Contact Mechanics

Contact Mechanics

Edited by

M. Raous

Laboratoire de Mécanique et d'Acoustique—CNRS
Marseille, France

and

M. Jean and J. J. Moreau

Laboratoire de Mécanique et Génie Civil
Montpellier, France

Springer Science+Business Media, LLC

Library of Congress Cataloging-in-Publication Data

Contact Mechanics International Symposium (2nd : 1994 : Carry-le
-Rouet, France)
Contact mechanics / edited by M. Raous, M. Jean, and J.J.Moreau.
p. cm.
Includes bibliographical references and index.
ISBN 978-0-306-45065-5 ISBN 978-1-4615-1983-6 (eBook)
DOI 10.1007/978-1-4615-1983-6
1. Contact mechanics--Congresses. I. Raous, M. II. Jean, M.
III. Moreau, J. J. (Jean Jacques), 1923- . IV. Title.
TA353.C667 1995
621--dc20 95-34062
CIP

Proceedings of the 2nd Contact Mechanics International Symposium,
held September 19–23, 1994, in Carry-Le-Rouet, France

ISBN 978-0-306-45065-5

Originally published by Plenum Press, New York in 1995

10 9 8 7 6 5 4 3 2 1

PREFACE

This proceedings volume contains 66 papers presented at the second "Contact Mechanics International Symposium" held in Carry-Le-Rouet, France, from September 19th to 23rd, 1994, attended by 110 participants from 17 countries. This symposium was the continuation of the first CMIS held in 1992 in Lausanne, of the Symposium *Euromech 273* "Unilateral Contact and Dry Friction" held in 1990 in La Grande Motte, France, and of the series of "Meetings on Unilateral Problems in Structural Analysis" organized in Italy, every other year, during the eighties.

The primary purpose of the symposium was to bring specialists of contact mechanics together in order to draw a representative picture of the state of the art and to identify new trends and new features in the field.

In view of the contributions made, one may assert that the mechanics of contact and friction has now reached a stage where the foundations are clear both from the mathematical and from the computational standpoints. Some of the difficulties met may be identified by saying that frictional contact is governed by resistance laws that are nonsmooth and whose flow rule is not associated with the yield criterion through the traditional normality property.

However, most large finite element softwares, in their recent releases, are able to take Coulomb friction and the unilaterality of contact constraints into account, even in the context of finite deformations. They at least work for quasistatic evolutions and also some dynamical situations now begin to be tractable. A subject of wide interest is collisions between bodies whether their global deformation is entered into the computation or whether, on the contrary, they are treated as rigid at the macroscopic scale of observation. In that line, the dynamical treatment of systems involving a large number of bodies provides effective insight into the behaviour of granular media. Recent progress in the analysis of the instabilities connected with friction makes us expect a better understanding of friction-induced vibrations and noise. Constant effort is applied towards a more precise description of friction than the Coulomb law. A better connection between tribology and general mechanics is sure to be productive. For instance, new models taking into account the analysis of asperities are available for numerical use.

The volume is divided into four sections: *Mathematical formulations*, *Numerical aspects*, *Friction modelling*, and *Dynamics of rigid and deformable bodies*. Experimental aspects are to be found in the last two sections. In each section, papers are as often as possible grouped according to their subtopic affinity. Lectures, communications and poster presentations may be identified through the table of contents.

Efforts from several parts have made the symposium possible. We would like first to thank all the authors for their contributions and the members of the Scientific Committee for their patronage and their assistance in selecting papers. We acknowledge the effectiveness of the Organizing Committee, made of members of the Laboratoire de Mécanique et d'Acoustique, with special gratitude to Elaine Pratt and Patrick Chabrand for the elaboration of the final form of this volume. We are pleased to thank Annie Fornacciari, Secretary of the Symposium, for her strong contribution in the preparation and the organization. We would like to associate also all the PhD students of the group of Mechanics and Modelling of Contact in Marseille for their very active help during the meeting.

We are furthermore indebted to the following sponsors, who have marked their interest in contact mechanics and in the research activity in Marseille by their financial support:

Association Universitaire de Mécanique (AUM)
Conseil Général des Bouches-du-Rhône
Centre National de la Recherche Scientifique (CNRS)
Calcul Scientifique Appliqué à la Mécanique pour l'Industrie (CSAMI)
Electricité de France (EDF)
Groupe Français de Mécanique des Matériaux (MECAMAT)
International Science Foundation (ISF)
Laboratoire de Mécanique et d'Acoustique (LMA)
Marseille-Provence-Métropole
Ministère de l'Enseignement Supérieur et de la Recherche
Renault
Société Française de Tribologie (SFT)
Université de Provence (Aix-Marseille I)

Michel Raous
LMA, Marseille
Chairman

Jean Jacques Moreau
LMGC, Montpellier
Co-Chairman

Michel Jean
LMGC, Montpellier
Co-Chairman

CONTENTS

MATHEMATICAL FORMULATIONS

Analysis and Optimization

Existence and Uniqueness Results

Lagrangian and Mixed Formulations

Algorithms

FRICTION MODELLING

Rough Surface Effects

Adhesion and Wear

DYNAMICS OF RIGID AND DEFORMABLE BODIES

Rigid Bodies and Collection of Particles

Impact and Vibrations

SHAPE OPTIMIZATION OF ELASTO-PLASTIC BODIES IN CONTACT

J. Haslinger

Department of Metal Physics
Charles University, Prague
Czech Republic

INTRODUCTION

The aim of this contribution is to present the mathematical analysis of optimal shape design problems of bodies being in mutual contact, which are made of a physically non-linear material, obeying the non-linear Hooke's law. This is an extension of previous results, dealing with elastic materials, discussed in [Haslinger, Neittaanmäki (1988)]. In Section 1, the state problem for physically non-linear materials is formulated and the sufficient conditions, guaranteeing the existence and the uniqueness of the solution are presented. In Section 2, the optimal shape design problem is formulated. Next Section deals with the existence of at least one optimal shape. A sketch of the existence proof is presented. In Section 4, the approximation of the optimal shape design problem is shortly described. The detailed analysis of this topic can be found in [Haslinger, Mäkinen].

SETTING OF THE STATE PROBLEM

Let $\Omega \subset \mathbb{R}^2$ be a bounded domain, the lipschitz boundary $\partial\Omega$ of which is decomposed as follows:

$$\partial\Omega = \bar{\Gamma}_u \cup \bar{\Gamma}_P \cup \bar{\Gamma}_c .$$

On each part different boundary conditions will be prescribed. Next, we shall assume that $\bar{\Gamma}_u \neq \emptyset$. A deformable body will be represented by the infinite cylinder $Q \subset \mathbb{R}^3$:

Contact Mechanics, Edited by M. Raous *et al.*
Plenum Press, New York, 1995

$$Q = \{(x_1,x_2,x_3) | (x_1,x_2) \in \Omega,\ x_3 \in \mathbb{R}^1\} = \Omega \times \mathbb{R}^1$$

with the boundary $\partial Q = \bar{\Theta}_u \cup \bar{\Theta}_P \cup \bar{\Theta}_c$, where

$$\Theta_j = \Gamma_j \times \mathbb{R}^1, \qquad j = u, P, c.$$

The body will be subjected to body forces $F = (F_1,F_2,F_3)$ and surface tractions $P = (P_1,P_2,P_3)$ acting on Θ_P.

Next we shall assume:

(j) material properties of Q do not depend on the x_3 - coordinate;

(jj) $F_i = F_i(x_1,x_2)$, $i = 1,2$; $F_3 \equiv 0$;
$P_i = P_i(x_1,x_2)$, $i = 1,2$; $P_3 \equiv 0$;

(jjj) $u_3 \equiv 0$ on Θ_u.

If (j) - (jjj) are satisfied, then $u_3 \equiv 0$ and $u_i = u_i(x_1,x_2)$, $i = 1,2$. Consequently, we may analyse a plane problem for the cross section Ω, only.

Next we shall assume that Ω is made of a material, obeying the theory of small elasto - plastic deformations, (see [Washizu (1968)]). The corresponding non - linear Hooke's law is given by (summation convention is used)

$$\begin{cases} \sigma_{ij} \equiv \sigma_{ij}(u) = \kappa\varepsilon_{ll}\delta_{ij} + 2\mu(\Gamma)(\varepsilon_{ij} - 1/3\ \delta_{ij}\varepsilon_{ll}) \\ \varepsilon_{ij} \equiv \varepsilon_{ij}(u) = 1/2\ (\partial u_i/\partial x_j + \partial u_j/\partial x_i), \quad i,j = 1,2. \end{cases} \tag{1.1}$$

Here $u=(u_1,u_2)$ denotes a displacement field, δ_{ij} is the Kronecker symbol. Symbols κ and μ stand for the bulk and shear moduli, respectively. The stress tensor $\sigma(u)=(\sigma_{ij}(u))^2_{i,j=1}$, is related to the strain tensor $\varepsilon(u)=(\varepsilon_{ij}(u))^2_{i,j=1}$ by means of a non-linear relation, the physical justification of which can be found in [Washizu (1968)].

We assume that the shear moduli μ is a function of the invariant

$$\Gamma = 1/\sqrt{3}[(\varepsilon_{11}-\varepsilon_{22})^2 + \varepsilon_{11}^2+\varepsilon_{22}^2+ 6\varepsilon_{12}^2]^{1/2}.$$

In the following, we shall assume that $\kappa = \kappa(x)$, $\mu = \mu(t,x)$, $t \geq 0$, $x \in \Omega$ are continuous functions of their variables and μ is continuously differentiable with respect to t:

$$\kappa \in C(\bar{\Omega}), \mu \in C(\mathbb{R}^1_+ \times \bar{\Omega}), \partial\mu/\partial t \in C(\mathbb{R}^1_+ \times \bar{\Omega}).$$

Moreover

$$0<\kappa_0 \le \kappa(x) \le \kappa_1 \tag{1.2}$$
$$0<\mu_0 \le \mu(t,x) \le 3/2\kappa(x) \qquad \forall x \in \Omega,\ t > 0 \tag{1.3}$$
$$0<\tau_0 \le \mu(t,x) + 2\partial\mu(t,x)/\partial t . t \le \tau_1, \qquad \forall x \in \Omega, \forall t>0, \tag{1.4}$$

where $\kappa_0, \kappa_1, \mu_0, \tau_0, \tau_1$ are given positive constants.

Now, we shall formulate the *boundary conditions*. Let Ω be unilaterally supported by a rigid halfplane S along the part Γ_c (contact part). For the sake of simplicity we shall assume that

$$S = \{(x_1,x_2) \in \mathbb{R}^2 | x_2 \le 0\} \equiv \mathbb{R}^2_-$$

and Γ_c is given by the graph of a non-negative, lipschitz continuous function α, defined on [a,b]. The contact conditions on Γ_c are the following:

$$\left\{ \begin{array}{ll} u_2(x_1,\alpha(x_1)) \ge -\alpha(x_1) & x_1 \in (a,b) \\ T_2 \equiv \sigma_{2j}(u)n_j \ge 0, & T_2(u_2+\alpha) = 0 \text{ on } \Gamma_c \\ T_1 \equiv \sigma_{1j}(u)n_j = 0 & \text{on } \Gamma_c. \end{array} \right. \tag{1.5}$$

On the remaining parts Γ_u and Γ_P, the body is supposed to be fixed and to be subjected to surface tractions P, respectively:

$$u_i = 0 \text{ on } \Gamma_u, \ i = 1,2; \tag{1.6}$$
$$T_i \equiv \sigma_{ij}(u)n_j = P_i \text{ on } \Gamma_P, \ i = 1,2. \tag{1.7}$$

By a *classical solution*, describing the equilibrium state we mean any displacement field $u = (u_1,u_2)$, satisfying the non-linear Hooke's law (1.1), the system of boundary conditions (1.5)-(1.7) and the equilibrium equations

$$\partial\sigma_{ij}/\partial x_j + F_i = 0 \text{ in } \Omega, \ i = 1,2. \tag{1.8}$$

In order to give the *variational formulation* of this problem, we introduce the following notations:

$$V=\{v \in (H^1(\Omega))^2 | v_i = 0 \text{ on } \Gamma_u,\ i = 1,2\},$$

$$K=\{v \in V|\ v_2(x_1,\alpha(x_1)) \ge -\alpha(x_1), x_1 \in (a,b)\},$$

$$a_\Omega(u,v) \equiv (\sigma_{ij}(u),\varepsilon_{ij}(v))_{0,\Omega} \equiv \int_\Omega \sigma_{ij}(u)\varepsilon_{ij}(v)dx$$

$$L_\Omega(v) \equiv (F_i,v_i)_{0,\Omega} + (P_i,v_i)_{0,\Gamma_P} \equiv \int_\Omega F_i v_i dx + \int_{\Gamma_P} P_i v_i ds$$

By a *variational solution* of the problem under consideration we mean any $u \in K$ such that

$$a_\Omega(u,v-u) \ge L_\Omega(v-u) \quad \forall v \in K. \tag{$\mathcal{P}$}$$

Using the Green's theorem and a suitable choice of test functions in ($\mathcal{P}$), one can deduce that the conditions (1.5)-(1.7) as well as the equilibrium equations (1.8) are satisfied. Therefore, if a variational solution is smooth enough then it is a classical one, as well. It is not difficult to show that the potential of ($\mathcal{P}$) is given by

$$\Phi_\Omega(v) = 1/2 \int_\Omega (\kappa\varepsilon_{11}^2(v) + \int_0^{\Gamma^2(u)} \mu(t)dt)dx - L_\Omega(v)$$

and the problem ($\mathcal{P}$) is equivalent to

$$\begin{cases} \text{find } u \in K \text{ such that} \\ \Phi_\Omega(u) \le \Phi_\Omega(v)\ \forall\ v \in K\ . \end{cases} \tag{$\mathcal{P}'$}$$

If (1.2)-(1.4) are satisfied, then Φ_Ω is *strictly convex, coercive* and *differentiable* in V. Consequently, ($\mathcal{P}'$)(and also ($\mathcal{P}$)) has a unique solution u. The formulation ($\mathcal{P}$) is nothing else then the necessary and sufficiant condition for u, to be a minimiser of Φ_Ω on K. The proof can be found in [Nečas, Hlaváček (1981)].

SHAPE OPTIMIZATION - FORMULATION OF THE PROBLEM

Until now, we assumed that the shape of Ω is given. In shape optimization, the boundary $\partial\Omega$ (or a part of it) plays the role of the *control variable*. The goal is to find such a shape, for which the structure has à-priori given properties (or properties, which are as close as possible to a given target).

The problem can be formulated, using the tools of the optimal control theory. Denote by $\mathcal{O}$ a family of *admissible domains*, among which we search for and optimal one:

$$\mathcal{O} = \{\Omega \subset \mathbb{R}^2 \mid \emptyset \neq C \subseteq \Omega \subseteq D, \partial\Omega \text{ is lipschitz continuous}\},$$

where C,D are given domains. On each $\Omega \in \mathcal{O}$ we formulate the problem ($\mathcal{P}$) (or($\mathcal{P}'$)), introduced in the last Section for a particular choice of Ω. In order to emphasize the dependence of the state problem on the choice of $\Omega \in \mathcal{O}$, we shall write $(\mathcal{P}(\Omega))$, $(\mathcal{P}'(\Omega))$, $V(\Omega)$, $K(\Omega)$, $u(\Omega)$ ect. in what follows.

Let $\hat{\Omega} \subset \mathbb{R}^2$ be a bounded domain, such that $\hat{\Omega} \supset \Omega\ \forall\Omega \in \mathcal{O}$. In order to guarantee the existence and the uniqueness of the solution of $(\mathcal{P}(\Omega))$, we shall suppose that the material functions κ, μ are defined on $\hat{\Omega}$ and satisfy (1.2)-(1.4) in $\mathbb{R}^1_+ \times \hat{\Omega}$. Also $F \in (L^2(\hat{\Omega}))^2$ and $P \in (L^2(\Gamma_P))^2$ for any $\Omega \in \mathcal{O}$ is supposed.

Finally, let $J\colon (\Omega,y) \to \mathbb{R}^1$, $\Omega \in \mathcal{O}$, $y \in V(\Omega)$ be a *cost functional. The optimal shape design problem* in an abstract setting reads as follows:

$$\left\{\begin{array}{l} \text{find } \Omega^* \in \mathcal{O} \text{ such that} \\ J(\Omega^*, u(\Omega^*)) \le J(\Omega, u(\Omega))\ \forall\ \Omega \in \mathcal{O} \end{array}\right. \qquad (\mathbb{P})$$

where $u(\Omega) \in K(\Omega)$ solves $(P(\Omega))$.

AN EXISTENCE RESULT FOR ($\mathbb{P}$)

The aim of the present Section is to establish the existence of at least one solution of ($\mathbb{P}$). To avoid the technical difficulties, we shall consider a simple family of admissible domains $\mathcal{O}$, consisting of all "rectangular" domains $\Omega(\alpha)$ of the form:

$$\Omega(\alpha)=\{(x_1,x_2) \in \mathbb{R}^2 \mid x_1 \in (a,b),\ \alpha(x_1) \le x_2 \le \gamma\},$$

where $\gamma>0$ is a given number, α is a non-negative lipschitz continuous function, defining the contact surface $\Gamma_c(\alpha) = \{(x_1,x_2) \mid x_2 = \alpha(x_1),\ x_1 \in (a,b)\}$, along which the contact conditions (1.5) will be prescribed. The contact part $\Gamma_c(\alpha)$ will be the object of the optimization. The *design variable* α, determining $\Gamma_c(\alpha)$ is assumed to belong to the set U_{ad}, defined by

$$U_{ad} = \{\alpha \in C^{0,1}([a,b]) \mid 0 \le \alpha(x_1) \le C_o, |\alpha'(x_1)| \le C_1 \text{ a.e. in } (a,b), \text{ meas } \Omega(\alpha) = C_2\},$$

i.e. U_{ad} contains all uniformly bounded and uniformly continuous functions in (a,b), preserving the area of $\Omega(\alpha)$. The constants C_0, C_1 and C_2 are choosen in such a way that U_{ad} is non-empty. On the remaining part of the boundary $\partial\Omega(\alpha)$, zero displacements (on Γ_u) and surface tractions P (on Γ_P) will be prescribed. We suppose also that $\text{meas}_1\, \Gamma_u \ge \delta > 0$ for any $\alpha \in U_{ad}$, where $\delta > 0$ is a given constant and the symbol $\text{meas}_1\, \omega$ stands for 1-dimensional Lebesgue measure of $\omega \subset \partial\Omega(\alpha)$. There is one-to-one correspondence between $\mathcal{O}$ and U_{ad}:

$$\mathcal{O} = \{\Omega(\alpha),\ \alpha \in U_{ad}\}.$$

Therefore instead of $(\mathcal{P}(\Omega))$, $V(\Omega)$, $K(\Omega)$, ... we shall write $(\mathcal{P}(\alpha))$, $V(\alpha)$, $K(\alpha)$, ... in what follows.

In order to prove the existence of at least one solution of $(\mathbb{P})$, the *continuous dependence* of the solution $u(\alpha)$ of $(\mathcal{P}(\alpha))$ on the design variable α has to be proved.

This required property now follows from:

Lemma 3.1 *Let $\alpha_n \rightrightarrows \alpha$ (uniformly) in [a,b], $\alpha_n, \alpha \in U_{ad}$, and let $u_n \equiv u(\alpha_n)$ be solutions of $(\mathcal{P}(\alpha_n))$. Then there exist: a subsequence of $\{u_n\}$(denoted by the same symbol) and a function $u \in K(\alpha)$ such that*

$$u_n \rightharpoonup u \text{ in } (H^1(G_m))^2,\ n \to \infty \tag{3.1}$$

for any m integer, sufficiently large, where

$$G_m = \{(x_1,x_2) \in \mathbb{R}^2 \mid x_1 \in (a,b),\ \alpha(x_1) + 1/m < x_2 < \gamma\}. \tag{3.2}$$

Moreover, $u \equiv u(\alpha)$ solves $(\mathcal{P}(\alpha))$.

In order to prove the existence of at least one solution of $(\mathbb{P})$, the following *lower semicontinuity* of J will be required:

$$(\mathbb{A}) \quad \left\{ \begin{array}{l} \alpha_n \rightrightarrows \alpha \text{ (uniformly) in } [a,b],\ \alpha_n,\ \alpha \in U_{ad} \\ y_n \rightharpoonup y \text{ in } (H^1(G_m))^2\ \forall m,\ y_n \in V(\alpha_n),\ y \in V(\alpha) \end{array} \right. \Longrightarrow$$

$$\Longrightarrow \liminf J(\alpha_n, y_n) \geq J(\alpha, y), \ n \to \infty$$

Theorem 3.1 Let the cost functional J satisfy ($\mathbb{A}$). *Then* ($\mathbb{P}$) *has at least one solution.*

Proof: the result follows immediately from Lemma 3.1 and the fact that U_{ad} is compact in C([a,b]) - norm.

Remark 3.1 Examples of cost functionals, satisfying ($\mathbb{A}$):

$$J(\alpha, u(\alpha)) = \| u(\alpha) - z_d \|^2_{0,\Omega(\alpha)},$$

$$J(\alpha, u(\alpha)) = \Phi_{\Omega(\alpha)}(u(\alpha)) .$$

□

It is well-known that the mapping $\alpha \to J(\alpha, u(\alpha))$ is not continuously differentiable, in general (see [Haslinger, Neittaanmäki(1988)], [Sokolowski, Zolesio (1992)]. A posible way, how to overcome this difficulty, is to "regularize" the state problem $(\mathcal{P}(\alpha))$ by means of a suitable penalty approach. Denote by

$$B_\alpha(u,v) \equiv \int_a^b [(u_2(x_1,\alpha(x_1))+\alpha(x_1))^-]^2 v_2(x_1,\alpha(x_1)dx_1$$

$$\equiv \int_a^b ((u_2(\alpha)+\alpha)^-)^2 v_2(\alpha)dx_1, \quad u,v \in V(\alpha)$$

the penalty functional (the symbol $(a)^-$ means the negative part of a real number a). It is easy to see that

$$u \in K(\alpha) \Leftrightarrow u \in V(\alpha) \text{ and } B_\alpha(u,v) = 0 \ \forall\, v \in V(\alpha).$$

Instead of $(\mathcal{P}(\alpha))$ we shall consider a family of the non-linear elliptic problems:

$$a_{\Omega(\alpha)}(u_\varepsilon(\alpha), v) + 1/\varepsilon B_\alpha(u_\varepsilon, v) = L_{\Omega(\alpha)}(v) \ \forall\, v \in V(\alpha), \qquad (\mathcal{P}(\alpha))_\varepsilon$$

where $\varepsilon \to 0+$ is the penalty parameter. The problem ($\mathbb{P}$), will be replaced by

$$\begin{cases} \text{find } \alpha_\varepsilon^* \in U_{ad} \text{ such that} \\ J(\alpha_\varepsilon^*, u_\varepsilon(\alpha_\varepsilon^*)) \le J(\alpha, u_\varepsilon(\alpha)) \ \forall\, \alpha \in U_{ad}, \end{cases} \qquad (\mathbb{P})_\varepsilon$$

with $u_\varepsilon(\alpha) \in V(\alpha)$ being the solution of $((\mathcal{P}(\alpha))_\varepsilon$.

Remark 3.2 The main reason, why we replace $(\mathbb{P})$ by $(\mathbb{P})_\varepsilon$ is the fact, that the function $\alpha \to J(\alpha, u_\varepsilon(\alpha))$ is continuously differentiable, provided the cost functional J is smooth enough. Consequently, classical gradient type minimization methods can be used, when realizing $(\mathbb{P})_\varepsilon$ numerically. □

Using the same ideas as in Lemma 3.1, one can show that for any $\varepsilon > 0$ there exists at least one solution α_ε^* of $(\mathbb{P})_\varepsilon$. A natural question arises, namely if there is some relation between $(\mathbb{P})$ and $(\mathbb{P})_\varepsilon$, when $\varepsilon \to 0+$. Let $\{\varepsilon_k\}$ be a sequence of the penalty parameters tending to zero. Then it is possible to prove

Theorem 3.2 *Let $\alpha_k^* \in U_{ad}$ be a solution of $(\mathbb{P})_{\varepsilon_k}$ and $u_k^*(\alpha_k^*)$ the corresponding state. Then there exist: subsequences $\{\alpha_{k_j}^*\} \subset \{\alpha_k^*\}, \{u_{k_j}^*(\alpha_{k_j}^*)\} \subset \{u_k^*(\alpha_k^*)\}$ and elements $\alpha^* \in U_{ad}$, $u^*(\alpha^*) \in K(\alpha^*)$ such that*

$$\alpha_{k_j}^* \rightrightarrows \alpha^* \text{ (uniformly) in } [a,b],$$
$$u_{k_j}^*(\alpha_{k_j}^*) \to u^*(\alpha^*) \quad \text{(weakly) in } (H^1(G_m))^2$$

for any m integer, sufficiently large, where $G_m = \{(x_1,x_2) \in \mathbb{R}^2 | x_1 \in (a,b),\ \alpha^*(x_1) + 1/m < x_2 < \gamma\}$.
Moreover α^* is a solution of $(\mathbb{P})$ and $u^*(\alpha^*)$ solves $(\mathcal{P}(\alpha^*))$.

Proof can be done by the same technique as in [Haslinger, Neittaanmäki, Tiihonen (1986)].

APPROXIMATION OF $(\mathbb{P})$

Instead of the system of admissible domains $\mathcal{O}$, we shall assume its *approximation* $\mathcal{O}_h$, containing domains, the shape of which is determined by a finite number of parameters only. Assuming $\mathcal{O}$ as in Section 3, one can take

$$\mathcal{O}_h = \{\Omega(\alpha_h) | \ \alpha_h \in U_{ad}^h\},$$

where

$$\mathcal{U}^h_{ad} = \{\alpha_h \in C^0([a,b]) \mid \alpha_h \text{ are piecewise linear in } [a,b]\} \cap \mathcal{U}_{ad}.$$

This means that $\Omega(\alpha_h)$ is a *polygonal* domain and standard finite elements can be used. Let $\mathcal{T}(h,\alpha_h)$ denote a triangulation of $\bar{\Omega}(\alpha_h)$. Besides of usual requirements on the mutual position of triangles from $\mathcal{T}(h,\alpha_h)$, we shall suppose that for any $h > 0$ fixed, the triangulations $\{\mathcal{T}(h,\alpha_h),\ \alpha_h \in \mathcal{U}^h_{ad}\}$ are *topologically equivalent*, i.e. they have the same number of the nodes and the nodes have the same neighbours for any choice of $\alpha_h \in \mathcal{U}^h_{ad}$. Moreover, for any $h > 0$ fixed, the triangulations $\mathcal{T}(h,\alpha_h)$ depend continuously on $\alpha_h \in \mathcal{U}^h_{ad}$.

Let $\alpha_h \in \mathcal{U}^h_{ad}$ be given and let $\mathcal{T}(h,\alpha_h)$ be a triangulation of $\bar{\Omega}(\alpha_h)$, satisfying the requirements, formulated above. With any $\mathcal{T}(h,\alpha_h)$ we associate the finite dimensional space

$$V_h(\alpha_h) = \{v_h \in C(\bar{\Omega}(\alpha_h)) \mid v_h \text{ is piecewise linear over } \mathcal{T}(h,\alpha_h),\ v_h = 0 \text{ on } \bar{\Gamma}_u\}$$

and the closed convex subset $K_h(\alpha_h) = K(\alpha_h) \cap V_h(\alpha_h)$.

Instead of $(\mathcal{P}(\alpha))$ we assume its finite element approximation

$$\left\{\begin{array}{l} \text{find } u_h \equiv u_h(\alpha_h) \in K_h(\alpha_h) \text{ such that} \\ a_{\Omega(\alpha_h)}(u_h, v_h - u_h) \ge L_{\Omega(\alpha_h)}(v_h - u_h)\ \forall\, v_h \in K_h(\alpha_h). \end{array}\right. \qquad (\mathcal{P}_h(\alpha_h))$$

Finally, let $J_h : (\alpha_h, y_h) \to \mathbb{R}^1$, $\alpha_h \in \mathcal{U}^h_{ad}$, $y_h \in V_h(\alpha_h)$ be *an approximation* of J.

The approximation of the optimal shape design problem $(\mathbb{P})$ now reads as follows:

$$\left\{\begin{array}{l} \text{find } \alpha^*_h \in \mathcal{U}^h_{ad} \text{ such that} \\ J_h(\alpha^*_h, u_h(\alpha^*_h)) \le J_h(\alpha_h, u_h(\alpha_h)) \quad \forall\ \alpha_h \in \mathcal{U}^h_{ad} \end{array}\right. \qquad (\mathbb{P})_h$$

It is possible to show that under "reasonable" assumptions, $(\mathbb{P})$ and $(\mathbb{P})_h$ are closed on subsequences when $h \to 0+$.

Next, we present the matrix formulation of $(\mathcal{P}_h(\alpha_h))$ and $(\mathbb{P})_h$. Let $\alpha \in \mathbb{R}^D$ be the discrete design variable associated with $\alpha_h \in \mathcal{U}^h_{ad}$ (x_2-coordinates of the nodes of α_h in our special case). By $\mathcal{U}$ we denote the set of all admissible discrete design variables, i.e. $\alpha \in \mathcal{U}$ iff $\alpha_h \in U^h_{ad}$.

Let $\alpha \in \mathcal{U}$ be given. Then $(\mathcal{P}_h(\alpha_h))$ expressed in the matrix form reads as follows:

$$\begin{cases} \text{find } q \equiv q(\alpha) \in \mathcal{K}(\alpha) \text{such that} \\ (K(\alpha,q)q, x-q)_{R_n} \geq (F(\alpha), x-q)_{R_n} \ \forall\, x \in \mathcal{K}(\alpha), \end{cases} \qquad (\bar{P}(\alpha))$$

where q is the unknown, containing the nodal displacements, $K(\alpha,q)$ is the stifness matrix corresponding to the non-linear Hooke's law, depending on α and q, $F(\alpha)$ is the force vector. The symbol $(\ ,\ )_{R_n}$ stands for the scalar product in $\mathbb{R}^n$ and $\mathcal{K}(\alpha) \subset \mathbb{R}^n$ is the closed convex subset isomorphic with $K_h(\alpha_h)$, $\alpha_h \in U^h_{ad}$. The problem $(\bar{P}(\alpha))$ can be equivalently written as the minimization problem

$$\begin{cases} \text{find } q \equiv q(\alpha) \in \mathcal{K}(\alpha) \text{ such that} \\ \Phi(\alpha, q(\alpha)) \leq \Phi(\alpha, x) \quad \forall\, x \in \mathcal{K}(\alpha), \end{cases} \qquad (\bar{P}'(\alpha))$$

where $\Phi(\alpha,q)$ is a strictly convex, differentiable function of $(\alpha,q) \in \mathcal{U} \times \mathbb{R}^n$ and such that $\nabla_q\Phi(\alpha,q) = K(\alpha,q)q$. Finally, the optimal shape design problem expressed in the matrix form reads as follows:

$$\begin{cases} \text{find } \alpha^* \in \mathcal{U} \text{ such that} \\ C(\alpha^*, q(\alpha^*)) \leq C(\alpha, q(\alpha)) \quad \forall\, \alpha \in \mathcal{U},. \end{cases} \qquad (\bar{\mathbb{P}})$$

where $q(\alpha) \in \mathcal{K}(\alpha)$ solves $(\bar{P}'(\alpha))$ and C is the matrix expression of J_h

The practical realization of $(\bar{\mathbb{P}})$ is presented in [Haslinger, Mäkinen (1992)].

REFERENCES

Washizu, K.(1968): Variational methods in elasticity and plasticity, Oxford: Pergamon Press

Nečas, J, Hlaváček, I. (1981): Mathematical theory of elastic and elasto-plastic bodies: An introduction, 1981, Elsevier, Amsterdam

Haslinger, J, Neittaanmäki, P. Tiihonen, T.(1986): Shape optimization in contact problems based on penalization of the state inequality, Apl. Mat. 31 (1986), 54-77

Haslinger, J; Neittaanmäki, P.(1988): Finite element approximation for optimal shape design: Theory and Applications, J. Wiley & Sons, 1988

Hlaváček, I.(1989): Inequalities of Korn's type, uniform with respect to a class of domains, Apl. Mat. 34, (1989), 105-112

Haslinger, J., Mäkinen, R. (1992): Shape optimization of elasto-plastic bodies under plane strains: sensitivity analysis and numerical implementation, Struct. Optim. 4, 133-141 (1992)

Sokolowski, J., Zolesio, J.P. (1992): Introduction to Shape optimization, Shape sensitivity analysis, Springer-Verlag, 1992

Haslinger, J., Mäkinen: Contact shape optimization for physically non-linear materials (in preparation)

ANALYSIS OF AN INCREMENTAL FORMULATION FOR FRICTIONAL CONTACT PROBLEMS

Marius Cocu[1], Elaine Pratt[1] and Michel Raous[2]

[1]U.F.R. - M.I.M., Université de Provence
Laboratoire de Mécanique et d'Acoustique - C.N.R.S.
[2]Laboratoire de Mécanique et d'Acoustique - C.N.R.S.
31, chemin Joseph Aiguier
13402 Marseille Cedex 20 France

INTRODUCTION

This work concerns a quasistatic Signorini's problem with a nonlocal Coulomb's friction law for a linear elastic body. When unilateral contact and friction are considered, both displacement (normal contact) and velocity (friction) are involved. The static (displacement) formulation cannot describe evolutive situations, however introducing the unilateral condition $u_N \leq 0$ into the dynamic formulation is not so simple as $\dot{u}_N$, in some situations, may be either positive or negative.

The classical formulation of this quasistatic problem is given by problem P_1. We give a new variational formulation of this problem which consists of two coupled variational inequalities. The unilateral contact conditions are described by a classical variational inequality and all the other equations and boundary conditions of the problem including the friction law are contained in a separate inequality which has a differential form.The existence of a solution for quasistatic frictional contact problems with a normal compliance law was proved by Andersson (1991) using incremental formulations and, in the presence of a time regularization, by Klarbring et al. (1991) in a different manner.

Having set the problem under a variational form (problem P_2) we prove the existence of a solution under the hypothesis that the friction coefficient is sufficiently small.

The general outline of the proof is as follows. We begin by considering different equivalent incremental formulations of the problem which may be considered as a time dicretization of problem P_2. These incremental formulations, if the friction coefficient is small, possess a unique solution so that we are able to construct a sequence of mappings belonging to $L^2(0,T;V)$ and defined by using the solutions of one of the incremental formulations.We then obtain bounds on these incremental solutions and on the difference between two consecutive solutions which enable us to show that this sequence possesses a sub-sequence which converges weakly in $L^2(0,T;V)$. This weak limit is then shown to belong to $W^{1,2}(0,T;V)$. The final part of the proof consists in showing that the weak limit is in fact a solution of problem P_2.

The incremental formulations have the advantage of providing the numerical approach we have used for computing the example given in the final section which illustrates the importance of taking the velocity into account where friction is concerned even for quasistatic loadings.

Contact Mechanics, Edited by M. Raous *et al.*
Plenum Press, New York, 1995

THE QUASISTATIC PROBLEM

We consider an elastic body occupying a domain Ω of $\mathbf{R}^d$, $d = 2, 3$, with a Lipschitz boundary $\Gamma = \bar{\Gamma}_1 \cup \bar{\Gamma}_2 \cup \bar{\Gamma}_3$, where Γ_1, Γ_2 and Γ_3 are open and disjoint parts of Γ which do not depend on time. The displacements are prescribed on Γ_1 with $\text{mes}(\Gamma_1) > 0$ and an evolutive force density φ_2 is applied on Γ_2. The solid is initially in contact with a rigid fixed support on Γ_3 and the evolution of the displacements on this part of the boundary is restricted by a unilateral condition of non penetration into the support and submitted to friction forces when contact occurs. On Ω a volume force density φ_1 is applied.

We suppose that the given forces φ_1 and φ_2 are sufficiently smooth in space and in time and that we may neglect the acceleration terms.

We denote by $\sigma=(\sigma_{ij})$ the stress tensor, $e=(e_{ij})$ the strain tensor, $u=(u_i)$ the displacement field and E the elasticity tensor of the material with the usual properties of ellipticity and symmetry. On Γ_3 we use the following notations for the normal and tangential components of the displacement vector and stress vector :

$u_N = u_i\, n_i, \quad u_T = u - u_N\, n$

$\sigma_N = \sigma_{ij}\, n_i\, n_j, \quad \sigma_T = \sigma.n - \sigma_N\, n,$

where $n = (n_i)$ is the outward normal unit vector to Γ.

The classical quasi static formulation is

Problem P_1: Find u such that $\forall\, t \in [0,T]$

$$\text{div}\, \sigma = -\varphi_1 \text{ in } \Omega, \tag{1}$$

$$\sigma = Ee, \quad e = \text{grad}_s\, u \text{ in } \Omega, \tag{2}$$

$$u(0) = u_0 \text{ in } \Omega, \tag{3}$$

$$u = 0 \text{ on } \Gamma_1, \tag{4}$$

$$\sigma.n = \varphi_2 \text{ on } \Gamma_2, \tag{5}$$

$$u_N \le 0, \; \sigma_N \le 0, \; u_N\, \sigma_N = 0 \text{ on } \Gamma_3, \tag{6}$$

$$|\sigma_T| \le \mu |R\sigma_N| \quad \text{and} \begin{cases} |\sigma_T| < \mu |R\sigma_N| & \Rightarrow \dot{u}_T = 0 \\ |\sigma_T| = \mu |R\sigma_N| & \Rightarrow \exists \lambda \ge 0 \;, \dot{u}_T = -\lambda \sigma_T \end{cases} \quad \text{on } \Gamma_3, \tag{7}$$

where μ is the friction coefficient and $R\sigma_N$ is a regularization of the normal contact force introduced by Duvaut (1980).

VARIATIONAL FORMULATION

We shall adopt the following notations:

$V = \{ v \in (H^1(\Omega))^d \,;\, v = 0 \text{ a.e. on } \Gamma_1 \}$

$\| \;\|$ shall denote the norm induced by $(\,,\,)$, the scalar product on V

$K = \{ v \in V;\, v_N \le 0 \text{ a.e. on } \Gamma_3 \}$

$W = W^{1,2}(0,T;V)$, see for example Cazenave and Haraux (1990)

$\langle\, ,\, \rangle$ shall denote the duality pairing on $H^{\frac{1}{2}}(\Gamma_3) \times H^{-\frac{1}{2}}(\Gamma_3)$.

We suppose that $\varphi_1 \in W^{1,2}(0,T;(L^2(\Omega))^d)$, $\varphi_2 \in W^{1,2}(0,T;(L^2(\Gamma_2))^d)$, $\mu \in C(0,T;L^\infty(\Gamma_3))$ with $\mu \ge 0$ a.e. on Γ_3 for $\forall\, t \in [0,T]$ and that $R: H^{-\frac{1}{2}}(\Gamma_3) \to L^2(\Gamma_3)$ is a linear and completely continuous operator.

Throughout this paper we consider only the Lebesgue mesure and we shall therefore omit the integration variable whenever there is no ambiguity (for example we shall set $\int_\Omega f$ for $\int_\Omega f(x)dx$).

Using Green's formula, (1), (2), the boundary conditions (4), (5) and a test function v (homogeneous to a velocity) in V we have, if u is a sufficiently regular solution: $\forall v \in V$

$$a(u, v-\dot{u}) - \int_\Omega \varphi_1(v-\dot{u}) - \int_{\Gamma_2} \varphi_2(v-\dot{u}) - \int_{\Gamma_3} \sigma_N(v_N-\dot{u}_N) - \int_{\Gamma_3} \sigma_T(v_T-\dot{u}_T) = 0 \tag{8}$$

with $a(u,v) = \int_\Omega \sigma_{ij}(u)e_{ij}(v)$.

A variational formulation of the Coulomb law (7) is given by the following inequality, see Duvaut-Lions (1972):

$$\forall v \in V \quad \sigma_T(v_T - \dot{u}_T) + j(u,v) - j(u,\dot{u}) \geq 0 \tag{9}$$

with $j(u,v) = \int_{\Gamma_3} \mu |R\sigma_N(u)| \, |v_T|$.

The following properties on a and j hold, see Duvaut-Lions (1972):

$$\exists m > 0 \text{ such that } \forall u \in V \quad a(u,u) \geq m\|u\|^2 \tag{10}$$

$$\exists M > 0 \text{ such that } \forall u \in V \; \forall v \in V \quad a(u,v) \leq M\|u\| \, \|v\| \tag{11}$$

$$\exists C > 0 \text{ such that } \forall (u_1, u_2, v) \in V^3 \quad |j(u_1,v) - j(u_2,v)| \leq \bar{\mu}C\|u_1 - u_2\| \|v\| \tag{12}$$

with $\bar{\mu} = \|\mu\|_{L^\infty(\Gamma_3)}$,

$$\forall (u, v_1, v_2) \in V^3 \quad j(u, v_1) - j(u, v_2) \leq j(u, v_1 - v_2), \tag{13}$$

where inequalities (12) and (13) are true for all $t \in [0,T]$.

From (8) and (9) we obtain the following implicit variational inequality : $\forall t \in [0,T]$

$$\forall v \in V \quad a(u, v-\dot{u}) - (f, v-\dot{u}) + j(u,v) - j(u,\dot{u}) - \langle \sigma_N(u), v_N - \dot{u}_N \rangle \geq 0 \tag{14}$$

with $(f,v) = \int_\Omega \varphi_1 v + \int_{\Gamma_2} \varphi_2 v$.

Another relation on σ_N is needed. The following weak formulation of the unilateral condition (6) can be given:

$$\forall z \in K \quad \langle \sigma_N(u), z_N - u_N \rangle \geq 0.$$

If $\sigma_N(u)$ belongs to $L^2(\Gamma_3)$, then the duality pairing coincides with the scalar product on $L^2(\Gamma_3)$.

We may now give the following weak formulation of problem P_1.

Problem P_2: Find $u \in W^{1,2}(0,T;V)$ such that $\forall t \in [0,T]$ a.e. $u(t) \in K$ and

$$a(u(t), v-\dot{u}(t)) + j(u(t), v) - j(u(t), \dot{u}(t)) \geq (f(t), v-\dot{u}(t)) + \langle \sigma_N(u(t)), v_N - \dot{u}_N(t) \rangle \quad \forall v \in V \tag{15}$$

$$\langle \sigma_N(u(t)), z_N - u_N(t) \rangle \geq 0 \quad \forall z \in K \tag{16}$$

$u(0) = u_0$, where $u_0 \in K$ is given. (17)

The initial u_0 cannot be totally arbitrary in K, in fact we must suppose that u_0 satisfies the following compatibility condition:

$$a(u_0, w-u_0) + j(u_0, w-u_0) \geq (f^0, w-u_0) \quad \forall w \in K, \text{ where } f^0 = f(0) \tag{18}$$

The proof of the formal equivalence, i.e. if u is sufficiently regular, is straightforward:
equation (1) and condition (5) are easily obtained by choosing $v = \dot{u} \pm \varphi$ on Ω with $\varphi \in (D(\Omega))^d$, then $v = \dot{u} \pm \psi$ on Γ_2 with $\psi \in V$ and $v = \dot{u}$ on Γ_3. Inequality (9) then holds which establishes the friction relations (7). The choice of u in K insures the unilateral condition $u_N \leq 0$ a.e. on Γ_3. By choosing $z_N = 0$ and $z_N = 2\, u_N$ in (16) we obtain:

$$\int_{\Gamma_3} \sigma_N u_N = 0. \tag{19}$$

It follows that $\int_{\Gamma_3} \sigma_N z_N \geq 0 \quad \forall z \in K$ which implies $\sigma_N \leq 0$ and finally, using (19), (6) holds.

INCREMENTAL FORMULATIONS

Let us consider a time discretization by taking $n \in \mathbf{N}^*$ and setting $\Delta t = T/n$. We then consider a sequence $(P_i^n)_{i=0...n-1}$ of variational inequalities defined by:

Problem P_i^n: Find $u^{i+1} \in K$ such that

$$a(u^{i+1}, v - \frac{u^{i+1}-u^i}{\Delta t}) + j(u^{i+1}, v) - j(u^{i+1}, \frac{u^{i+1}-u^i}{\Delta t}) \geq (f^{i+1}, v - \frac{u^{i+1}-u^i}{\Delta t}) + \left\langle \sigma_N(u^{i+1}), v_N - \frac{u_N^{i+1}-u_N^i}{\Delta t} \right\rangle \quad \forall v \in V \tag{20}$$

$$\left\langle \sigma_N(u^{i+1}), z_N - u_N^{i+1} \right\rangle \geq 0 \quad \forall z \in K \tag{21}$$

where $u^0 = u_0$ and $f^{i+1} = f((i+1).\Delta t)$.

Existence of a Solution of Problem P_i^n

In order to show that there exists a solution of each problem P_i^n we shall construct equivalent problems.

Let us first consider the following equivalent problems obtained by multiplying the first inequality of P_i^n by Δt and by setting $w = \Delta t.v + u^i$.

Problem Q_i^n: Find $u^{i+1} \in K$ such that

$$a(u^{i+1}, w-u^{i+1}) + j(u^{i+1}, w-u^i) - j(u^{i+1}, u^{i+1}-u^i) \geq (f^{i+1}, w-u^{i+1}) + \left\langle \sigma_N(u^{i+1}), w_N - u_N^{i+1} \right\rangle \quad \forall w \in V$$

$$\left\langle \sigma_N(u^{i+1}), z_N - u_N^{i+1} \right\rangle \geq 0 \quad \forall z \in K.$$

Problems Q_i^n are equivalent to the following quasivariational inequalities:

Problem S_i^n: Find $u^{i+1} \in K$ such that

$$a(u^{i+1}, w-u^{i+1}) + j(u^{i+1}, w-u^i) - j(u^{i+1}, u^{i+1}-u^i) \geq (f^{i+1}, w-u^{i+1}) \quad \forall w \in K$$

Remark 1. Inequalities such as S_i^n have been considered by Cocu (1984), for example, and have been shown to possess a unique solution if the friction coefficient which appears in j is sufficiently small. To be more precise the friction coefficient must satisfy the following inequality $\bar{\mu} \leq \mu_1 < \frac{m}{C}$ where m and C have been defined in (10) and (12). We shall from now on suppose that the friction coefficient is small enough to ensure that the above inequality is true.

Remark 2. The compatibility condition (18) will ensure that u^1 is equal to u^0 if f^1 is equal to f^0.

Bounds for the Solution of P_i^n

We shall exhibit bounds for the incremental solution u^i and also for the difference $u^{i+1} - u^i$. By setting in inequalities S_i^n, for $i = 0 \ldots n-1$, $w = 0$ we obtain

$$-a(u^{i+1}, u^{i+1}) + j(u^{i+1}, -u^i) - j(u^{i+1}, u^{i+1}-u^i) \geq -(f^{i+1}, u^{i+1}) \qquad \forall i = 0 \ldots n-1$$

Therefore

$$a(u^{i+1}, u^{i+1}) \leq (f^{i+1}, u^{i+1}) + j(u^{i+1}, -u^i) - j(u^{i+1}, u^{i+1}-u^i) \qquad \forall i = 0 \ldots n-1$$

and by (13)

$$a(u^{i+1}, u^{i+1}) \leq (f^{i+1}, u^{i+1}) + j(u^{i+1}, -u^{i+1}) \qquad \forall i = 0 \ldots n-1.$$

Finally by (10) and (12) we obtain :

$$m\|u^{i+1}\|^2 \leq \|f^{i+1}\| \, \|u^{i+1}\| + \bar{\mu}C\|u^{i+1}\|^2 \qquad \forall i = 0 \ldots n-1$$

so that if $\bar{\mu}C < m$ we have :

$$\|u^{i+1}\| \leq \frac{\|f^{i+1}\|}{m - \bar{\mu}C} \qquad \forall i = 0 \ldots n-1 \ .$$

By setting $w = 0$ in condition (18) we finally obtain :

$$\|u^i\| \leq \frac{\|f^i\|}{m - \bar{\mu}C} \qquad \forall i = 0 \ldots n \ . \tag{22}$$

Let us now set $\Delta u^i = u^{i+1} - u^i$ and $\Delta f^i = f^{i+1} - f^i$ for $i = 0...n-1$. Considering inequality S_i^n in which we set $w = u^i$ and adding it to inequality S_{i-1}^n in which we set $w = u^{i+1}$ (if $i = 0$ we set $w = u^1$ in condition (18)) we obtain the following inequality: $\forall i = 0 \dots n-1$

$$a(\Delta u^i, \Delta u^i) \le (\Delta f^i, \Delta u^i) + j(u^i, \Delta u^i + \Delta u^{i-1}) - j(u^i, \Delta u^{i-1}) - j(u^{i+1}, \Delta u^i)$$

And by (10) and (12) we obtain the following bound for Δu^i

$$\|\Delta u^i\| \le \frac{\|\Delta f^i\|}{m - \bar{\mu} C} \qquad \forall i = 0 \dots n-1 \ . \tag{23}$$

EXISTENCE OF A QUASI-STATIC SOLUTION

We define, for all $n \in \mathbf{N}^*$, a function f_n: $[0,T] \to V$ by $f_n(0) = f(0)$ and $f_n(t) = f(t_{i+1})$ for all $t \in]t_i, t_{i+1}]$ where $t_i = i.\Delta t$ and $\Delta t = T/n$. We then consider inequalities S_i^n with $f^i = f(t_i)$, for $i = 0...n$, in order to define a function $u_n \in L^2(0,T;V)$ by

$u_n(0) = u_0$ and

$u_n(t) = u^{i+1}$, the solution of S_i^n , for $t \in]t_i, t_{i+1}]$.

The proofs of the following results of convergence will be given in a forthcoming paper.

Lemma 1 - There exists a subsequence $(u_{n_p})_p$ of (u_n) such that for all $t \in [0,T]$ $(u_{n_p}(t))_p$ converges weakly in V to u(t), where u is an element of $L^2(0,T;V)$.

Define the following function:

$$\begin{cases} \tilde{u}_n(t) = u^i + \dfrac{(t - t_i)}{\Delta t}(u^{i+1} - u^i) & \forall t \in]t_i, t_{i+1}[\\ \tilde{u}_n(0) = u_0 & \end{cases}$$

where n belongs to the subset of $\mathbf{N}$ corresponding to the subsequence of (u_n) one has exhibited in Lemma 1.

Lemma 2 - There exists a subsequence of $(\tilde{u}_n)$ that converges weakly towards u in $W^{1,2}(0,T;V)$.

In the following, we shall still denote a subsequence by its initial sequence.

As $\left\langle \sigma_N(u^{i+1}), u_N^{i+1} \right\rangle = 0$, $\sigma_N(u^{i+1}) \le 0$ and $u_N^i \le 0$ we have

$$\left\langle \sigma_N(u^{i+1}), v_N - \frac{u_N^{i+1} - u_N^i}{\Delta t} \right\rangle \ge \left\langle \sigma_N(u^{i+1}), v_N \right\rangle$$

and therefore u^{i+1} shall satisfy for $i = 0...n-1$

$$a(u^{i+1}, v - \frac{u^{i+1} - u^i}{\Delta t}) + j(u^{i+1}, v) - j(u^{i+1}, \frac{u^{i+1} - u^i}{\Delta t}) \ge$$
$$(f^{i+1}, v - \frac{u^{i+1} - u^i}{\Delta t}) + \left\langle \sigma_N(u^{i+1}), v_N \right\rangle$$

Thus for all $v \in L^2(0,T;V)$

$$\int_0^T (a(u_n(t), v(t) - \frac{d}{dt}\tilde{u}_n(t)) + j(u_n(t), v(t)) - j(u_n(t), \frac{d}{dt}\tilde{u}_n(t)))dt \geq \int_0^T (f_n(t), v(t) - \frac{d}{dt}\tilde{u}_n(t))dt + \int_0^T \langle \sigma_N(u_n(t)), v_N(t) \rangle dt \tag{24}$$

Lemma 3 - We have for all $v \in L^2(0,T;V)$

$$\lim_{n\to\infty} \int_0^T (f_n(t), v(t) - \frac{d}{dt}\tilde{u}_n(t))dt = \int_0^T (f(t), v(t) - \dot{u}(t))dt,$$

$$\lim_{n\to\infty} \int_0^T j(u_n(t), v(t))dt = \int_0^T j(u(t), v(t))dt,$$

$$\lim_{n\to\infty} \int_0^T a(u_n(t), v(t))dt = \int_0^T a(u(t), v(t))dt.$$

Lemma 4 - For some subsequences of (u_n) and $(\tilde{u}_n)$

$$\liminf_{n\to\infty} \int_0^T a(u_n(t), \frac{d}{dt}\tilde{u}_n(t))dt \geq \int_0^T a(u(t), \dot{u}(t))dt$$

$$\liminf_{n\to\infty} \int_0^T j(u_n(t), \frac{d}{dt}\tilde{u}_n(t))dt \geq \int_0^T j(u(t), \dot{u}(t))dt$$

Thus the existence of a quasistatic solution follows immediately from Lemmas 3 and 4.

THEOREM - The weak limit u of (u_n) satisfies the following inequalities:

$$\int_0^T (a(u(t), v(t) - \dot{u}(t)) + j(u(t), v(t)) - j(u(t), \dot{u}(t)))dt$$

$$\geq \int_0^T (f(t), v(t) - \dot{u}(t))dt + \int_0^T \langle \sigma_N(u(t)), v_N(t) - \dot{u}_N(t) \rangle dt$$

for all $v \in L^2(0,T;V)$ and $\forall t \in [0,T]$ a.e. $u(t) \in K$ such that

$$\langle \sigma_N(u(t)), z_N - u_N(t) \rangle \geq 0 \quad \forall z \in K.$$

A NUMERICAL EXAMPLE

We consider a symmetric linear elastic bloc set on a plane surface submitted to a vertical force density f = - 5 daN/mm^2 and to a horizontal force density F = 10 daN/mm^2. Unilateral contact with quasistatic Coulomb's friction law and $\mu = 0.5$ are assumed. We consider plane strain elasticity with a Young's modulus E = 13 000 daN/mm^2 and a Poisson's ratio $\nu = 0.2$. The problem, considered by Raous et al.(1988), Licht et al.(1991), has been solved by a finite element method, using a 230 node mesh with 32 contact nodes. If one applies f before F (fig.1) one obtains a different solution from the one obtained by applying F first (fig.2).

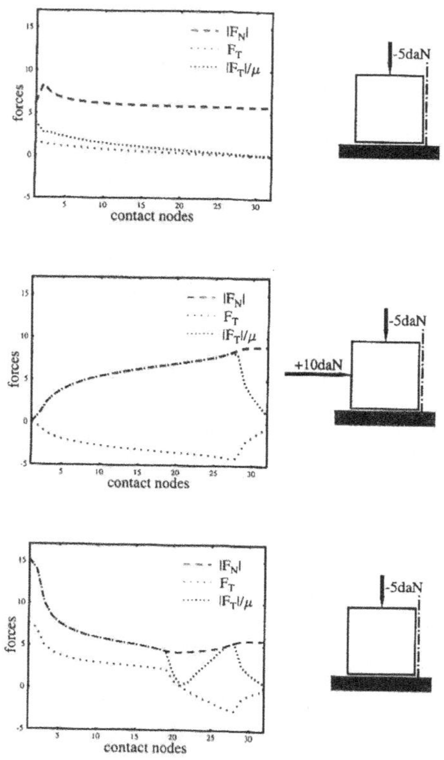

Figure 1. Normal and tangential forces for sequence 1.

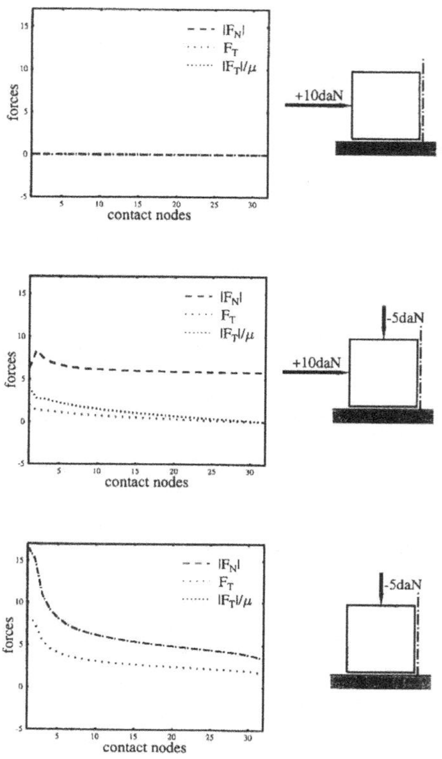

Figure 2. Normal and tangential forces for sequence 2.

REFERENCES

Andersson, L.E., 1991, A quasistatic frictional problem with normal compliance, *Nonlinear Analysis T.M.A.* 16:347.

Cazenave, T., and Haraux, A., 1990, "Introduction aux Problèmes d'Evolution Sémi-linéaires" Mathématiques et Applications, Ellipses, Paris.

Cocu, M., 1984, Existence of solutions of Signorini problems with friction, *Int. J. Engng. Sci.* 22:567.

Duvaut, G., 1980, Equilibre d'un solide élastique avec contact unilatéral et frottement de Coulomb, *C. R. Acad. Sci. Paris*, 290, série A:263.

Duvaut, G., and Lions, J.-L., 1972, "Les Inéquations en Mécanique et en Physique", Dunod, Paris.

Klarbring, A., Mikelic, A., and Shillor, M., 1991, A global existence result for the quasistatic frictional contact problem with normal compliance, in "International Series of Numerical Mathematics" 101:85, Birkhäuser Verlag, Basel.

Licht, C., Pratt, E., and Raous, M., 1991, Remarks on a numerical method for unilateral contact including friction, in "International Series of Numerical Mathematics" 101:129, Birkhäuser Verlag, Basel.

Raous, M., Chabrand, P., and Lebon, F., 1988, Numerical methods for frictional contact problems and applications, *J. M. T. A., Special issue to vol.* 7:111.

LINEAR INDENTERS UNDER CONDITIONS OF PARTIAL SLIP

D.A. Hills[1], A. Sackfield[2] and C.E. Truman[2]

[1]Department of Engineering Science
Oxford University, England, OX1 3PJ
[2]Department of Mathematics
Nottingham Trent University, England. NG1 4BU

INTRODUCTION

Two experimental problems demanding well defined, clearly understood contacts are the use of elastic indentation testing to assess the brittle fracture strength of ceramics and other brittle materials (Ostojic and McPherson, 1987), and the quantification of fretting fatigue damage (Hills and Nowell, 1994). In such cases the use of *incomplete* contacts, ie. those where the contact size increases with applied load, are very much to be preferred over *complete* contacts, such as that arising under a flat ended punch, as the resulting indentation is much less sensitive to manufacturing errors, is more straightforward to analyse, and the resulting stress field is devoid of singularities. The most obvious configuration to choose is therefore the Hertzian contact, developed by pressing a sphere into the substrate material, and indeed this has wide application to each of the problems described. However, there are many cases where an alternative shape of indenter has much to recommend it, at least in part because a different contact stress field may be used to explore the influence of the state of stress on surface crack propulsion. The simplest alternative shapes which may be used are the wedge and cone. These clearly have the immediate disadvantage that a singular point exists at the apex, so that there will inevitably be a small enclave of plasticity present there. However, providing that the size of this zone is small in comparison with the half width or radius, a, of the contact patch, the problem will be dominated by the elastic response of the substrate, and a wholly elastic solution will give a very fair represenstation of the stress and displacement field, except near the plastic zone itself. In most practical problems the cracks actually nucleate near the *edge* of the contact patch, and hence this is not a serious difficulty.

CONTACT PROBLEM

In this brief paper we are not able to develop the full solutions to the problems posed,

Contact Mechanics, Edited by M. Raous *et al.*
Plenum Press, New York, 1995

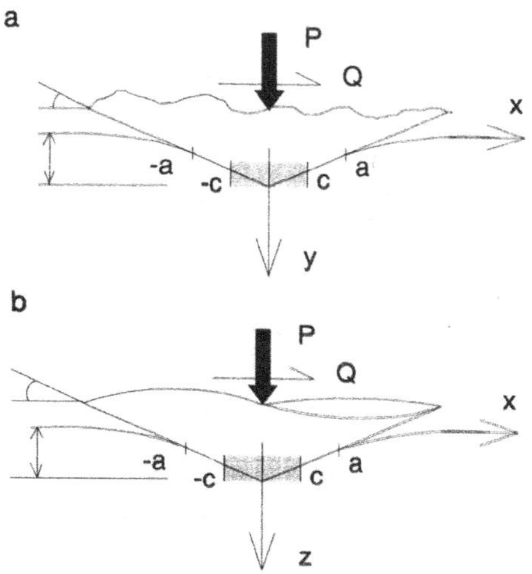

Figure 1. Wedge and Cone Geometries

and so only the results will be given here. The underlying theory may be found in Hills, Nowell and Sackfield (1993); each uses a formulation appropriate to a half-space, and we also assume throughout the paper that, if the contacting bodies are elastically dissimilar, the influence of the shearing tractions on the surface normal displacement may be neglected. The geometry of both the plane and axi-symmetric cases is depicted in Fig.1. In order to permit a unified treatment of the plane and axi-symmetric geometries, we have deliberately used a complementary nomenclature for the pair of problems. Further details of the evaluation of the contact law and pressure distribution may be found in Truman, Sackfield and Hills (1994). The results found are given in table A. where P is the applied load, R is the radius of the cylinder or sphere, and A is the composite compliance of the material pair, defined by

$$A = \frac{\kappa_1 + 1}{4\mu_1} + \frac{\kappa_2 + 1}{4\mu_2} \tag{1}$$

$$\kappa_i = 3 - 4\nu_i \qquad \text{in plane strain}$$

$$\kappa_i = (3 - \nu_i)/(1 + \nu_i) \qquad \text{in plane stress}$$

μ_i is the modulus of rigidity of material i, and ν_i its Poisson's ratio. The normalized contact pressure distributions are plotted out in fig.2. It is interesting to note how the *average* contact pressure varies with the geometry of the indenter for each of the problems under consideration; if we consider the Hertz problem first, in which either a cylinder or sphere indents the half-space, it is clear that the mean pressure $\bar{p}$ varies as $R^{-1/2}$ and $R^{-2/3}$ respectively. By contrast, in the case of the wedge or cone indentation, the contact pressure is, in each case, linearly proportional to the external angle ϕ, fig.1.

INTERIOR STRESS FIELD

The interior stress field obtaining beneath a sliding conical indenter has been found by Hanson(1992,1993), using a formulation initially for the more general case when the substrate possesses only transverse isotropy. The interior stress field for the isotropic case may be obtained more directly using the Green, Collins and Barber method (see

Table 1. **Contact Law**

1. *Pressure Distribution*		**Linear**	**Parabolic**
Plane	$\frac{ap}{P}$	$\frac{1}{\pi}\cosh^{-1}\left(\frac{a}{\lvert x\rvert}\right)$	$\frac{2}{\pi}\sqrt{1-\left(\frac{r}{a}\right)^2}$
Axi-Symmetric	$\frac{a^2p}{P}$	$\frac{1}{\pi}\cosh^{-1}\left(\frac{a}{r}\right)$	$\frac{3}{2\pi}\sqrt{1-\left(\frac{r}{a}\right)^2}$

2. *Contact Patch Size*	**Linear**	**Parabolic**
Plane	$a = \frac{PA}{2\theta}$	$a^2 = \frac{2PAR}{\pi}$
Axi-Symmetric	$a^2 = \frac{PA}{\pi\theta}$	$a^3 = \frac{3PAR}{8}$

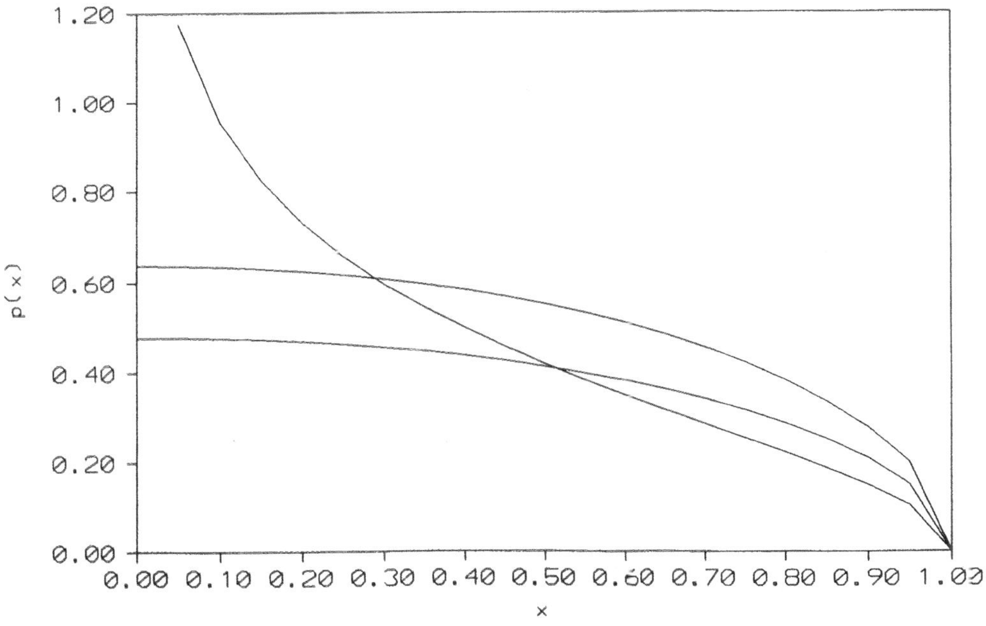

Figure 2. Normalised Pressure Distributions

Table 2. **Partial Slip Response**

Stick Zone Size		**Linear**	**Parabolic**
Plane	$\frac{c}{a}$	$1 - \frac{Q}{fP}$	$\sqrt{1 - \frac{Q}{fP}}$
Axi-Symmetric	$\frac{c}{a}$	$\sqrt{1 - \frac{Q}{fP}}$	$\sqrt[3]{1 - \frac{Q}{fP}}$

Hills, Nowell and Sackfield, 1993). The stress field resulting from indentation by a wedge may be found most readily by using a Muskhelishvili representation of the state of stress (Truman, Sackfield and Hills, 1994). Results will not be reproduced here, and we turn now to the question of the response of the contacts in question to the development of conditions under which partial slip obtains.

For brevity here only the results for partial slip problems between similar materials with an applied tangential force less than or equal to that necessary to cause sliding are shown in table 2 (the Mindlin and Cattaneo problem). For further details Truman, Sacfield and Hills (1994) may be consulted.

ACKNOWLEDGMENT

One of us (C.E.T.) gratefully acknowledges the support of the Electro-Mechanical Technology Advancing Foundation, Tokyo, and the European Research Centre of NSK-RHP Bearings during this work.

REFERENCES

Hanson, M.T., 1992, The elastic field for conical indentation including sliding friction for transverse isotropy, *J. Appl. Mech.*, 59:S123

Hanson, M.T., 1993, The elastic field for a sliding conical punch on an isotropic half-space, *J. Appl. Mech.*, 60:557.

Hills, D.A. and Nowell, D., 1994, "Mechanics of Fretting Fatigue", Kluwer Academic Press, Dordrecht, Netherlands.

Hills, D.A., Nowell, D. and Sackfield, A., 1993, "Mechanics of Elastic Contacts", Butterworth Heinemann, Oxford.

Ostojic, P. and McPherson, R., 1987, A review of indentation theory:its development, principles and limitations, *Int. J. Fract.*, 33:297.

Truman, C.E., Sackfield, A. and Hills, D.A., 1994, Contact mechanics of wedge and cone indenters, *Int. J. Mech. Sci.*, in press.

DYNAMICAL CONTACT PROBLEMS FOR VISCOELASTIC BODIES

Jiří Jarušek

Institute of Inform. Theory and Autom., Czech. Acad. Sci.
Prague, Czech Republik

INTRODUCTION

The aim of this contribution is to present a surway about author's recent results in the field concerning the existence and regularity of solutions. It is well known that it is very hard to solve dynamical contact problem (cf. Amerio and Prouse, 1975, Bamberger and Schatzmann, 1983, Maruo, 1985, Schatzmann, 1983 etc., for a global information Cabannes and Citrini, 1987 etc.) due to its hyperbolic character for which the convenient constraint formulation is in velocities while, due the obvious physical reality, the constraint should be formulated in displacements. This yields the necessity to put displacement into the variational inequality which gives the opposite sign at the velocity term after the integration by parts. The viscoelasticity approach can overcome such problem, since it enables to prove the strong convergence of velocities. In this context we study two types of models:

1. the material has a certain singular memory of the Hencky type,

2. the stress tensor depends on the gradient of velocities.

We note that the viscoelasticity "parabolizes" the problem in some sense (this phenomenon is more remarkable at the second approach).

Another reason to involve the viscoelasticity occurs when the friction in the framework of the Coulomb law must be treated. In the dynamical case the velocity arises in the friction boundary term, therefore some volume gradient term on the left-hand side of the a priori estimate is needed. Of course, for such problems only the second approach yields some results.

As the viscoelastic approach seems to be based on a good physical reasoning, it can give for suitable materials interesting results also in cases, where the existence of a solution of the corresponding problems neglecting the viscoelasticity remain still open.

It is necessary to note that the existence of energy conserving solutions and the unicity of problems remains mostly open.

We remark that in the sequel τ is used for the time variable. The dot denotes the time derivative, ∇ the space gradient and the prime a derivative of a function defined on a subset of $\mathbb{R}$. W_p^k denotes the appropriate Sobolev space (the order of the derivative k could be fractional) and $W_2^k \equiv H^k$. B_0 denotes the space of bounded maps with the sup-norm. The above described notation is also used for the Bochner-type spaces.

Contact Mechanics, Edited by M. Raous *et al.*
Plenum Press, New York, 1995

THE MATERIALS WITH SINGULAR MEMORY

We start from the simple linear case of a membrane. We look for the solution u of the following problem

$$\begin{aligned} \ddot{u} - \operatorname{div}\sigma(u) - f = g \text{ on } Q_T = I_T \times \Omega, \\ u \geq 0,\ g \geq 0,\ ug = 0 \text{ on } Q_T, \\ u(0,\cdot) = u_0,\ \dot{u}(0,\cdot) - u_1 \geq 0,\ (\dot{u}(0,\cdot) - u_1)u_0 = 0 \text{ on } \Omega, \\ \sigma(u)n \geq 0,\ u \geq 0,\ u(\sigma(u)n) = 0 \text{ on } S_T,\quad S_T = I_T \times \partial\Omega, \end{aligned} \tag{1}$$

where Ω is a domain in $I\!R^N, N \in I\!N$, having the boundary $C_{1,1}$–smooth (the Lipschitz gradient), n is the unit outer normal vector and $I_T \equiv (0,T)$ a time interval. Here

$$\begin{aligned} \sigma(u)(x,\tau) = \sigma^I(u)(x,\tau) + \sigma^V(u)(x,\tau),\quad \sigma^I(u) \equiv \nabla u \quad \text{and} \\ \sigma^V(u)(x,\tau) \equiv \textstyle\int_{-\infty}^{t} K(s-\tau)\nabla(u(x,s) - u(x,\tau))\,ds \equiv K \otimes \nabla u(x,\tau), \\ K(\tau) = \begin{cases} 0, & \tau \leq 0, \\ c(\tau)\tau^{-2\alpha} + a(\tau), & \tau > 0, \end{cases} \end{aligned} \tag{2}$$

Both a and c are sufficiently smooth, nonnegative and nonincreasing functions on $I\!R_+$ with $c(\tau) > 0$ for τ from a neighbourhood of the origin. Let $f \in L_2(Q_T)$, $u_0 \in H^1(\Omega)$, $u_0 \geq 0$ a.e. in Ω, $u_1 \in H^1(\Omega)$. Put

$$C = \{v \in H^1(Q_T);\ v \geq 0 \text{ a.e. in } Q_T\}. \tag{3}$$

Then a function $u \in C$ such that $\nabla u \in H^\alpha(I_T;\ L_2(\Omega; I\!R^N))$, $\dot{u} \in L_\infty(I_T;\ L_2(\Omega))$ and $\dot{u}(T,\cdot) \in L_2(\Omega)$ will be a weak solution to (1), if for every $v \in C$

$$\begin{aligned} \textstyle\int_{Q_T}(\nabla u,\ \nabla(v-u)) + (K \otimes \nabla u,\ \nabla(v-u)) - \dot{u}(\dot{v}-\dot{u}) - f(v-u)\,dx\,d\tau + \\ \textstyle\int_\Omega(\dot{u}(v-u))(T,\cdot) - u_1(v(0,\cdot) - u_0)\,dx \geq 0. \end{aligned} \tag{4}$$

In Jarušek (1994) the following theorem was proved (the result for $N = 1$ is straightforward):

THEOREM 1. *Under assumptions concerning Ω, u_0 u_1, let α from (2) satisfy*

$$\alpha \in (1 - 2/N, 1/2),\ N \geq 2,\quad \alpha \in (0, 1/2),\ N = 1, \tag{5}$$

and let $||a||_{L_\infty(I\!R)}, ||c||_{L_\infty(I\!R)}$ be so small that for some $k < 1$ and every $w \in L_2(I_T;\ H^1(\Omega))$

$$\int_{Q_T}(K \otimes \nabla w, \nabla w)\,dx\ dt \leq k\ ||\nabla w||^2_{L_2(Q_T;\ R^N)}. \tag{6}$$

Then there exists a weak solution to the problem (1).

We remark that, particularly, for $N \geq 4$ the interval in (5) is empty.

The method of its proof is common to all the theorems in this and the next section. Taking a penalty function h in the form

$$h(r) = (r^-)^2,\ r \in I\!R,\ \text{where } r^- = \min(0,r),\ r \in I\!R, \tag{7}$$

we introduce the penalized problem having the classical formulation

$$\ddot{u}_\varepsilon - \operatorname{div}\sigma(u_\varepsilon) - f + \tfrac{1}{\varepsilon}h'(u_\varepsilon) = 0 \text{ on } Q_T, \qquad (8)$$
$$u_\varepsilon(0,.) = u_0,\ \dot{u}_\varepsilon(0,.) = u_1 \text{ on } \Omega, \quad \sigma(u_\varepsilon)n = 0 \text{ on } S_T.$$

Its variational formulation follows: $u_\varepsilon \in H^1(Q_T)$ having $\dot{u}_\varepsilon \in B_0(I_T;\ L_2(\Omega))$, $\nabla u_\varepsilon \in B_0\left(I_T;\ L_2(\Omega; I\!R^N)\right) \cap H^\alpha\left(I_T;\ L_2(\Omega; I\!R^N)\right)$, $K \otimes \nabla u_\varepsilon \in L_2(Q_T;\ I\!R^N)$ and $\dot{u}_\varepsilon(T,\cdot) \in L_2(\Omega)$ will be a weak solution to (8), if for every $v \in L_2\left(I_T;\ H^1(\Omega)\right)$ the equation

$$\int_{Q_T} (\nabla u_\varepsilon, \nabla v) + (K \otimes \nabla u_\varepsilon, \nabla v) + \ddot{u}_\varepsilon v - fv + \frac{1}{\varepsilon}h'(u_\varepsilon)v \,dx\, d\tau = 0 \qquad (9)$$

holds. We put in the standard way $v = \dot{u}_\varepsilon$ into (9). By computation we easily prove

LEMMA 1. *Let B be a Hilbert space and let v: $I\!R \to B$ satisfy $v, \dot{v} \in (L_1 \cap B_0)(I\!R_+;\ B)$ and let v be extended such that $v(\tau) = v(0)$ for $\tau \leq 0$. Then*

$$\int_0^{+\infty} (K \otimes v, \dot{v})_B \,ds = \frac{1}{2}\int_0^{+\infty}\int_0^{+\infty} -K'(s)|v(\tau) - v(\tau - s)|_B^2 \,d\tau\, ds. \qquad (10)$$

Using this and the assumptions to K, a and c, we arrive at a priori estimate

$$\sup_{\tau \in I_T}\left(\|\dot{u}_\varepsilon(\tau,\cdot)\|^2_{L_2(\Omega)} + \|\nabla u_\varepsilon(\tau,\cdot)\|^2_{L_2(\Omega;\ I\!R^N)} + \tfrac{1}{\varepsilon}\|u_\varepsilon^-(\tau,\cdot)\|^2_{L_2(\Omega)}\right) + \qquad (11)$$
$$\|\nabla u_\varepsilon\|^2_{H^\alpha(I_T;\ L_2(\Omega; I\!R^N))} + \|u_\varepsilon\|^2_{L_2(Q_T)} \leq c_0 \equiv c_0(\mathcal{J})$$
$$\text{with } \mathcal{J} \equiv [\|u_0\|_{H^1(\Omega)}, \|u_1\|_{L_2(\Omega)}, \|f\|_{L_2(Q_T)}] \in I\!R^3,$$

where the memory gives the fractional norm in the described way. An additional but very important a priori estimate

$$\frac{1}{\varepsilon}\int_{Q_T} h'(u_\varepsilon)\,dx\,d\tau = \frac{2}{\varepsilon}\|u_\varepsilon^-\|_{L_1(Q_T)} < c_1 \equiv c_1(\mathcal{J}) \qquad (12)$$

is a consequence of putting $v = 1$ into (9).

The standard Galerkin procedure gives the existence of a solution of the equation (9). The a priori estimates (11) and (12) together with the standard imbedding theorem used to the adjoint operator yield the boundedness of $\{\ddot{u}_\varepsilon;\ \varepsilon > 0\}$ in $L_1\left(I_T;\ (H^{\frac{N}{2}+\eta}(\Omega))^*\right)$ for every $\eta > 0$. Then, under assumption (5), the interpolation technique based on results of Běsov, Il'jin and Nikol'skij (1975), Schmeisser (1987) and Schmeisser and Triebel (1987) yields the relative compactness of $\{\dot{u}_\varepsilon;\ \varepsilon > 0\}$ in $L_1(I_T;\ H^1(\Omega))$. (We interpolate the estimate for $\ddot{u}_\varepsilon$ with the estimate of the fractional norm in (11)). The estimate (11) yields simultaneously the boundedness of $\{\dot{u}_\varepsilon;\ \varepsilon > 0\}$ in $L_\infty(I_T;\ H^1(\Omega))$, From this we obtain (via the Hölder inequality and the convergence of norms) the strong convergence of velocities for some $\varepsilon_k \to 0$. The strong convergence of gradients follows from (6). (Denote the limit by u, put $v = u_{\varepsilon_k} - u$ into (9), add $\int_{Q_T} (\nabla u, \nabla(u - u_{\varepsilon_k})) + (K \otimes \nabla u, \nabla(u - u_{\varepsilon_k}))\,dx\,d\tau$ to both sides of the equality and estimate the appropriate L_2–norm employing their weak convergence) Then it is easy to see that u is a weak solution to (1). □

We remark that some "fractional" regularity $u \in H^{1+\delta}(Q_T)$ for any $\delta \in (0, \frac{\alpha}{4})$, $N = 2$, or any $\delta \in \frac{3}{10}(0, \alpha - \frac{1}{3})$, $N = 3$, can be proved by means of the localization and shift technique described in Jarušek (1983) in all details. Moreover, the proof technique of Theorem 1 can be used also for the Dirichlet boundary value condition replacing the Signorini one in (1).

For details see Jarušek (1994). The method of proofs of the existence and regularity of solutions can also be used for polyharmonic problems (including those for plates — cf. Jarušek[1].

In the contact problem of **a body** the same problems with the velocity occurs in spite of the boundary character of the contact (cf. Jarušek[2]). Here we show the possibility to use the approach to nonlinear models. Let Ω, I_T, Q_T and S_T have the same sense and properties as in the preceding case. For a displacement u the stress tensor $\tilde{\sigma}(u) \equiv (\sigma_{ij}(u))$ is assumed to have the form $\left(\sigma^I_{ij}(u)\right) + \left(\sigma^V_{ij}(u)\right)$, where $\sigma^I_{ij}(u) = \frac{\partial W}{\partial e_{ij}}(\cdot, \tilde{e}(u))$, $i,j = 1,\dots,N$, and

$$\sigma^V_{ij}(u) = -\int_0^\tau K(\tau-\ell)\frac{\partial V}{\partial e_{ij}}(\cdot, \tilde{e}(u(\tau,\cdot)) - \tilde{e}(u(\ell,\cdot)))\,d\ell \tag{13}$$

with the small strain tensor $\tilde{e}(u) \equiv (e_{ij}(u)) = \left(\frac{1}{2}\left(\frac{\partial u_i}{\partial x_j} + \frac{\partial u_j}{\partial x_i}\right)\right)$. The inviscid stored energy function $W\colon I\!R^{N+N^2} \to I\!R$ is assumed to be C_2–smooth and to have the partial Hess matrix strongly elliptic and bounded, i.e. there are $\beta_0^W, \beta_1^W > 0$ such that

$$\beta_0^W \xi_{ij}\xi_{ij} \le \frac{\partial^2 W}{\partial e_{ij}\partial e_{kl}}(x,\omega)\,\xi_{ij}\xi_{kl} \quad \text{and} \quad \frac{\partial^2 W}{\partial e_{ij}\partial e_{kl}}(x,\omega)\xi_{ij}\zeta_{kl} \le \beta_1^W\sqrt{\xi_{ij}\xi_{ij}}\sqrt{\zeta_{kl}\zeta_{kl}} \tag{14}$$

for all symmetric $N\times N$ matrices ξ,ζ,ω and all $x \in \Omega$ and satisfying $W(\cdot,0) = 0$, $\frac{\partial W}{\partial e_{ij}}(\cdot,0) = 0$. The viscous stored energy function V is assumed to have the same properties as W, particularly the constants corresponding to those in (14) will be denoted by β_0^V, β_1^V. Here and in the sequel the summation convention is consequently used. The kernel K is assumed in the form defined in (2) with the same properties of the functions. The classical formulation of the problem is

$$\begin{gathered} \ddot{u}_i - \frac{\partial}{\partial x_j}\sigma_{ij}(u) = f_i,\ i=1,\dots,N \quad \text{on} \quad Q_T \equiv I_T\times\Omega, \\ u_n \le 0,\ T_n(u) \le 0,\ T_n(u)\,u_n = 0, \quad T_t(u) = 0 \quad \text{on} \quad S_{c,T} \equiv I_T \times \partial\Omega_c, \\ T = T_0 \quad \text{on } S_{T,T} \equiv I_T \times \partial\Omega_T, \qquad u(0,\cdot) = u_0,\ \dot{u}(0,\cdot) = u_1 \quad \text{on} \quad \Omega. \end{gathered} \tag{15}$$

Here the boundary of Ω is divided into two disjoint measurable parts: a contact part $\partial\Omega_c$, where no friction occurs, and the remaining part $\partial\Omega_T$ where the stress is prescribed. We remark that the existence of a part of the boundary, where the displacement is given, is also admissible, but we avoid technicalities. T denotes the boundary stress vector ($T_i(u) \equiv \sigma_{ij}(u)n_j$, $i=1,\dots,N$). For a vector function $w\colon \partial\Omega \to I\!R^N$ we denote $w_n \equiv w_i n_i$ its normal component and $w_t \equiv w - w_n n$ its tangential component.

Denoting

$$\mathcal{K} := \{v \in H^1(\Omega; I\!R^N);\ v_n \le 0 \text{ a.e. in } \partial\Omega_c\}, \tag{16}$$

we can introduce the variational formulation of the problem: a weak solution to (15) will be a function $u \in B_0(I_T;\ H^1(\Omega; I\!R^N))$ for which $u(\tau,\cdot) \in \mathcal{K}$ for a.e. $\tau \in I_T$, $\nabla u \in H^\alpha(I_T;\ L_2(\Omega; I\!R^{N^2}))$, $\dot{u} \in L_\infty(I_T;\ L_2(\Omega; I\!R^N))$, $\dot{u}(\mathcal{T},\cdot) \in L_2(\Omega; I\!R^N)$ and for all $v \in H^1(Q_T; I\!R^N)$ such that $v(\tau,\cdot) \in \mathcal{K}$ a.e. in I_T the following inequality holds:

$$\begin{gathered} \int_{Q_T} \sigma_{ij}(u)e_{ij}(v-u) - \dot{u}_i(\dot{v}_i - \dot{u}_i)\,dx\,d\tau + \int_\Omega (\dot{u}_i(v_i - u_i))(\mathcal{T},\cdot)\,dx \ge \\ \int_\Omega (u_1)_i(v_i(0,\cdot) - (u_0)_i)\,dx + \int_{Q_T} f_i(v_i - u_i)\,dx\,d\tau + \int_{S_{T,T}} T_{0,i}(v_i - u_i)\,dx\,d\tau. \end{gathered} \tag{17}$$

We solve our problems under the assumption

$$u_0 \in \mathcal{K},\ u_1 \in L_2(\Omega; I\!R^N),\ f \in L_2(Q_T; I\!R^N), \text{ and } T_0 \in H^1(I_T;\ (H^{\frac{1}{2}}(\partial\Omega_T; I\!R^N))^*). \tag{18}$$

Our aim is to prove

THEOREM 2. *Let the assumptions concerning* Ω, *its boundary, the functions* V, W *the assumptions* (5) *and* (18) *and*

$$2\beta_1^V \int_{I\!R_+} K(s)\,\mathrm{d}s < \beta_0^W \tag{19}$$

hold. Then there exists a weak solution to the contact problem (15).

To prove it, we introduce the penalized problem whose variational formulation follows: u_ε will be the weak solution of the penalized problem, if $u_\varepsilon \in B_0(I_T;\ H^1(\Omega; I\!R^N))$ for which $\dot{u}_\varepsilon \in B_0(I_T;\ L_2(\Omega; I\!R^N))$ and $\ddot{u}_\varepsilon \in L_2(I_T;\ (H^1(\Omega; I\!R^N))^*)$, the initial condition in (15) is satisfied and the following equation

$$\int_{Q_T} (\ddot{u}_\varepsilon)_i v_i + \sigma_{ij}(u_\varepsilon) e_{ij}(v)\,\mathrm{d}x\,\mathrm{d}\tau + \int_{S_{cT}} \frac{1}{\varepsilon}\tilde{h}'((u_\varepsilon)_n) v_n\,\mathrm{d}x\,\mathrm{d}\tau = \int_{Q_T} f_i v_i\,\mathrm{d}x\,\mathrm{d}\tau + \int_{S_{T,T}} T_{0,i} v_i\,\mathrm{d}x\,\mathrm{d}\tau \tag{20}$$

holds for all $v \in L_2(I_T;\ H^1(\Omega; I\!R^N))$. The penalty function $\tilde{h}$ has here the form

$$\tilde{h}\colon z \mapsto (z^+)^2 \text{ with } z^+ \equiv \max(0, z),\ z \in I\!R. \tag{21}$$

In fact, the penalized problem consists in replacing the Signorini boundary value condition on S_c in (15) by the condition $T_n(u_\varepsilon) = -\frac{1}{\varepsilon}\tilde{h}'((u_\varepsilon)_n)$.

Putting $v = \dot{u}_\varepsilon$ on $I_\tau = (0, \tau)$, $v = 0$ on $I_T \setminus I_\tau$ into (20) denoting $Q_\tau \equiv I_\tau \times \Omega$, $\tau \in I_T$, and using the same notation for S_τ, $S_{T,\tau}$, $S_{c,\tau}$ we obtain

$$\begin{aligned}
&\int_\Omega \left(\tfrac{1}{2}|\dot{u}_\varepsilon|^2 + W(\cdot, \tilde{e}(u_\varepsilon))\right)(\tau, \cdot)\,\mathrm{d}x - \int_{Q_\tau}\int_0^\tau K'(s-\ell) V(\cdot, \tilde{e}(u_\varepsilon(s, \cdot)) - \tilde{e}(u_\varepsilon(\ell, \cdot))\,\mathrm{d}\ell\,\mathrm{d}s\,\mathrm{d}x \\
&\quad + \int_{\partial\Omega_c} \tfrac{1}{\varepsilon}|(u_\varepsilon)_n^+|^2(\tau, \cdot)\,\mathrm{d}x + \int_0^\tau\int_\Omega K(\tau-\ell) V(\cdot, \tilde{e}(u_\varepsilon(\tau, \cdot)) - \tilde{e}(u_\varepsilon(\ell, \cdot)))\,\mathrm{d}\ell\,\mathrm{d}x \\
&\quad = \int_\Omega \tfrac{1}{2}|u_1|^2 + (W(\cdot, \tilde{e}(u_0))\,\mathrm{d}x + \int_{Q_\tau} f_i(\dot{u}_\varepsilon)_i\,\mathrm{d}x\,\mathrm{d}t \\
&\quad + \int_{S_{T,\tau}} -\dot{T}_{0,i}(u_\varepsilon)_i\,\mathrm{d}x\,\mathrm{d}s + \int_{\partial\Omega_T} T_{0,i}(\tau, \cdot)(u_\varepsilon)_i(\tau, \cdot) - T_{0,i}(0, \cdot) u_{0,i}\,\mathrm{d}x.
\end{aligned} \tag{22}$$

Let us denote by $\mathcal{R}$ the finite-dimensional kernel of $\tilde{e}$ (the space of shifts and rotations of body being rigid and undeformable) and by Y its orthogonal complement in $H^1(\Omega; I\!R^N)$ in the $L_2(\Omega; I\!R^N)$-scalar product. The assumptions to V and W yield their almost quadratic growth and the strong monotonicity of their gradients with respect to $\tilde{e}$. This and the Korn inequality (Duvaut and Lions 1972, Chapter III, Theorem 3.1) yields the a priori estimate

$$\begin{aligned}
&\sup_{\tau \in I_T}\left(\|\dot{u}_\varepsilon(\tau, \cdot)\|^2_{L_2(\Omega; I\!R^N)} + \|\nabla u_\varepsilon(\tau, \cdot)\|^2_{L_2(\Omega;\ I\!R^{N^2})} + \tfrac{1}{\varepsilon}\|(u_\varepsilon)_n^+(\tau, \cdot)\|^2_{L_2(\partial\Omega_c)}\right) + \\
&\|\nabla \pi_Y u_\varepsilon\|^2_{H^\alpha(I_T;\ L_2(\Omega; I\!R^{N^2}))} \le c_0 \equiv c_0(\mathcal{J}) \quad \text{with } \mathcal{J} \equiv \Big[\beta_0^W, \beta_1^W, \beta_0^V, \beta_1^V, \\
&\|u_0\|_{H^1(\Omega; I\!R^N)}, \|u_1\|_{L_2(\Omega; I\!R^N)}, \|f\|_{L_2(Q_T; I\!R^N)}, \|T_0\|_{H^1(I_T;\ (H^{-\frac{1}{2}}(\partial\Omega_T; I\!R^N))^*)}\Big],
\end{aligned} \tag{23}$$

where π_Y is the projection $H^1(\Omega; I\!R^N) \to Y$ and the parameter α is from (2). An additional a priori estimate

$$\frac{1}{\varepsilon}\|(u_\varepsilon)_n^+\|_{L_1(S_{c,T})} < const. \tag{24}$$

is obtained by putting $\tilde{v} \in H^1(Q_T; I\!R^N)$ such that $\dot{\tilde{v}} \in B_0(I_T;\ L_2(\Omega; I\!R^N))$, and $\tilde{v}_t \equiv 0$, $\tilde{v}_n \equiv 1$ both on S_{cT} into (20).

Due to nonlinearities and the possible non-triviality of $\mathcal{R}$ we must be here more careful even in the Galerkin approximation to (20). We use a L_2-orthogonal basis of $H^1(\Omega;\mathbb{R}^N)$ whose first components create a basis of $\mathcal{R}$. To estimate $\ddot{u}_\varepsilon$ we use (24) and the set of imbeddings

$$L_1(S_{c,T}) \hookrightarrow L_1(I_T;\, L^*_\infty(\partial\Omega_c)) \hookrightarrow L_1\Big(I_T;\, (H^{\frac{N-1}{2}+\vartheta}(\partial\Omega_c))^*\Big), \tag{25}$$

being valid with $\vartheta > 0$ arbitrary. This yields the boundedness of $\{T_n(u_\varepsilon);\, \varepsilon > 0\}$ in the last space in (25) and leads (after the use of (23) and of the Green theorem) to the uniform estimate of $\ddot{u}_\varepsilon$, $\varepsilon > 0$ in $L_1\Big(I_T;\, (H^{\frac{N}{2}+\vartheta}(\Omega;\mathbb{R}^N))^*\Big)$ for every $\vartheta > 0$. This estimate can be "decomposed" into the uniform estimate of their projections onto Y and $\mathcal{R}$. The interpolation technique giving the strong $L_1(I_T; H^1(\Omega;\mathbb{R}^N))$-compactness works here for $\pi_Y\dot{u}$, while the compactness of their projections onto $\mathcal{R}$ follows from (23), (24) and from the fact that for all elements of $\mathcal{R}$ all their space derivatives of the order greater than 1 are equal to zero. Such an idea is employed both in the Galerkin procedure and in the convergence procedure with $\varepsilon_k \to 0$. The strong convergence of gradients is based on the strong monotonicity of the operator $\tilde{e}(u) \mapsto \tilde{\sigma}(u)$ which holds due to the mentioned properties of V and W and due to the assumption (19). Then its proof is similar to the membrane case. □

The regularity of the solution away from $S_{c,T}$ is obvious. Along $S_{c,T}$ some regularity of the above mentioned type can be proved.

MATERIALS WITH VISCOELASTICITY OF THE TYPE $\nabla\dot{u}$

The **mebrane** case is here easier than in the preceding section. We consider the problem (1), where $\sigma^V(u) = \nabla\dot{u}$ and solve it via the same penalization. In the a priori estimate (11) the fractional-derivative norm of ∇u is replaced by $\|\nabla\dot{u}\|_{L_2(Q_T;\mathbb{R}^N)}$. The additional estimate (12) remains valid. This enables to use directly the Hilbert-space interpolation with the help of the Fourier transformation based on results described e.g. in Lions and Magenes (1968). Thus we prove $\{\dot{u}_\varepsilon,\, \varepsilon > 0\}$ is bounded in $H^\beta(I_T; L_2(\Omega))$ for any $\beta \in (0, \frac{1}{N+2})$. This yields the strong convergence of velocities and Theorem 1 remains valid. More space regularity holds here (the proof starts from putting $v = \Delta u_\varepsilon$ into the appropriate version of (9)). The remark to polyharmonic case remains also true. For all details see Jarušek, Málek, Nečas and Šverák (1992).

In the **body** case we assume here the strain–stress relation as above, where the relation

$$\tilde{\sigma}^V \equiv (\sigma^V_{ij}(u)) = A\tilde{e}(\dot{u}) \equiv (a_{ijkl}e_{kl}(\dot{u})) \tag{26}$$

replaces the relation (13). The usual symmetries of coefficients, the continuity and the ellipticity of the quadratic form corresponding to A are assumed. Instead of zero friction in (15) the Coulomb friction law with given friction force $-G$ is considered:

$$|T_t(u)| \le G,\; T_t(u)\dot{u}_t + G|\dot{u}_t| = 0,\; |T_t(u)| = G \ne 0 \Longrightarrow \underset{\lambda\le 0 \text{ on } S_{c,T}}{\exists} \quad \dot{u}_t = \lambda T_t \text{ on } S_{c,T}. \tag{27}$$

The variational inequality to weak formulation of this contact problem with given friction is

$$\begin{aligned}\int_{Q_T} \sigma_{ij}(u)e_{ij}(v-u) - \dot{u}_i(\dot{v}_i - \dot{u}_i)\,\mathrm{d}x\,\mathrm{d}\tau + \int_{S_{c,T}} G(|v_t + \dot{u}_t - u_t| - |\dot{u}_t|)\,\mathrm{d}x\,\mathrm{d}\tau +\\ \int_\Omega (\dot{u}_i(v_i - u_i))(T,\cdot)\,\mathrm{d}x \ge \int_\Omega u_{1,i}(v_i(0,\cdot) - u_{0,i})\,\mathrm{d}x +\\ \int_{Q_T} f_i(v_i - u_i)\,\mathrm{d}x\,\mathrm{d}\tau + \int_{S_{T,T}} T_{0,i}(v_i - u_i)\,\mathrm{d}x\,\mathrm{d}\tau.\end{aligned} \tag{28}$$

with the same assumptions to u and v as in the preceding section. Here the following theorem holds (where the assumption to T could be weakened and the existence of a boundary part with a prescribed displacement could be involved):

THEOREM 3. *Let the assumptions concerning Ω, its boundary, the operator A, the function W, and the assumption* (18) *be fulfilled. Let $G \in L_2\left(I_T;\, (H^{\frac{1}{2}-\alpha}(\partial\Omega_c))^*\right)$ for some $\alpha \in (0, \frac{1}{2})$, $G \geq 0$ in the dual sense and let $\operatorname{supp} G \subset \operatorname{Int} \partial\Omega_c$ (in the topology of $\partial\Omega$). Then there exists a weak solution to the contact problem with given friction* [(15), (27)].

The occurence of the friction term complicates its proof. To treat it, we use the penalty function (21) together with smoothing convex functions

$$K_\eta: x \mapsto \left\langle \begin{array}{ll} |x|, & |x| \geq \eta \\ -\frac{1}{8\eta^3}|x|^4 + \frac{3}{4\eta}|x|^2 + \frac{3}{8}\eta, & |x| \leq \eta \end{array} \right., \quad x \in I\!R^N, \text{ for } \eta > 0 \tag{29}$$

for $K_0: x \mapsto |x|$, $x \in I\!R^N$. For arbitrary $\eta, \beta > 0$ we have

$$K_\eta \in C_2(I\!R^N), \quad 0 \leq K_\eta \text{ are Lipschitz with the constant 1 on } I\!R^N, \tag{30}$$
$$\operatorname{supp}(K_\eta - K_0) \subset \{x \in I\!R^N;\ |x| \leq \eta\} \quad \text{and} \quad \|K_\eta - K_0\|_{C_{1-\beta}(I\!R^N)} \leq const\, \eta^\beta,\ \beta \in (0,1).$$

For $\varepsilon > 0$ and $\eta > 0$ we define that $u_{\varepsilon,\eta}$ is the weak solution of the penalized problem, iff $u_{\varepsilon,\eta} \in B_0(I_T;\, H^1(\Omega; I\!R^N))$ for which $\dot{u}_{\varepsilon,\eta} \in B_0(I_T;\, L_2(\Omega; I\!R^N)) \cap L_2(I_T;\, H^1(\Omega; I\!R^N))$ and $\ddot{u}_{\varepsilon,\eta} \in L_2(I_T;\, (H^1(\Omega; I\!R^N)^*)$, the initial conditions in (15) are satisfied and the equation

$$\int\limits_{Q_T} (\ddot{u}_{\varepsilon,\eta})_i v_i + \sigma_{ij}(u_{\varepsilon,\eta}) e_{ij}(v)\, dx\, d\tau + \int\limits_{S_{c,T}} G K_\eta'((\dot{u}_{\varepsilon,\eta})_t) v_t + \tfrac{1}{\varepsilon}\tilde{h}'((u_{\varepsilon,\eta})_n) v_n\, dx\, d\tau =$$
$$\int\limits_{Q_T} f_i v_i\, dx\, d\tau + \int\limits_{S_{T,T}} T_{0,i} v_i\, dx\, d\tau \tag{31}$$

holds for any $v \in L_2(I_T;\, H^1(\Omega; I\!R^N))$.

The a priori estimate is derived like in the corresponding problem in Sec. 2 and has the form (23) for $u_{\varepsilon,\eta}$, where the term $\|\nabla \pi_Y u_{\varepsilon,\eta}\|^2_{H^\alpha(I_T;\, L_2(\Omega; I\!R^{N^2}))}$ is replaced by $\|\nabla \dot{u}_{\varepsilon,\eta}\|^2_{L_2(Q_T;\, I\!R^{N^2})}$ and $\mathcal{J}$ depends also on $\|G\|_{L_2\left(I_T;\, (H^{\frac{1}{2}}(\partial\Omega_c))^*\right)}$. From it we proceed in a similar way as in the proof of the preceding theorem. We derive the additional a priori estimate (24). In the convergence procedures (in the Galerkin method and for $\eta \to 0$ and $\varepsilon \to 0$) we need the strong convergence of the friction terms to obtain the strong convergence of gradients and this requires the better regularity of G together with a very precise use of the imbedding theorem and the trace theorem. The main point of the proof of the convergence of the smoothed friction term to the original one is the estimate

$$\|(K_\eta - K_0)(w)\|^2_{H^{\frac{1}{2}-\theta}(\partial\Omega_c; I\!R^N)} \leq \|K_\eta - K_0\|^2_{C_0(I\!R)} (\operatorname{mes}_{N-1} \partial\Omega_c)^2 + \tag{32}$$
$$\|K_\eta - K_0\|^2_{C_{1-\beta}(I\!R)} \int\limits_{\partial\Omega_c} \int\limits_{\partial\Omega_c} \frac{|w(x) - w(y)|^{2-2\beta}}{|x-y|^{N-2\theta}}\, dx\, dy,$$

being valid for any $\theta \in (0, \frac{1}{2})$ and each $\beta \in (0,1)$, and the compact imbedding $H^{\frac{1}{2}}(\partial\Omega_c) \hookrightarrow W_{2-2\beta}^{\frac{1-2\theta}{2-2\beta}}(\partial\Omega_c)$, $\beta \in (0, 2\theta)$, based again on the results of Schmeisser and Triebel (1987), Chapter 2, Sec. 4. and on (30). The details of the proof will be published in Jarušek[3]. □

REMARK. More interesting than the described problem it seems to be the original problem with friction, where $G = \mathcal{F}|T_n(u)|$ and only the coefficient of friction $\mathcal{F}$ is prescribed. The unique up-to-now known working method is the fixed-point one used for the quasistatic case e.g. in Nečas, Jarušek and Haslinger (1980), Jarušek (1983) and Jarušek (1984). Here the problem seems to be the limited time regularity of the solution which is the same as in the corresponding membrane case. More promising is the case with the cone in velocities (its version with given friction was solved under particular assumptions in Duvaut and Lions (1972) and in general in Jarušek (1990). Here the author is close to definite results.

REFERENCES

Amerio L. and Prouse G., 1975, Study of the motion of a string vibrating against an obstacle, *Rend. Mat.* VI, 8:563.

Bamberger A. and Schatzmann M., 1983, New results on the vibrating string with a continuous obstacle, *SIAM J. Math. Anal.* 14:560.

Běsov O.V., Il'jin V.P. and Nikol'skij S.M., 1975, "Integral Transformations of Functions and Imbedding Theorems" (in Russian), Nauka, Moskava.

Cabannes H. and C. Citrini (eds.), 1987, "Vibrations with Unilateral Constraints", Proc. conf. Euromech 209 held at Como (June 1986), Tecnoprint.

Duvaut G. and Lions J.L., 1972, "Les inéquations en mécanique et en physique", Dunod, Paris.

Jarušek J., 1983 Contact problems with bounded friction, Coercive case, *Czech. Math. J.* 33:237.

Jarušek J., 1984, Contact problems with bounded friction, Semicoercive case, *Czech. Math. J.* 34:619.

Jarušek J., 1990, Contact problems with given time-dependent friction force in linear viscoelasticity, *Comm. Math. Univ. Carolinae* 31:257.

Jarušek J., Málek J., Nečas J.and Šverák V., 1992 Variational inequality for a viscous drum vibrating in the presence of an obstacle, *Rend. Mat.* VII, 12:943.

Jarušek J., 1994, Solvability of the variational inequality for a drum with a memory vibrating in the presence of an obstacle, *Boll. Unione Mat. Ital.* **7** (8–A):113.

Jarušek[1] J., Solvability of unilateral hyperbolic problems involving viscoelasticity via penalization, to appear in "Proc. Int. conf. diff. eq. & math. modelling" held at Varenna 1992 (R. Salvi, ed.).

Jarušek[2] J., Dynamical contact problems for bodies with a singular memory, to appear in *Boll. Unione Mat. Ital.*

Jarušek[3] J., Dynamical contact problems with given friction for viscoelastic bodies (in review).

Lions J.L. and Magenes E., 1968, "Problèmes aux limites non-homogènes et applications", Dunod, Paris.

Maruo K., 1985, Existence of solutions of some nonlinear wave equations, *Osaka Math. J.* 22:31.

Nečas J., Jarušek J. and Haslinger J., 1980, On the solution of the variational inequality to the Signorini problem with small friction, *Boll. Unione Mat. Ital.* **5** (17 B):796.

Schatzmann M., 1980, A hyperbolic problem of second order with unilateral constraints: the vibrating string with a concave obstacle, *J. Math. Anal. Appl.* 73:183.

Schmeisser H.-J., 1987, Vector-valued Sobolev and Besov spaces. In: "Seminar of the Karl–Weierstrass–Institute 1985/86", Teubner Texte Math. Vol. 96, Teubner Vg., Leipzig.

Schmeisser H.-J. and Triebel H., 1987, "Topics in Fourier Analysis and Function Spaces", Akad. Vg. Geest & Portig, Leipzig.

CONTACT - FRICTION OF FUZZY TYPE. CONTACT - FRICTION OF FRACTAL TYPE

E.S. Mistakidis, P.D. Panagiotopoulos and O.K. Panagouli

Institute of Steel Structures, Aristotle University
GR-54006 Thessaloniki, Greece

INTRODUCTION

The aim of this paper is to introduce fuzzy and fractal friction laws. Fuzzy superpotential laws were first considered by the second author in Panagiotopoulos (1993). The phenomenon is described in terms of a nonconvex superpotential which is appropriately defined and a hemivariational inequality is obtained.

In the sequel, fractal type friction laws are considered. These laws are approximated by appropriately defined fractal interpolation functions. The numerical treatment of each arizing nonmonotone problem is accomplished by an advanced solution method (see e.g. Mistakidis and Panagiotopoulos (1994)), which approximates the nonmonotone problem by a sequence of monotone problems.

FUZZY LAWS

In this section an appropriate superpotential law is obtained in order to describe fuzzy phenomena in contact problems. Let l be an open subset of the real line $\mathbf{R}$ and let M be a measurable subset of l such that for every open and nonempty subset I of l, $\mathrm{mes}(I \cap M)$ and $\mathrm{mes}(I \cap (l - M))$ are positive. Let

$$g(u) = \begin{cases} +b_1 \text{ if } u \in M \\ -b_2 \text{ if } u \notin M \end{cases} \tag{1}$$

and

$$f(u) = \int_0^u g(u)du. \tag{2}$$

Then f is Lipschitzian and it can be verified that

$$\bar{\partial} f(u) = [-b_2, b_1] \tag{3}$$

Contact Mechanics, Edited by M. Raous *et al.*
Plenum Press, New York, 1995

for every $u \in l$, i.e. we obtain an infinite number of jumps in l. Suppose now the friction boundary conditions depicted in Fig. 1a. More specifically let us assume that in the small interval $-\varepsilon < [u_T] < \varepsilon$ the traction $-S_T$ may take any value between $-\alpha$ and α. We may write this law in the form:

$$\text{if} \quad [u_T] < -\varepsilon \quad -S_T = \alpha \tag{4}$$
$$\text{if} \quad -\varepsilon \leq [u_T] \leq \varepsilon \quad -\alpha \leq -S_T \leq \alpha \tag{5}$$
$$\text{if} \quad [u_T] > \varepsilon \quad -S_T = -\alpha \tag{6}$$
$$\tag{7}$$

or equivalently in the form

$$-S_T \in \partial j([u_T]) + \bar{\partial} f([u_T]), \tag{8}$$

where j is convex and results from the maximal monotone graph $ABCDEF$ and f is given by the relations (1)-(3) with $b_1 = b_2 = \alpha$ and $l = (-\varepsilon, \varepsilon)$.

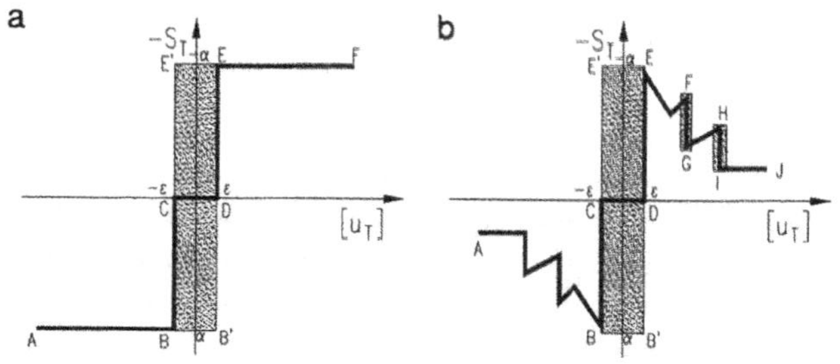

Figure 1. Fuzzy frictional laws.

Analogously is written the law of Fig. 1b. Now $\partial j([u_T])$ is replaced in (8) by $\bar{\partial} j([u_T])$, where j is nonconvex and results from the nonmonotone graph $ABCDEFGHIJ$. Physically the above diagrams correspond to monotone and nonmonotone interface friction laws with a nonfully determined behaviour around the adhesive friction region (fuzzy adhesive behaviour). Such nonfully determined regions may be considered around the complete vertical segments of the nonmonotone law (see the shaded areas around FG and HI in Fig. 1b). The expression of such nonfully determined laws in terms of nonconvex superpotentials permits the formulation of a variational theory for this class of problems.

FRACTAL FRICTION LAWS

Fractal type friction laws have been registered lately by various experiments (see e.g. Feder (1991)). All these laws are nonmonotone - nonsmooth so it is difficult to incorporate the description of such laws into the standard formalism which usually operates only with simple functions.

In the present work the basic idea for the mathematical description of these laws is the use of some fractal functions, the so called "fractal interpolation functions". In this case a friction law ϕ is the fractal graph of a continuous function g which interpolates a given set of data $\{(x_i, y_i), i = 0, 1, \ldots, N\}$. More specifically if C^0 is the set of the continuous functions $g : [x_0, x_N] \rightarrow \mathbf{R}$, then the sequence of functions $g_{i+1}(x) = (Tg_i)(x)$, where the operator

$T: C^0 \to C^0$, defined by

$$(Tg)(a_i x + e_i) = c_i x + d_i g(x) + f_i \quad i = 1, 2, \ldots, N, \tag{9}$$

converges to a curve ϕ. This curve is the unique attractor of the hyperbolic iterated function system (I.F.S.) $\{\mathbf{R}^2; w_i, i = 1, 2, \ldots, N\}$ (Barnsley (1988)) defined by the shear transformation

$$\{x, y\} \to w_i \left\{ \begin{array}{c} x \\ y \end{array} \right\} = \left[\begin{array}{cc} a_i & 0 \\ c_i & d_i \end{array} \right] \left\{ \begin{array}{c} x \\ y \end{array} \right\} + \left\{ \begin{array}{c} e_i \\ f_i \end{array} \right\} \quad i = 1, \ldots, N. \tag{10}$$

The factors d_i are free parameters of the problem and satisfy $0 \leq d_i < 1$. Moreover the remaining coefficients are given by the formulas

$$a_i = \frac{(x_i - x_{i-1})}{(x_N - x_0)}, \qquad e_i = \frac{(x_N x_{i-1} - x_0 x_i)}{(x_N - x_0)} \tag{11}$$

$$c_i = \frac{(y_i - y_{i-1})}{(x_N - x_0)} - d_i \frac{(y_N - y_0)}{(x_N - x_0)}, \quad f_i = \frac{(x_N y_{i-1} - x_0 y_i)}{(x_N - x_0)} - d_i \frac{(x_N y_0 - y_N x_0)}{(x_N - x_0)}. \tag{12}$$

Since ϕ is the fixed point of the given transformation T i.e.

$$\phi = T\phi \text{ and } \phi_{i+1} = T\phi_i, \quad \phi_i \to \phi, \tag{13}$$

we are led, in the case of fractal type friction laws, to the following formulation of the unilateral contact and friction problem: Let $\Omega \subset \mathbf{R}^2$ be a two-dimensional linear elastic structure coming in contact with another elastic or rigid body through an interface Γ_S, and suppose that in the tangential to the interface direction a fractal law holds between the tangential traction $-S_T$ and the tangential displacement u_T. It is assumed that on the boundary part Γ_U the displacements are prescribed whereas on the boundary part Γ_F the loading is given. According to Panagiotopoulos (1985),(1993), an equilibrium position of Ω is characterized by the following hemivariational inequality problem: Find $u \in V_0$ such as to satisfy the inequality

$$\alpha(u, v - u) + \int_{\Gamma_S} j^0(u_T, v_T - u_T) d\Gamma \geq (l, v - u) \ \forall v \in V_0 \tag{14}$$

where V_0 is the kinematically admissible set, $\alpha(\,.\,,\,.\,)$ the linear elastic strain energy, (l, u) the work of the external forces, $j(.)$ is the nonconvex superpotential of the fractal law and j^0 is the directional derivative of Clarke. In order to solve the problem, instead of the initial fractal law, a sequence of problems is solved, where in every step we use an approximation ϕ_i of the fractal law. The arizing nonmonotone law problems are treated with the method described in Mistakidis and Panagiotopoulos (1994). The solution method is demonstrated in the example that follows. We consider the linear elastic body of Fig. 2 and we assume that on the interface a fractal type friction law holds. The initial points $\{(x_i, y_i), i = 1, 2, 3\}$ are: (0.00,0.75), (0.50,1.00), (0.50,0.20) and (1.00,0.45). The free parameters are taken to have the values $d_1 = d_2 = d_3 = 0.2$, $d_1 = d_2 = d_3 = 0.3$ and $d_1 = d_2 = d_3 = 0.4$ respectively. For each set of values d_i $i = 1, 2, 3$ a fractal law is derived as it is shown in Fig.3, where every fractal law is approximated by five iterations. The friction coefficient μ is 0.2 for all the cases. In figures 4a,b,c the differences of the frictional forces along the interface between two consecutive approximations for the three different fractal type frictions laws are given. It is observed that we have convergence to the solution of the problem after the third approximation of the fractal law in the first case where $d_1 = d_2 = d_3 = 0.2$, whereas this convergence is achieved after the fourth approximation of the fractal law in the case where $d_1 = d_2 = d_3 = 0.4$.

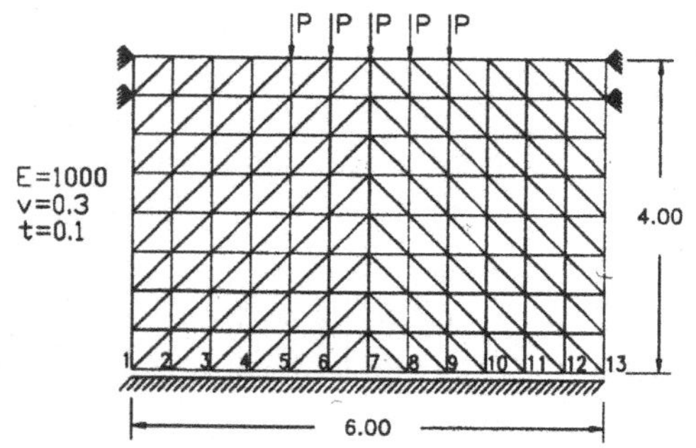

Figure 2. The analysed structure.

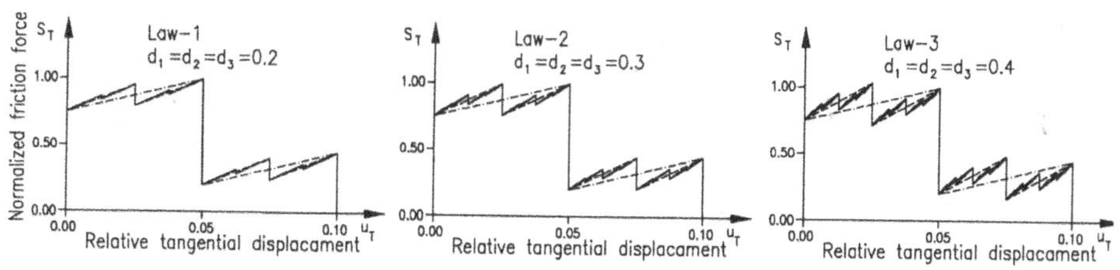

Figure 3. The adopted nonmonotone fractal laws.

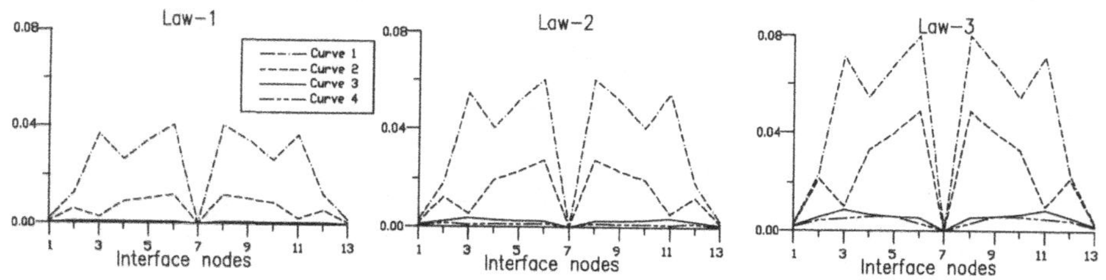

Figure 4. Differences of the friction forces between two consecutive approximations of the fractal law (curve i corresponds to the differences between the i and $i+1$ iterations).

It is important to notice here that the convergence to the final solution of the problem depends strongly on the fractal dimension of ϕ. As this dimension becomes larger (i.e. as d_i become larger), a larger number of approximations is needed in order to have convergence (see also on this context Panagiotopoulos et al. (1994)).

REFERENCES

Barnsley, M., 1988, "Fractals Everywhere," Academic Press, Boston-New York.

Feder, H.J.S. and Feder, J., 1991, Self-organized criticality in a stick-slip process, *Phys. Rev. Lett.* 66:2669.

Mistakidis, E.S. and Panagiotopoulos, P.D., 1994, On the Approximation of Nonmonotone Multivalued Problems by Monotone Subproblems, *Computer Methods in Applied Mechanics and Engineering* 114:55.

Panagiotopoulos, P.D., 1985, "Inequality Problems in Mechanics and Applications. Convex and Nonconvex Energy Functions," Birkhäuser, Boston-Basel. Russian translation (MIR, Moscow, 1989).

Panagiotopoulos, P.D., 1993, "Hemivariational Inequalities. Applications in Mechanics and Engineering," Springer, Berlin-Heidelberg-New York.

Panagiotopoulos, P.D., Panagouli, O.K and Mistakidis, E.S., 1994, Fractal geometry and fractal material behaviour in solids and structures, *Archive of Applied Mechanics* 63:1.

UPPER AND LOWER SOLUTION BOUNDS IN UNILATERAL CONTACT ELASTOSTATICS UNDER SECOND-ORDER GEOMETRIC EFFECTS

A.A. Liolios

Institute of Structural Mechanics, Democritus University of Thrace
GR-67100 Xanthi, Greece

ABSTRACT

The paper deals with a procedure, which provides upper and lower bounds for solution quantities, either global or local, for inequality problems of structural elastostatics, when second-order geometric effects are taken into account.

INTRODUCTION

Two-sided solution bounds can be used either to check the numerical results or to predict an interval of values for a solution quantity in practical engineering problems, which in most cases can be solved numerically only. Methods of deriving such two-sided solution bounds for elliptic linear equality boundary value problems of mathematical physics and especially of elastostatics are already well-known, see e.g. Synge (1957) and Stumpf(1970). A generalization of these bounding procedures for inequality problems has been also obtained (Liolios, 1982, 1985; Sewell, 1977).

The aim of the present paper is to extend further these methods to inequality problems of structural elastostatics when second-order geometric effects have to be taken into account. So, a procedure based on some functional analysis concepts is presented, which provides upper and lower bounds, either global or local, for solution quantities of the above problem when it is piece-wise linearized.

PROBLEM CONDITIONS

The problem formulation of Liolios (1982) is followed herein. So the problem is discussed for simplicity in matrix operator notation. First the structure is discretized into finite elements of the "natural" type (Maier, 1971) and the unilateral behaviour is considered

Contact Mechanics, Edited by M. Raous *et al.*
Plenum Press, New York, 1995

as piece-wise linearized. Thus the constitutive conditions for the unassembled structure are

$$\boldsymbol{e} = \boldsymbol{\varepsilon} + \boldsymbol{\theta} + \boldsymbol{\mu}; \quad \boldsymbol{\sigma} = \boldsymbol{E}\boldsymbol{\varepsilon} \text{ or } \boldsymbol{\varepsilon} = \boldsymbol{E}^{-1}\boldsymbol{\sigma}, \tag{1); (2a, b}$$

$$\boldsymbol{\mu} = \boldsymbol{B}\boldsymbol{v}, \quad \boldsymbol{t} = \boldsymbol{B}^T\boldsymbol{\sigma} - \boldsymbol{A}\boldsymbol{v} - \boldsymbol{r}, \quad \boldsymbol{r} \geq \boldsymbol{0}, \tag{3), (4a, b}$$

$$\boldsymbol{t} \leq \boldsymbol{0}, \quad \boldsymbol{v} \geq \boldsymbol{0}, \quad \boldsymbol{v}^T\boldsymbol{t} = 0, \tag{5a, b, c}$$

where $\boldsymbol{e}, \boldsymbol{\varepsilon}, \boldsymbol{\theta}, \boldsymbol{\mu}$ are the total, elastic, imposed and unilateral strain vectors, respectively; $\boldsymbol{\sigma}$ is the stress vector; $\boldsymbol{t}, \boldsymbol{v}, \boldsymbol{r}$ are the stress, deformation and resistence vectors of the unilateral constraints, respectively; $\boldsymbol{B}$ is the unilateral-transformation matrix; and $\boldsymbol{E}$, $\boldsymbol{A}$ are the "natural" stiffness and unilateral-interaction matrices, respectively, both symmetric and positive definite.

Further, compatibility and equilibrium for the assembled structure are expressed by the following conditions

$$\mathrm{e} = \boldsymbol{G}^T\boldsymbol{u}, \tag{6}$$

$$\boldsymbol{G}\boldsymbol{\sigma} = \boldsymbol{p} - \boldsymbol{D}\boldsymbol{u}. \tag{7}$$

Here $\boldsymbol{G}$ is the rectangular equilibrium matrix and $\boldsymbol{G}^T$ its transposed; $\boldsymbol{p}$ is the load vector; $\boldsymbol{u}$ is the displacement vector; and $\boldsymbol{D}$ is the symmetric, positive definite, constant geometric stiffness matrix, depending linearly on preexisting constant stresses (Maier, 1971). Thus, through the term $\boldsymbol{D}\boldsymbol{u}$ alone the geometry changes affect the equilibrium (second-order geometric effects).

So, if $\boldsymbol{E}, \boldsymbol{G}, \boldsymbol{A}, \boldsymbol{B}$ and $\boldsymbol{D}$ are known, then the so-formulated problem consists in finding the set $\{\boldsymbol{u}, \boldsymbol{v}; \boldsymbol{\sigma}, \boldsymbol{t}; \boldsymbol{e}, \boldsymbol{\varepsilon}, \boldsymbol{\mu}\}$ that corresponds to the given action set $\{\boldsymbol{p}, \boldsymbol{\theta}\}$ and satisfies (1)-(7). Solution existence, uniqueness and regularity questions for this inequality problem can be treated by the variational inequality approach, see e.g. Panagiotopoulos (1985). Here we assume that a unique solution exists, for which two-sided solution bounds are required.

THE HILBERT SPACE OF THE STRUCTURE STATES AND SOME BASIC INEQUALITIES

The ordered set $\{\boldsymbol{\sigma}, \boldsymbol{t}; \boldsymbol{u}, \boldsymbol{v}\}$ is said to be the state $\boldsymbol{S}$ of the structure. From relations (1)-(7) it follows that such a structure state can be fully determined by either the pair $\{\boldsymbol{\sigma}, \boldsymbol{t}\}$ or the pair $\{\boldsymbol{u}, \boldsymbol{v}\}$. Further, with each state $\boldsymbol{S}$ we associate an element of a real Hilbert space $\mathcal{H}$. For this space we define as scalar product of two states $\boldsymbol{S}_1$ and $\boldsymbol{S}_2$ the bilinear form

$$\boldsymbol{S}_1 \cdot \boldsymbol{S}_2 = \boldsymbol{\varepsilon}_1^T\boldsymbol{E}\boldsymbol{\varepsilon}_2 + \boldsymbol{v}_1^T\boldsymbol{A}\boldsymbol{v}_2 + \boldsymbol{u}_1^T\boldsymbol{D}\boldsymbol{u}_2. \tag{8}$$

This symmetric form is positive definite, because the matrices $\boldsymbol{E}$, $\boldsymbol{A}$ and $\boldsymbol{D}$ are so. According to definition (8), the squared norm $||\boldsymbol{S}||^2$ is the double of the generalized energy of the structure.

Now we introduce two sets of admissible states. A state $\boldsymbol{S}_\sigma$ determined by the stress pair $\{\boldsymbol{\sigma}_\sigma, \boldsymbol{t}_\sigma\}$ is said to be a stress admissible state, iff this state satisfies (2)-(4), (5a) and (7). On the other hand, we define the kinematically admissible states $\boldsymbol{S}_k$ as those which are determined by the pairs $\{\boldsymbol{u}_k, \boldsymbol{v}_k\}$ and satisfy (1)-(4), (5b) and (6). Taking into account these definitions of $\boldsymbol{S}_\sigma$ and $\boldsymbol{S}_k$ and the rels. (1)-(7), we have for the solution state $\boldsymbol{S}_0$ and any pair $\{\boldsymbol{S}_\sigma, \boldsymbol{S}_k\}$

$$\boldsymbol{v}_0^T\boldsymbol{t}_\sigma \leq 0, \quad \boldsymbol{v}_k^T\boldsymbol{t}_0 \leq 0, \quad \alpha = -\boldsymbol{v}_k^T\boldsymbol{t}_\sigma \geq 0 \tag{9a, b, c}$$

$$(\boldsymbol{S}_0 - \boldsymbol{S}_\sigma) \cdot (\boldsymbol{S}_0 - \boldsymbol{S}_k) = +\boldsymbol{v}_k^T \boldsymbol{t}_0 + \boldsymbol{v}_0^T \boldsymbol{t}_\sigma - \boldsymbol{v}_k^T \boldsymbol{t}_\sigma. \tag{10}$$

Due to (9), equality (10) is transformed to the following basic inequalities:

$$(\boldsymbol{S}_0 - \boldsymbol{S}_\sigma) \cdot (\boldsymbol{S}_0 - \boldsymbol{S}_k) \leq 0 \tag{11}$$

$$(\boldsymbol{S}_0 - \boldsymbol{S}_\sigma)^2 + (\boldsymbol{S}_0 - \boldsymbol{S}_k)^2 \leq -2\boldsymbol{v}_k^T \boldsymbol{t}_\sigma + (\boldsymbol{S}_\sigma - \boldsymbol{S}_k)^2 + 2\boldsymbol{v}_0^T \boldsymbol{t}_\sigma, \tag{12}$$

$$(\boldsymbol{S}_0 - \boldsymbol{S}_\sigma)^2 + (\boldsymbol{S}_0 - \boldsymbol{S}_k)^2 \leq -2\boldsymbol{v}_k^T \boldsymbol{t}_\sigma + (\boldsymbol{S}_\sigma - \boldsymbol{S}_k)^2 + 2\boldsymbol{v}_0^T \boldsymbol{t}_0, \tag{13}$$

GLOBAL TWO-SIDED SOLUTION BOUNDS

Applying now the procedures of Liolios(1982, 1985), we derive from inequalities (11)-(13) the following upper and lower global bounds for the solution state $\boldsymbol{S}_0$:

$$[\boldsymbol{S}_0 - \boldsymbol{C}_0]^2 \leq R_0^2, \quad (||\boldsymbol{C}_0|| - R_0) \leq ||\boldsymbol{S}_0|| \leq (||\boldsymbol{C}_0|| + R_0), \tag{14a, b}$$

$$-\Pi_c(\boldsymbol{S}_\sigma) \leq -\Pi_c(\boldsymbol{S}_0) = \Pi_p(\boldsymbol{S}_0) \leq \Pi_p(\boldsymbol{S}_k), \tag{15}$$

where

$$\boldsymbol{C}_0 = \frac{1}{2}(\boldsymbol{S}_\sigma + \boldsymbol{S}_k), \quad R_0^2 = \alpha + \frac{||\boldsymbol{S}_\sigma - \boldsymbol{S}_k||^2}{4} = [\alpha + \Pi_p(\boldsymbol{S}_k) + \Pi_c(\boldsymbol{S}_\sigma)]/2, \tag{16a, b}$$

$$\Pi_p(\boldsymbol{S}) = \Pi_p(u, v) = \frac{1}{2}[\boldsymbol{\varepsilon}^T \boldsymbol{E} \boldsymbol{\varepsilon} + \boldsymbol{v}^T \boldsymbol{A} \boldsymbol{v} + \boldsymbol{u}^T \boldsymbol{D} \boldsymbol{u}] - \boldsymbol{q}^T \boldsymbol{u} + \boldsymbol{r}^T \boldsymbol{v}, \tag{17}$$

$$\Pi_c(\boldsymbol{S}) = \Pi_c(\boldsymbol{\sigma}, \boldsymbol{t}) = \frac{1}{2}[\boldsymbol{\sigma}^T \boldsymbol{E}^{-1} \boldsymbol{\sigma} + \boldsymbol{v}^T \boldsymbol{A} \boldsymbol{v} + \boldsymbol{u}^T \boldsymbol{D} \boldsymbol{u}] + \boldsymbol{\sigma}^T \boldsymbol{\theta}. \tag{18}$$

LOCAL TWO-SIDED SOLUTION BOUNDS

Further, to obtain local bounds for a solution quantity $F(\boldsymbol{\xi})$ at the point ξ of the structure, we introduce (Liolios, 1982) a suitable Green state $\check{\boldsymbol{S}}$ with corresponding stress, $\check{\boldsymbol{S}}_\sigma$, and kinematically, $\check{\boldsymbol{S}}_k$, admissible states. So we obtain finally

$$|F(\boldsymbol{\xi}) - F_\alpha(\boldsymbol{\xi})| \leq R_0 \check{D}, \tag{19a}$$

where

$$F_\alpha(\boldsymbol{\xi}) = (\boldsymbol{q}^T \check{\boldsymbol{u}}_k - \check{\boldsymbol{\sigma}}_\sigma^T \boldsymbol{\theta}) + \frac{1}{2}(\boldsymbol{S}_\sigma + \boldsymbol{S}_k) \cdot (\check{\boldsymbol{S}}_\sigma - \check{\boldsymbol{S}}_k), \tag{19b}$$

$$\check{D} = ||\check{\boldsymbol{S}}_\sigma - \check{\boldsymbol{S}}_k||. \tag{19c}$$

NUMERICAL REALIZATION OF THE BOUNDS AND CONCLUDING REMARKS

Upper and lower bounds in terms of the admissible states $\boldsymbol{S}_\sigma$, $\boldsymbol{S}_k$, $\check{\boldsymbol{S}}_\sigma$ and $\check{\boldsymbol{S}}_k$ are provided by (14), (15) and (19) for the solution state $\boldsymbol{S}_0$. Improvements of these bounds can be obtained by minimizing the bounds-gaps R_0 and $\check{D}$. This minimization leads to quadratic programming problems. Moreover, (15) expresses two dual minimum principles, the first one of the generalized potential energy (17) and the second of the generalized complementary energy (18). So, the required admissible states $\boldsymbol{S}_\sigma$ and $\check{\boldsymbol{S}}_\sigma$ (resp. $\boldsymbol{S}_k$ and $\check{\boldsymbol{S}}_k$) can be numerically constructed as solutions of convex optimization problems by using stress diffusive (resp. displacement conforming) finite element models (Fraeijs de Veubeke, 1965).

Thus, the bounding procedure presented herein can be realized numerically by using available programmes of the finite element method and of nonlinear mathematical programming (Bathe,1982; Panagiotopoulos, 1985). Finally, this approach can be extended and applied to various other piece-wise linearized inequality problems of mechanics, see e.g. Liolios (1989).

REFERENCES

Bathe, K.-J., 1982, "Finite Element Procedures in Engineering Analysis," Prentice-Hall, New Jersey.

Fraeijs de Veubeke, B., 1965, Displacement and equilibrium models in the finite element method, in: "Stress Analysis," Zienkiewicz, O. C., and Holister, G. S., Eds., Wiley, London.

Liolios, A. A., 1982, Upper and lower bounds in unilateral and nonlinear Structural elastostatics, *Jnl Appl. Math. Mech. (ZAMM)*, 62:T138.

Liolios, A. A., 1985, A direct formulation of dual extremum principles in unilateral elastostatics by the generalization of the hypercircle method, *Jnl Appl. Math. Mech. (ZAMM)*, 65:T348.

Liolios, A. A., 1989, A linear complementarity approach to the nonconvex dynamic problem of unilateral contact with friction between adjacent structures, *Jnl Appl. Math. Mech. (ZAMM)*, 69:T420.

Maier, G.,1971, Incremental elastoplastic analysis in the presence of large displacements and physical instabilizing effects, *Int. Jnl Solids and Structures*, 7:345.

Panagiotopoulos, P.D., 1985, "Inequality Problems in Mechanics and Applications. Convex and Nonconvex Energy Functions," Birkhäuser Verlag, Boston,Basel, Stuttgart, (Russian translation: MIR Publ. Moscow 1989).

Sewell, M.J., 1987, "Maximum and Minimum Principles," Cambridge University Press, Cambridge.

Stumpf, H.,1970, "Eingrenzungsverfahren in der Elastomechanik," Westdeutscher Verlag, Köln.

Synge, J. L.,1957, "The Hypercircle in Mathematical Physics," Cambridge University Press.

STIFFNESS MAXIMIZATION OF GENERAL STRUCTURE IN SIGNORINI-TYPE CONTACT

Joakim Petersson

Department of Mechanical Engineering
Division of Mechanics
Linköping Institute of Technology
S-581 83 Linköping
SWEDEN

INTRODUCTION

Structural optimization concerns the aim of making a particular type of structure sustain its loads as efficiently as possible, without violating a set of *a priori* determined design constraints. In this article, we focus on a general Signorini problem, corresponding to linearly elastic structures in unilateral contact with rigid and frictionless supports. The object desired from the design is a maximum of equilibrium potential energy. This well-known objective was treated as early as 1968 by Prager and Taylor and 1973 by Hemp. It is primarily, but not solely, a measure of structural stiffness.

Design constraints will be presumed to consist of, at least, a so-called isoperimetric constraint, accounting for a fixed accessible amount of material. In any case, the design is allowed to reach zero values, so the problem can be regarded as a type of topology optimization.

The strain energy of the structure is assumed to depend linearly on the design. Trusses, sheets and sandwich-type beams and plates are examples of possible structures in accordance with this assumption, and the first two will be subject to a more detailed examination.

The contents of the report is the following. First the general assumptions and the state problem are dealt with. Then the structural optimization problem is treated in a general form. The objective functional is shown to yield, in a sense, small contact pressures and displacements, and the strain energy density will be evenly distributed.

The design problem is identified with the dual problem connected to a saddle point in the potential energy as a function of displacement and design. The solvability of the design problem is shown from a saddle point theorem.

The paper ends with two particular examples, one discrete structure and one continuum structure, that fit into the abstract formulations. Optimum structures, of these particular kinds, are shown to be (at least nearly) uniformly stressed.

ABSTRACT STATE PROBLEM AND ASSUMPTIONS

We assume that the problem of finding the state of equilibrium for the structure can be written as

$$(S)_h \quad \text{Find } \tilde{u} \in \mathcal{U} : \; J_h(\tilde{u}) \leq J_h(u) \quad \forall u \in \mathcal{U}$$

where the total potential energy functional

$$J_h(u) = \frac{1}{2} a_h(u, u) - L(u)$$

is quadratic.The parameter h is referred to as the design variable, and it belongs to the set of permissible designs $\mathcal{H}$. We assume that for any h in $\mathcal{H}$, $a_h(\cdot, \cdot)$ is a bounded, symmetric and bilinear form, and $L(\cdot)$ is a bounded linear functional on a Hilbert space V. The former represents (twice of) the strain energy stored in the structure and the latter the work of the external loads.

$\mathcal{U} \subset V$ is the set of kinematically admissible displacements and we postulate that it can be written as

$$\mathcal{U} = \{u \in V \mid C(u) \leq g\} \neq \emptyset,$$

where $C(\cdot)$ is a linear trace map (into a trace space U), that for each displacement gives its outward normal[1] component in the contact region, g, which is the initial gap to a rigid obstacle, belongs to U and $\leq$ denotes a suitable partial ordering on U such that $\mathcal{U}$ is closed and convex in V.

Let W denote a normed vector space and W' its dual. An important assumption that we make here is that $h \mapsto J_h(u)$ is linear. More precisely, we suppose that the bilinear form can be written as

$$\frac{1}{2} a_h(u, u) = \langle \, h, A(u) \, \rangle, \tag{1}$$

where $\langle \cdot, \cdot \rangle$ denotes the duality pairing on $W' \times W$ and

$$V \ni u \longmapsto A(u) \in W$$

is an elastic potential map that for each displacement gives the corresponding distribution of strain energy density. In order to comply with (1), we must assume that $\mathcal{H}$ is a subset of W'. Furthermore, $\mathcal{H}$ is presumed to be non-empty, weakly* compact, convex and such that $a_h(\varphi, \varphi)$ is non-negative for all h in $\mathcal{H}$ and V-elliptic for at least one $h = h_0$ in $\mathcal{H}$.

In the upcoming chapter we will formulate a structural optimization problem in which permissible designs will "compete" in order to maximize a stiffness measure. These permissible designs must be such that they require the same amount of material, *Vol*. Hence, all h in $\mathcal{H}$ satisfy the isoperimetric constraint

$$\langle \, h, 1 \, \rangle = Vol \tag{2}$$

reflecting this fact. Here $1 \in W$ is such that it assigns the value one to all parts of the structure.

[1]or inward normal to obstacle

Benedict (1982) proposed to choose the *equilibrium* potential energy as a measure of stiffness for structures possibly in contact. We will adhere to this efficiency measure and see what it means through Clapeyron's theorem for unilateral contact, which we derive below.

The state problem $(S)_h$ is constrained, but can be transformed into a formulation free of constraints on u, through a Lagrangian technique. We form the Lagrangian

$$\mathcal{L}_h(u,p) = J_h(u) + [\,p, C(u) - g\,], \tag{3}$$

in which $[\cdot\,,\cdot]$ denotes the duality pairing on $U' \times U$, and p is a non-negative multiplier for the non-penetration constraint $C(u) \leq g$. It represents (generalized) contact pressure and is (in equilibrium) non-zero only if the constraint is active, i.e. $C(u) = g$; hence

$$[\,p_h, C(u_h) - g\,] = 0, \tag{4}$$

where p_h and u_h are contact pressure and displacement in equilibrium for the design h.

Requiring stationarity of the functional in (3) with respect to u, i.e. taking the Gateaux derivative with respect to u (at u_h) in the direction φ and setting this equal to zero, one gets

$$a_h(u_h, \varphi) - L(\varphi) + [\,p_h, C(\varphi)\,] = 0 \quad \forall \varphi \in V. \tag{5}$$

Choosing $\varphi = u_h$ and substituting the result in the definition of $J_h(u)$, one has with (4) that

$$J_h(u_h) = -\frac{1}{2}\{[\,p_h, g\,] + L(u_h)\}, \tag{6}$$

which is the desired form of Clapeyron's theorem. The quantity in (6) can also according to $(S)_h$ be written as $\inf_{u \in \mathcal{U}} J_h(u) \equiv \varphi(h)$, and the general stiffness design problem can be formulated as

$$(d) \quad \text{Find } \tilde{h} \in \mathcal{H} : \ \varphi(\tilde{h}) \geq \varphi(h) \quad \forall h \in \mathcal{H}.$$

The meaning of (d) is clear from (6): *Maximizing the equilibrium potential energy means minimizing a sum of weighted measures of contact pressures and displacements.*

Saddle Point Formulation and Existence

In (d) the infimum function is maximized; let us formulate the converse, namely minimizing the supremum function:

$$(p) \quad \text{Find } \tilde{u} \in \mathcal{U} : \ \psi(\tilde{u}) \leq \psi(u) \quad \forall u \in \mathcal{U},$$

where $\psi(u) \equiv \sup_{h \in \mathcal{H}} J_h(u)$. The problems (p) and (d) are referred to as *primal* and *dual* problems, and they are naturally linked together through a saddle point:

$$(SJ) \quad \text{Find } (\tilde{u}, \tilde{h}) \in \mathcal{U} \times \mathcal{H} : J_h(\tilde{u}) \leq J_{\tilde{h}}(\tilde{u}) \leq J_{\tilde{h}}(u) \quad \forall (u,h) \in \mathcal{U} \times \mathcal{H}.$$

The dual problem can be viewed as a classical nested approach of structural optimization, in which the state variables are eliminated, and the primal one has turned out to be practicable to treat in computations; cf. Ben-Tal et al. (1993).

The following has been shown to hold (Ekeland and Temam, 1976):

Theorem 1 *Suppose $(\tilde{u}, \tilde{h}) \in \mathcal{U} \times \mathcal{H}$ solves (SJ). Then $\tilde{u}$ solves (p) and $\tilde{h}$ solves (d). Conversely, if $\tilde{u}$ and $\tilde{h}$ are any two solutions to (p) and (d), then, provided the optimal objective values coincide (as they do when (SJ) is solvable), $(\tilde{u}, \tilde{h})$ solves (SJ).*

We are in a position to show existence of solutions to our general problem (SJ) through a saddle point theorem below. It looks very much like Proposition 2.4, chapter VI, in Ekeland and Temam (1976), but does not assume reflexivity of W'. In order to prove it, it is sufficient to start from Corollary 1 in Ky Fan (1964) and then proceed as in the proof of the main existence statement by Petersson (1993).

Theorem 2 *Suppose that J maps $\mathcal{U} \times \mathcal{H}$ into $\mathbf{R}$, and V is a Hilbert space and W a normed vector space. $\mathcal{U} \subset V$ is non-empty, closed and convex. $\mathcal{H} \subset W'$ is non-empty, weakly* compact and convex. J is weakly* upper semi-continuous and concave in h for all u in $\mathcal{U}$, and lower semi-continuous and convex in u for all h in $\mathcal{H}$. Finally, if*

$$\exists\, h_0 \in \mathcal{H} : \lim_{\substack{u \in \mathcal{U} \\ \|u\|_V \to +\infty}} J_{h_0}(u) = +\infty \quad (\textit{coercivity}),$$

then there exists at least one solution to (SJ).

From (1) we have that J is linear, and also weakly* continuous in h. In fact, if

$$h_n \xrightarrow{w^*} h \iff \langle\, h_n, \varphi \,\rangle \longrightarrow \langle\, h, \varphi \,\rangle \quad \forall \varphi \in W,$$

then by choosing $\varphi = A(u)$ one arrives at

$$J_{h_n}(u) = \langle\, h_n, A(u) \,\rangle - L(u) \longrightarrow \langle\, h, A(u) \,\rangle - L(u) = J_h(u).$$

Trivially, then, J is weakly* upper semi-continuous and concave in h. Furthermore J is lower semi-continuous and convex in u since $a_h(\varphi, \varphi) \geq 0$ for all $h \in \mathcal{H}$. Finally, since $a_{h_0}(\cdot\,,\cdot)$ is V-elliptic $J_{h_0}(\cdot)$ is coercive, hence all the prerequisites of Theorem 2 are satisfied, and therefore, from Theorems 1 and 2, *the general stiffness optimization problem (d) has at least one solution.*

Let us note the following important facts. As opposed to $(S)_{h_0}$, $(S)_h$ need not be solvable, since $J_h(\cdot)$ need not be coercive. This has as a consequence that one cannot define a design-to-state mapping as is common in optimal control theory.

That (SJ) serves its purpose can be seen directly without using (d). Indeed, let $h \in \mathcal{H}$ be any design with an existing equilibrium displacement u_h. Then, $(S)_h$ and the left inequality in (SJ) imply

$$J_h(u_h) \leq J_h(\tilde{u}) \leq J_{\tilde{h}}(\tilde{u}),$$

which, by virtue of the right inequality in (SJ), states that the "saddle point design" generates the largest value of the equilibrium potential energy.

Optimality Condition

From the left inequality in (SJ) it is easy to obtain a commonly recognized optimality condition. If $(\tilde{u}, \tilde{h})$ solves (SJ), then $\tilde{h}$ maximizes the functional $h \mapsto \frac{1}{2} a_h(\tilde{u}, \tilde{u}) = \langle\, h, A(\tilde{u}) \,\rangle$, over the set $\mathcal{H}$. Assuming (2) is the only active constraint among the ones defining $\mathcal{H}$, we treat the maximization problem with a Lagrangian technique, by forming the following functional:

$$\mathcal{L}(h, \lambda) = \lambda(\langle\, h, 1 \,\rangle - Vol) - \langle\, h, A(\tilde{u}) \,\rangle, \tag{7}$$

where $\lambda \in \mathbf{R}$. Requiring stationarity with respect to h:

$$\lambda\langle \varphi, 1 \rangle - \langle \varphi, A(\tilde{u}) \rangle = \langle \varphi, \lambda 1 - A(\tilde{u}) \rangle = 0 \quad \forall \varphi \in W'.$$

By virtue of a lemma of the Hahn-Banach theorem, this implies

$$A(\tilde{u}) = \lambda 1. \tag{8}$$

Hence, if the isoperimetric constraint is the only active constraint, then *an optimal design has a uniform distribution of strain energy density.* This condition will often suffice to guarantee an evenly stressed structure.

PARTICULAR EXAMPLES

In this chapter we will exemplify situations that fit into the general setting described in the second chapter. Possible examples are: a one-dimensional rod in which the cross-sectional area is the design variable, a beam in which the width is the design, and a finite element discretization of a two-dimensional continuum in which the thickness for each finite element is subject to change. The first two examples are distributed parameter systems and the last a discrete. We will here have a closer look at two examples, one of each type, namely, truss and sheet. The first one has in the non-contact case been theoretically and numerically studied by e.g. Bendsøe (1992). In 1970 Céa and Malanowski gave an existence result corresponding to the second one, excluding both contact and complete material removal. In this treatment zero values of the design variable are allowed, and (d) is a type of *topology* optimization problem.

Some of the abstract concepts used previously will in the upcoming examples take the concrete forms shown in Table 1.

Table 1. Particular forms of some sets and mappings in the two examples.

	V	U	W	W'	$A(u)$	$C(u)$
truss	$\mathbf{R}^n$	$\mathbf{R}^r$	$\mathbf{R}^m$	$\mathbf{R}^m$	$\frac{1}{2}\{u^T K_1 u, \cdots, u^T K_m u\}^T$	Cu
sheet	$(H_0^1(\Omega))^2$	$H^{\frac{1}{2}}(\Gamma_c)$	$L^1(\Omega)$	$L^\infty(\Omega)$	$\frac{1}{2} u_{i,j} E_{ijkl} u_{k,l}$	$(u \cdot n)\vert_{\Gamma_c}$

Truss Structure

Consider a truss with m bars and n unknown nodal displacements. If the ith bar has the length L_i, then its (assembled) element stiffness matrix (per unit bar volume) can be written

$$K_i = \frac{E}{L_i^2}\gamma_i\gamma_i^T \tag{9}$$

where E is Young's modulus and $\gamma_i \in \mathbf{R}^n$ is a vector of direction cosines for the bar's orientation. If h_i denotes the volume of the ith bar, the structural stiffness matrix can be written as a sum of element stiffness matrices:

$$K(h) = \sum_{i=1}^{m} h_i K_i. \tag{10}$$

$K(h)$ is symmetric and, with non-negative bar volumes, positive semi-definite, as is seen from (9) and (10).

The functionals defined previously are here

$$a_h(u, v) = u^T K(h) v \tag{11}$$

and

$$L(u) = f^T u, \tag{12}$$

where $f \in \mathbf{R}^n$ contains external nodal forces.

The set of permissible designs takes the following form:

$$\mathcal{H} = \left\{ h \in \mathbf{R}^m \,\middle|\, \sum_{i=1}^{m} h_i \equiv h^T \mathbf{1}_m = Vol, \quad 0 \le h_i \quad \forall i \right\}$$

and the set of kinematically admissible displacements:

$$\mathcal{U} = \{ u \in \mathbf{R}^n \mid Cu \le g \},$$

where C is a $r \times n$-matrix whose rows are the vectors of direction cosines of the inward obstacle normals.

In this example, properties in weak, weak* and strong senses coincide. $\mathcal{H}$ is bounded ($\| h \| \le Vol$) and closed, and hence weakly* compact by Heine-Borel's theorem.

Assuming there are sufficient prescribed displacements, $h_0 := \frac{Vol}{m}\mathbf{1}_m \in \mathcal{H}$ gives a positive definite stiffness matrix, hence $a_{h_0}(\cdot\,,\cdot)$ is V-elliptic. More details (verifying the presumptions in the second chapter) can be found in Petersson and Klarbring (1993). Cf. also Table 1.

Finally, from (8) we have that the strain energy density $\frac{1}{2}\tilde{u}^T K_i \tilde{u}$ is λ in any non-removed bar (i.e. when $h_i \ge 0$ is passive). But this quantity can trivially also be written as $\sigma_i^2/2E$ where σ_i is the stress in the ith bar. Hence

$$|\sigma_i| = \sqrt{2E\lambda} \tag{13}$$

for all non-removed bars. One concludes that *all bars in an optimum truss are equally stressed*, (in tension or compression).

Sheet Structure

Consider a sheet with a mid-surface occupying a domain Ω in $\mathbf{R}^2$, and with a thickness $h(x)$ at the point x in Ω. The boundary is partitioned into the parts Γ_d, Γ_t and Γ_c, the first two denoting parts where zero displacements and tractions are prescribed, and the last meaning candidate contact boundary.

Each point x in Ω has an in-plane displacement $u(x)$, and the corresponding field belongs to

$$V = (H_0^1(\Omega))^2 = \left\{ u = (u_1, u_2) \in (H^1(\Omega))^2 \mid u_i = 0 \text{ on } \Gamma_d,\ i = 1, 2 \right\},$$

and to the convex set

$$\mathcal{U} = \{ u \in V \mid u_i n_i - g \le 0 \text{ on } \Gamma_c \},$$

where n_i are the components of the outward unit normal vector to the boundary $\partial\Omega$, and the summation convention from 1 to 2 is used.

The linear and bilinear forms are

$$L(u) = \int_\Omega f \cdot u \, d\Omega + \int_{\Gamma_t} t \cdot u \, d\Gamma$$

and

$$a_h(u,v) = \int_\Omega h u_{i,j} E_{ijkl} v_{k,l} d\Omega,$$

where E_{ijkl} are the elasticity constants satisfying the usual ellipticty, symmetry and boundedness features. $L(\cdot)$ depends on body forces $f \in (L^2(\Omega))^2$ and tractions $t \in (L^2(\Gamma_t))^2$.

The set of permissible designs can be defined as

$$\mathcal{H} = \left\{ h \in L^\infty(\Omega) \,\middle|\, h_{min} \le h \le h_{max} \text{ a.e. in } \Omega, \int_\Omega h \, d\Omega = Vol \right\},$$

where h_{min} and h_{max} are bound functions satisfying certain properties in such a way that $\mathcal{H}$ is non-empty, convex and weakly* compact. Details concerning this can be found in the licentiate thesis by Petersson (1994).

The "coercive design" h_0 can be constructed as a suitable convex combination of h_{max} and h_{min}. Note that it is admissible to choose the latter to be zero. The former is needed for compactness, unlike for a truss when *Vol* is an implicit upper bound on the design variable.

Finally, we show that something similar to the uniformly stressed optimum truss holds in the present case.

Let us assume isotropy and plane stress. Then, if σ_i denotes principal stresses and E and ν are Young's modulus and Poisson's ratio, the strain energy density in terms of stresses, A_σ, and von Mises equivalent stress, σ_M, can be written as

$$A_\sigma = \frac{1}{2E}(\sigma_1^2 + \sigma_2^2 - 2\nu\sigma_1\sigma_2), \quad \sigma_M = (\sigma_1^2 + \sigma_2^2 - \sigma_1\sigma_2)^{1/2}, \tag{14}$$

see for instance Saada (1974). By using the non-negativity of $(\sigma_1 + \sigma_2)^2(1/2 - \nu)$ and $(\sigma_1 - \sigma_2)^2(1/2 - \nu)$ one obtains

$$(1+\nu)\sigma_M^2 \le 3EA_\sigma, \quad (1-\nu)\sigma_M^2 \ge EA_\sigma, \tag{15}$$

respectively. Rewriting (15), one has

$$\frac{1}{1-\nu}EA_\sigma \le \sigma_M^2 \le \frac{3}{1+\nu}EA_\sigma. \tag{16}$$

From (8) and (16) one concludes that, in case of homogeneous material, *the distribution of von Mises equivalent stress is constant, or nearly constant, for optimum sheets.* In fact, for rubber-like materials with $\nu = 0.5$ we have equality signs in (16), and for $\nu = 0.3$, like for many metals, (16) reads approximately

$$1.2\sqrt{EA_\sigma} \le \sigma_M \le 1.5\sqrt{EA_\sigma}. \tag{17}$$

ACKNOWLEDGEMENTS

This work was financially supported by the Swedish Research Council for Engineering Sciences (TFR).

REFERENCES

Bendsøe, M.P., 1992, "Aspects of Topology Optimization of Discrete and Continuum Structures," The Technical University of Denmark, Lyngby.

Benedict, R.L., 1982, Maximum stiffness design for elastic bodies in contact, *J. Mech. Design* 104:825.

Ben-Tal, A., Kočvara, M., and Zowe, J., 1993, Two non-smooth approaches to simultaneous geometry and topology design of trusses, *in*: "Topology Design of Structures," M.P. Bendsøe and C.A. Mota Soares, ed., Kluwer Academic Publishers, Dordrecht.

Céa, J., and Malanowski, K., 1970, An example of a max-min problem in partial differential equations, *SIAM J. Control* 8:305.

Ekeland, I., and Temam, R., 1976, "Convex Analysis and Variational Problems," North Holland, Amsterdam.

Hemp, W.S., 1973, "Optimum Structures," Oxford University Press, Bath.

Ky Fan, M., 1964, Sur un théorème minimax, *C.R. Acad. Sc. Paris* 259:3925.

Petersson, J., submitted 1993, On stiffness maximization of variable thickness sheet with unilateral contact, to appear in *Quart. Appl. Math.*

Petersson, J., 1994, "Stiffness Optimization in Unilateral Contact Problems," Department of Mechanical Engineering, Linköping.

Petersson, J., and Klarbring, A., submitted 1993, Saddle point approach to stiffness optimization of discrete structures including unilateral contact, to appear in *Control and Cybernetics.*

Prager, W., and Taylor, J.E., 1968, Problems of optimal structural design, *J. Appl. Mech.* 35:102.

Saada, A.S., 1974, "Elasticity, Theory and Applications," Krieger, Malabar.

DYNAMIC EVOLUTION OF AN ELASTIC BEAM IN FRICTIONAL CONTACT WITH AN OBSTACLE

Kevin T. Andrews, M. Shillor, and S. Wright

Department of Mathematical Sciences
Oakland University
Rochester, MI 48309-4401, USA

INTRODUCTION

Problems involving contact and friction phenomena have received a great deal of attention in recent years and by now there is a considerable body of engineering literature devoted to this subject. In contrast, there are relatively few general mathematical results available in this area, due to the substantial difficulties encountered in establishing existence results for initial-boundary value problems that model these phenomena. Moreover, in both cases, most of the existing literature deals with static situations, or, occasionally, with a sequence of static problems, which arise from the time discretization of an evolution problem. Modeling and mathematical analysis of such problems can be found in Duvaut and Lions [DL], Moreau et al. [MPS], Kikuchi and Oden [KO], and Telega [Te1], and the references therein (see also Curnier [Cu]). There are, however, some recent results on quasistatic and dynamic behavior in Andersson [An], Telega [Te2], Klarbring et al. [KMS2] and Oden and Martins [OM].

The classic approach to modelling the frictional contact of elastic bodies employs Signorini's nonpenetration condition and Coulomb's law of dry friction. But, as has been pointed out in [OM, DL, Te1, Te2, KMS1, KMS2] and elsewhere, there are both physical and mathematical difficulties associated with the presence of these conditions in models of dynamic contact phenomena. From the physical point of view, unilateral contact conditions seem unrealistic except for very smooth surfaces, as they assert that there is no mutual penetration of the contacting bodies. Mathematically, the Signorini-Coulomb conditions lead to initial-boundary value problems for which the existence of solutions has only been shown in some special cases, e.g., ([NJH], [J1], [J2] and the references in [MO]). The work of Duvuat and Lions [DL, Chapter III] especially illustrates the mathematical difficulties that may be encountered in handling dynamic problems in this area, even in situations where the friction bound is treated as a prescribed function.

In an effort to overcome these difficulties, which often are reflected in the behavior of numerical solutions obtained from algorithms that are based on these two conditions (see, e.g., Raous et al. [RCL]), some investigators have introduced alternative

Contact Mechanics, Edited by M. Raous *et al.*
Plenum Press, New York, 1995

models for the contact interface. Oden and Martins [OM], for example, have proposed a model where the contact interface has a normal compliance characterized by a power-law relationship between the normal pressure and the penetration. A similar generalization of Coulomb's law was also proposed. The computational, theoretical and experimental justification for these conditions has been developed in a series of papers (see, e.g., [JK, KMS1, KMS2, KT, MO, OP, WO]) and there is by now a considerable body of work devoted to this topic.

In a recent paper, [ASW], we incorporated these normal compliance conditions into a model for the dynamic vibrations of an Euler-Bernoulli beam that is in frictional contact with an obstacle. The model takes into account both horizontal and vertical displacements of locations along the beam. Well-posedness results are obtained for a weak formulation of an initial-boundary value problem containing a constitutive relation that includes Kelvin-Voigt damping. More surprisingly, we can also show the existence and uniqueness of a problem that employs the Signorini-Coulomb conditions and Kelvin-Voigt damping. In both cases it is possible to separate the problem of finding the horizontal displacement from the problem of finding the vertical displacement. The horizontal displacement problem is solved first and then the normal contact stress so obtained enters the friction functional in the vertical displacement problem. It is this latter fact that presents the greatest mathematical difficulties, particularly in the problem involving the Signorini-Coulomb conditions, where the normal contact stress possesses minimal regularity. Nevertheless, we can show the existence of a weak solution to the vertical displacement problem, using a crucial fractional Sobolev space estimate derived from the Fourier transform and interpolation. In the absence of damping we can also show the existence of a solution to an Euler-Bernoulli beam problem that employs the Signorini-Coulomb conditions and a regularized normal contact stress.

THE MODELS AND STATEMENTS OF RESULTS

In this section, we present models for the dynamic evolution of a viscoelastic beam in frictional contact with a rigid obstacle. The beam is attached to a wall at its left end, but its right end is free to come into frictional contact with a rigid obstacle situated some distance to the right. The physical setting is depicted in Figure 1.

We assume that the area-center of gravity of the beam in its (stress free) reference configuration coincides with the interval $0 \leqslant x \leqslant 1$. We let $g > 0$ denote the initial gap between the end $x = 1$ and the obstacle. For $T > 0$, we set $\Omega_T = (0,1) \times (0,T)$, and let $u = u(x,t)$ and $v = v(x,t)$, $(x,t) \in \Omega_T$ represent the horizontal and vertical displacements of the beam at location x and time t.

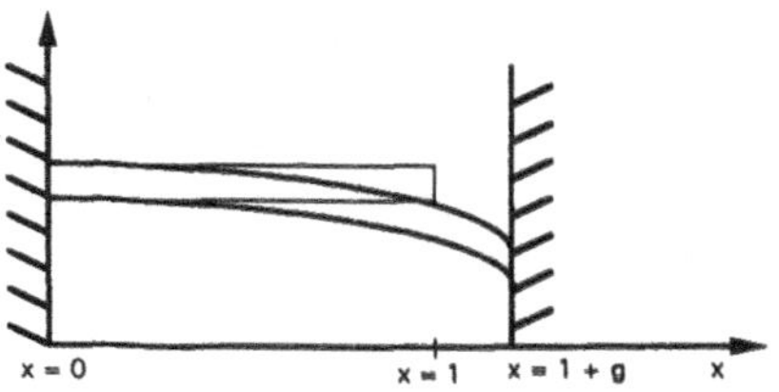

Figure 1. The reference configuration and the deflected beam

Let $\sigma_N = \sigma_N(x,t)$ be the contact pressure at (x,t) and let $\sigma_T = \sigma_T(x,t)$ be the shear stress at (x,t). Then the equations of motion in nondimensional units take the form

$$u_{tt} - (\sigma_N)_x(x,t) = k, \tag{1}$$
$$v_{tt} - (\sigma_T)_x(x,t) = f, \tag{2}$$

where f and k denote the vertical and horizontal applied forces, respectively. We take for our constitutive relationships the Kelvin-Voigt viscoelasticity laws

$$\sigma_N(x,t) = cu_x(x,t) + bu_{xt}(x,t), \quad (x,t) \in \Omega_T, \tag{3}$$

and

$$\sigma_T(x,t) = -(v_{xxx}(x,t) + dv_{xxxt}(x,t)), \quad (x,t) \in \Omega_T. \tag{4}$$

Here b, c, and d are nonnegative material coefficients. In fact, b and d represent the viscosities in the horizontal and vertical directions, respectively. To complete the model we must include appropriate intial and boundary conditions. The initial conditions take the form

$$u(x,0) = u_0(x), \qquad u_t(x,0) = u_1(x), \quad 0 \leqslant x \leqslant 1, \tag{5}$$
$$v(x,0) = v_0(x), \qquad v_t(x,0) = v_1(x), \quad 0 \leqslant x \leqslant 1, \tag{6}$$

where u_0, u_1, v_0 and v_1 are prescribed functions. Since we suppose that the beam is rigidly attached at its left end, we set

$$u(0,t) = 0, \tag{7}$$
$$v(0,t) = 0, \quad v_x(0,t) = 0, \quad 0 \leqslant t \leqslant T. \tag{8}$$

For the boundary conditions at the free end we consider several alternative conditions. For the horizontal displacement the first of these is the classic Signorini nonpenetration conditions

$$u(1,t) \leqslant g, \quad \sigma_N(1,t) \leqslant 0, \quad \text{and} \quad \sigma_N(1,t)(g - u(1,t)) = 0, \quad 0 \leqslant t \leqslant T. \tag{9}$$

As was previously mentioned there are both physical and mathematical difficulties associated with the inclusion of this condition in mathematical models of frictional contact. Consequently, following [MO] and [KMS1], we shall also consider the alternative normal compliance condition

$$\sigma_N(1,t) = -c_N(u(1,t) - g)_+^{m_N}, \quad 0 \leqslant t \leqslant T, \tag{10}$$

where c_N and m_N are two positive constants, and $(u(1,t)-g)_+ = \max\{(u(1,t)-g), 0\}$ represents the positive part of the function $u(1,t) - g$. We note that in both (9) and (10) that when there is no contact, i.e., when $u(1,t) < g$, then $\sigma_N(1,t) = 0$. However, (10) permits the contacting end to penetrate the obstacle, i.e, it permits $u(1,t) > g$. (9) may be thought of as a limiting case of (10) as c_N tends to infinity and for this reason it is typical to choose c_N large (see, e.g., [KT]).

For the vertical displacement problem we assume that the sum of the moments acting on the free end is zero, i.e.,

$$v_{xx}(1,t) + dv_{xxt}(1,t) = 0, \quad 0 \leqslant t \leqslant T, \tag{11}$$

and we also include a friction law of the form

$$|\sigma_T(1,t)| \leqslant h(t), \tag{12}$$

$$\text{if } |\sigma_T(1,t)| = h(t) \quad \text{then} \quad v_t(1,t) = -\lambda\sigma_T(1,t), \text{ for some } \lambda \geqslant 0, \tag{13}$$

$$\text{if } |\sigma_T(1,t)| < h(t) \quad \text{then} \quad v_t(1,t) = 0, \tag{14}$$

for $0 \leqslant t \leqslant T$. Physically, the function $h(t)$ may be thought of as a friction bound and the conditions (12)-(14) interpreted in the following way: when the shear stress σ_T equals $\pm h$ then the shear will be in the direction opposite to the slip, and when the shear is strictly less (in absolute value) than the friction bound, then the end sticks to the obstacle.

The function h can be chosen in at least three different ways. One way is to treat h as a prescribed function. With this choice, equations (2),(4),(6),(8),(11)-(14) become an independent model for the transverse vibrations of a viscoelastic beam in frictional contact with a rigid obstacle. In [ASW] we proved that such problems have weak solutions provided h is in $H^{-\epsilon}(0,T)$, for some $0 < \epsilon < (1/8)$. A second way of treating h is to let

$$h(t) = -\mu\sigma_N(1,t), \quad 0 \leqslant t \leqslant T, \tag{15}$$

where μ is a positive constant, called the coefficient of friction. With this choice, the conditions (12)-(15) constitute the classic Coulomb law of dry friction. Of course this law may be used in combination with either the Signorini condition (9) or the normal compliance condition (10). However, in the latter case, we may also allow for a more general h of the form

$$h(t) = c_T(u(1,t) - g)_+^{m_T}, \quad 0 \leqslant t \leqslant T, \tag{16}$$

where c_T and m_T are two positive constants which may be chosen independently of c_N and m_N. This completes our description of the various boundary conditions we will consider.

The alert reader will have noticed that equations (1),(3), (5),(7) and either (9) or (10) constitute an initial-boundary value problem for the horizontal displacement u that can be solved independently of the vertical displacement v. Indeed, problems of this type have already been considered in [Ki1, Ki2, KS, LS, MO, SB].

Now, as is well known, a friction law of the form (12)–(14) imposes a regularity ceiling which, generally, precludes the existence of classical solutions to problems containing this boundary condition. Thus, it is natural to consider weak, or variational inequality, formulations of the above equations. We will give two such formulations in this section. Toward that end we introduce the following spaces and notation. (For definitions of any unexplained notation we refer the reader to [LM] or [LSU]).

Let $H = L^2(0,1)$, $E = \{w \in H^1(0,1) : w(0) = 0\}$ and $V = \{w \in H^2(0,1) : w(0) = w'(0) = 0\}$. Clearly we have

$$V \subseteq E \subseteq H = H' \subseteq E' \subseteq V',$$

where E' and V' are, respectively, the topological duals of E and V.

Our first formulation incorporates the normal compliance condition (10) and the generalized Coulomb condition (15) and so it is convenient to introduce

$$j_N(u,w) = c_N \int_0^T (u(1,t)-g)_+^{m_N} w(1,t)dt, \tag{17}$$

the *"normal compliance"* functional, and

$$j_T(u,w) = c_T \int_0^T (u(1,t)-g)_+^{m_T} |w(1,t)|dt, \tag{18}$$

the *"friction"* functional. Note that if $m_N \geqslant 1$ and $m_T \geqslant 1$, then both functionals are defined and convex on $L^2(0,T;E) \times L^2(0,T;E)$ but that, unlike $j_N(u,\cdot)$, $j_T(u,\cdot)$ is not Gâteaux differentiable.

We can now give our first weak formulation, which is obtained in the usual way by multiplying (1)-(2) by suitable test functions and integrating by parts.

Definition 1. A pair of functions $(u,v) \in L^2(0,T;E) \times L^2(0,T;V)$ is said to be a weak solution to (1)-(8),(10), provided that

$$u_t \in L^2(0,T;E), \quad u_{tt} \in L^2(0,T;E'), \quad u(\cdot,0) = u_0, \quad u_t(\cdot,0) = u_1, \tag{19}$$

$$v_t \in L^2(0,T;V), \quad v_{tt} \in L^2(0,T;V'), \quad v(\cdot,0) = v_0, \quad v_t(\cdot,0) = v_1, \tag{20}$$

and for each $\varphi \in L^2(0,T;E)$ and $w \in L^2(0,T;V)$

$$\int_0^T \langle u_{tt},\varphi\rangle dt + c\int_0^T (u_x,\varphi_x)dt + b\int_0^T (u_{xt},\varphi_x)dt + j_N(u,\varphi) = \int_0^T \langle k,\varphi\rangle dt, \tag{21}$$

$$\begin{aligned} &\int_0^T \langle v_{tt}, w - v_t\rangle dt + \int_0^T (v_{xx}, w_{xx} - v_{xxt})dt + d\int_0^T (v_{xxt}, w_{xx} - v_{xxt})dt \\ &\quad + j_T(u,w) - j_T(u,v_t) \\ &\geqslant \int_0^T (f, w - v_t)dt. \end{aligned} \tag{22}$$

In the integrands contained in (21) and (22), $\langle\cdot,\cdot\rangle$ denotes respectively the duality pairing between E and E' and between V and V' while in both cases $(\cdot,\cdot)$ denotes the inner product in H.

We have the following existence and uniqueness result for the above problem.

Theorem 2. *Let $b \geqslant 0$, $c > 0$, and $d > 0$. Let $k, k_t \in L^2(0,T;E')$, $f \in L^2(\Omega_T)$, $u_0 \in E$, $v_0 \in V$, u_1, $v_1 \in H$, and $m_N \geqslant 1$, $m_T \geqslant 1$. Then there exists a unique solution to problem (19)-(22), provided that in the case when $b = 0$, the requirement in (21) that $u_t \in L^2(0,T;E)$ is replaced by $u_t \in L^2(0,T;H)$.*

As previously indicated, the unique solvability of (19) and (21) when $b = 0$ is established in [MO, Theorem 4.1]. However, the proof given there can be modified in obvious ways to obtain the result for $b > 0$.

We have, in addition, the following stability result.

Theorem 3. *Let (y_i, z_i), $i = 1, 2$, be two solutions to (19)-(21) corresponding to initial data $(y_{i0}, y_{i1}) \in E \times H$ and $(z_{i0}, z_{i1}) \in V \times H$ and applied forces $(f_i, k_i) \in L^2(\Omega_T) \times L^2(\Omega_T)$. Then there is an absolute constant $C > 0$, and a constant $C_1 > 0$ which only depends boundedly on the norms of the data for the horizontal problem, such that*

$$\begin{aligned}\|y_1 - y_2\|_{H^1(\Omega_T)} &\leqslant C_1(\|k_1 - k_2\|_{L^2(\Omega_T)} + \|y'_{10} - y'_{20}\|_H + \|y_{11} - y_{21}\|_H),\\ \|z_1 - z_2\|_{W^{2,1}(\Omega_T)} &\leqslant C_1(\|k_1 - k_2\|_{L^2(\Omega_T)} + \|y'_{10} - y'_{20}\|_H + \|y_{11} - y_{21}\|_H)\\ &\quad + C(\|f_1 - f_2\|_{L^2(\Omega_T)} + \|z''_{10} - z''_{20}\|_H + \|z_{11} - z_{21}\|_H).\end{aligned}$$

Our second formulation incorporates the Signorini condition (9) and the friction law (12)-(14). For this purpose it is convenient to set

$$K = \{f \in E : f(1) \leqslant g\}.$$

We also recall the scale of Hilbert spaces of distributions $H^\epsilon(0,T)$, $\epsilon \in (-\infty, \infty)$, as defined in [LM, Vol. I] and introduce, for $\epsilon \geqslant 0$, the functional

$$J(z) = \langle h, |z(1,\cdot)|\rangle_{-\epsilon,\epsilon},$$

where $h \in H^{-\epsilon}(0,T)$ and $\langle\cdot,\cdot\rangle_{-\epsilon,\epsilon}$ denotes the duality pairing between $H^\epsilon(0,T)$ and $H^{-\epsilon}(0,T)$.

The variational formulation of the problem with Signorini and Coulomb boundary conditions is as follows.

Definition 4. Let $\epsilon \geqslant 0$, and suppose $h \in H^{-\epsilon}(0,T)$. A pair of functions $(u, v) \in L^2(0,T;K) \times L^2(0,T;V)$ is said to be a weak solution to (1)-(9),(11)-(14) provided that

$$u_t \in L^2(0,T;E),\ u_{tt} \in L^2(0,T;E'),\ u(\cdot,0) = u_0,\ u_t(\cdot,0) = u_1, \tag{23}$$

$$v_t \in L^2(0,T;V),\ v_{tt} \in H^{-\epsilon}(0,T;V'),\ v(\cdot,0) = v_0,\ v_t(\cdot,0) = v_1, \tag{24}$$

and for each $\varphi \in L^2(0,T;K)$ and $w \in H^\epsilon(0,T;V)$,

$$\int_0^T \langle u_{tt}, \varphi - u\rangle dt + c\int_0^T (u_x, \varphi_x - u_x)dt + b\int_0^T (u_{xt}, \varphi_x - u_x)dt \geqslant \int_0^T (k, \varphi - u)dt, \tag{25}$$

$$\begin{aligned}\int_0^T \langle v_{tt}, w - v_t\rangle dt + \int_0^T (v_{xx}, w_{xx} - v_{xxt})dt + d\int_0^T (v_{xxt}, w_{xx} - v_{xxt})dt&\\ + J(w) - J(v_t)&\\ \geqslant \int_0^T (f, w - v_t)dt.&\end{aligned} \tag{26}$$

We have the following existence and uniqueness result.

Theorem 5. *Let $b \geqslant 0$, $c > 0$ and $d > 0$. Let $k \in L^2(\Omega_T)$, $f \in L^2(\Omega_T)$, $u_0 \in K$, $v_0 \in V$ and u_1, $v_1 \in H$. If $h = -\mu\sigma_N(1,\cdot)$, where μ is a positive constant and σ_N is given by (3), then there exists a unique solution to problem (23)–(26), provided that in the case when $b = 0$ the requirement in (21) that $u_t \in L^2(0,T;E)$ is replaced by $u_t \in L^2(0,T;H)$.*

Once again the solution to the horizontal problem (23) and (25) can be found in [SB], [Ki1], or [Ki2] for the case when $b = 0$ and in [KS] for the case when $b > 0$.

Hence, the main interest lies in establishing the solution to the vertical displacement problem (24) and (26).

Finally, we consider a problem which uses the standard Signorini-Coulomb conditions without Kelvin-Voigt damping. This is the kind of model for frictional contact most frequently employed in engineering applications, despite the difficulties discussed above. In this setting we find it necessary to introduce a positive regularization operator, i.e, an operator $R : L^2(0,T) \to H_L^2(0,T)$ such that $R(\nu) \geqslant 0$ if $\nu \geqslant 0$ and such that there exists an absolute constant $C > 0$ so that

$$\|R(\nu)\|_{H_L^2(0,T)} \leqslant C\|\nu\|_{L^2(0,T)}$$

for each $\nu \in L^2(0,T)$. Here $H_L^2(0,T) = \{w \in H^2(0,T) : w(0) = 0\}$. One way to construct such an operator is by using the convolution with a positive C^∞ kernel (see, e.g., [G] or [B]).

We now have the following result.

Theorem 6. *Let $b = d = 0$ and $c > 0$. Let $k \in L^2(\Gamma_T)$, $f \in H^1(0,T;H)$, $u_0 \in K$, $u_1 \in H$, $v_0 \in H^4(0,1)$ and $v_1 \in V$ with $v_0(0) = v_0'(0) = v_0''(1) = v_0'''(1) = 0$. If $h = R(-\mu\sigma_N(1,\cdot))$, where μ is a positive constant and σ_N is given by (3), then there exists a solution to problem (23)–(26). Moreover, v can be chosen so that $v \in W^{1,\infty}(0,T;V)$ and $v_{tt} \in L^\infty(0,T;H)$.*

The full details can be found in [ASW]. Some numerical simulations are in progress and will be presented elsewhere.

REFERENCES

[An]. L. E. Andersson, *A quasistatic frictional problem with normal compliance*, Nonlin. Anal. TMA **16**(4) (1991), 347-370.

[ASW]. Kevin T. Andrews, M. Shillor, and S. Wright, *On the dynamic vibrations of an elastic beam in frictional contact with a rigid obstacle*, preprint.

[B]. G. Bonfanti, *A noncoercive friction problem with tangential applied forces in three dimensions*, BUMI **7** (1993), 149-165.

[Cu]. A. Curnier, *A theory of friction*, Int. J. Solids Structures **20** (1984), no. 7, 637-647.

[DL]. G. Duvuat and J. L. Lions, *Inequalities in Mechanics and Physics*, Springer-Verlag, Berlin, 1976.

[G]. F. Gastaldi, *Remarks on a noncoercive contact problem with friction in elastostatics*, IAN Publication 650 (1988).

[J1]. J. Jarusek, *Contact problems with bounded friction, Semicoercive case*, Czech. Math. J. **34** (1984), 619-629.

[J2]. J. Jarusek, *Contact problems with given time-dependent friction force in linear viscoelasticity*, Comment. Math. Univ. Carol. **31** (1990), no. 2, 257-262.

[JK]. L. Johansson and A. Klarbring, *Thermoelastic frictional contact problems: Modelling, finite element approximation and numerical realization*, Comp. Meth. Appl. Mech. Engrg. **105** (1993), 181-210.

[KO]. N. Kikuchi and J.T. Oden, *Contact Problems in Elasticity*, Philadelphia, SIAM, 1988.

[Ki1]. J. U. Kim, *A boundary thin obstacle problem for a wave equation*, Commun. Partial Diff. Eq. **14** (1989), no. 8&9, 1011-1026.

[Ki2]. J. U. Kim, *A one-dimensional dynamic contact problem in linear visco-elasticity*, Math. Meth. Appl. Sci. **13** (1990), 55-79.

[KMS1]. A. Klarbring, A. Mikelić and M. Shillor, *Frictional contact problems with normal compliance*, Int. J. Engng Sci. **26**(8) (1988), 811-832.

[KMS2]. A. Klarbring, A. Mikelić, and M. Shillor, *A global existence result for the quasistatic frictional contact problem with normal compliance*, in Unilateral Problems in Structural Mechanics IV (Capri, 1989) (G. DelPiero and F. Maceri, eds.), Birkhauser, Boston, 1991, pp. 85-111.

[KT]. A. Klarbring and B. Torstenfelt, *A Newton method for contact problems with friction and interface compliance,* preprint.

[KS]. K. Kuttler and M. Shillor, *A dynamic contact problem in one-dimensional thermoviscoelasticity,* preprint.

[LSU]. O. A. Ladyzenskaja, V. A. Solonnikov and N. N. Uralceva, *Linear and Quasilinear Equations of Parabolic Type*, American Mathematical Society, Providence, 1968.

[LM]. J. L. Lions and E. Magenes, *Nonhomogeneous Boundary Value Problems and Applications*, Vol. I&II, Springer, New York, 1972.

[LS]. G. Lebeau and M. Schatzman, *A wave problem in a half-space with a unilateral condition at the boundary,* J. Diff. Eqns. **53** (1984), 309-361.

[MO]. J.A.C. Martins and J.T. Oden, *Existence and uniqueness results for dynamic contact problems with nonlinear normal and friction interface laws*, Nonlin. Anal. TMA **11**(3) (1987), 407-428.

[MPS]. J.J. Moreau, P.D. Panagiotopoulos, and G. Strang (eds.), *Topics in Nonsmooth Mechanics*, Birkhauser, Basel, 1988.

[NJH]. J. Necas, J. Jarusek, and J. Haslinger, *On the solution of the variational inequality to the Signorini problem with small friction*, Boll. Un. Mat. Ital. **5** (1980), no. 17-B, 796-811.

[OM]. J.T. Oden and J.A.C. Martins, *Models and computational methods for dynamic friction phenomena*, Comput. Meth. Appl. Mech. Engrg. **52** (1985), 527-634.

[OP]. J.T. Oden and E. Pires, *Nonlocal and nonlinear friction laws and variational principles for contact problems in elasticity*, J. Appl. Mech. **50** (1983), 67-76.

[RCL]. M. Raous, P. Chabrand, and F. Lebon, *Numerical methods for frictional contact problems and applications*, J. Theoretical Appl. Mechanics **7**(1) (1988), 111-128.

[SB]. M. Schatzman and M. Bercovier, *Numerical approximation of a wave equation with unilateral constraints*, Math. Compt. **53** (1989), no. 187, 55-79.

[Te1]. J.J. Telega, *Topics on Unilateral Contact Problems of Elasticity and Inelastisity*, preprint.

[Te2]. J. J. Telega, *Quasi-static Signorini's contact problem with friction and duality*, in Unilateral Problems in Structural Mechanics IV (Capri, 1989) (G. DelPiero and F. Maceri, eds.), Birkhauser, Boston, 1991, pp. 199-214.

[WO]. L. White and J.T. Oden, *Dynamics and Control of Viscoelastic Solids with Contact and Friction Effects*, Nonlin. Anal. TMA **13** (1989), 459-474.

NUMERICAL TREATMENT OF NONMONOTONE QUASI-STATIC FRICTIONAL CONTACT PROBLEMS VIA D.C. ENERGY DECOMPOSITION AND MULTIPHASE METHODS

Georgios E. Stavroulakis

Faculty of Mathematics and Physics
Aachen University of Technology, RWTH
D-52062 Aachen, F.R. Germany

INTRODUCTION

Theoretical and numerical study of quasistatic numerical contact problems with nonmonotone Coulomb-like friction is considered through difference convex (d.c.) decomposition of the nonconvex superpotential energy function. Separate handling of convexity and concavity, whenever this information is explicitly (d.c. Hiriart-Urruty, 1985) or implicitly (quasidifferentials Demyanov, 1989) available, is currently considered to be a reasonable approach for the study of nonconvex optimization problems, since results of convex analysis and optimization can be fully utilized. The impact of this approach on the development of numerical algorithms for frictional contact problems is studied here.

Following J.J. Moreau (1968) a quasistatic frictional contact joint can be formulated by introducing the subdifferential of a nonsmooth (absolute value) convex superpotential. Thus variational inequality problems arise. In the case of nonmonotone friction laws, the superpotential is nonconvex as well. If the generalized subdifferential of F.H. Clarke is used hemivariational inequality problems (Panagiotopoulos, 1994) arise.

A difference convex decomposition of the nonconvex superpotential is used here on the original nonconvex and nondifferentiable problem and gives rise to a two-field finite element formulation that involves elements of both monotone friction and correction (descending) tractions. Thus a system of variational inequalities (Stavroulakis et al., 1993, Panagiotopoulos et al., 1992) arise. Multiphase numerical algorithms can be constructed from this theory using results of quasidifferentiable optimization and d.c. duality and optimization where classical frictional contact problems with monotone laws arise as subproblems (Auchmuty, 1989, Tao et al., 1988, Stavroulakis et al., 1993). The advantage gained by using existing software is obvious. The method is classified as the analogon of sub (or super) gradient iterative optimization algorithms for the nonconvex potential energy optimization problem (Tao et al., 1988).

SYSTEMS OF VARIATIONAL INEQUALITIES

Let a two dimensional linear elastic structure in $\Omega \subset R^2$ be considered with a

Contact Mechanics, Edited by M. Raous *et al.*
Plenum Press, New York, 1995

boundary composed of the complementary and nonoverlapping parts: support boundary part Γ_U, loaded boundary part Γ_F and frictional contact boundary Γ_S.

In order to demonstrate the d.c. decomposition concept the friction problem with given normal traction will be considered first. For monotone, static, Coulomb friction law it gives rise to the following *potential energy minimization problem* : Find $u \in R^n$

$$\Pi(u) \;=\; \min \{ \; \Pi(v) \mid v \; \in R^n \} \tag{1}$$

with

$$\Pi(v) \;=\; \frac{1}{2}\alpha(v,v) \;-\; (f,v) \;-\; < F_i, v_i >_{\Gamma_F} \;-\; < C_N, v_N >_{\Gamma_S}$$
$$+ \int_{\Gamma_S} \mu \mid C_N \mid (\mid v_T \mid) \; d\Gamma \;=\; \Pi_1(v) \;+\; \Phi_1(v) \tag{2}$$

An equivalent *variational inequality* problem can be written or a *convex multivalued equation* : Find $u \in R^n$ such that

$$\mathbf{0} \in \partial(\Pi_1(\mathbf{u}) + \Phi_1(\mathbf{u})) \tag{3}$$

A nonmonotone friction law will be considered analogously by using the nonconvex and nonsmooth *potential energy* defined by

$$\Pi(u) \;=\; \Pi_1(u) \;+\; \int_{\Gamma_S} G_T(u_T; C_N, \mu) \; d\Gamma_S \tag{4}$$

where $S_T = g_T(u_T; S_N, \mu)$ is the nonmonotone, possibly multivalued friction law and

$$G_T(u_T; C_N, \mu) \;=\; \int_0^{u_T} g_T(v; c_T, \mu) dv \tag{5}$$

Let a *d.c. decomposition* of the superpotential $G_T(u_T; C_T, \mu)$ be given:

$$G_T(u_T; C_N, \mu) \;=\; \phi_1(u_T; C_N, \mu) \;-\; \phi_2(u_T; C_N, \mu) \tag{6}$$

where $\phi_1(u_T; C_N, \mu)$ and $\phi_2(u_T; C_N, \mu)$ are convex, possibly nondifferentiable constituents. Then $\Phi(v)$ term in (4) is decomposed as:

$$\Phi(u) \;=\; \int_{\Gamma_S} G_T(u_T) d\Gamma_S \;=\; \int_{\Gamma_S} \phi_1(u_T) d\Gamma_S \;-\; \int_{\Gamma_S} \phi_2(u_T) d\Gamma_S \;=\; \Phi_1(u) \;-\; \Phi_2(u) \tag{7}$$

Thus the *d.c. potential energy* arises:

$$\Pi(v) \;=\; \Pi_1(v) \;+\; \Phi_1(v) \;-\; \Phi_2(v) \;=\; \overline{\Phi_1}(v) \;-\; \overline{\Phi_2}(v) \tag{8}$$

Exploitation of the necessary condition for (8) to attain a minimum, i.e. Find $u \in R^n$ such that

$$\overline{\Phi_2}(u) \subset \overline{\Phi_1}(u) \tag{9}$$

leads to the following *system of variational inequalities*: Find $u \in R^n$ and a field of frictional correcting tractions $w_2 \in W$ such as to satisfy:

$$\alpha(u, v-u) \;-\; (f, v-u) \;-\; < F, v-u >_{\Gamma_F} \;-\; < C_N, v_N - u_N >_{\Gamma_S}$$
$$+ \; \Phi_1(v_T) \;-\; \Phi_1(u_T) \;-\; < w_2, v_N - u_N >_{\Gamma_S} \;\geq\; 0, \; \forall v \in R^n \tag{10}$$

and for each $w_2 \in W$ such that

$$\Phi_2(v_T) \;-\; \Phi_2(u_T) \;\geq < w_2, v_N - u_N >_{\Gamma_S}, \; \forall v \in V_0 \tag{11}$$

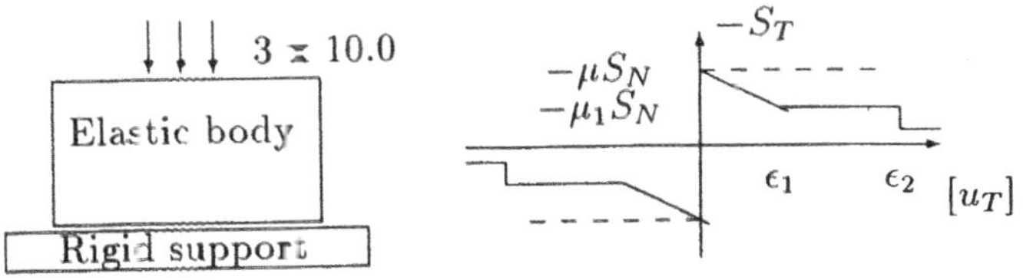

Figure 1. Configuration of a stamp problem and nonmonotone friction law

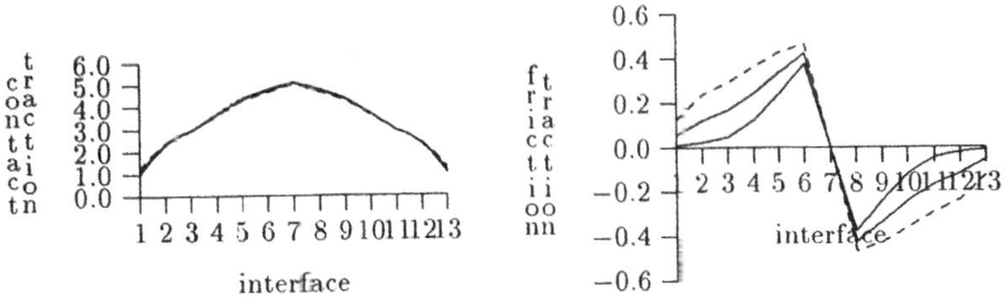

Figure 2. Interface stresses (for $\mu = 0.1, \epsilon_1 = 0.01, \epsilon_2 = 0.05$)

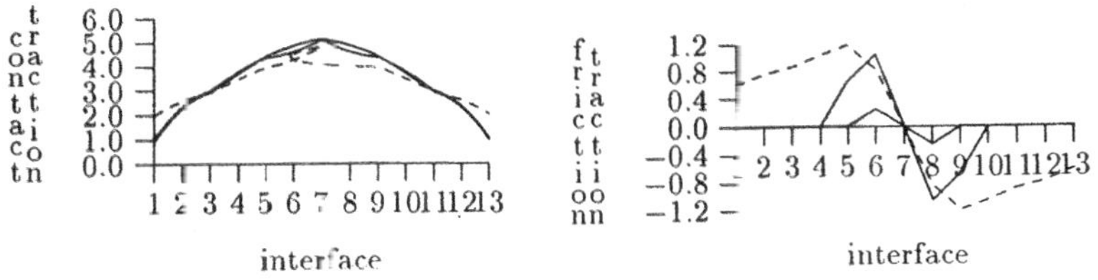

Figure 3. Interface stresses (for $\mu = 0.3, \epsilon_1 = 0.003, \epsilon_2 = 0.005$)

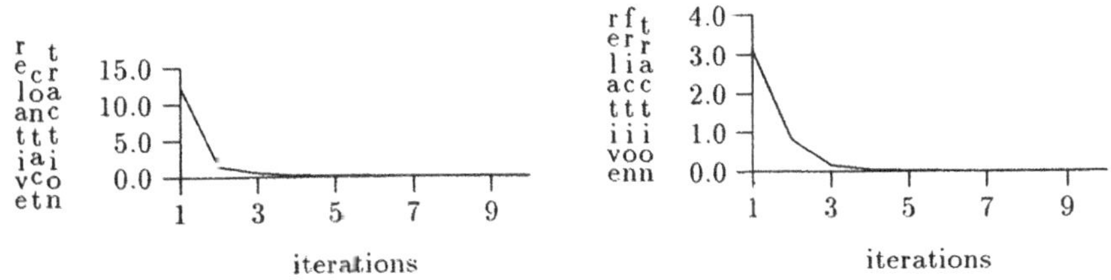

Figure 4. Convergence of the iterative scheme in terms of relative norms of interface contact and friction tractions (case of Fig.3)

Variables w_2 may be considered as hidden or artificial boundary tractions that account for the descending branches in the nonmonotone friction relation (see Fig.1,b). A generalization of the above approach within the quasidifferential optimization theory is presented in Stavroulakis et al. (1993). Note that both d.c. decomposition and quasidifferentials are not uniquely defined (see e.g. Demyanov, 1989, Hiriart-Urruty, 1985).

For the numerical solution a method proposed by Auchmuty (1989) is used here, where first the Lagrangian function $L_{II}(u, w_2) \; : \; V_0 \; \times \; W \; \rightarrow R^1$ is constructed by applying partial convex conjugation on $\overline{\Phi_2}(.)$:

$$L_{II}(u, w_2) \; = \; \overline{\Phi_1}(u) \; + \; \overline{\Phi_2}^c(w_2) \; - \; < u, w_2 > \tag{12}$$

Since $L_{II}(u, w_2)$ is convex in each one of its arguments when the other is considered as constant the multiphase algorithm is defined::

$$L_{II}(u, w_2) \; \leq \; L_{II}(v, w_2), \; \forall v \in R^n \text{ and } L_{II}(u, w_2) \; \leq \; L_{II}(v, \overline{w}_2), \; \forall \overline{w}_2 \in W \tag{13}$$

The first subproblem in (13) concerns an elastic structure with frictional boundary with given normal traction and initial (imposed) tangential loading w_2 along the boundary Γ_S. The correcting terms w_2 are updated in the second subproblem (see Stavroulakis et al., 1993 for details). Algorithm (13) has been proposed in Auchmuty (1989) for the calculation of critical points of the potential energy function. Various extensions are discussed in Tao et al. (1988).

NUMERICAL EXAMPLE

The stamp problem of Fig. 1 is solved with the nonmonotone friction laws shown in Fig.1b. For the elastic plane stress structure $E = 1000.0, \nu = 0.3$ and $t = 1.0$. The interface nodal stresses are shown on Figs. 2 and 3, for two different frictional coefficients. Here $\mu_1 = \mu$, $\mu_1 = 0.5\mu$ and $\mu_1 = 0.1\mu$ are considered for parametric investigation. Convergence has been achieved after 4 and 10 (worst case) iterations of algorithm (12) resp. The change of the quadratic norm of interface stresses during the iterations are given on Fig.4.

ACKNOWLEDGEMENTS

Continuous support from Prof. P.D. Panagiotopoulos, Aristotle University, Thessaloniki and RWTH Aachen is gratefully acknowledged. Financial support of the Alexander von Humboldt Foundation, Bonn and kind hospitality of Prof. Dr.Ing. J. Ballmann and Prof. Dr. H. Th. Jongen, RWTH Aachen are also acknowledged.

REFERENCES

Auchmuty, G., 1989, Duality algorithms for nonconvex variational principles. *Num. Funct. Anal. and Optimization*, 10:211-264.

Demyanov, V.F., 1989, Smoothness of nonsmooth functions, *in* : "Nonsmooth Optimization and Related Topics", F.H. Clarke, V.F. Demyanov and F. Giannessi eds., Plenum Press, New York, London.

Hiriart-Urruty, J.-B., 1985, Generalized differentiability, duality and optimization for problems dealing with differences of convex functions, *in* : "Convexity and Duality in Optimization", J. Ponstein ed., Lect Notes in Econ. and Math. Systems Vol. 256, Springer 1985.

Moreau, J.J., 1968, La notion de sur-potentiel et les liaisons unilaterales en elastostatique. *C.R. Acad. Sc., Paris*, 267A:954-957.

Panagiotopoulos, P.D. and Stavroulakis, G.E., 1992, New type of variational principles based on the notion of quasidifferentiability, *Acta Mechanica* 94:171-194.

Panagiotopoulos, P.D., 1994, "Hemivariational Inequalities. Applications in Mechanics and Engineering", Springer Verlag, New York, Berlin.

Stavroulakis, G.E. and Panagiotopoulos, P.D., 1993 Convex multilevel decomposition algorithms for non-monotone problems, *Intern. Journal of Numerical Methods in Engineering* 36(11): 1945-1961.

Tao, P.D. and Souad, El.B., 1988, Duality in D.C. optimization. Subgradient methods, *in* : "Intern. Series of Numerical Mathematics" Vo9l:276-294, Birkhäuser Verlag, Basel.

FRICTIONAL CONTACT PROBLEM IN ELASTOSTATICS : REVISITING THE UNIQUENESS CONDITION

Pierre Alart , Frédéric Lebon, François Quittau and Karine Rey

Laboratoire de Mécanique et Génie Civil
Université Montpellier 2
Pl. E. Bataillon, 34095 Montpellier Cedex 5
France

INTRODUCTION

The purpose of this paper is to enlight, using a simple-minded approach, the existence and uniqueness of the solution of frictional contact problems in elastostatics.

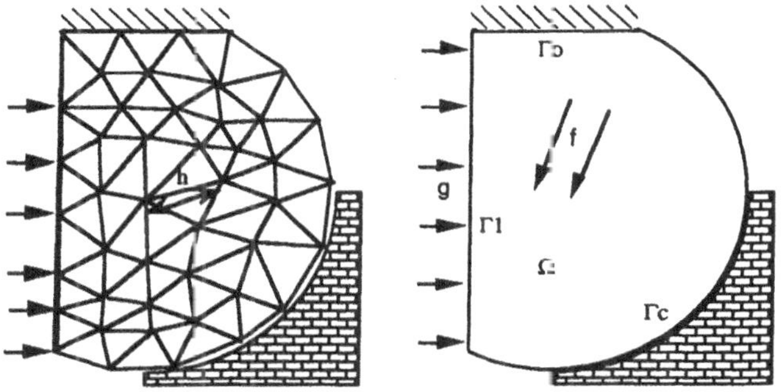

$$\mu^{h}_{crit} \overset{?}{\rightarrow} \mu_{crit}$$

Figure 1. Does an asymptotic critical friction coefficient exist?

It is recalled that such a uniqueness condition may be obtained by considering a non local (or regularized) friction law (Duvaut, 1982; Cocu, 1984). There exists a unique solution if the friction coefficient is smaller than a critical value. Unfortunately, it is a sufficient but not necessary condition. Moreover, this value depends on the regularisation parameter and tends to zero when the regularisation vanishes.

About the non regularized problem, some existence results have been proved (Necas et al., 1980). In this case, analytical expressions of the critical value for simple examples are given in particular when the medium occupies an infinite layer.

In an other way, uniqueness conditions have been exhibited for discret (or discretized) problems using fixed point theorems (Licht et al., 1990), complementary methods (Klarbring,

Contact Mechanics, Edited by M. Raous *et al.*
Plenum Press, New York, 1995

1985) or non smooth analysis (Alart, 1993). For discrete problems examples of non-uniqueness like the Klarbring's truss example (1985) are available. Such examples are not known for continuous media.

The condition of Alart (1993) is necessary and sufficient. The purpose of this paper is to investigate the critical friction value for the uniqueness of a discretized problem and its limit when the mesh becomes finer (figure 1). Unfortunately the results of Alart (1993) are not easy to handle. The computation of many matrices determinants is needed.

In the first section, we review the useful mathematical background. We then provide in the second section, results concerning a discretized medium occuping a square (square test). For that we use at first a symbolic computation software with coarse meshes and then a numerical computation program to study the influence of refinement. The third section is devoted to interpreting some results on rectangular tests obtained by a symbolic computation software. We close by discussing the meaning and the relevance of the notions introduced in the previous sections.

This research provides an insight into the mechanism of non uniqueness due to friction. In this way, it would be interesting to have a non uniqueness example about a continuous problem.

THEORETICAL BACKGROUND

Frictional contact operator

We are interested in the problem of a discretized elastic body in frictional contact with a rigid obstacle. We write directly the expressions concerning the discret or discretized case. Deriving from an implicit augmented Lagrangian (Alart and Curnier, 1991), the frictional contact operator noted F depends on the generalized displacement vector u and the generalized contact force vector λ (composed by all nodal contact force vectors):

$$F(u,\lambda) = \sigma_n^- \; n + \mathrm{proj}_{C(\sigma_n^-)}[\,\sigma_t\,] \text{ with} \tag{1}$$

$$\sigma_n = \lambda_n + ru_n \; ; \; \sigma_t = \lambda_t + ru_t \; ; \; x^- = \min(0,x) \; ;$$

where n is the outward normal unit vector to the obstacle, λ_n and λ_t denote the normal and tangential components of λ, d_n and δ_t represent the normal distances and the tangential slip increments between the nodes of the eventual contact boundary and the obstacle and r is a positive real number characteristic of the penalty-duality methods (Alart and Curnier, 1991). $C(\lambda_n)$ denotes the section of the Coulomb's cone at the level λ_n. This operator has to be written within a system of equations gouverning the equilibrium of the discretized elastic body in frictional contact :

$$H(u,\lambda) = 0 \quad : \left\{ \begin{array}{lcl} Ku - F_{ext} + F(u,\lambda) & = & 0 \\ -\frac{1}{r}\,(\,\lambda - F(u,\lambda)\,) & = & 0 \end{array} \right. \tag{2}$$

Readers can refer to Lebon, Alart and Doudet (1994) about solving this system by Newton methods. The contact operator presents interesting properties regarding existence and uniqueness: it is Conewise Linear in 2D discretisation and Raywise Linear in 3D. This last notion has been introduced by Alart (1993) for a general purpose.

Theorem. The system (2) has a unique solution if and only if the determinant is positive for each tangent matrix of H associated to a cone of linearity.

This result is not obvious as shown by Alart (1993). The proof uses a property of the set of all tangent matrices specific to the frictional contact operator : for two matrices of this set, there exists a sequence of matrices $(M_i)_i$ of this set such that $M_{i+1} - M_i$ is a rank-1 matrix. Using a weak notion of differentiability, this set, noted δH, is the *base of the generalized Jacobian* at the vertex of the cones (Clarke, 1982; Alart, 1993). Each determinant is a polynomial function of degree p (p is the number of nodes candidate for contact) with respect to the friction coefficient μ. Moreover, the Poisson's modulus ν is a parameter of each polynomial. For a fixed value of ν, we define the *critical friction coefficient* as the greatest value of μ such that each determinant is non negative.

For better understanding, we present a simple example representing a square shaped domain discretized using four triangular elements (figure 2). The upper side is fixed but the lower side is in eventual contact with a flat obstacle. Consequently, the stiffness matrix K is of order 6. It is a part of the 10 order tangent matrix of the system of equations H mentionned above. We have four status (gap, stick, foreward slip and backward slip) for each nodal contact. With two contact nodes, the system of nonlinear equations has 16 (i.e. 4^2) cones of linearity (endowed with a determinant), called *global status*. The base of the generalized Jacobian is then defined as follows:

$$\delta H = \underline{K} + \prod_{i=1}^{p} \{ \underset{\text{gap}}{\underset{\uparrow}{0}},\ \underset{\text{stick}}{\underset{\uparrow}{n_i n_i^T + t_i t_i^T}},\ \underset{\text{slip +}}{\underset{\uparrow}{(n_i - \mu t_i) n_i^T}},\ \underset{\text{slip -}}{\underset{\uparrow}{(n_i + \mu t_i) n_i^T}} \} \tag{3}$$

where $\underline{K} = \begin{bmatrix} K & 0 \\ 0 & -\frac{I}{r} \end{bmatrix}$ and $\underline{n_i} = \begin{Bmatrix} \sqrt{r}\, n_i \\ \frac{1}{\sqrt{r}} n_i \end{Bmatrix}$

(gap , slip+) or (slip - , gap):

$$\mu_{crit} = \frac{24\nu^3-80\nu^2+80\nu-25}{-24\nu^2+38\nu-15}$$

(stick , slip+) or (slip - , stick):

$$\mu_{crit} = \frac{-96\nu^3+216\nu^2-160\nu+39}{16\nu^2-24\nu-9}$$

(slip+ , slip+) or (slip - , slip -):

$$\mu_{crit} = \frac{((8\nu^3-13\nu^2+5\nu)(3-4)^3)^{1/2}}{16\nu^3-24\nu^2+9\nu}$$

(slip - , slip+): $\mu_{crit} = 1$

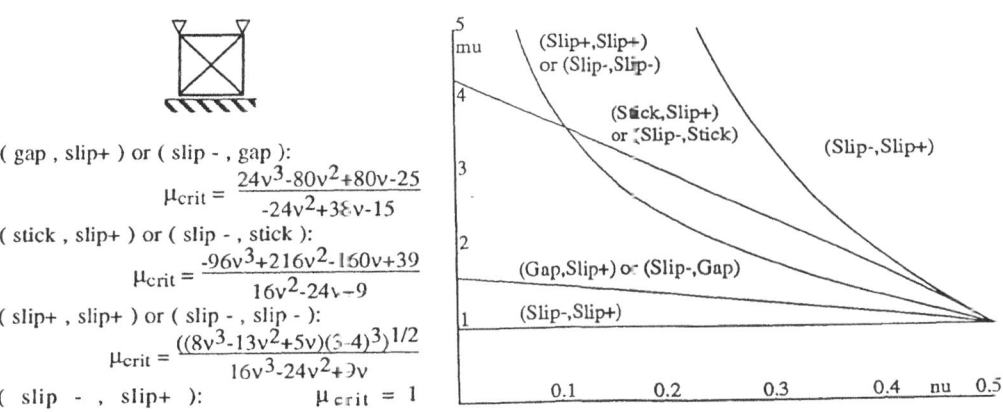

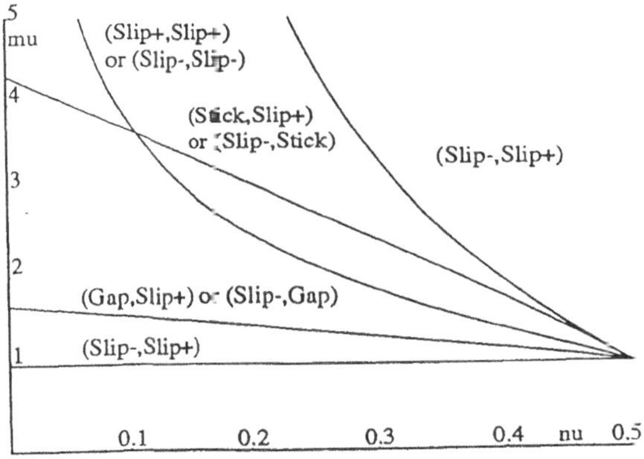

Figure 2. A simple example: seven status give a condition on the friction coefficient but only one of them determines the critical friction coefficient (equal to 1) for ν between zero and 0.5.

SQUARE TEST

In this section, we keep the previous geometry but we study the influence of the mesh. We have to evaluate 4^p determinants if p denotes the number of eventual contact nodes. In a first study, we used a symbolic computation software (MAPLE) to select the optimal global status, that is to say the global status which determine the critical friction coefficient.

Influence of the mesh type (coarse meshes)

This study is limited to coarse meshes because the cost of the symbolic computation of determinant increases fastly with the size of the system. Moreover, the software cannot give usually formal expressions of the roots of a polynomial of degree greater than 5 (as well as mathematicians) with exceptions.

The table 1 shows that only one global status defines the critical friction coefficient. This status, which we call *standard status*, can be extended to finer discretisation of the contact region: the nodes of the left half-part of the contact surface slip backward while the nodes of the other part slip forward; the status of the eventual node in the center is unimportant. We can imagine a distribution of contact regions for the initial problem involving a continuous medium as a limit of such a standard status associated to a discretised contact surface : two opposite slip regions separated by an adherent zone in the middle.

Table 1. Influence of the mesh type for some coarse meshes (Quittau, 1993).

Nodal contact number	Mesh	Optimal global status	Critical friction coefficient (value)	Critical friction coefficient (variation) ν : 0 .... 0.5
2		← →	1	1 → 1
2		← →	1	1 → 1
3		← ← → ← → →	$\frac{48\nu^2-78\nu+31}{40\nu^2-70\nu+29}$	1.069 ↓ 1
3		← ← → ← → →	$\frac{18\nu-17}{16\nu-18}$	0.944 ↓ 0.8
3		← ← → ← → →	too complex	0.998 ↓ 0.8
3		← ← → ← → →	$\frac{120\nu^3-352\nu^2+332\nu-101}{80\nu^3-272\nu^2+282\nu-91}$	0.998 ↓ 0.8
4		← ← → →	too complex	1.109 ↓ 1
5		← ← ← → → ← ← → → →	too complex	1.03 ↓ 0.855
5	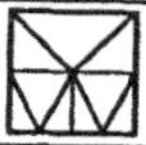	← ← ← → → ← ← → → →	too complex	1.017 ↓ 0.825

Moreover the kind of mesh has a weak influence on the value of the critical friction coefficient. Howeover, for p equal to 4 and 5, the only 2^p global slip status (associated to the p local slip status) have been examinated when all the 4^p have been taken into account for p equal to 2 and 3. A recent work of Rey (1994) shows results of the same kind for p equal to 7, 9 and 11.

Influence of the mesh refinement

For finer meshes, it is necessary to use a numerical software, but we limit the study to the standard status selected by the symbolic computation approach in the previous section. We choose as reference the most complex mesh investigated above. We call it Base5 and we obtain finer meshes either by local refinement (Loc9, Loc17, Loc33) or by global refinement (Glo9 and Glo17). The number at the end of the name denotes the number of contact nodes. On the other hand, we consider a set of regular meshes (Reg5, Reg9 and Reg17). The situation is summarized in the figure 3.

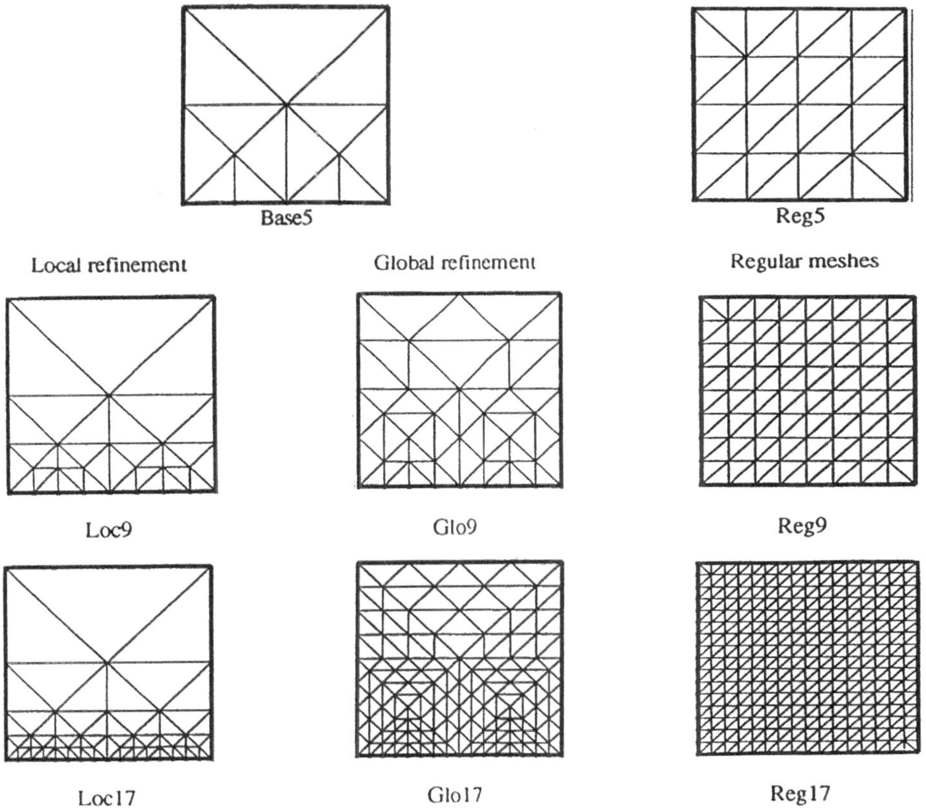

Figure 3. Meshes investigated by the numerical approach

Figures 4 and 5a show that all refinements lead to a smaller critical friction coefficient. But it is less decreasing for the local refinement (figure 4a): the critical value is almost determined by the initial coarse mesh Base5. In figures 4b and 5a, we have a global refinement with two different initial coarse meshes. The gaps between the curves with 17 nodes and 9 nodes (Glo17-Glo9 or Reg17-Reg9) is smaller than the one between the curves with 9 nodes and 5 nodes (Glo9-Base5 or Reg9-Reg5). Consequently, we may hope that the curves tend to a positive lower limit when the discretisation step increases. This limit might be an upper bound for the initial problem with a continuous medium.

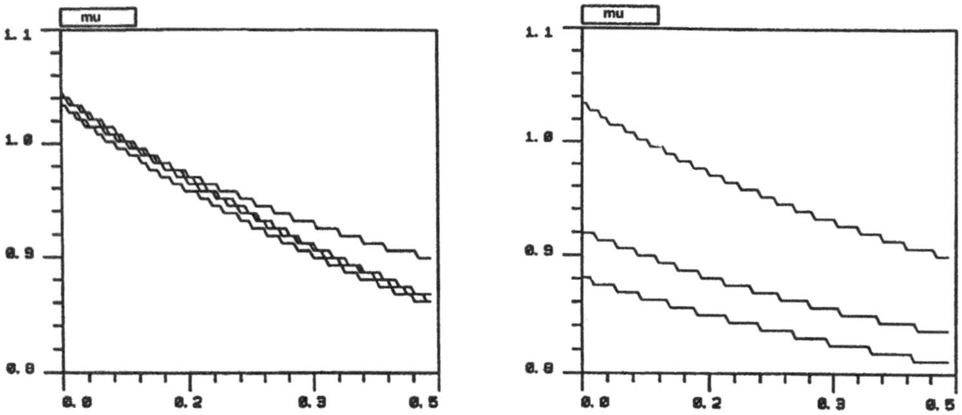

a) Local refinement (base5, loc9, loc17 and loc33)

b) Global refinement (base5, glo9 and glo17)

Figure 4. Influence of the refinement type. The curves are named between parenthesis successively when they are met reading the graph from the upper part to the v axis.

Finally with the global refinement the influence of the initial coarse mesh disappears because the curves Glo17 and Reg17 coincide in the figure 5b. On the contrary, the local refinement give an over-estimated critical coefficient if the initial mesh is too coarse. This practice is often used in computational contact mechanics. Is this method well-founded?

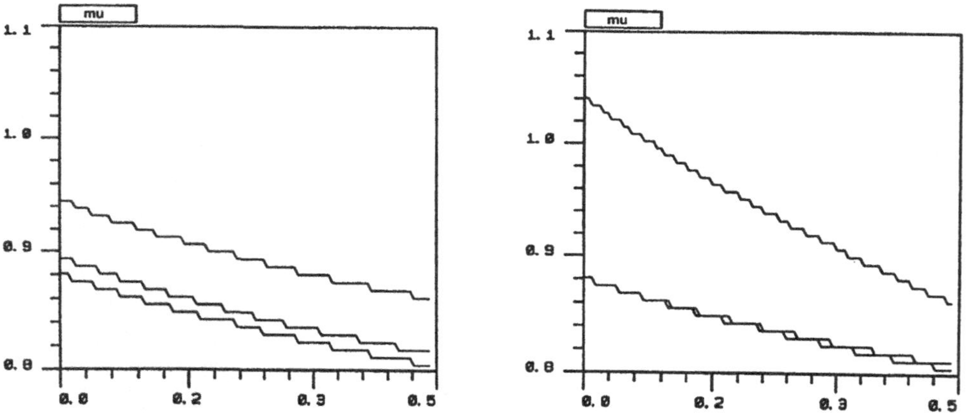

a) Regular meshes (reg5, reg9 and reg17)

b) Comparison between the finest meshes (loc17, glo17 and reg17)

Figure 5. Investigation of regular meshes and finest meshes.

RECTANGULAR TEST

To study the influence of the geometry on the uniqueness condition, we consider a set of rectangles. We introduce an aspect ratio, noted e, defined by the length over the height. We keep the same boundary and contact conditions as for the previous square test. Only symbolic computations have been done.

The results are less easy to interpret because the situation is more complicated even for coarse meshes. Firstly, we have several optimal status which determine the critical friction coefficient. For e equal to 2, we recover the standard status for ν between 0 and 0.3 and for ν between 0.3 and 0.5 we get new status which we call *soapy status* because all contact nodes slide in the same way (figure 6). For e equal to 3, the standard status is obtained for a large range of ν, but a third status appears, which can be identified to a *fractal status*. As a matter of fact, we can distinguish an elementary pattern made by two successive contact nodes with opposite sliding directions except an eventual middle node.

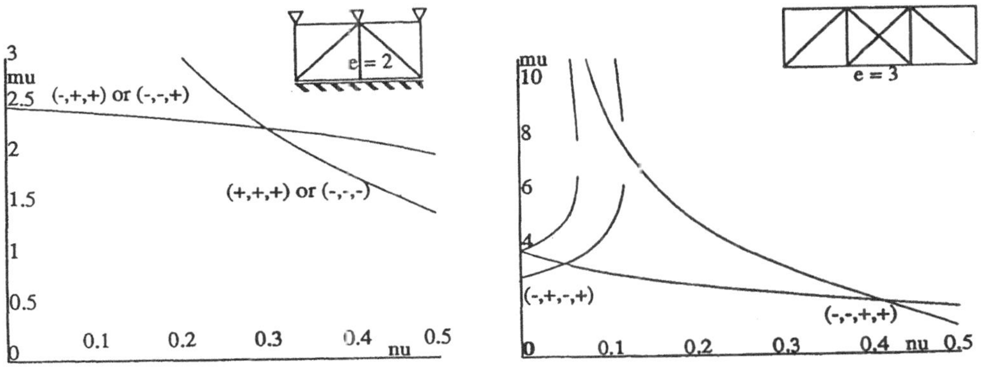

a) 2 optimal status: standard and soapy b) 2 optimal status: standard and fractal

Figure 6. Critical friction coefficient for e = 2 and e = 3. "S" denotes slip- or slip+.

For e equal to 4, the three status are optimal according to an interval of ν; but we can note that the range of influence of the fractal status is larger than for the previous case (figure 7). To confirm the importance of the fractal status, only this status and the standard one are presented for e equal to 6. Indeed, the complete study needs 4096 (4^6) determinant computations! Moreover the symbolic software cannot give always the roots of the polynomial of degree 7 and the cross-ruled zone represents then the negative part of the polynomial in the figure 7b.

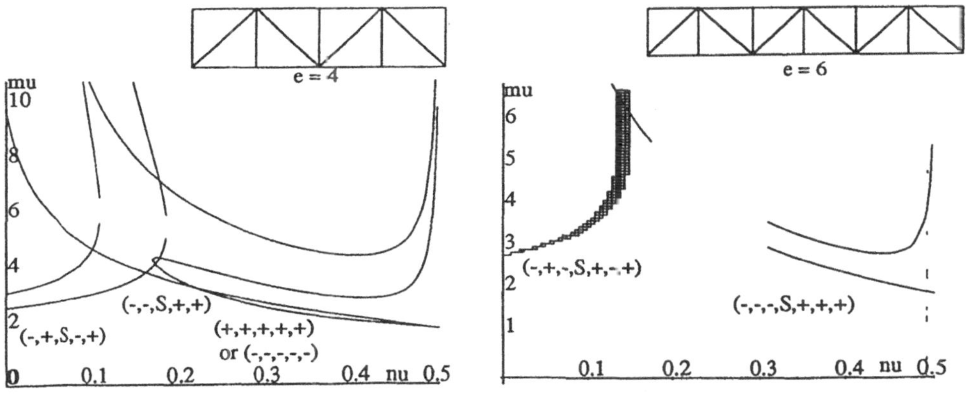

a) 3 optimal status: standard, soapy and fractal b) 2 status are represented: standard and fractal

Figure 7. Critical friction coefficient for e = 4 and e = 6.

CONCLUSION

This first study shows encouraging results concerning the square test. Only one status determines the critical friction coefficient. This status has a physical meaning and the numerical approach could lead us to an eventual asymptotic critical coefficient to compare with mathematical estimation for the continuous problem.

The results obtained for the rectangular tests lead us to define the fractal status. Nevertheless, what is the meaning of such a status? We do not know if this status depends only on the geometry or aswell as on the discretisation level. That is, for a fixed aspect ratio, can we define a stable asymptotic status, i.e. a finite distribution of contact zones for the continuous problem (figure 8)? In this case, this status would not be really fractal! On the contrary, if the status depends on the discretisation step with the elementary pattern defined above, we cannot imagine an asymptotic finite distribution of contact zones. Moreover, does it exist a loading which gives such a pathological solution? The answer to these questions needs substantial symbolic computations.

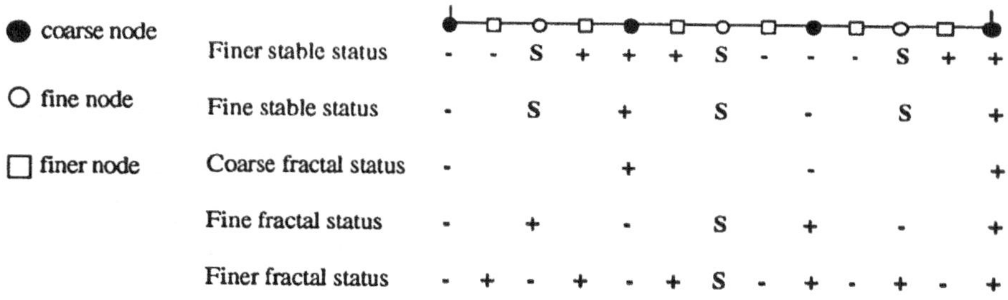

Figure 8. Two possible evolutions of the fractal status with respect to the discretisation level for e equal to 3: a stable status and a really fractal one depending on the discretisation level.

REFERENCES

Alart, P., 1993, Critères d'injectivité et de surjectivité pour certaines applications de Rn dans lui-même; application à la mécanique du contact, *RAIRO, Mod. Math. et An. Nu.*, 27, 203:222.

Alart, P., and Curnier, A., 1991, A mixed formulation for frictional contact problems prone to Newton like solution methods, *Comp. Meth. in Appl. Mech. and Eng.*, 92, 353:375.

Cocu, M. , 1984, Existence of solutions of Signorini problems with friction, Int. J. Engng. Sci., 22, 567:575.

Clarke, F. H., 1983, "Optimization and Nonsmooth Analysis", Wiley, New York.

Duvaut, G., 1982, Loi de frottement non locale, *J. Th. Appl. Mech.*, special issue, 73:78.

Klarbring, A., 1985, Contact problems in linear elasticity, Ph. D. thesis n°133, University of Linköping.

Lebon, F., Alart, P., and Doudet, P., 1994, Friction and preconditioners, this volume.

Licht, C., Pratt, E., and Raous M., 1990, Remarks on a numerical method for unilateral contact includuing friction, *in* "Unilateral Problems in Structural Analysis", CISM, Capri.

Necas, J., Jarusek, J., and Haslinger, J. , 1980, On the solution of the variational inequality to the Signorini problem with small friction, *Bolletino U.M.I.*, 5, 796:811.

Quittau, F., 1993, Calcul formel et frottement, recherche d'une condition d'existence et d'unicité pour le problème discret de contact unilatéral avec frottement, D.E.A. report, Montpellier University.

Rey, K., 1994, Contact unilatéral avec frottemental : influence du maillage sur la condition d'unicité (calcul formel et numérique), D.E.A. report, Montpellier University.

AN EXISTENCE RESULT FOR A CLASS OF LIMIT STATE PROBLEMS

Lars-Erik Andersson[1] and Anders Klarbring[2]

[1]Department of Mathematics
[2]Department of Mechanical Engineering
Linköping University
S-581 83 Linköping, Sweden

INTRODUCTION

The present work is concerned with frictional joints treated from the point of view of linear elasticity. As a model of such a joint we consider a linear elastic body in frictional contact with a rigid support. The objective is to analyze limit states of frictional joints, i.e. states where a small perturbation may introduce a large change of configuration. As a more precise definition of a limit state, the requirement that the forces and the static deformation are such that a time constant rigid body velocity field can be added to the deformation, is used. That is, the elastic body moves (or slides) in a steady state fashion relative to the rigid support. To get insight into the nature of such a definition two examples are briefly discussed from a qualitative point of view:

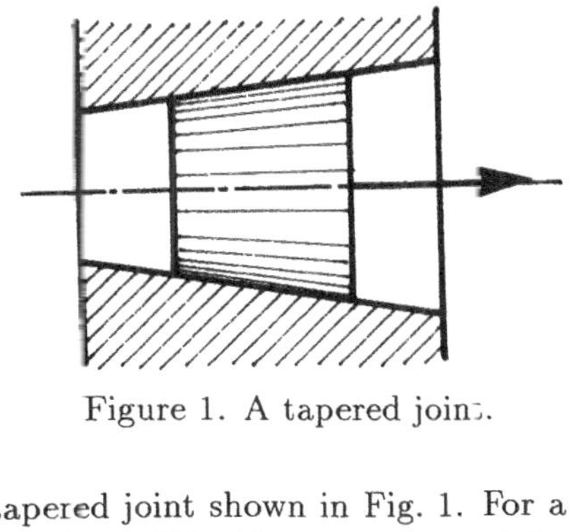

Figure 1. A tapered joint.

First, consider the tapered joint shown in Fig. 1. For a friction coefficient that is large enough to make the joint self-locking, any force F from zero to infinity may be necessary to take the joint apart, depending on the previous load history. It is also clear that, unless the force is zero, the constant limit velocity field required by the definition will not be present. Rather, the joint will unassemble in a dynamic fashion.

Contact Mechanics, Edited by M. Raous *et al.*
Plenum Press, New York, 1995

As a second example, consider the shrink-fitted shaft and bushing assemblage in Fig. 2. Here it is clear that the definition makes sense for small friction coefficients.

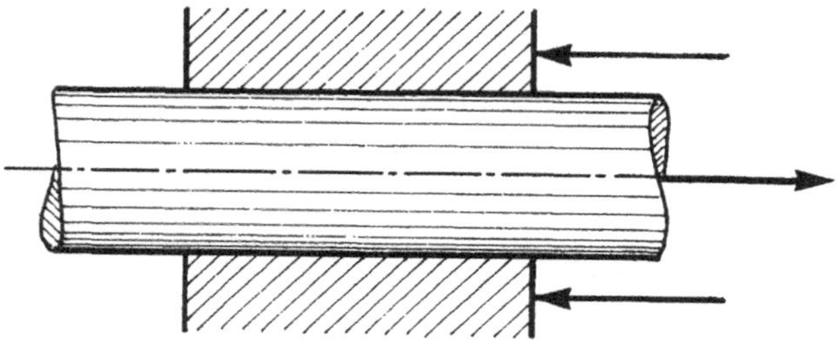

Figure 2. A shrink-fitted shaft and bushing assemblage.

However, for large friction coefficients we may still have a "violent" limit state: consider the uncorking of a bottle.

Thus, the definition above covers a subclass of the situations which one would like to call limit states of frictional joints. Nevertheless, for this subclass some exact mathematical statements, which may form a starting point for a more complete theory, are given in this paper. The problem of finding the forces and the static deformation corresponding to a given limit velocity is formulated. The existence of a solution is shown under restrictions on the magnitude of the friction coefficient and the direction of the external force. This result is in agreement with our intuitive understanding of the problem.

THE QUASI-STATIC FRICTIONAL CONTACT PROBLEM

Since the interest is in analyzing constant velocity states, it is sufficient to start from the so-called quasi-static formulation, where dynamic forces are neglected.

Consider a linear elastic body that occupies a region Ω of $\mathbf{R}^3$. The boundary of Ω consists of disjoint parts S_t and S_c. The body is subjected to body forces $\alpha\mathbf{f} = \alpha(f_1, f_2, f_3)$ over Ω and surface tractions $\alpha\mathbf{t} = \alpha(t_1, t_2, t_3)$ over S_t, where α is a scalar load parameter and $\mathbf{f}$ and $\mathbf{t}$ are given vector fields. The following classical equations of linear elasticity are valid:

$$\frac{\partial \sigma_{ij}}{\partial x_j} + \alpha f_i = 0 \quad \text{in} \quad \Omega, \tag{1}$$

$$\sigma_{ij} = E_{ijkl}\frac{\partial u_k}{\partial x_l} \quad \text{in} \quad \Omega, \tag{2}$$

$$\sigma_{ij} n_j = \alpha t_i \quad \text{on} \quad S_t. \tag{3}$$

Here $\mathbf{u} = (u_1, u_2, u_3)$ is the displacement vector, $\boldsymbol{\sigma} = \{\sigma_{ij}\}$ is the stress tensor and n_j are the components of the outward unit normal vector. E_{ijkl} are elasticity constants which satisfy the usual symmetry and ellipticity conditions. Furthermore, $i, j, k, l = 1, 2, 3$, the summation convention is used and $(0, x_1, x_2, x_3)$ is the cartesian reference frame.

The boundary part S_c is the contact boundary where the body may come into contact with a rigid support. To state the laws of contact and friction we decompose the displacement and traction vectors on S_c into normal and tangential components:

$$\sigma_N = \sigma_{ij} n_i n_j, \qquad \sigma_{Ti} = \sigma_{ij} n_j - \sigma_N n_i,$$

$$u_N = u_i n_i, \qquad u_{Ti} = u_i - u_N n_i.$$

The contact law is taken to be the classical one of Signorini, i.e.

$$\sigma_N \leq 0, \quad u_N - g \leq 0, \quad \sigma_N(u_N - g) = 0 \quad \text{on } S_c, \tag{4}$$

where g is the initial gap between the body and the rigid support. Note that there is no sign restriction for g.

The friction law is that of Coulomb, which can be written as

$$\begin{aligned} |\boldsymbol{\sigma}_T| &\leq \mu|\sigma_N| & & \qquad (5)\\ \text{if } |\boldsymbol{\sigma}_T| &< \mu|\sigma_N| \text{ then } \dot{\mathbf{u}}_T = \mathbf{0} & \text{on } S_c, & \qquad (6)\\ \text{if } |\boldsymbol{\sigma}_T| &= \mu|\sigma_N| > 0 \text{ then } \dot{\mathbf{u}}_T = -\lambda\boldsymbol{\sigma}_T,\ \lambda \geq 0 & & \qquad (7) \end{aligned}$$

where μ is the friction coefficient, a superposed dot denotes time derivative, $\boldsymbol{\sigma}_T = (\sigma_{T1}, \sigma_{T2}, \sigma_{T3})$ and $\mathbf{u}_T = (u_{T1}, u_{T2}, u_{T3})$.

In conclusion, (1) through (7) constitute the quasi-static frictional contact problem.

A limit state of the mechanical system is a state where a constant rigid velocity field w can be superposed onto a static deformation.

As a particular case of a limit state problem we will consider a situation when the rigid velocity field $\mathbf{w} = \mathbf{e}_1 = (1,0,0)$ is superimposed onto a static deformation $\mathbf{u}^\circ$ (the limit state). Substituting the displacement $\mathbf{u} = \mathbf{u}^\circ + \tau\mathbf{e}_1$, where $\mathbf{u}^\circ$ is independent of time, into (2) and (4) we obtain

$$\sigma_{ij} = E_{ijkl}\frac{\partial u_k^\circ}{\partial x_l} \quad \text{in } \Omega, \tag{8}$$

$$\sigma_N \leq 0, \quad u_N^\circ - g \leq 0, \quad \sigma_N(u_N^\circ - g) = 0 \quad \text{on } \Gamma_c. \tag{9}$$

Furthermore, the friction law (5) through (7) implies that

$$\boldsymbol{\sigma}_T = -\mu|\sigma_N|\mathbf{e}_1 \quad \text{on } \Gamma_c. \tag{10}$$

Here Γ_c is the part of S_c where $w_N = 0$. The problem to find the fields $\boldsymbol{\sigma}$ and $\mathbf{u}^\circ$ and the scalar α such that (1), (3) and (8) through (10) are satisfied constitutes a limit state problem that will be considered in subsequent sections. The given data are $\mathbf{f}, \mathbf{t}, E_{ijkl}, n_i$ and μ.

VARIATIONAL FORMULATION

The problem of the previous section may be formulated as a variational inequality in the following way. Let

$$a(\mathbf{u}, \mathbf{v}) = \int_\Omega E_{ijkl}\frac{\partial u_i}{\partial x_j}\frac{\partial v_k}{\partial x_l}dx.$$

denote the energy functional. The convex set of admissable deformations is defined as

$$\hat{K} = \{\mathbf{u} \in \hat{V} |\ u_N - g_N \le 0 \text{ on } \Gamma_c\}.$$

For technical reasons the initial gap is here defined by means of a function $\mathbf{g} = (g_1, g_2, g_3) \in \hat{V}$ such that $g = g_N|_{\Gamma_c} = g_i n_i|_{\Gamma_c}$. Then we may write

$$\hat{K} = \mathbf{g} + \hat{K}_\circ$$

where

$$\hat{K}_\circ = \{\mathbf{u} \in \hat{V} |\ u_N \le 0 \text{ on } \Gamma_c\}$$

is a convex cone with vertex in the origin. Assuming sufficient regularity and using Green's formula the limit state problem defined by (1), (3), (8), (10) takes the form

$$(\hat{\mathrm{P}}) \qquad \begin{cases} \text{Find } \hat{\mathbf{u}} \in \hat{K}_\circ \text{ and } \alpha \in \mathbf{R} \text{ such that for all } \mathbf{v} \in \hat{K}_\circ \\ a(\hat{\mathbf{u}}, \mathbf{v} - \hat{\mathbf{u}}) - \int_{\Gamma_c} \mu\sigma_N(\hat{\mathbf{u}})(v_1 - \hat{u}_1) ds \\ \ge \alpha\langle \mathbf{F}, \mathbf{v} - \hat{\mathbf{u}}\rangle + \langle \ell_g, \mathbf{v} - \hat{\mathbf{u}}\rangle - a(\mathbf{g}, \mathbf{v} - \hat{\mathbf{u}}), \end{cases}$$

where

$$\langle \mathbf{F}, \mathbf{v}\rangle = \int_\Omega f_i v_i dx + \int_{S_t} t_i v_i ds, \quad \langle \ell_g, \mathbf{v}\rangle = \int_{\Gamma_c} \mu\sigma_N(\mathbf{g}) v_1 ds$$

and $\sigma_N(\mathbf{u}^\circ)$, $\sigma_N(\hat{\mathbf{u}})$ and $\sigma_N(\mathbf{g})$ are defined by (8). The solutions $\mathbf{u}^\circ$ and $\hat{\mathbf{u}}$ are related as $\mathbf{u}^\circ = \hat{\mathbf{u}} + \mathbf{g}$. The force ℓ_g may be interpreted as being due to the 'shrink-fitting' (if $g < 0$).

DECOUPLED VARIATIONAL FORMULATIONS

Let $\mathcal{L}$ denote the 1-dimensional subspace

$$\mathcal{L} = \{\mathbf{v} \in \hat{V} |\ \mathbf{v} = k\mathbf{e}_1 \ \text{ for some } k \in \mathbf{R}\}$$

of $\hat{V}$ and $\mathcal{L}^\perp$ the orthogonal complement with respect to the L^2-norm (or equivalently to the H^2-norm), i.e.,

$$\mathcal{L}^\perp = \{\mathbf{v} \in \hat{V} |\ \int_\Omega v_1 dx = 0\}.$$

Each element $\mathbf{u} \in \hat{V}$ may then be decomposed as $\mathbf{u} = \bar{\mathbf{u}} + r\mathbf{e}_1$ with

$$\bar{\mathbf{u}} = (u_1 - \frac{1}{|\Omega|}\int_\Omega u_1 dx)\mathbf{e}_1 + u_2\mathbf{e}_2 + u_3\mathbf{e}_3 = \bar{u}_i\mathbf{e}_i \in \mathcal{L}^\perp$$

and

$$r = \frac{1}{|\Omega|}\int_\Omega u_1 dx.$$

For F and ℓ_g in the dual space of $\hat{V}$ we have similar decompositions, $F = \bar{F} + r\mathbf{e}_1$ and $\ell_g = \bar{\ell}_g + s\mathbf{e}_1$ with $\langle \bar{F}, \mathbf{e}_1\rangle = \langle \bar{l}_g, \mathbf{e}_1\rangle = 0$. It is a decomposition of the force into a part which tends to push the body in $\mathbf{e}_1$-direction and a part which is 'orthogonal' to this direction. Now it also follows that

$$\langle \bar{\mathbf{F}}, \bar{\mathbf{v}}\rangle = \langle \mathbf{F}, \bar{\mathbf{v}}\rangle \quad \forall \bar{\mathbf{v}} \in \hat{K} \cap \mathcal{L}^\perp$$

$$\langle \mathbf{F} - \bar{\mathbf{F}}, \mathbf{e}_1 \rangle = \langle \mathbf{F}, \mathbf{e}_1 \rangle = \int_\Omega f_1 dx + \int_{S_t} t_1 ds.$$

and similarly for ℓ_g.

We now substitute the fields $\hat{\mathbf{u}} = \bar{\mathbf{u}} + r\mathbf{e}_1$ and $\mathbf{v} = \bar{\mathbf{v}} + s\mathbf{e}_1$, where $\bar{\mathbf{v}}, \bar{\mathbf{u}} \in \mathcal{L}^\perp$, into problem $(\hat{\mathrm{P}})$. Since $a(\hat{\mathbf{u}}, \mathbf{v}) = a(\bar{\mathbf{u}}, \bar{\mathbf{v}})$ and $\sigma_N(\hat{\mathbf{u}}) = \sigma_N(\bar{\mathbf{u}})$ one then obtains

$$\begin{aligned} a(\bar{\mathbf{u}}, \bar{\mathbf{v}} - \bar{\mathbf{u}}) &- \int_{\Gamma_c} \mu\sigma_N(\bar{\mathbf{u}})(\bar{v}_1 - \bar{u}_1) ds - (s - r) \int_{\Gamma_c} \mu\sigma_N(\bar{\mathbf{u}}) ds \\ &\geq \alpha\langle \bar{\mathbf{F}}, \bar{\mathbf{v}} - \bar{\mathbf{u}} \rangle + \alpha(s - r)\langle \mathbf{F} - \bar{\mathbf{F}}, \mathbf{e}_1 \rangle + \langle \bar{\ell}_g, \bar{\mathbf{v}} - \bar{\mathbf{u}} \rangle \\ &+ (s - r)\langle \ell_g - \bar{\ell}_g, \mathbf{e}_1 \rangle - a(\mathbf{g}, \bar{\mathbf{v}} - \bar{\mathbf{u}}). \end{aligned}$$

Here we make a further *assumption on the geometry of the problem.* We assume that $\hat{K}_\circ \cap Q = \mathcal{L}$, where Q denotes the finite-dimensional space of rigid body motions. Then s is arbitrary and we find that a solution of problem $(\hat{\mathrm{P}})$ can be constructed from a solution of the following problem:

$$(\hat{\mathrm{P}})_d \quad \begin{cases} \text{Find } \bar{\mathbf{u}} \in \hat{K}_\circ \cap \mathcal{L}^\perp \text{ and } \alpha \in \mathbf{R} \text{ such that} \\ -\int_{\Gamma_c} \mu\sigma_N(\bar{\mathbf{u}}) ds = \alpha\langle \mathbf{F} - \bar{\mathbf{F}}, \mathbf{e}_1 \rangle + \langle \ell_g - \bar{\ell}_g, \mathbf{e}_1 \rangle \\ \text{and for all } \bar{\mathbf{v}} \in \hat{K}_\circ \cap \mathcal{L}^\perp \\ a(\bar{\mathbf{u}}, \bar{\mathbf{v}} - \bar{\mathbf{u}}) - \int_{\Gamma_c} \mu\sigma_N(\bar{\mathbf{u}})(\bar{v}_1 - \bar{u}_1) ds \\ \geq \alpha\langle \bar{\mathbf{F}}, \bar{\mathbf{v}} - \bar{\mathbf{u}} \rangle + \langle \bar{\ell}_g, \bar{\mathbf{v}} - \bar{\mathbf{u}} \rangle - a(\mathbf{g}, \bar{\mathbf{v}} - \bar{\mathbf{u}}). \end{cases}$$

If $(\bar{\mathbf{u}}, \alpha)$ is a solution of problem $(\hat{\mathrm{P}})_d$, $(\hat{\mathbf{u}} = \bar{\mathbf{u}} + r\mathbf{e}_1, \alpha)$ is a solution of problem $(\hat{\mathrm{P}})$ for any $r \in \mathbf{R}$. The equation of $(\hat{\mathrm{P}})_d$ simply expresses global equilibrium in the $\mathbf{e}_1$-direction. Note that a certain non-uniqueness of solutions has appeared. The problem is indifferent with respect to a rigid body displacement in the $\mathbf{e}_1$-direction.

A further reformulation of the problem will show useful for our existence proof. That is, the equality of $(\hat{\mathrm{P}})_d$ can be merged into the inequality to result in

$$(\hat{\mathrm{P}}\mathrm{V}) \quad \begin{cases} \text{Find } (\mathbf{u}, \alpha) \in (\hat{K}_\circ \cap \mathcal{L}^\perp) \times \mathbf{R} \text{ such that for all } (\mathbf{v}, \beta) \in (\hat{K}_\circ \cap \mathcal{L}^\perp) \times \mathbf{R} \\ a(\mathbf{u}, \mathbf{v} - \mathbf{u}) - \alpha\langle \bar{\mathbf{F}}, \mathbf{v} - \mathbf{u} \rangle - \int_{\Gamma_c} \mu\sigma_N(\mathbf{u})(v_1 - u_1) ds \\ +\alpha\langle \mathbf{F} - \bar{\mathbf{F}}, \mathbf{e}_1 \rangle(\beta - \alpha) + \int_{\Gamma_c} \mu\sigma_N(\mathbf{u}) ds(\beta - \alpha) \\ -\langle \bar{\ell}_g, \mathbf{v} - \mathbf{u} \rangle + a(\mathbf{g}, \mathbf{v} - \mathbf{u}) + \langle \ell_g - \bar{\ell}_g, \mathbf{e}_1 \rangle(\beta - \alpha) \geq 0. \end{cases}$$

Note that we have dropped the bar-sign for elements $\mathbf{u}, \mathbf{v} \in \hat{K}_\circ \cap \mathcal{L}^\perp$.

FUNCTIONAL SETTING

In order to formulate and prove existence results we must now be more specific about the choice of function spaces and about various assumptions.

First $\Omega \subset \mathbf{R}^3$ is assumed to be an open bounded Lipschitz domain, and S_t and Γ_c relatively open subsets of $\partial\Omega$ with $\bar{S}_t \cap \bar{\Gamma}_c = \emptyset$. For the volume forces and surface traction fields $\mathbf{f}$ and $\mathbf{t}$ we require that $f_i \in L^2(\Omega)$, $t_i \in H^{-1/2}(\partial\Omega)$ and supp $\mathbf{t} \subset \bar{S}_t$. Also let $E_{ijkl} \in L^\infty(\Omega)$ and $E_{ijkl}\xi_{ij}\xi_{kl} \geq c_\circ \xi_{ij}\xi_{ij}$ for some $c_\circ > 0$ and for all symmetric ξ_{ij}.

The space of smooth functions $\hat{V}$ is replaced by the Hilbert space $V = [H^1(\Omega)]^3$ with norm denoted by $\|\cdot\|_V$. $\mathcal{L}$ and $\mathcal{L}^\perp$ are defined as previously, but with $\hat{V}$ replaced by V. For the gap function $\mathbf{g}$ we assume that $\mathbf{g} \in V$ and in addition that $\partial\sigma_{ij}(\mathbf{g})/\partial x_j \in L^2(\Omega)$. The sets $\hat{K}$ and $\hat{K}_\circ$ are replaced by

$$K = \{\mathbf{u} \in V |\ (u_N - g_N)|_{\Gamma_c} \leq 0\}, \quad K_\circ = \{\mathbf{u} \in V |\ u_N|_{\Gamma_c} \leq 0\},$$

where the gap function $g_N \in H^{1/2}(\partial\Omega) := W \subset L^1(\partial\Omega)$. Then K and $K_\circ$ are closed convex cones in V and $K = \mathbf{g} + K_\circ$. Under these assumptions it turns out that problem $\hat{\mathrm{P}}\mathrm{V}$ should be replaced by

$$(\mathrm{PV}) \begin{cases} \text{Find } (\mathbf{u}, \alpha) \in (K_\circ \cap \mathcal{L}^\perp) \times \mathbf{R} \text{ such that for all } (\mathbf{v}, \beta) \in (K_\circ \cap \mathcal{L}^\perp) \times \mathbf{R} \\ a(\mathbf{u}, \mathbf{v} - \mathbf{u}) - \alpha\langle \bar{\mathbf{F}} - \bar{\mathbf{L}}_{f,\mu}, \mathbf{v} - \mathbf{u}\rangle_{V',V} + j_0(\mathbf{u}, \mathbf{v} - \mathbf{u}) \\ +\alpha\langle \mathbf{F} - \bar{\mathbf{F}} - (\mathbf{L}_{f,\mu} - \bar{\mathbf{L}}_{f,\mu}), \mathbf{e}_1\rangle_{V',V}(\beta - \alpha) - j_0(\mathbf{u}, \mathbf{e}_1)(\beta - \alpha) \\ -\langle \bar{\ell}_g, \mathbf{v} - \mathbf{u}\rangle_{V',V} + a(\mathbf{g}, \mathbf{v} - \mathbf{u}) + \langle \ell_g - \bar{\ell}_g, \mathbf{e}_1\rangle_{V',V}(\beta - \alpha) \geq 0. \end{cases}$$

Here F, $\bar{F}$, $L_{f,\mu}$, $\bar{L}_{f,\mu} \in V'$ and $j_0 : V \times V \to \mathbf{R}$ is a bounded, bilinear (non-symmetric) form. $L_{f,\mu}$ and j_0 originate from the term $\int_{\Gamma_c} \mu\sigma_N(\bar{\mathbf{u}})(\bar{v}_1)ds$ in the following way. $\sigma_N(\mathbf{u}) := \sigma_N(\pi_{\alpha\mathbf{f}}\mathbf{u})$ where the mapping $\pi_{\alpha\mathbf{f}} : V \to V(\alpha\mathbf{f}) = \{\mathbf{v} \in V |\ \frac{\partial\sigma_{ij}(\mathbf{v})}{\partial x_j} + \alpha f_i = 0\}$ is the orthoprojection onto the closed linear manifold $V(\alpha\mathbf{f}) \subset V$. Further $\pi_{\alpha\mathbf{f}}\mathbf{u} = \alpha\pi_{\mathbf{f}}\mathbf{0} + \pi_{\mathbf{0}}\mathbf{u}$, $\sigma_N(\pi_{\alpha\mathbf{f}}\mathbf{u}) = \alpha\sigma_N(\pi_{\mathbf{f}}\mathbf{0}) + \sigma_N(\pi_{\mathbf{0}}\mathbf{u})$ giving the terms $L_{f,\mu} = -\mu\sigma_N(\pi_f(\mathbf{0}) \in V'$ and

$$j_0(u, v) = -\langle \mu_N(\pi_{\mathbf{0}}\mathbf{u}), \psi v_1\rangle_{H^{-1/2}(\Omega), H^{1/2}(\Omega)}.$$

Here $\psi \in C_0^\infty(\mathbf{R}^3$ is a function such that $\psi \equiv 1$ on a neighbourhood of S_c and $\psi \equiv 0$ on a neighbourhood of S_t. It should be noted that the obtained solution $\mathbf{u}$ is independent of the particular choice of ψ and that $\pi_{\alpha\mathbf{f}}\mathbf{u} = u$.

Introducing the notation $\mathcal{U} = (\mathbf{u}, \alpha)$, $\mathcal{V} = (\mathbf{v}, \beta)$ and

$$\begin{aligned} \mathcal{A}(\mathcal{U}, \mathcal{V}) &= a(\mathbf{u}, \mathbf{v}) + j_0(\mathbf{u}, \mathbf{v}) - \alpha\langle \bar{\mathbf{F}} - \bar{\mathbf{L}}_{f,\mu}, \mathbf{v}\rangle_{V',V} \\ &+ \alpha\langle \mathbf{F} - \bar{\mathbf{F}} - (\mathbf{L}_{f,\mu} - \bar{\mathbf{L}}_{f,\mu}), \mathbf{e}_1\rangle_{V',V}\beta - j_0(\mathbf{u}, \mathbf{e}_1)\beta, \end{aligned}$$

$$\mathcal{L}(\mathcal{V}) = \langle \bar{\ell}_g, \mathbf{v}\rangle_{V',V} - a(\mathbf{g}, \mathbf{v}) - \langle \ell_g - \bar{\ell}_g, \mathbf{e}_1\rangle_{V',V}\beta,$$

problem (PV) takes the form

$$(\mathrm{PV}) \begin{cases} \text{Find } \mathcal{U} \in (K_\circ \cap \mathcal{L}^\perp) \times \mathbf{R} \text{ such that for all } \mathcal{V} \in (K_\circ \cap \mathcal{L}^\perp) \times \mathbf{R} \\ \mathcal{A}(\mathcal{U}, \mathcal{V} - \mathcal{U}) \geq \mathcal{L}(\mathcal{V} - \mathcal{U}). \end{cases}$$

It is clear that $\mathcal{A}$ is a bilinear, nonsymmetric and continuous functional on the closed convex cone $(K_\circ \cap \mathcal{L}^\perp) \times \mathbf{R}$ and that $\mathcal{L}$ is a bounded linear functional on $V \times \mathbf{R}$.

AN EXISTENCE RESULT

We now state the following lemma.

Lemma 1 *There exists a constant $c_k > 0$ such that*

$$a(\mathbf{v}, \mathbf{v}) \geq c_k \|\mathbf{v}\|_V^2 \tag{11}$$

holds for all $\mathbf{v} \in K_\circ \cap \mathcal{L}^\perp$ as well as for all $\mathbf{v} \in Q^\perp$, where $Q^\perp$ is the orthogonal complement of Q in the H^1-norm.

Proof The sets $K_\circ \cap \mathcal{L}^\perp \subset V$ and $Q^\perp \subset V$ are both convex closed cones of V, with vertex in the origin, and $(K_\circ \cap \mathcal{L}^\perp) \cap Q = \{\mathbf{0}\}$, $Q^\perp \cap Q = \{\mathbf{0}\}$ Thus, the lemma follows from Korn's inequality, see Nečas and Hlaváček (1981). ∎

Using the definition of that j_0 one can prove that

$$|j_0(\mathbf{u},\mathbf{u})| \leq A_1 \|\mu\|_{Lip} \|\mathbf{u}\|_V^2. \tag{12}$$

where A_1 depends only on the geometry of Ω and where $\|\mu\|_{Lip}$ denotes the Lipschitz-norm of the coefficient of friction. Similarly,

$$|j_0(\mathbf{u},\mathbf{e}_1)| \leq A_1 |\Omega| \|\mu\|_{Lip} \|\mathbf{u}\|_V \tag{13}$$

with $|\Omega|$ the volume of Ω. Let us introduce the shorter notation $A = A_1\|\mu\|_{Lip}$. Then we obtain the following estimate for the bilinear functional $\mathcal{A}$.

$$\begin{aligned} \mathcal{A}(\mathcal{U},\mathcal{U}) &\geq (c_k - A)\|\mathbf{u}\|_V^2 - |\alpha|\|\mathbf{u}\|_V \|\bar{\mathbf{F}} - \bar{\mathbf{L}}_{f,\mu}\|_{V'} \\ &+ \alpha^2 \langle \mathbf{F} - \bar{\mathbf{F}} - (\mathbf{L}_{f,\mu} - \bar{\mathbf{L}}_{f,\mu}), \mathbf{e}_1 \rangle_{V',V} - A|\Omega|\|\mathbf{u}\|_V \alpha \end{aligned}$$

or with shorter notation

$$\mathcal{A}(\mathcal{U},\mathcal{U}) \geq c_{11}\|\mathbf{u}\|_V^2 + c_{22}\alpha^2 - 2c_{12}|\alpha|\|\mathbf{u}\|_V. \tag{14}$$

Now, a necessary and sufficient condition that there exist an $\epsilon > 0$ such that

$$c_{11}\|\mathbf{u}\|^2 + c_{22}\alpha^2 - 2c_{12}|\alpha|\|\mathbf{u}\|_V \geq \epsilon\{\|\mathbf{u}\|^2 + \alpha^2\}$$

for all $\|\mathbf{u}\|_V$, α, is that $c_{11} > 0$, $c_{22} > 0$ and $c_{11}c_{22} - c_{12}^2 > 0$, i.e. that

$$c_k - A > 0, \quad \langle \mathbf{F} - \bar{\mathbf{F}} - (\mathbf{L}_{f,\mu} - \bar{\mathbf{L}}_{f,\mu})\ \mathbf{e}_1 \rangle_{V',V} > 0, \tag{15}$$

$$(c_k - A)\langle \mathbf{F} - \bar{\mathbf{F}} - (\mathbf{L}_{f,\mu} - \bar{\mathbf{L}}_{f,\mu}), \mathbf{e}_1 \rangle_{V',V} > \frac{1}{4}\left\{\|\bar{\mathbf{F}} - \bar{\mathbf{L}}_{f,\mu}\|_{V'} + A|\Omega|\right\}^2. \tag{16}$$

Under the assumptions (15) and (16) we therefore have, for some $\epsilon > 0$, the following *coercivity property* for the non-symmetric bilinear functional $\mathcal{A}$,

$$\mathcal{A}(\mathcal{U},\mathcal{U}) \geq \epsilon\|\mathcal{U}\|_{V\times\mathbf{R}}^2 \qquad \text{for all } \mathcal{U} \in (K_\circ \cap \mathcal{L}^\perp) \times \mathbf{R}. \tag{17}$$

We are now ready to formulate the main result of our paper, stating that problem (PV) has at least one solution.

Theorem 1 *Assume that our parameters satisfy the inequalities (15) and (16). Then there is at least one vector* $\mathcal{U} = (\mathbf{u},\alpha) \in (K_\circ \cap \mathcal{L}^\perp) \times \mathbf{R}$ *such that*

$$\mathcal{A}(\mathcal{U}, \mathcal{V} - \mathcal{U}) \geq \mathcal{L}(\mathcal{V} - \mathcal{U})$$

for all $\mathcal{V} = (\mathbf{v},\beta) \in (K_\circ \cap \mathcal{L}^\perp) \times \mathbf{R}$.

The proof of this theorem follows from general results for abstract inequalities in Banach spaces, see for example Cocu (1984) or, Fan (1972) and Brezis et. al. (1972) for more general results. A crucial part of the proof is to verify that the functional

$$(K_\circ \cap \mathcal{L}^\perp) \times \mathbf{R} \ni \mathcal{U} \mapsto \mathcal{A}(\mathcal{U},\mathcal{U}) \in \mathbf{R}$$

is weakly lower semicontinuous, i.e., that

$$\mathcal{A}(\mathcal{U},\mathcal{U}) \leq \liminf_{n\to\infty} \mathcal{A}(\mathcal{U}_n,\mathcal{U}_n)$$

for every sequence $\{\mathcal{U}_n\}_{n=1}^{\infty} \subset (K_\circ \cap \mathcal{L}^\perp) \times \mathbf{R}$ converging weakly towards $\mathcal{U} \in V \times \mathbf{R}$. Since $\mathcal{A}$ is non-symmetric this is not completely trivial but may be proved, using Lemma 1 and the fact that Q is finite-dimensional.

From (15), (16) and the fact that A contains the factor $\|\mu\|_{Lip}$ and that $\|\mathbf{L}_{f,\mu} - \bar{\mathbf{L}}_{f,\mu}\|_{V'} \leq C\|\mu\|_{Lip}$ we get the following corollary

Corollary 1 *If* $\langle \mathbf{F} - \bar{\mathbf{F}}, \mathbf{e}_1\rangle_{V',V} > 0$ *and* $c_k\langle \mathbf{F} - \bar{\mathbf{F}}, \mathbf{e}_1\rangle_{V',V} > \frac{1}{4}\|\bar{\mathbf{F}} - \bar{\mathbf{L}}_{f,\mu}\|_V^2$ *then there exists a* $\delta > 0$ *such that the problem (PV) has a solution whenever* $0 \leq \|\mu\|_{Lip} < \delta$.

Some important comments relating to this corollary are as follows: (i) the fact that a 'small' friction coefficient is needed is in agreement with previous results on frictional contact, see for instance Klarbring (1990). In fact, for large frictional coefficients it can be anticipated that the limit state corresponds to a chattering motion of stick-slip type instead of the constant velocity state considered here. (ii) The condition $c_k\langle \mathbf{F} - \bar{\mathbf{F}}, \mathbf{e}_1\rangle_{V',V} > \frac{1}{4}\|\bar{\mathbf{F}} - \bar{\mathbf{L}}_{f,\mu}\|_V^2$, which implies $\langle \mathbf{F} - \bar{\mathbf{F}}, \mathbf{e}_1\rangle_{V',V} > 0$, means, firstly, that there must be a force resultant in the $\mathbf{e}_1$-direction in order to have a sliding in this direction and, secondly, it is a condition on the direction of the forces: for given geometry and constitutive constants a large "tangential" component $\mathbf{F} - \bar{\mathbf{F}}$ promotes satisfaction of the condition, while a large "normal" component $\bar{\mathbf{F}}$ counteracts it.

REFERENCES

Brezis H., Nirenberg L. and Stampacchia G., 1972, A remark on Ky Fan's minimax principle, *Bolletino U.M.I.*, **6** *293(4), 293-300*

Cocu M., 1984, Existence of solutions of Signorini problems with friction, *Int. J. Engng. Sci.*, **22***(5), 567-575*

Fan K., 1972, A minimax inequality and applications, *Inequalities III (ed. O. Shisha), Academic Press, 103-113*

Klarbring A., 1990, Examples of non-uniqueness and non-existence of solutions to quasistatic contact problems with friction, *Ing. Arch.* **60**, *529-542*

Nečas J. and Hlaváček I., 1981, Mathematical theory of elastic and elasto-plastic bodies: an introduction, Elsevier, Amsterdam

THE INFLUENCE OF THE FRICTION COEFFICIENTS ON THE UNIQUENESS OF THE SOLUTION OF THE UNILATERAL CONTACT PROBLEM

I. Doudoumis, E. Mitsopoulou and N. Charalambakis

School of Engineering
Aristotle University of Thessaloniki, Greece

ABSTRACT

In the paper the quasistatic two dimensional unilateral contact frictional problem between elastic bodies is studied. The problem is formulated as an incremental Linear Complementarity Problem. Based on the theorems of the mathematical theory of linear complementarity, a sufficient criterion is given for the friction coefficient's values by which it is assured that the frictional-unilateral contact problem has a unique solution for the given initial contact state. That is we define quantitatively whether the values of the friction coefficients belong to a class of "sufficiently small" coefficients.

FORMULATION OF THE PROBLEM

The quasistatic two dimensional unilateral contact frictional problem between elastic bodies is considered, under the assumption of infinitesimal strains and displacements. The Coulomb's law of dry friction is assumed to hold at the contact area. The elastic bodies are discretized by the finite element method and thus the contact area consists of pairs of distinct points i (s. figure 1). The Coulomb's coefficient of friction $\mu_i = \tan \phi_i$ may take different values at the different points.

Since the solution of this problem depends on the loading history (path dependent), an incremental formulation is used. That is we define the unknown increments of the displacements and stresses due to an infinitesimal increment of the external loads (load increment n), when the initial conditions are known. The initial conditions are known since we have already determined the terminal equilibrium contact state of the previous load increment $(n-1)$.

At the pairs of discrete contact points fictitious rigid bonds with infinitesimal size are introduced. The stresses $\mathbf{s}_i = (s_T, s_N)_i$ of the bond i represent the tangential and

Contact Mechanics, Edited by M. Raous *et al.*
Plenum Press, New York, 1995

normal contact reactions, and the strains $\boldsymbol{\varepsilon}_i = (\varepsilon_T, \varepsilon_N)_i$ represent the corresponding relative displacements of the nodal pair i.

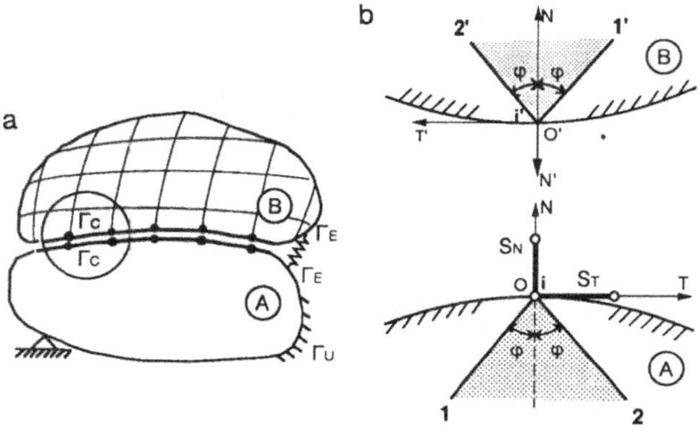

Figure 1. a) Discretization of the bodies and the contact area
b) Coulomb's angular region and interface bond at the nodal pair i

The incremental contact law which governs the behaviour of each bond i depends on the initial contact state of the bond which may be one of four discrete types (Klarbring 1985, Doudoumis 1991):

a. The initial contact state is sticking contact (the current contact stress lies in the interior of Coulomb's angular region):

$$s_{Ni} < 0 \, , \quad f_{Ti} = \mu_i s_{Ni} + \mid s_{Ti} \mid < 0 \, , \quad \varepsilon_{Ni} = 0 \tag{1}$$

Incremental law:

$$\delta \mathbf{s}_i \in R_2 \quad \text{and} \quad \delta \boldsymbol{\varepsilon}_i = 0 \, . \tag{2}$$

b. The initial contact state is separation (zero contact stress):

$$\mathbf{s}_i = 0 \, , \quad \boldsymbol{\varepsilon}_{Ni} > 0 \, . \tag{3}$$

Incremental law:

$$\delta \mathbf{s}_i = 0 \quad \text{and} \quad \delta \boldsymbol{\varepsilon}_i \in R_2 \tag{4}$$

c. The initial contact state is sliding contact with $s_{Ni} < 0$ (the current contact stress lies on Coulomb's angular line):

$$s_{Ni} < 0 \, , \quad f_{Ti} = \mu_i s_{Ni} + \mid s_{Ti} \mid = 0 \, , \quad \varepsilon_{Ni} = 0 \, . \tag{5}$$

Incremental law:

$$\begin{array}{c} \delta \mathbf{f}_i^c = \mathbf{N}_i^c \cdot \delta \mathbf{s}_i^c \, , \quad \delta \varepsilon_i = \mathbf{V}_i^c \cdot \delta \boldsymbol{\lambda}_i^c \\ -\delta \mathbf{f}_i^c \geq \mathbf{0} \, , \quad \delta \boldsymbol{\lambda}_i^c \geq 0 \, , \quad -\delta \mathbf{f}_i^{cT} \cdot \delta \boldsymbol{\lambda}_i^c = 0 \end{array} \tag{6}$$

where

$$\begin{array}{l} \mathbf{N}_i^c = [\cos \theta_i \, , \mu_i] \\ \mathbf{V}_i^c = [\cos \theta_i \, , 0]^T \\ \delta \boldsymbol{\lambda}_i^c = \mid \delta \varepsilon_{T_i} \mid \end{array} \tag{7}$$

($\theta_i = 180^o$ when $s_{T_i} < 0$, $\theta_i = 0^o$ when $s_{T_i} > 0$).

d. The initial contact state is sliding contact with $s_{Ni} = 0$ (the contact stress lies on the vertex of Coulomb's angular region):

$$s_i = 0\,, \quad \varepsilon_{Ni} = 0\,, \quad (f_{Ti} = \mu_i s_{Ni} + \mid s_{Ti} \mid = 0)\,. \tag{8}$$

Incremental law:

$$\begin{aligned} &\delta \mathbf{f}_i^d = \mathbf{N}_i^d \cdot \delta \mathbf{s}_i^d\,, \quad \delta \varepsilon_i = \mathbf{V}_i^d \cdot \delta \boldsymbol{\lambda}_i^d \\ &-\delta \mathbf{f}_i^d \geq 0\,, \quad \delta \boldsymbol{\lambda}_i^d \geq 0\,, \quad -\delta \mathbf{f}_i^{dT} \cdot \delta \boldsymbol{\lambda}_i^d = 0 \end{aligned} \tag{9}$$

where

$$\mathbf{N}_i^d = \begin{bmatrix} -1 & \mu_i \\ 1 & \mu_i \\ 0 & 1 \end{bmatrix}\,, \quad \mathbf{V}_i^d = \begin{bmatrix} -1 & 1 & 0 \\ 0 & 0 & 1 \end{bmatrix}\,, \quad \delta \boldsymbol{\lambda}_i = \begin{bmatrix} \mid \delta \varepsilon_T^- \mid_i \\ \mid \delta \varepsilon_T^+ \mid_i \\ \mid \delta \varepsilon_N \mid_i \end{bmatrix}\,. \tag{10}$$

For the total number m_c of the unilateral bonds with initial contact state c, the incremental law is written in matrix form:

$$\begin{aligned} &\delta \mathbf{f}^c = \mathbf{N}^c \cdot \delta \mathbf{s}^c\,, \quad \delta \varepsilon = \mathbf{V}^c \cdot \delta \boldsymbol{\lambda}^c \\ &-\delta \mathbf{f}^c \geq 0\,, \quad \delta \boldsymbol{\lambda}^c \geq 0\,, \quad -\delta \mathbf{f}^{cT} \cdot \delta \boldsymbol{\lambda}^c = 0 \end{aligned} \tag{11}$$

where

$$\begin{aligned} \delta \mathbf{f}^c &= \{\delta \mathbf{f}_1^c, \delta \mathbf{f}_2^c \cdots \delta \mathbf{f}_{m_c}^c\} \\ \delta \boldsymbol{\lambda}^c &= \{\delta \boldsymbol{\lambda}_1^c, \delta \boldsymbol{\lambda}_2^c \cdots \delta \boldsymbol{\lambda}_{m_c}^c\} \\ \mathbf{N}^c &= diag\{\mathbf{N}_1^c, \mathbf{N}_2^c, \cdots \mathbf{N}_{m_c}^c\} \\ \mathbf{V}^c &= diag\{\mathbf{V}_1^c, \mathbf{V}_2^c, \cdots \mathbf{V}_{m_c}^c\} \end{aligned} \tag{12}$$

For the total number m_d of the unilateral bonds with initial contact state d, the incremental law is written in matrix form:

$$\begin{aligned} &\delta \mathbf{f}^d = \mathbf{N}^d \cdot \delta \mathbf{s}^d\,, \quad \delta \varepsilon = \mathbf{V}^d \cdot \delta \boldsymbol{\lambda}^d \\ &-\delta \mathbf{f}^d \geq 0\,, \quad \delta \boldsymbol{\lambda}^d \geq 0\,, \quad -\delta \mathbf{f}^{dT} \cdot \delta \boldsymbol{\lambda}^d = 0 \end{aligned} \tag{13}$$

where

$$\begin{aligned} \delta \mathbf{f}^d &= \{\delta \mathbf{f}_1^d, \delta \mathbf{f}_2^d \cdots \delta \mathbf{f}_{m_d}^d\} \\ \delta \boldsymbol{\lambda}^d &= \{\delta \boldsymbol{\lambda}_1^d, \delta \boldsymbol{\lambda}_2^d \cdots \delta \boldsymbol{\lambda}_{m_d}^d\} \\ \mathbf{N}^d &= diag\{\mathbf{N}_1^d, \mathbf{N}_2^d, \cdots \mathbf{N}_{m_d}^d\} \\ \mathbf{V}^d &= diag\{\mathbf{V}_1^d, \mathbf{V}_2^d, \cdots \mathbf{V}_{m_d}^d\} \end{aligned} \tag{14}$$

Thus, for the total number $m = m_c + m_d$ of unilateral bonds with initial conditions of type c and d the following relations hold:

$$\delta \mathbf{f} = \mathbf{N} \cdot \delta \mathbf{s} \quad \delta \boldsymbol{\varepsilon} = \mathbf{V} \cdot \delta \boldsymbol{\lambda} \quad -\delta \mathbf{f} \geq 0\,, \quad \delta \boldsymbol{\lambda} \geq 0 \quad -\delta \mathbf{f}^T \cdot \delta \boldsymbol{\lambda} = 0 \tag{15}$$

where:

$$\delta \mathbf{f} = \begin{bmatrix} \delta \mathbf{f}^c \\ \delta \mathbf{f}^d \end{bmatrix}, \delta \boldsymbol{\lambda} = \begin{bmatrix} \delta \boldsymbol{\lambda}^c \\ \delta \boldsymbol{\lambda}^d \end{bmatrix}$$

$$\delta \mathbf{s} = \begin{bmatrix} \delta \mathbf{s}^c \\ \delta \mathbf{s}^d \end{bmatrix}, \delta \boldsymbol{\varepsilon} = \begin{bmatrix} \delta \boldsymbol{\varepsilon}^c \\ \delta \boldsymbol{\varepsilon}^d \end{bmatrix} \tag{16}$$

$$\mathbf{N} = \begin{bmatrix} \mathbf{N}^c & 0 \\ 0 & \mathbf{N}^d \end{bmatrix}, \mathbf{V} = \begin{bmatrix} \mathbf{V}^c & 0 \\ 0 & \mathbf{V}^d \end{bmatrix} .$$

Furthermore, $2m$ stiffness equilibrium equations hold, in terms of the strains and stresses of the $m = m_c + m_d$ interface contact bonds:

$$\begin{bmatrix} \delta \mathbf{s}^c \\ \delta \mathbf{s}^d \end{bmatrix} = \begin{bmatrix} \delta \mathbf{s}_0^c \\ \delta \mathbf{s}_0^d \end{bmatrix} + \begin{bmatrix} \mathbf{K}^{cc} & \mathbf{K}^{cd} \\ \mathbf{K}^{dc} & \mathbf{K}^{dd} \end{bmatrix} \begin{bmatrix} \delta \boldsymbol{\varepsilon}^c \\ \delta \boldsymbol{\varepsilon}^d \end{bmatrix}$$

$$\text{or} \quad \delta \mathbf{s} = \delta \mathbf{s}_0 + \mathbf{K} \delta \boldsymbol{\varepsilon} . \tag{17}$$

Here $\mathbf{K}$ is a "stiffness" matrix of influence coefficients, which expresses the reactions of the bonds of a bilateral structure (which is derived from the given "unilateral" structural system), due to unit fictitious strains at the bonds. It is supposed that the structural system does not possess rigid body's degrees of freedom, and thus the matrix $-\mathbf{K}$ is always positive definite (P.D.). The vector $\delta \mathbf{s}_0$ is the known vector of the increments of the stresses of the bonds at the bilateral structure, due to a given increment of the external loading. In the above relations the unilateral bonds, with initial contact state of type a and b have not indroduced, because $\delta \boldsymbol{\varepsilon}^i$ and $\delta \mathbf{s}^i$ respectively are a priori known (equal to zero).

The relations (15),(17) lead to the following incremental Linear Complementarity Problem: find $\delta \mathbf{f}$, $\delta \boldsymbol{\lambda}$ such that

$$-\delta \mathbf{f} = \mathbf{M} \cdot \delta \boldsymbol{\lambda} + \delta \mathbf{q} , \quad -\delta \mathbf{f} \geq 0 , \quad \delta \boldsymbol{\lambda} \geq 0 , \quad -\delta \mathbf{f}^T \cdot \delta \boldsymbol{\lambda} = 0 \tag{18}$$

where $\mathbf{M} = -\mathbf{N} \cdot \mathbf{K} \cdot \mathbf{V}$ is a given nonsymmetric square matrix and $\delta \mathbf{q} = -\mathbf{N} \cdot \delta \mathbf{s}_0$ is a known vector.

The numerical solution of the above incremental LCP can be achieved by using direct or iterative solution methods (Murty 1988), e.g. Lemke's pivoting algorithm (Doudoumis, Mitsopoulou and Nikolaidis 1992).

ON THE UNIQUENESS OF THE SOLUTION

The uniqueness of the solution of the L.C.P. (18) depends on the nature of the matrix $\mathbf{M}$. Since this matrix depends on the values of the friction coefficients μ_i of the nodal pairs of unilateral contact, we attempt to define a set of values of μ_i (sufficient conditions) for which the matrix $\mathbf{M}$ has the appropriate characteristics to give a unique solution (Doudoumis 1991).

The terms of matrix $\mathbf{K}$ in (17) (and respectively the terms of matrices $\mathbf{N}$ and $\mathbf{V}$) are rearranged in order that the $m = m_c + m_d$ tangential components are written first and then the m normal components, and the matrices are written :

$$\mathbf{K} = \left[\begin{array}{cc|cc} \mathbf{K}_{TT}^{c} & \mathbf{K}_{TT}^{cd} & \mathbf{K}_{TN}^{c} & \mathbf{K}_{TN}^{cd} \\ \mathbf{K}_{TT}^{cd} & \mathbf{K}_{TT}^{d} & \mathbf{K}_{TN}^{cd} & \mathbf{K}_{TN}^{d} \\ \hline \mathbf{K}_{NT}^{c} & \mathbf{K}_{NT}^{cd} & \mathbf{K}_{NN}^{c} & \mathbf{K}_{NN}^{cd} \\ \mathbf{K}_{NT}^{cd} & \mathbf{K}_{NT}^{d} & \mathbf{K}_{NN}^{cd} & \mathbf{K}_{NN}^{d} \end{array} \right] , \tag{19}$$

$$\mathbf{N} = \left[\begin{array}{c|c|c|c} \mathbf{e}^c & \mathbf{0} & \boldsymbol{\mu}^c & \mathbf{0} \\ \hline \mathbf{0} & \mathbf{e}^d & \mathbf{0} & \boldsymbol{\mu}^d \\ \hline \mathbf{0} & \mathbf{0} & \mathbf{0} & \mathbf{I} \end{array}\right], \tag{20}$$

$$V = \left[\begin{array}{c|c|c} \mathbf{e}^{cT} & \mathbf{0} & \mathbf{0} \\ \hline \mathbf{0} & \mathbf{e}^{dT} & \mathbf{0} \\ \hline \mathbf{0} & \mathbf{0} & \mathbf{0} \\ \hline \mathbf{0} & \mathbf{0} & \mathbf{I} \end{array}\right] \tag{21}$$

where

$$\begin{aligned} \mathbf{e}^c &= diag[e_1^c\ e_2^c \cdots e_{m_c}^c] \\ e_i^c &= -1 \quad \text{or} \quad 1 \\ \boldsymbol{\mu}^c &= diag[\mu_1^c\ \mu_2^c \cdots \mu_{m_c}^c] \end{aligned} \tag{22}$$

and

$$\begin{aligned} \mathbf{e}^d &= diag[\mathbf{e}_1^d\ \mathbf{e}_2^d \cdots \mathbf{e}_{m_d}^d] \\ \boldsymbol{\mu}^d &= diag[\boldsymbol{\mu}_1^d\ \boldsymbol{\mu}_2^d \cdots \boldsymbol{\mu}_{m_d}^d] \end{aligned} \tag{23}$$

$$\mathbf{e}_i^d = \begin{bmatrix} -1 \\ 1 \end{bmatrix} \quad \text{and} \quad \boldsymbol{\mu}_i^d = \begin{bmatrix} \boldsymbol{\mu} \\ \boldsymbol{\mu} \end{bmatrix} .$$

In the case of initial contact state d, for each nodal pair i among the possible subsequent contact states there are two opposite sliding directions (see subvectors $\mathbf{e}^d$, $\boldsymbol{\mu}^d$), and the m_d (from the $2m_d$) rows of matrix $\mathbf{N}$ and the m_d (from the $2m_d$) columns of matrix $\mathbf{V}$ are linearly dependent. Thus the matrix $\mathbf{M} = -\mathbf{N}\cdot\mathbf{K}\cdot\mathbf{V}$ is row adequate (det $|\ \mathbf{M}\ | = 0$). Since the sliding of a nodal pair in two opposite directions cannot occur simultaneously, we define,whithin the needs of the present investigation, for each nodal pair i only one sliding direction (which may be -1 or 1), and all possible combinations for these sliding directions of the nodal pairs must be taken into account. Thus the diagonal supermatrices $\mathbf{e}^d = diag[\mathbf{e}_1^d\mathbf{e}_2^d...\mathbf{e}_{md}^d]$, and $\boldsymbol{\mu}^d = diag[\boldsymbol{\mu}_1^d\boldsymbol{\mu}_2^d...\boldsymbol{\mu}_{md}^d]$, reduce to diagonal matrices, because the order (2×1) of the vector components $\mathbf{e}_i^d$, and $\boldsymbol{\mu}_i^d$ is reduced to (1×1) ($\mathbf{e}_i^d = e_i^d = -1$ or 1, and $\boldsymbol{\mu}_i^d = \mu$).

In view of the above, the $(m + m_d) \times (m + m_c + m_d)$ matrix $\mathbf{N}$ and the $(m + m_c + m_d) \times (m + m_d)$ matrix $\mathbf{V}$ can be written as

$$\mathbf{N} = \left[\begin{array}{cc|cc} \multicolumn{2}{c|}{\mathbf{e}} & \multicolumn{2}{c}{\boldsymbol{\mu}} \\ \hline \mathbf{0} & \mathbf{0} & \mathbf{0} & \mathbf{I} \end{array}\right], \tag{24}$$

and

$$\mathbf{V} = \left[\begin{array}{c|c} \mathbf{e} & \mathbf{0} \\ \hline \mathbf{0} & \mathbf{0} \\ \hline \mathbf{0} & \mathbf{I} \end{array}\right], \tag{25}$$

where

$$\mathbf{e} = diag[e_1^c\ e_2^c \cdots e_{m_c}^c\ e_1^d\ e_2^d \cdots e_{m_d}^d] \tag{26}$$

$$\boldsymbol{\mu} = diag[\mu_1^c\ \mu_2^c \cdots \mu_{m_c}^c\ \mu_1^d\ \mu_2^d \cdots \mu_{m_d}^d] . \tag{27}$$

Now the diagonal matrix $\boldsymbol{\mu}$ may be written in the form:

$$\boldsymbol{\mu} = \rho\ diag[\mu_1^{*c}\ \mu_2^{*c}\ \mu_{m_c}^{*c}\ \mu_1^{*d} \cdots m_{m_d}^{*d}] = \rho \cdot \boldsymbol{\mu}^* \quad (\rho \geq 0) . \tag{28}$$

where $diag[\mu_1^{*c}\ \mu_2^{*c}\ \mu_{m_c}^{*c}\ \mu_1^{*d} \cdots m_{m_d}^{*d}]$ is a matrix of the relative values of the coefficients μ_i, and ρ is a nonnegative multiplication factor.

The matrix $\mathbf{M} = -\mathbf{N} \cdot \mathbf{K} \cdot \mathbf{V}$ after the above (rel. (19)(24)(25)) can be written as a linear function of ρ :

$$\mathbf{M}(\rho) = \mathbf{H} - \rho\mathbf{G} \tag{29}$$

where

$$\mathbf{H} = \left[\begin{array}{c|c} \mathbf{eK}_{TT}\mathbf{e} & \mathbf{eK}^*_{TN} \\ \hline \mathbf{K}^*_{NT}\mathbf{e} & \mathbf{K}^d_{NN} \end{array}\right] , \tag{30}$$

$$\mathbf{G} = \mu^* \left[\begin{array}{c|c} \mathbf{K}_{NT}\mathbf{e} & \mathbf{K}^*_{NN} \\ \hline \mathbf{0} & \mathbf{0} \end{array}\right] , \tag{31}$$

and

$$\begin{aligned} \mathbf{K}^*_{NT} &= [\ \mathbf{K}^{cd}_{NT}\ \ \mathbf{K}^d_{NT}\ ] , \\ \mathbf{K}^*_{TN} &= \mathbf{K}^{*\,T}_{NT} \\ \mathbf{K}^*_{NN} &= \begin{bmatrix} \mathbf{K}^{cd}_{NN} \\ \mathbf{K}^d_{NN} \end{bmatrix} . \end{aligned} \tag{32}$$

The matrix $\mathbf{H}$ in relation (30) is an alternative form of matrix $-\mathbf{K}$. That is some of the sings of the K_{ij} terms have been changed in order to be consistent with the corresponding strains ε_{jT} (which are always considered positive). Thus the matrix $\mathbf{M}(0) = \mathbf{H}$ is positive definite, because the matrix $-\mathbf{K}$ has been considered P.D. It should be noted also that the P.D. symmetric matrix $\mathbf{M}(0) = \mathbf{H}$ is the matrix of the frictionless contact case ($\mu = 0$) which, as it is known, has a unique solution.

In order to define a sufficient condition for the values of μ_i for which the problem has a unique solution the following theorems (Murty 1988) of linear complementarity are used:

a. A sufficient condition for the solution to exist and to be unique (for any increment of the external loading), is the positive definiteness of matrix $\mathbf{M}$.

b. The nonsymmetric matrix $\mathbf{M}(\rho)$ is P.D. if and only if the symmetric matrix:

$$\mathbf{D}(\rho) = (\mathbf{M} + \mathbf{M}^T)/2 = (\mathbf{H} + \mathbf{H}^T)/2 - \rho(\mathbf{G} + \mathbf{G}^T)/2 = \mathbf{H} - \rho(\mathbf{G} + \mathbf{G}^T)/2 \tag{33}$$

is P.D.

c. The square symmetric matrix: $\mathbf{D}(\rho)$ is P.D. if and only if all its main principal subdeterminants are strictly positive.

As we have already proved

$$\det \mid \mathbf{M}(0) \mid \equiv \det \mid \mathbf{D}(0) \mid > 0 . \tag{34}$$

For increasing values of $\rho > 0$ the matrix $\mathbf{D}(\rho)$ may stop being positive definite. When this fact will take place, a certain main principal subdeterminant (suppose $\det \mid \mathbf{D}^*(\rho) \mid$) will not be positive any more. Since all the subdeterminants of $\mathbf{D}(\rho)$ are continuous funtions of ρ this determinant will become zero for a value of ρ:

$$\det \mid \mathbf{H}^* - \rho(\mathbf{G}^* + \mathbf{G}^{*T})/2 \mid = 0 \tag{35}$$

That is we are searching for the smallest value of ρ for which rel. (35) holds. The above problem is an eigenvalue problem and the smallest eigenvalue ρ_1 has to be defined.

But according to a property of symmetric square matrices (Bathe and Wilson 1976) the smallest eigenvalue of all the main principal subdeterminants of a square symmetric matrix is the smallest eigenvalue ρ_1 of the determinant of the square matrix and rel.(35) is written as:

$$\det \mid \mathbf{H} - \rho(\mathbf{G} + \mathbf{G}^T)/2 \mid = 0 \tag{36}$$

or (since $\det \mid \mathbf{H} \mid > 0$ then $\mathbf{H}^{-1}$ exists):

$$\det \mid \mathbf{H}^{-1}(\mathbf{G} + \mathbf{G}^T)/2 - \mathbf{I}w \mid = 0 \tag{37}$$

where $w = 1/\rho$.

The largest root of equation (37) gives the required limit value ρ_1 of ρ $(0 \leq \rho < \rho_1)$ for which the matrix $\mathbf{M}$ is positive definite. Thus we have defined a quantitative sufficient criterion for the values $\boldsymbol{\mu} = \rho\{\mu_1^* \; \mu_2^* \cdots \mu_m^*\}$ of the friction coefficients, by which it is assured that the frictional unilateral contact problem has a unique solution.

NUMERICAL EXAMPLE

The uniqueness of the solution of the frame shown in figure 2 is studied here. The frame is supported by 3 unilateral frictional supports on a rigid foundation. The geometry and data of the problem are indicated in the figure.

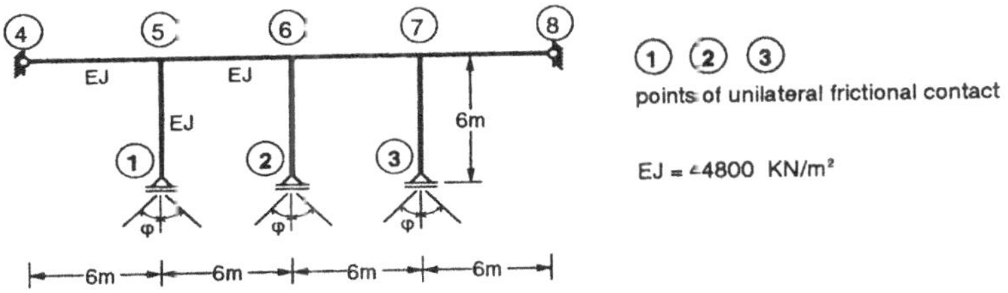

Figure 2. Symmetric plane frame with 3 unilateral frictional supports

We suppose first that the initial contact state is sliding with $s_N < 0$ (of type c) and $\mathbf{e}^c = diag[1 \;\; -1 \;\; 1]$ (the nodes 1 and 3 are slipping to the right while 2 is slipping to the left. For these initial conditions the 3×3 $\mathbf{H}$ and $\mathbf{G}$ matrices are formed and the eigenvalues of the problem are found through the solution of (37). The smallest eigenvalue obtained is equal to 0.826 which means that a sufficient condition for the solution to exist and be unique is $\mu < 0.826$ or $\phi < 39{,}60^0$.

If the initial contact condition is sliding with $s_N = 0$ (of type d) state $(\mathbf{s}^i \;, \varepsilon = 0)$, for the three contact points $2^3 = 8$ or due to symmetry only 3 different values of $\mathbf{e}^d$ must be considered. Thus three different 6×6 eigenvalue problems have to be solved:

a. for $\mathbf{e}^d = diag[1 \; 1 \; 1]$ it is obtained $\rho_1 = 0.462$
b. for $\mathbf{e}^d = diag[1 \;\; -1 \;\; -1]$ it is obtained $\rho_1 = 0.352$
c. for $\mathbf{e}^d = diag[1 \;\; -1 \; 1]$ it is obtained $\rho_1 = 0.315$

Since the smallest value of ρ_1 is that of case 3 the maximum value of μ is given by this value or $0 \leq \mu < 0.315$ and $0 \leq \phi < 17.5^0$.

REFERENCES

Bathe, K.J. and Wilson, E.L., 1976: "Numerical Methods in Finite Element Analysis". Prentice-Hall, Inc. Englewood Cliff, New Jersey.

Doudoumis, I.N. , Mitsopoulou, E.N., Nikolaidis, G.N., 1992: A Comparative numerical study on the unilateral contact problem with friction, Proceedings of the 1st National Congress of Computational Mechanics, Athens.

Doudoumis, I.N. , 1991: Modelling "infill finite elements" with unilateral contact frictional interface conditions and elastic or inelastic material laws, Ph.D. Thesis, Aristotle University of Thessaloniki, Thessaloniki.

Klarbring, A. , 1985: Contact problems with friction -using a finite dimensional description and the theory of linear complementarity, Linkoping Studies in Science and Technology, Thesis No 20.

Murty K.G. , 1988: "Linear Complementarity, Linear and Nonlinear Programming", Heldermann Verlag, Berlin.

Panagiotopoulos, P.D. , 1993: "Hemivariational Inequalities. Applications in Mechanics and Engineering", Springer Verlag, Berlin, N. York.

A FUNCTIONAL FRAMEWORK FOR THE SIGNORINI PROBLEM WITH COULOMB'S FRICTION

Charles Naéjus,[1,2] Alain Cimetière[1] and Alain Léger [2]

[1]UFR-Sciences Poitiers, Laboratoire de Mécanique Théorique
40 avenue du Recteur Pineau, 86022 Poitiers cedex (France)
[2]EDF-DER, Mécanique et Modèles Numériques
1 avenue du Général De Gaulle, 92141 Clamart cedex (France)

INTRODUCTION

In this lecture we go back over the mathematical formulation of the Signorini problem with Coulomb's friction. We essentially aim at reconsidering the functional framework for three reasons. Our intention is, first, to give a precise meaning to the normal contact stress (see for example Duvaut, 1982), then to give a weak formulation of Coulomb's rule, finally to take into account the time dependency and general coefficients of friction.

Indeed, the previous formulations did not take into account the quasistatic evolution and were restricted to the case of particular coefficients of friction .

In the first part of the paper, we discuss the time dependency of the problem, by studying a finite dimensional problem. We give an variational inequality of the quasistatic evolution, which takes into account both the unilateral contact and the Coulomb's friction law. As an application we get the time discretization of the problem by a "trapezoidal integration scheme".

Then we focus on the functional framework of the equilibrium problem which could correspond to one increment of the discretization in time for a continuous elastic medium. This constitutes the main part of the lecture. In particular, special attention is paid to some properties of the traces spaces, thanks to the notion of Dirichlet space.

Finally, a quasivariational formulation follows, first for the equilibrium problem, then for the initial quasistatic evolution problem.

A VARIATIONAL FORMULATION OF THE QUASISTATIC PROBLEM

Let P be a point moving in a region of $\mathbb{R}^3$ limited by an obstacle. Let $u(t)$, $t \in [0,T]$ be the displacement of P from its initial position, $\dot{u}$ its (right-)velocity. We denote by K the set of all the possible displacements of P. We suppose that there exists

Contact Mechanics, Edited by M. Raous *et al.*
Plenum Press, New York, 1995

a regular enough scalar function g such that any displacement $\mathbf{u}(t)$ of P belongs to K, if and only if $g(\mathbf{u}(t)) \geq 0$. Let then $V_K(u(t))$ be the set of admissible velocities :

$$V_K(\mathbf{u}(t)) = \text{ closure of } \left\{\mathbf{v}\, ; \nabla g(\mathbf{u}(t)) \cdot \mathbf{v} \in \mathbb{R}^+ + \mathbb{R}.g(\mathbf{u}(t))\right\} = \overline{\left\{\bigcup_{\epsilon>0} \frac{K-\mathbf{u}}{\epsilon}\right\}}.$$

We will note that $\dot{\mathbf{u}}(t) \in V_K(\mathbf{u}(t))$. At a contact point an index N or T will classically hold for the normal or the tangential component either for the velocity $\dot{\mathbf{u}}$ or for the reaction $\boldsymbol{R}$ of the obstacle. The contact between the point P and the obstacle is unilateral :

$$g(\mathbf{u}(t)) \geq 0, \quad R_N(\mathbf{u}(t)) \geq 0, \quad g(\mathbf{u}(t)) \cdot R_N(\mathbf{u}(t)) = 0, \tag{1}$$

with Coulomb's friction law :

$$\begin{cases} |\boldsymbol{R}_T| \leq \mu |R_N| & \\ |\boldsymbol{R}_T| < \mu |R_N| & \Rightarrow \quad \dot{\mathbf{u}}_T = 0 \\ |\boldsymbol{R}_T| = \mu |R_N| \neq 0 & \Rightarrow \quad \exists \lambda \geq 0 \ such\ \dot{\mathbf{u}}_T = -\lambda \boldsymbol{R}_T \end{cases} \tag{2}$$

where μ is the coefficient of friction.

An Equivalent Formulation for Unilateral Contact with Coulomb's Friction

It is known, from Duvaut and Lions(1972) or Moreau(1973) that the following variational inequality is strictly equivalent to the law (2) :

$$\boldsymbol{R}_T(t) \cdot (\mathbf{v}_T - \dot{\mathbf{u}}_T(t)) + \mu |R_N(t)| (|\mathbf{v}_T| - |\dot{\mathbf{u}}_T(t)|) \geq 0 \ , \forall \mathbf{v} \in V_K(u(t))$$

Let now $\mathbf{u} \in W^{1,1}(0,T,\mathbb{R}^3)$. It has been established by Moreau(1988) that the condition (1) holds for all $t \in [0,T]$ if and only if :

$$\begin{cases} \mathbf{u}(0) \in K \ and, \\ \boldsymbol{R}(t) \cdot (\mathbf{v} - \dot{\mathbf{u}}(t)) \geq 0, \quad \forall \mathbf{v} \in V_K(\mathbf{u}(t)) \ \text{ a.e. } t \in [0,T]. \end{cases}$$

Then we deduce the following proposition :

Proposition 1. Let $\mathbf{u} \in W^{1,1}(0,T,\mathbb{R}^3)$; then the condition of unilateral contact with Coulomb's friction, (1) and (2) holds for all $t \in [0,T]$ if and only if

$$\begin{cases} \mathbf{u}(0) \in K \ and \\ \left.\begin{array}{l} \boldsymbol{R}(t) \cdot (\mathbf{v} - \dot{\mathbf{u}}(t)) + \mu |R_N(t)| (|\mathbf{v}_T| - |\dot{\mathbf{u}}_T(t)|) \geq 0, \\ \forall \mathbf{v} \in V_K(\mathbf{u}(t)) \end{array}\right\} \text{ a.e. } t \in [0,T] \end{cases}$$

That means variational inequalities can be written for the analysis of the quasistatic evolution of bodies under unilateral contact with Coulomb's friction. Let's look for instance at a simple three degrees of freedom system, such as the model used by Curnier and Alart(1988). The point P, is now submitted both to a force $\boldsymbol{f}(t)$ and to a spring system whose stiffness matrix is denoted M. At $t = 0$ we suppose that the system is balanced with $\boldsymbol{f}(0) = \boldsymbol{f}_0$ and has a displacement $\mathbf{u}(0) = \mathbf{u}_0$. The corresponding quasistatic problem reads as follows :

$$\forall t \in [0,T], \quad \begin{cases} M\mathbf{u}(t) = \boldsymbol{f}(t) + \mathbf{R}(t) \\ \mathbf{u}(t) \textit{ and } \mathbf{R}(t) \textit{ verify conditions } (1) \textit{ and } (2) \end{cases}$$

Thanks to proposition 1, one of its variational formulation is :

$$\begin{cases} \mathbf{u} \in W^{1,2}(0,T;\mathbb{R}^3); \quad \mathbf{u}(0) = \mathbf{u}_0 \\ \left. \begin{array}{l} M\mathbf{u}(t) \cdot (\mathbf{v} - \dot{\mathbf{u}}(t)) + \mu R_N(\mathbf{u}(t))(|\mathbf{v}_T| - |\dot{\mathbf{u}}_T(t)|) \geq \boldsymbol{f}(t) \cdot (\mathbf{v} - \dot{\mathbf{u}}(t)) \\ \forall \mathbf{v} \in V_K(\mathbf{u}(t)) \end{array} \right\} \text{ a.e. } t \in [0,T] \end{cases}$$

An Implicit Incremental Scheme

For the time discretization, the first useful step consists in integrating the above inequality on the interval [0,T] that leads to the following equivalent formulation.

$$\begin{cases} \mathbf{u} \in W^{1,2}(0,T;\mathbb{R}^3); \quad \mathbf{u}(0) = \mathbf{u}_0 \\ \displaystyle\int_0^T M\mathbf{u}(t) \cdot (\dot{\mathbf{v}}(t) - \dot{\mathbf{u}}(t))dt + \int_0^T \mu R_N(\mathbf{u}(t))(|\dot{\mathbf{v}}_T(t)| - |\dot{\mathbf{u}}_T(t)|)dt \geq \\ \qquad\qquad\qquad\qquad \displaystyle\int_0^T \boldsymbol{f}(t) \cdot (\dot{\mathbf{v}}(t) - \dot{\mathbf{u}}(t))dt \\ \forall \mathbf{v} \in W^{1,2}(0,T;\mathbb{R}^3), \ \dot{\mathbf{v}}(t) \in V_K(u(t)), \text{ a.e. } t \in [0,T] \end{cases} \tag{3}$$

Let us then consider an ϵ–subdivision of the interval [0,T] :

$$[0,T[= \bigcup_{k=1,N_\epsilon} [t^\epsilon_{k-1}, t^\epsilon_k[, \quad N_\epsilon \epsilon = T, \quad t^\epsilon_k = k\epsilon$$

and let W^ϵ the space of continuous functions on $[0,T]$, affine on each interval $[t^\epsilon_{k-1}, t^\epsilon_k[$. Any element $\mathbf{v}^\epsilon$ in W^ϵ, with $\mathbf{v}^\epsilon_k = \mathbf{v}^\epsilon(t^\epsilon_k)$ can consequently be written as

$$\mathbf{v}^\epsilon(t) = \mathbf{v}^\epsilon_k + \frac{\Delta \mathbf{v}^\epsilon_k}{\epsilon}(t - t^\epsilon_k), \qquad \text{for any } t \in [t^\epsilon_k, t^\epsilon_{k+1}[, \quad \Delta \mathbf{v}^\epsilon_k = \mathbf{v}^\epsilon_{k+1} - \mathbf{v}^\epsilon_k$$

This leads to the following implicit incremental scheme for the time discretization of problem (3)

Find $\mathbf{u}^\epsilon \in W^\epsilon$ such that

$$\begin{cases} \mathbf{u}^\epsilon_0 = \mathbf{u}_0, \textit{ and for } k = 0, \ldots, N_\epsilon - 1 \\ M(\mathbf{u}^\epsilon_k + \dfrac{\Delta \mathbf{u}^\epsilon_k}{2}) \cdot (\Delta \mathbf{v}^\epsilon_k - \Delta \mathbf{u}^\epsilon_k) + \mu R_N(\mathbf{u}^\epsilon_k + \dfrac{\Delta \mathbf{u}^\epsilon_k}{2})(|\Delta \mathbf{v}^\epsilon_{kT}| - |\Delta \mathbf{u}^\epsilon_{kT}|) \geq \\ \qquad\qquad\qquad\qquad (\mathbf{f}^\epsilon_k + \dfrac{\Delta \mathbf{f}^\epsilon_k}{2}) \cdot (\Delta \mathbf{v}^\epsilon_k - \Delta \mathbf{u}^\epsilon_k) \\ \forall \mathbf{v}^\epsilon \in W^\epsilon, \quad \Delta \mathbf{v}^\epsilon_k \in K - \mathbf{u}^\epsilon_k \end{cases} \tag{4}$$

Let $\mathbf{u}^\epsilon$ a solution of (4) then if K is a convex set we have $\mathbf{u}^\epsilon(t) \in K$, $t \in [0,T]$, else the condition $\mathbf{u}^\epsilon(t_k^\epsilon) \in K$ only is guaranted.

The previous analysis can easily be extented to more general problems with a finite number of degrees of freedom. When continuous media are considered, the functional framework has to be specified, which is the aim of the following section.

THE STATIC SIGNORINI PROBLEM WITH COULOMB'S FRICTION

We now analyse the equilibrium of an elastic solid occupying a regular domain Ω of $\mathbb{R}^3$, clamped on a part Γ_D of its boundary, submitted to given forces $\boldsymbol{F}$ on another part Γ_F, and to unilateral contact conditions with dry Coulomb friction against a rigid obstacle on the last part Γ_C of its boundary.

The solid is assumed to be linearly elastic, that is

$$\sigma_{ij}(\mathbf{u}) = a_{ijkl}\varepsilon_{kl}(\mathbf{u}), \quad \varepsilon_{kl}(\mathbf{u}) = \frac{1}{2}\left(\frac{\partial u_k}{\partial x_l} + \frac{\partial u_l}{\partial x_k}\right),$$

where the coefficients a_{ijkl} verify the usual ellipticity, symmetry and regularity conditions. Let $\boldsymbol{n}$ be the normal to Γ and denote σ_N, σ_T, u_N, and u_T the normal and tangential components of the stresses and displacements on Γ.

Then the equilibrium problem classically reads as follows :

$$\begin{cases} (a) \begin{bmatrix} -\operatorname{div}\sigma(\mathbf{u}) & = & \boldsymbol{f} & \text{in} & \Omega \\ \mathbf{u} & = & 0 & \text{on} & \Gamma_D \\ \sigma\boldsymbol{n} & = & \boldsymbol{F} & \text{on} & \Gamma_F \end{bmatrix} \\ \left.\begin{array}{l} (b)\ [\sigma_N(\mathbf{u}) \le 0,\ u_N \le 0,\ \sigma_N u_N = 0 \\ (c) \begin{bmatrix} |\sigma_T| \le \mu|\sigma_N| & & \\ |\sigma_T| < \mu|\sigma_N| & \Rightarrow & \mathbf{u}_T = 0 \\ |\sigma_T| = \mu|\sigma_N| \neq 0 & \Rightarrow & \exists\ \lambda \ge 0 \text{ such } \mathbf{u}_T = -\lambda\sigma_T \end{bmatrix} \end{array}\right\} \text{on } \Gamma_C \end{cases} \qquad (5)$$

Some Functional Spaces and Definitions

We use the classical functional spaces L^2, H^1, $H^{1/2}$ and denote $\boldsymbol{L}^2$, $\boldsymbol{H}^1$, $\boldsymbol{H}^{1/2}$ the space $(L^2)^3$, $(H^1)^3$, $(H^{1/2})^3$.

Then it is natural from problem (5) to introduce the following Hilbert spaces :

$$\boldsymbol{V} = \{\mathbf{v} \in \boldsymbol{H}^1(\Omega);\ \mathbf{v} = 0 \text{ sur } \Gamma_D\}, \quad \boldsymbol{V}_{\text{div}} = \{\mathbf{v} \in \boldsymbol{V}, \operatorname{div}\sigma(\mathbf{v}) \in \boldsymbol{L}^2(\Omega)\}$$

with the norms $\|\mathbf{v}\|_{\boldsymbol{V}} = \|\mathbf{v}\|_{\boldsymbol{H}^1(\Omega)}$ and $\|\mathbf{v}\|_{\boldsymbol{V}_{\text{div}}} = \|\mathbf{v}\|_{\boldsymbol{V}} + \|\operatorname{div}\sigma(\mathbf{v})\|_{\boldsymbol{L}^2(\Omega)}$.

Let's now denote Γ_X a regular part of Γ, and $\tilde{v}$ the extension by zero of v outside of Γ_X in Γ. Let's then recall the definition of the space $H_{00}^{1/2}(\Gamma_X)$ given by Lions and Magenes(1968) :

$$H_{00}^{1/2}(\Gamma_X) = \left\{v \in H^{1/2}(\Gamma_X)\ /\ \tilde{v} \in H^{1/2}(\Gamma)\right\}$$

and note that this definition is equivalent to the following :

$$H_{00}^{1/2}(\Gamma_X) = \left\{v \in H^{1/2}(\Gamma_X),\ \delta^{-1/2}v \in L^2(\Gamma_X)\right\}$$

where δ holds for a $C^\infty(\overline{\Gamma_X})$ function equivalent to the distance to $\partial\Gamma_X$ function. It is known that $H_{00}^{1/2}(\Gamma_X)$ is a Hilbert space for the norm

$$\|v\|_{H_{00}^{1/2}(\Gamma_X)} = \|v\|_{H^{1/2}(\Gamma_X)} + \|\delta^{-1/2}v\|_{L^2(\Gamma_X)}$$

and $\langle\cdot,\cdot\rangle_{\frac{1/2}{00},\Gamma_X}$ denotes the duality pairing.

Let $\Sigma = \Gamma \setminus \overline{\Gamma_D}$. We note γ the trace operator on Γ, which is continuous and surjective from $\boldsymbol{V}$ onto $\boldsymbol{H}_{00}^{1/2}(\Sigma) = \left(H_{00}^{1/2}(\Sigma)\right)^3$ with

$$\gamma(\boldsymbol{v}) = \boldsymbol{v}_T + v_N, \quad \|\boldsymbol{v}_T\|_{\boldsymbol{H}_{00\,T}^{1/2}(\Sigma)} \leq \|\gamma(\boldsymbol{v})\|_{\boldsymbol{H}_{00}^{1/2}(\Sigma)} \leq c_1\|\boldsymbol{v}\|_{\boldsymbol{V}} . \tag{6}$$

From the unilateral constraint it becomes natural to introduce the convex $\boldsymbol{K}$:

$$\boldsymbol{K} = \{\boldsymbol{v} \in \boldsymbol{V};\ v_N \leq 0 \text{ on } \Gamma_C\}, \quad \boldsymbol{K}(\Gamma_C) = \gamma(\boldsymbol{K}) \text{ and}$$

$$K'(\Gamma_C) = \{\boldsymbol{r} \in \boldsymbol{H}_{00}^{1/2}(\Sigma)', \langle \boldsymbol{r}, \boldsymbol{v}\rangle \geq 0,\ \forall \boldsymbol{v} \in \boldsymbol{K}(\Gamma_C)\}.$$

Let's now define by a the usual $\boldsymbol{V}$ - elliptic bilinear form on $\boldsymbol{V} \times \boldsymbol{V}$:

$$a(\boldsymbol{u}, \boldsymbol{v}) = \int_\Omega a_{ijkl}\varepsilon_{kl}(\boldsymbol{u})\varepsilon_{ij}(\boldsymbol{v})d\boldsymbol{x} .$$

By the generalized Green's formula, any $\boldsymbol{u} \in \boldsymbol{V}_{\text{div}}$ is associated with its conormal derivative $\sigma(\boldsymbol{u}) \cdot \boldsymbol{n}$ such that :

$$\sigma(\boldsymbol{u}) \cdot \boldsymbol{n} = \sigma_N + \boldsymbol{\sigma}_T \in \boldsymbol{H}_{00}^{1/2}(\Sigma)' \text{ and}$$

$$\|\sigma_N(\boldsymbol{v})\|_{H_{00}^{1/2}(\Sigma)'} \leq \|\sigma(\boldsymbol{v})\boldsymbol{n}\|_{\boldsymbol{H}_{00}^{1/2}(\Sigma)'} \leq c_2\|\boldsymbol{v}\|_{\boldsymbol{V}_{\text{div}}} . \tag{7}$$

We finally recall the notions of restriction and extension in $H_{00}^{1/2}$ and $\left(H_{00}^{1/2}\right)'$. Let Γ_X now stand for a regular open part of Σ. For any $v \in H^{1/2}(\Gamma_X)$ we denote by $\tilde{v}$ its extension by zero onto Σ. On the other hand, for any $r \in H^{1/2}(\Sigma)'$ we denote by $r_{|\Gamma_X}$ its restriction to Γ_X. This restriction is continuous from $H_{00}^{1/2}(\Sigma)'$ into $H_{00}^{1/2}(\Gamma_X)'$ and is defined by

$$r_{|\Gamma_X}(v) = \langle r, \tilde{v}\rangle_{\frac{1/2}{00},\Sigma} .$$

Duvaut's Variational Formulation

Let's assume that $\boldsymbol{f}$ and $\boldsymbol{F}$ respectively belong to $\boldsymbol{L}^2(\Omega)$ and $\boldsymbol{H}^{-1/2}(\Gamma_F)$ and denote :

$$L(\boldsymbol{v}) = \int_\Omega \boldsymbol{f}dx + \langle \boldsymbol{F}, \gamma(\boldsymbol{v})\rangle_{\frac{1/2}{00}\Sigma} ,$$

We recall the following variational form of problem (5) introduced by Duvaut and Lions(1972)

$$\begin{cases} \textit{Find } \mathbf{u} \in \boldsymbol{K} \textit{ such that :} \\ a(\mathbf{u}, \mathbf{v} - \mathbf{u}) + \displaystyle\int_{\Gamma_C} \mu|\sigma_N(\mathbf{u})|(|\mathbf{v}_T| - |\mathbf{u}_T|)d\Gamma \geq L(\mathbf{v} - \mathbf{u}) \\ \textit{for any } \mathbf{v} \in \boldsymbol{K}. \end{cases} \tag{8}$$

This variational inequality has been widely used and discussed, and we just aim in this chapter at pointing to several difficulties coming from the mathematical meaning of the integral over Γ_C.

For the first point, it has already been noticed that $\sigma_N(u)$ is not defined for any $u \in H^1(\Omega)$.

Moreover $|\sigma_N(u)|$, the absolute value of $\sigma_N(u)$, does not generally exist for $\sigma_N(u)$ in $H_{00}^{1/2}(\Sigma)'$.

Another remark, according to the notions of restriction we just recalled, is that $\sigma_{N|\Gamma_C}$ does not generally belong to $H^{-1/2}(\Gamma_C)$ but only to $H_{00}^{1/2}(\Gamma_C)'$.

The last remark is that $\mu\sigma_N(u)$ does not exist if we only have $\sigma_N(u) \in H_{00}^{1/2}(\Sigma)'$ and μ bounded and semicontinuous on Γ_C. It is worth seeing that this remark holds for instance for the case for which μ is a piecewise constant function on Γ_C.

We now aim at answering these questions. It is then necessary to work a little more on the functional framework.

Dirichlet Space

The main tool we will use is the notion of Dirichlet space introduced by Beurling and Deny(1959). Let X a regular manifold of $\mathbb{R}^3$ and ξ a Radon measure, positive on X.

Then a Hilbert subspace H of $L^2(X, d\xi)$ is a Dirichlet space if the following properties hold :

(i) The injection of H into $L^2(X, d\xi)$ is continuous;

(ii) The intersection of H with the set $\mathcal{C}_c(X)$ of the continuous functions with compact support on X is dense both in H and in $\mathcal{C}_c(X)$.

(iii) The modulus contraction acts on H which means that :

$$\forall v \in H \begin{cases} |v| \in H \\ \| \, |v| \, \|_H \leq \|v\|_H \end{cases}$$

Then we recall a property of a Dirichlet space H, which will be used in the following.

Proposition 2. Let H'_+ be the positive cone of H' and $\mathcal{M}^+(X)$ the positive Radon measures defined on X. Then $H'_+ = H' \cap \mathcal{M}^+(X)$, and for any ρ in H'_+, we have :

(i) H is continuously embedded in $L^1(X, d\rho)$;

(ii) For any $v \in H$: $\langle \rho, v \rangle_{H',H} = \int_X v \, d\rho$.

We easily deduce the following propositions

Proposition 3. Let $r \in H' \cap \mathcal{M}(X)$ and μ a bounded and semicontinuous function on X then the absolute value of r, $|r|$, belongs to H' and we have $\mu|r| \in H' \cap \mathcal{M}(X)$.

Proposition 4. $H^{1/2}$ and $H_{00}^{1/2}$ are Dirichlet spaces.

The Weak Formulation of the Coulomb's Friction Law

Let's come back to the Signorini problem with Coulomb's friction and suppose now that μ is positive, bounded and semicontinuous on Γ_C. We can now establish the following results.

Proposition 5. Let $r \in K'(\Gamma_C)$, then $r_{|\Gamma_C} \in H_{00}^{1/2}(\Sigma)' \cap \mathcal{M}(\Sigma)$.

This proposition is a consequence of the propositions 2, 3 and of the monotonous convergence theorem from the integration theory.

Proposition 6. Let $\sigma_T \in H_{00\,T}^{1/2}(\Sigma)'$, $\sigma_N \in H_{00}^{1/2}(\Sigma)'$, $\mathbf{u} \in \mathbf{K}(\Gamma_C)$.

If $\sigma_{N|\Gamma_C} \in H_{00}^{1/2}(\Sigma)' \cap \mathcal{M}(\Sigma)$ then the two following assertions are equivalent :

(i) for any $\mathbf{v} \in \mathbf{K}(\Gamma_C)$, σ_N and σ_T :

$$\langle \sigma_{N|\Gamma_C}, v_N - u_N \rangle_{{}^{1/2}_{00},\Sigma} \geq 0,$$

$$\langle \sigma_T, \mathbf{v}_T - \mathbf{u}_T \rangle_{{}^{1/2}_{00},\Sigma} + \langle \mu|\sigma_{N|\Gamma_C}|, |\mathbf{v}_T| - |\mathbf{u}_T| \rangle_{{}^{1/2}_{00},\Sigma} \geq 0$$

(ii) There exists a positive measure $\varrho \in \mathcal{M}(\Gamma_C)$ which is equivalent to $\sigma_{N|\Gamma_C}$, and two functions $\mathbf{r}_T = (r_{T1}, r_{T2})$ and r_N such that $\sigma_{N|\Gamma_C} = r_N d\varrho$, $\sigma_{T|\Gamma_C} = \mathbf{r}_T d\varrho$ *and*

$$\left.\begin{array}{l} [u_N \leq 0,\ r_N \leq 0,\ r_N \cdot u_N = 0 \\ \\ \left[\begin{array}{lll} |\mathbf{r}_T| \leq \mu|r_N| & & \\ |\mathbf{r}_T| < \mu|r_N| & \Rightarrow & \mathbf{u}_T = 0 \\ |\mathbf{r}_T| = \mu|r_N| \neq 0 & \Rightarrow & \exists\, \lambda \geq 0 \text{ such } \mathbf{u}_T = -\lambda \mathbf{r}_T \end{array}\right. \end{array}\right\} d\varrho\text{-a.e. on } \Gamma_C$$

This result is a consequence of proposition 3, proposition 5 and of the Radon-Nikodim theorem.

Then an element $\mathbf{u}$ in V will be called a strong solution of problem (5), first if $\mathbf{u}$ satisfies the equilibrium equation in the distribution sense, secondly if $\mathbf{u}$ satisfies the boundary conditions on Γ_F and Γ_D in the trace sense, and thirdly if the unilateral contact with Coulomb's friction holds in the sense of point (ii) of proposition 6.

Let's note that such a formulation covers the classical one (5-c), when σ_N and $\mathbf{u}$ are regular enough on Γ_C.

The Quasivariational Inequality

Let's then define the set $\tilde{K}$ by

$$\tilde{\mathbf{K}} = \{\mathbf{v} \in \mathbf{K} \cap \mathbf{V}_{\text{div}},\ \sigma_N(\mathbf{v})_{|\Gamma_C} \in H_{00}^{1/2}(\Sigma)' \cap \mathcal{M}(\Sigma)\}.$$

then we can give the following result.

Theorem. The strong solutions of problem (5) are the same as those of the following quasivariational problem :

$$\begin{cases} \textit{Find } \mathbf{u} \in \tilde{K} \textit{ such that :} \\ a(\mathbf{u}, \mathbf{v} - \mathbf{u}) + \langle \mu |\sigma_N(\mathbf{u})_{|\Gamma_C}|, |\mathbf{v}_T| - |\mathbf{u}_T| \rangle_{\substack{1/2 \\ 00},\Sigma} \geq L(\mathbf{v} - \mathbf{u}) \\ \textit{for any } \mathbf{v} \in K. \end{cases} \tag{9}$$

Indeed, by standard techniques, for a solution of (9) or (5) we have

$$(\sigma_N(\mathbf{u}) - F_N)_{|\Gamma_C} = \sigma_{N|\Gamma_C} \ \in \ K(\Gamma_C),$$

then proposition 5 leads to

$$\sigma_{N|\Gamma_C} \in H_{00}^{1/2}(\Sigma)' \cap \mathcal{M}(\Sigma)$$

which allows the use of proposition 6.

APPLICATION TO THE QUASISTATIC PROBLEM

We can now give a variational formulation of the quasistatic Signorini problem with Coulomb's friction which reads as follows :

$$\begin{cases} \textit{Find } \mathbf{u} \in W^{1,2}(0,T;V),\ \mathbf{u}(t) \in \tilde{K} \textit{ such that :} \\ \left.\begin{array}{l} a(\mathbf{u}(t), \mathbf{v} - \dot{\mathbf{u}}(t)) + \langle \mu |\sigma_N(\mathbf{u}(t))_{|\Gamma_C}|, |\mathbf{v}_T| - |\dot{\mathbf{u}}_T(t)| \rangle_{\substack{1/2 \\ 00},\Sigma} \geq L(\mathbf{v} - \dot{\mathbf{u}}(t)) \\ \textit{for any } \mathbf{v} \in V_{\mathbf{K}}(\mathbf{u}(t)). \end{array}\right\} \text{ a.e. } t \in [0,T] \end{cases}$$

with

$$V_{\mathbf{K}}(\mathbf{u}(t)) = \overline{\mathbf{K} + \mathbb{R}\mathbf{u}(t)}.$$

REFERENCES

Beurling, A. and Deny, J.,1959, Dirichlet spaces, *Proc. Nat. Acad. Sci.*,208-215.

Curnier, A. and Alart, P.,1988, A generalized newton method for contact problems with friction, *J. Theor. Appl. Mech.*, sp. issue:7.

Duvaut, G., 1982, Loi de frottement non locale, *J. Méc. Théo. Appl.*,N. sp.

Lions, J.L. and Magenes, E.,1968,"Problèmes aux limites non homogènes. Vol.1," Dunod, Paris.

Moreau, J.J.,1973, On unilateral contraints, friction and plasticity, *in* : "New Variational Techniques in Mathematical physics", G. Capriz and G. Stampacchia,ed.,Edizioni Cremones, Roma.

Moreau, J.J.,1988, Unilateral contact and dry friction in finite freedoms dynamics, *in* : "Non Smooth Mechanics and Applications", J.J. Moreau and P. Panagiotopoulos, ed, Spriger-Verlag, Wien.

ON SOME SOURCES OF INSTABILITY/ILL-POSEDNESS IN ELASTICITY PROBLEMS WITH COULOMB'S FRICTION

J.A.C. Martins and F.M.F. Simões

C.M.E.S.T. - Instituto Superior Técnico
Av. Rovisco Pais, 1096 Lisboa Codex, Portugal

INTRODUCTION

The onset of instability and ill-posedness in evolution problems involving rate-independent solids has been the subject of numerous studies in recent years. A possible starting point for such studies is the (linearized) rate form of the equations of linear momentum balance that govern the propagation of velocity disturbances in a body. The objective of the analyses is to detect the first instance at which, as a result of some features of the tensor of the constitutive moduli, the square of one of the three speeds of propagation of plane waves ceases to be a strictly positive real number (Mandel, 1964). This may occur at some point in the interior of a body when the smaller square of a wave-speed vanishes (a stationary wave), which defines the onset of a *divergence instability*, or when the square of a wave speed becomes complex, a phenomenon termed *flutter instability* (Rice, 1976). The onset of a divergence instability can be interpreted also in a quasi-static context by analysing the rate form of the equilibrium equations (the rate form of the dynamical equations of motion with neglected inertia). Mathematically it corresponds to the loss of ellipticity of these equations and physically it corresponds to the onset of a stationary discontinuity: a strain localization phenomenon (Rice, 1976) responsible for instance for the formation of shear bands. Localization analysis has been extensively developed in the framework of the theories of plasticity and damage (Nguyen, 1993). Flutter instability may occur for elastic-plastic solids with non-associative flow rules whose constitutive moduli lack the major symmetry (Rice, 1976; Brannon and Drugan, 1993). It is an intrinsically dynamic phenomenon (a wave mode that tends to grow in time in an oscillatory manner): it occurs while the quasistatic rate equilibrium equations are still elliptic.

Similar situations may arise in finite or semi-infinite bodies due to the incorporation of the effects of the boundary conditions: at some point of the boundary of a body the speed of propagation of surface waves (known as

Contact Mechanics, Edited by M. Raous *et al.*
Plenum Press, New York, 1995

Rayleigh waves in the elastic case) may vanish (Needleman and Ortiz, 1991; Benallal et al., 1993) at the onset of a surface divergence instability, or may become complex at the onset of a surface flutter instability (Martins et al.; Loret et al.).

When classical rate-independent constitutive equations are used in the above studies, it happens that the wave length of the body or surface modes that correspond to the detected divergence or flutter instabilities is arbitrary, in particular, it has no minimum. This results from the fact that there is no characteristic length scale in those wave propagation problems. On the other hand it has the consequence that, as soon as the instability region is entered, it is not possible to proceed (as done in finite dimensional systems) by expressing a non-trivial solution as a linear combination of a finite number of linearly independent modes: the problem becomes (linearly) *ill-posed* in the sense of the definitions proposed by Schaeffer (1990), Benallal et al. (1993) and Benallal (1992). As soon as the range of ill-posedness is entered approximate numerical solutions exhibit a pathological mesh dependence which precludes any further meaningful analysis.

In order to cope with these difficulties various methods have been proposed in the literature (Pijaudier-Cabot and Bazant, 1987; Triantafyllidis and Aifantis, 1986; Needleman, 1988) which introduce a characteristic length scale into the constitutive model and the governing equations, so as to keep the problem well-posed.

In this paper we are interested in studying the circumstances in which divergence or flutter instabilities may arise in some frictional contact problems. Because we shall consider a linearly elastic body with the usual symmetry and ellipticity restrictions on the elasticity coefficients, no instabilities of the type above may arise in the interior of the body: only the occurrence of surface instabilities needs to be studied. On the other hand, the multivalued character of the classical frictional contact laws precludes the use of linearization techniques in arbitrary circumstances: we shall restrict our study to the most relevant branch of those laws: the "strict compression - strict sliding" branch, the one where the destabilizing effects of the non-associative Coulomb friction law take place.

In the next section of this paper we first recall the results of (Martins et al.) where the steady frictional sliding of an elastic half-space in contact with a rigid flat surface was considered: for sufficiently large friction coefficient and Poisson's ratio, a surface flutter instability develops such that no minimum wave length exists for the unstable surface modes. Then we show that surface divergence instability is excluded in the same problem.

In relation with this (non-existent) surface divergence instability, we discuss in the third section a power rate (second order work) criterion which gives a necessary condition for a divergence instability. Power rate criteria of the same type have been considered in related problems in elastoplasticity and rate independent solids by Hill (1958), Petryk (1993) and others and also, in the context of frictional contact problems, by Chateau and Nguyen (1991). In related problems where symmetry of the constitutive tensor holds, Petryk (1993) shows that a criterion of this type is also a sufficient condition for divergence instability. Some authors (Bazant and Cedolin (1991), Maier (1971), etc.) use some criteria of this type as a definition of instability even in problems where such symmetries do not hold. In the present paper where non-symmetry results from Coulomb's law, we construct surface type velocity fields that satisfy the above power rate necessary condition and which of course do not correspond to a surface divergence instability: no surface divergence instability exists.

In the spirit of (Pijaudier-Cabot and Bazant, 1987; Triantafyllidis and Aifantis, 1986; Needleman, 1988), we discuss in the final section of this paper the effect of introducing a characteristic length scale in the problem, by means of the earlier proposed non-local friction law (Oden and Pires, 1983) and normal compliance law (Oden and Martins, 1985). Of special interest will be the analysis of whether or not such modifications lead to minimum wave lengths for the detected surface flutter instabilities. Here we shall restrict the analysis to the case of incompressible bodies: for additional results see (Martins et al.; Simões and Martins).

DYNAMIC STABILITY OF THE STEADY-SLIDING EQUILIBRIUM STATES

Governing equations

Let us consider a semi-infinite, homogeneous and isotropic linearly elastic body in frictional contact with a semi-infinite rigid flat body, as shown in Fig. 1. Let ($\mathbf{e}_1$, $\mathbf{e}_2$, $\mathbf{e}_3$) be a fixed orthonormal reference frame with origin on the surface of the elastic body, with the $\mathbf{e}_2$ axis directed towards the interior of the elastic body. We denote by $\mathbf{x} = (x, y, z)$ the position vector of the particles of the elastic body. The rigid body has a prescribed, non-vanishing, time independent velocity with the direction of the negative $\mathbf{e}_1$ axis.

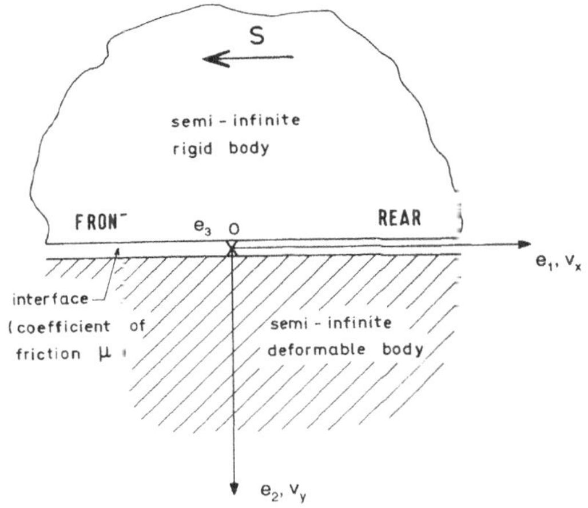

Figure 1. A semi-infinite elastic body in frictional contact with a moving semi-infinite rigid body.

In the analysis we shall be concerned with the propagation of small plane perturbations (orthogonal to $\mathbf{e}_3$) relative to a given steady-sliding equilibrium state of the elastic body: the steady-sliding slip velocity is different from zero and the corresponding normal stress on the interface $y = 0$ is assumed to be strictly negative. We further assume that for all times t considered in the analysis the displacements and velocities in the elastic body relative to its steady-sliding state are sufficiently small that: 1) the slip velocity on the interface does not vanish and has always the orientation of the steady-sliding one; 2) the amplitudes of the normal stress oscillations on the interface do not exceed the values of the corresponding steady-sliding normal stress, so that detachment never occurs.

The problem that we wish to solve is thus to find a plane velocity field $\mathbf{v}(x,y,t) = (v_x(x, y, t), v_y(x, y, t))$ in the half-plane $y \geq 0$ such that, for all times t, the rate form of the equations of linear elastodynamics

$$\frac{\partial^2 v_x}{\partial t^2} = \frac{G}{\rho}\left(\frac{\partial^2 v_x}{\partial x^2} + \frac{\partial^2 v_x}{\partial y^2}\right) + \frac{\lambda + G}{\rho}\left(\frac{\partial^2 v_x}{\partial x^2} + \frac{\partial^2 v_y}{\partial x \partial y}\right)$$

$$\frac{\partial^2 v_y}{\partial t^2} = \frac{G}{\rho}\left(\frac{\partial^2 v_y}{\partial x^2} + \frac{\partial^2 v_y}{\partial y^2}\right) + \frac{\lambda + G}{\rho}\left(\frac{\partial^2 v_x}{\partial x \partial y} + \frac{\partial^2 v_y}{\partial y^2}\right) \tag{1}$$

holds for $y < 0$ and the boundary conditions

$$v_y(x,0,t) = 0 \tag{2}$$

$$\dot{\sigma}_{yx}(x,0,t) + \mu\, \dot{\sigma}_{yy}(x,0,t) = 0 \tag{3}$$

hold on $y = 0$.

Surface solutions

In the present analysis we seek surface solutions to (1, 2, 3) in the form

$$\mathbf{v}(x, y, t) = \mathbf{V}(k\, y) \exp\,[i\, k\, (x - ct)] = (V_x(k\, y), V_y(k\, y)) \exp\,[i\, k\, (x - ct)] \tag{4}$$

where k is a positive real number which represents the angular frequency of the solutions along the $\mathbf{e}_1$ axis, i. e., $L = (2\,\pi\, /\, k)$ is the period of the solutions along that axis; $V_x(k\, y)$ and $V_y(k\, y)$ are sufficiently smooth functions defined on $[0, +\infty[$ with an appropriate decay as $y \to +\infty$, namely

$$\lim_{y \to +\infty} V_x(k\, y) = \lim_{y \to +\infty} V_y(k\, y) = 0\,; \tag{5}$$

c is a complex number such that Re(c) represents the speed at which the solutions (4) propagate along the $\mathbf{e}_1$ axis and k Im(c) represents the rate of exponential growth or decay of that solution in time.

Inserting (4) in (1, 2, 3), we obtain a system of linear and homogeneous second-order ordinary differential equations

$$\mathbf{A}\ \mathbf{V}'' + \mathbf{B}\ \mathbf{V}' + (\mathbf{C} - X^2\, \mathbf{I})\, \mathbf{V} = 0, \tag{6}$$

together with the boundary conditions at $y = 0$

$$V_y(0) = 0\ , \tag{7}$$

$$V_x'(0) + i\, V_y(0) + \mu\, [(\Lambda{+}2)\, V_y'(0) + i\, \Lambda\, V_x(0)] = 0\,. \tag{8}$$

In (6) **A**, **B** and **C** are 2x2 matrices defined componentwise by

$$A_{11} = -\,1,\quad A_{22} = -\,(\Lambda + 2),\quad A_{12} = A_{21} = 0,$$

$$B_{11} = B_{22} = 0, \quad B_{12} = B_{21} = -i\,(\Lambda + 1)\,,$$
$$C_{11} = \Lambda + 2, \quad C_{22} = 1, \quad C_{12} = C_{21} = 0\,;$$

$\Lambda = (\lambda / G)$, **I** is the 2×2 unit matrix, a superposed prime implies a derivative with respect to ky and

$$X = c / c_T\,. \tag{9}$$

Note that (6) is the dynamic counterpart of equation (IV.2.13) in (Benallal et al., 1993).

Solutions to system (6) are sought in the form of linear combinations of functions

$$\mathbf{V}(k\,y) = (V_x(k\,y), V_y(k\,y)) = (A, B)\exp(-\,b\,k\,y), \tag{10}$$

where A and B are undetermined amplitude constants and b is a complex number that satisfies

$$\mathrm{Re}(b) > 0\,. \tag{11}$$

Thus we allow the solution to oscillate along the $\mathbf{e}_2$ axis but, in accordance with (5), we require its amplitude to decay exponentially along this axis.

Inserting (10) in (6) and looking for non-trivial solutions we get the possible values for b

$$b^{(1)} = \sqrt{1 - \tau X^2}\,, \qquad b^{(2)} = \sqrt{1 - X^2}\,, \tag{12}$$

where

$$\tau = (c_T / c_L)^2 = \frac{1 - 2\nu}{2\,(1-\nu)} = \frac{1}{\Lambda + 2}\,, \tag{13}$$

so that, when $b^{(1)} \neq b^{(2)}$, the general form of the non-trivial surface modes is

$$\begin{aligned} V_x\,(ky) &= \left(A^{(1)}\, e^{-b^{(1)}ky} + A^{(2)}\, e^{-b^{(2)}ky} \right) \\ V_y\,(ky) &= \left(A^{(1)}\, i\, b^{(1)}\, e^{-b^{(1)}ky} + A^{(2)} \frac{i}{b^{(2)}}\, e^{-b^{(2)}ky} \right). \end{aligned} \tag{14}$$

Imposing now that (14) satisfies the boundary conditions (7, 8) and looking again for non-trivial solutions, the following characteristic equation is obtained

$$\mu\,(2 - 2\sqrt{1 - \tau X^2}\,\sqrt{1 - X^2} - X^2) = i\,X^2 \sqrt{1 - \tau X^2}\,. \tag{15}$$

Note that the values of k (or equivalently the wave lengths L) remain undetermined (there is no characteristic length in the problem). The relevance of this, particularly the absence of a minimum positive value for L, may be understood by observing that the exponentially growing time behavior of the surface solutions when Im(c) > 0 is unboundedly magnified for vanishing small values of the arbitrary wave lengths L:

$$\exp [k \, \mathrm{Im}\,(c)\, t] = \exp [2 \pi \frac{\mathrm{Im}(c)}{L} t].$$

Surface flutter instability

In (Martins et al.) the analysis of the nature of the solutions to (15) was performed. It was shown that if the pair of parameters (ν, μ) is in the interior of the shaded region represented in Fig. 2 or on the vertical line ($\nu = 0.5$, $\mu > 1$) then equation (15) has complex roots and the solutions that propagate towards the rear (Re(c) > 0) will exponentially grow with time (Im(c) > 0). Thus steady sliding is dynamically unstable. This *flutter instability* is due to the intrinsic non-symmetry of Coulomb's friction law and it manifests itself by growing surface oscillations which propagate from front to rear and which in a short time lead to situations of local loss of contact or stick.

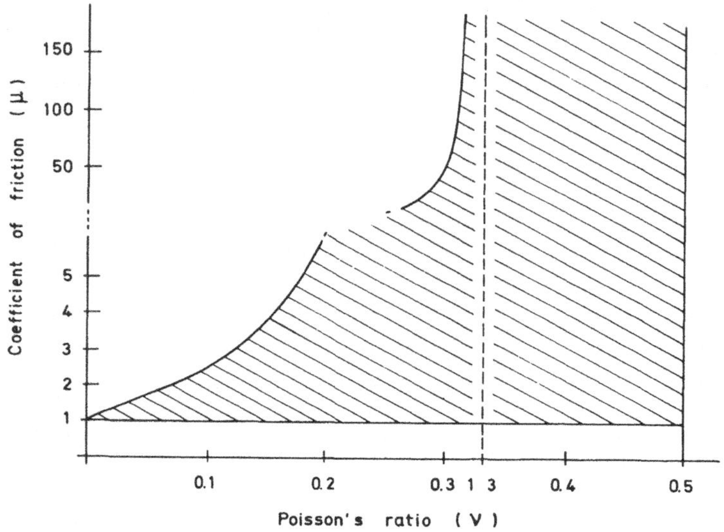

Figure 2. The region in the parameter plane (ν,μ) where equation (15) has complex roots.

Surface divergence instability

It also follows from the study in (Martins et al.) that *no solutions to* (5, 6, 7, 8) *exist with a strictly negative real value of* X^2. The possibility of finding a solution to (5, 6, 7, 8) with X = 0, i. e. a stationary surface wave, needs additional discussion. Since in this case $b^{(1)} = b^{(2)}$ [cf. (12)] the general solution to (6) is not a linear combination of modes of the form (10). Instead it can be shown to have the form

$$V_x(k\,y) = \left[A^{(1)} + A^{(2)}\left(-\frac{1}{\Lambda+1} + k\,y\right)\right] \exp(-k\,y)$$

$$V_y(k\,y) = i\left[A^{(1)} + A^{(2)}\left(\frac{\Lambda+2}{\Lambda+1} + k\,y\right)\right] \exp(-k\,y). \tag{16}$$

Then from the boundary conditions (7,8) we get $A^{(1)} = -(\Lambda + 2)(\Lambda + 1)^{-1} A^{(2)}$ and

$$(\Lambda + 2) + i\,\mu = 0. \tag{17}$$

Since (17) cannot hold, *the occurrence of stationary surface waves is excluded.*

POWER RATE NECESSARY CONDITION FOR DIVERGENCE INSTABILITY

Doing the (complex) inner product of the vector equation (6) with the vector function $\mathbf{V}(ky)$, integrating the resulting scalar equation between 0 and $+\infty$ (some terms are integrated by parts) and taking into account the boundary conditions (7,8) we get

$$P(\mathbf{V}, \mathbf{V}) = X^2 (\mathbf{V}, \mathbf{V}) \tag{18}$$

where

$$P(\mathbf{V}, \mathbf{V}) = \int_0^{+\infty} \left\{ \begin{aligned} &(\Lambda + 2)\left(V_x \bar{V}_x + V_y' \bar{V}_y'\right) + \left(\bar{V}_x' V_x' + V_y \bar{V}_y\right) \\ &+ i\left[\Lambda\left(V_x \bar{V}_y' - \bar{V}_x V_y'\right) + \left(V_y \bar{V}_x' - \bar{V}_y V_x'\right)\right] \end{aligned} \right\} d(ky)$$

$$- \mu\left[(\Lambda + 2)\, V_y'(0)\, \bar{V}_x(0) + i\,\Lambda\, V_x(0)\, \bar{V}_x(0)\right] \tag{19}$$

$$(\mathbf{V}, \mathbf{V}) = \int_0^{+\infty} (V_x \bar{V}_x + V_y \bar{V}_y)\; d(ky)\,. \tag{20}$$

A necessary condition for the occurrence of a surface divergence instability (X^2 real and ≤ 0) is thus

$$P(\mathbf{V}, \mathbf{V}) \text{ real and} \leq 0, \quad \text{for some } \mathbf{V} \neq \mathbf{0} \text{ such that } Vy(0) = 0, \tag{21}$$

or, equivalently,

$$\mathrm{Re}\,[P(\mathbf{V}, \mathbf{V})] \leq 0 \text{ and } \mathrm{Im}\,[P(\mathbf{V}, \mathbf{V})] = 0, \text{ for some } \mathbf{V} \neq \mathbf{0} \text{ such that } Vy(0) = 0. \tag{22}$$

Note that, when the above complex notation is not used and the physical (real) velocity fields for the present case of $\mathrm{Re}(c) = \mathrm{Re}(X) = 0$ are used, we have

$$\mathrm{Re}\,[\mathbf{v}(x, y, t)] = \mathbf{W}(x, y) \exp\,[k\, \mathrm{Im}(c)\, t] \tag{23}$$

where

$$W(x, y) = [\mathrm{Re}(V(ky)) \cos(kx) - \mathrm{Im}(V(ky)) \sin(kx)] . \tag{24}$$

Then the inequality in (22) holds iff the following (real) power rate statement holds:

$$P_L(\mathbf{W}, \mathbf{W}) = \int_0^L \int_0^{+\infty} \dot{\sigma}(\mathbf{W}) : \dot{\varepsilon}(\mathbf{W}) \; dy\, dx - \mu \int_0^L \dot{\sigma}_{yy}(\mathbf{W})\, W_x \, dx \leq 0, \tag{25}$$

because $P_L(\mathbf{W}, \mathbf{W}) = \pi G \,\mathrm{Re}\,[P(\mathbf{V}, \mathbf{V})]$. The (real) power rate (25) is the total power rate of the elastic forces and the friction stresses for a slice $\bar{\Omega}$ of the elastic body with dimension along x equal to the period L of the velocity field in that direction: $\bar{\Omega} = [0, L] \times [0, +\infty[$.

Here we illustrate with an example the fact that the necessary condition (22) may be easily satisfied by functions that do not solve the eigenproblem (6, 7, 8) with X^2 real and ≤ 0. We consider a function $\mathbf{V}$ given componentwise by

$$\begin{aligned} V_x(ky) &= C\,(\alpha\, e^{-b^{(1)}ky} + \beta\, e^{-b^{(2)}ky}) \\ V_y(ky) &= C\,(e^{-b^{(1)}ky} - e^{-b^{(2)}ky}), \end{aligned} \tag{26}$$

where C is an undetermined amplitude constant. For the material data $\Lambda = 1$ and $\mu = 0.46767$, the values $b^{(1)} = 1.6841 + 3\,i$, $b^{(2)} = 5 - 0.5\,i$, $\alpha = 1.06572 + 0.338055\,i$ and $\beta = 0.377622 - 1.51328\,i$ provide a surface velocity field of the form (26) such that (22) is satisfied with, actually, $\mathrm{Re}\,[P(\mathbf{V}, \mathbf{V})] = 0$. This velocity field of course is not a solution of (5 - 8) with $X = 0$.

This example serves to stress and further illustrate earlier statements (and finite dimensional friction examples) by Mandel (1964) that *power rate criteria of the type* (22, 25) *are not sufficient for dynamic (divergence) instability.*

FRICTION CONTACT MODELS WITH INTRINSIC LENGTH SCALES

We discuss the effect of introducing an intrinsic length scale in the problem by means of a non-local friction law (Oden and Pires, 1983) or a normal compliance law (Oden and Martins, 1985).

Non-local friction law

In this case the boundary condition (3) becomes

$$\dot{\sigma}_{yx}(x,0,t) + \mu\, S(\dot{\sigma}_{yy})(x,0,t) = 0, \tag{27}$$

where $S(\dot{\sigma}_{yy})(x,0,t)$ is the mollified normal stress rate which represents a weighted average of the normal stress rates in a certain neighbourhood of the point (x,0) and has the typical form

$$S(\dot{\sigma}_{yy})(x,0,t) = \int_{-\infty}^{+\infty} w(s)\, \dot{\sigma}_{yy}\, (x + s, 0, t)\, ds; \tag{28}$$

w(s) is a normalized weight function. Calculations are made with the normalized gaussian function

$$w(s) = \frac{1}{L_C\sqrt{2\,\pi}} \exp(-\frac{s^2}{2\,L_C^2}), \tag{29}$$

where L_C is a characteristic length of the interface.

Imposing that (14) satisfies the boundary conditions (2) and (27) we obtain the following characteristic equation

$$\mu\,(2 - 2\sqrt{1 - \tau\, X^2}\sqrt{1 - X^2} - X^2) \exp(-\frac{1}{2\,\alpha^2}) = i\, X^2 \sqrt{1 - \tau\, X^2}\ , \tag{30}$$

where

$$\alpha = \frac{1}{k\, L_C}. \tag{31}$$

Note that equation (15) is recovered when $\alpha \to \infty$ in equation (30).

In the case of an incompressible elastic body (ν = 0.5, $c_L = +\infty$) equation (30) becomes

$$\mu\,(2 - 2\sqrt{1 - X^2} - X^2) \exp(-\frac{1}{2\,\alpha^2}) - i\, X^2 = 0. \tag{32}$$

In addition to the trivial solution X = 0, we obtain by squaring this equation

$$X^2 = \frac{-4\, i\, \mu \exp(-\frac{1}{2\,\alpha^2})}{(1 - i\,\mu \exp(-\frac{1}{2\,\alpha^2}))^2}\ . \tag{33}$$

This solution does solve the original equation (30) and in order to satisfy restriction (11) we must have

$$\mu > \exp(\frac{1}{2\,\alpha^2}), \tag{34}$$

from which we obtain the infimum wave length of the flutter instability modes (for a given friction coefficient $\mu > 1$ and a given characteristic length $L_C > 0$)

$$L_{inf} = \frac{2\,\pi}{k} = \pi \sqrt{\frac{2}{\ln \mu}}\, L_C\ . \tag{35}$$

Thus *the characteristic length introduced by the non-local friction law prevents small wavelength modes of flutter instability.* The surface flutter instability region in the parameter plane (μ, α) is represented in Fig.3 .

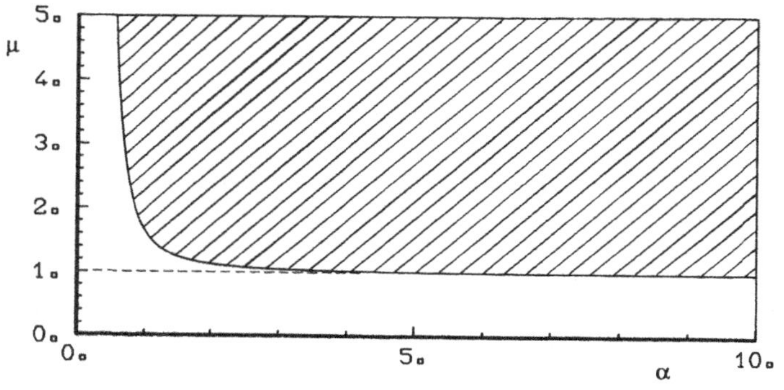

Figure 3. Surface flutter instability region in the parameter plane (μ,α) in the case of an incompressible elastic body.

Normal compliance law

Here we take into account the normal deformability of the interface between the elastic half-space and the flat surface. We denote by K_n the (linearized) normal stiffness of the interface at the steady sliding state. Thus, the boundary conditions are now (3) and

$$\dot{\sigma}_{yy}(x,0,t) = K_n \, v_y(x,0,t). \tag{36}$$

Imposing that the surface solutions (14) satisfy the boundary conditions (3, 36) we obtain the following characteristic equation

$$\sqrt{1 - \tau X^2}\ [2\sqrt{1 - X^2}\,(2 - \mu\,\alpha\,i) + \alpha\,X^2] = (2 - X^2)\,(2 - X^2 - \mu\,\alpha\,i), \tag{37}$$

where now

$$\alpha = \frac{1}{k\,L_C} \quad ; \quad L_C = \frac{G}{K_n} \ . \tag{38}$$

In the limit as $\alpha \to \infty$ in (37), equation (15) is recovered. In the case of an incompressible elastic body equation (37) becomes

$$2\sqrt{1 - X^2}\,(2 - \mu\,\alpha\,i) + \alpha\,X^2 = (2 - X^2)\,(2 - X^2 - \mu\,\alpha\,i) \tag{39}$$

and the search of the nature of its solutions leads to two transition lines in the parameter plane (μ,α):

$$\text{a)}\quad \mu^{(a)} = \frac{-2\sqrt{2}\sqrt{\alpha+4+\sqrt{(\alpha+4)^2-4(\alpha-1)}}}{\alpha\left(1+\frac{-(\alpha+4)-\sqrt{(\alpha+4)^2-4(\alpha-1)}}{2}\right)} \tag{40}$$

$$\text{b)}\quad \mu^{(b)} = \frac{2\sqrt{2}\sqrt{\alpha+4-\sqrt{(\alpha+4)^2-4(\alpha-1)}}}{\alpha\left(1+\frac{-(\alpha+4)+\sqrt{(\alpha+4)^2-4(\alpha-1)}}{2}\right)} . \tag{41}$$

The flutter region is defined by

$$\begin{cases} \mu > 0 & 0 < \alpha \leq 1 \\ \mu > \mu^{(a)} \vee \mu > \mu^{(b)} & 1 \leq \alpha \end{cases} . \tag{42}$$

In Fig.4 we represent the surface flutter instability region in the parameter plane (μ,α). We can see that, contrarily to what happens when a non-local friction law is used, the minimum value of the friction coefficient required for the occurrence of

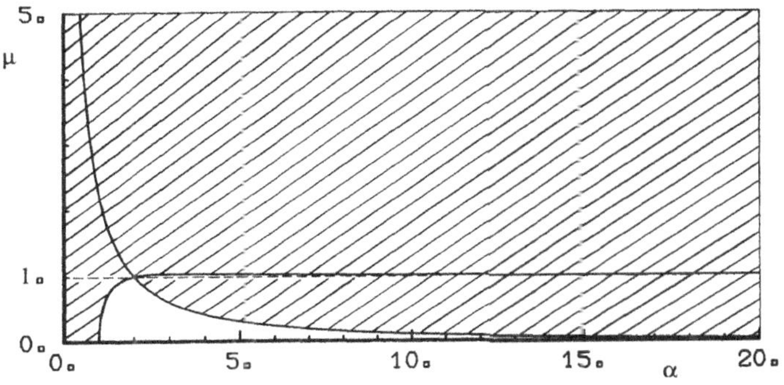

Figure 4. Surface flutter instability region in the parameter plane (μ,α) in the case of an incompressible elastic body.

surface flutter instability does not increase when $\alpha \to 0$. For instance, for $\alpha < 1$, surface flutter instability occurs for any positive friction coefficient. Thus *the characteristic length introduced by the normal compliance law does not prevent small wavelength modes of flutter instability.*

REFERENCES

Bazant, Z. and Cedolin, L., 1991, "Stability of Structures", Oxford U. Press.

Benallal, A., 1992, Ill-posedness and localization in solid structures, in: "Computational Plasticity", D.R.J. Owen et al , ed., Pineridge Press, Swansea.

Benallal, A., Billardon, R. and Geymonat, G., 1993, Bifurcation and localization in rate-independent materials. Some general considerations, in: "Bifurcation and Stability of Dissipative Systems", Q.S. Nguyen, ed., International Centre for Mechanical Sciences, Courses and Lectures N° 327, Springer-Verlag, Wien-New York.

Brannon, R.M. and Drugan, W.J., 1993, Influence of non-classical elastic-plastic constitutive features on shock wave existence and spectral solutions, *J. Mech. Phys. Solids* 41:297.

Chateau, X. and Nguyen, Q.S., 1991, Buckling of elastic structures in unilateral contact with or without friction, *Eur. J. Mech., A Solids* 10:71.

Hill, R., 1958, A general theory of uniqueness and stability in elastic-plastic solids, *J. Mech. Phys. Solids* 6:236.

Loret, B., Martins, J.A.C. and Simões, F.M.F., Surface boundary conditions trigger flutter instability in non-associative elastic-plastic solids. To appear in *Int. J. Solids Structures*.

Maier, G., 1971, Incremental plastic analysis in the presence of large displacements and physical instabilizing effects, *Int J. Solids Structures* 7:345.

Mandel, J., 1964, Conditions de stabilité et postulate de Drucker, in: "Rheology and Soil Mechanics", G. Kravtchenko and P. Sirieys, eds., IUTAM Symposium at Grenoble.

Martins, J.A.C., Faria, L.O. and Guimarães, J., Dynamic surface solutions in linear elasticity and viscoelasticity with frictional boundary conditions. To appear in *ASME J. Vibration and Acoustics*.

Needleman, A., 1988, Material rate dependence and mesh sensitivity in localization problems, *Comp. Meth. Appl. Mech. Eng.* 67:69.

Needleman, A. and Ortiz, M., 1991, Effect of boundaries and interfaces on shear-band localization, *Int. J. Solids Structures* 28:859.

Nguyen, Q.S., ed., 1993, "Bifurcation and Stability of Dissipative Systems", International Centre for Mechanical Sciences, Courses and Lectures N° 327, Springer-Verlag, Wien-New York.

Oden, J.T. and Pires, E.B., 1983, Nonlocal and nonlinear friction laws and variational principles for contact problems in elasticity, *J. Appl. Mech.* 50:67.

Oden, J.T. and Martins, J.A.C., 1985, Models and computational methods for dynamic friction phenomena, *Comp. Meth. Appl. Mech. Eng.* 52:527.

Petryk, H., 1993, General theory of bifurcation and stability in time-independent plasticity, in: "Bifurcation and Stability of Dissipative Systems", Q.S. Nguyen, ed., International Centre for Mechanical Sciences, Courses and Lectures N° 327, Springer-Verlag, Wien-New York.

Pijaudier-Cabot, G. and Bazant, Z.P., 1987, Nonlocal damage theory, *ASCE J. Eng. Mech.* 112:1512.

Rice, J.R., 1976, The localization of plastic deformation, in: "Theoretical and Applied Mechanics", W.T. Koiter, ed.,14th IUTAM Congress, Delft, North-Holland, Amsterdam.

Schaeffer, D. G., 1990, Instability and ill-posedness in the deformation of granular materials, *Int. J. Num. Anal. Meth. Geomech.* 14:253.

Simões, F.M.F. and Martins, J.A.C., Instability and ill-posedness in some friction problems. In preparation.

Triantafyllidis, N. and Aifantis, E.C., 1986, A gradient approach to localization of deformation - I. Hyperelastic materials, *J. Elasticity* 16:225.

FRICTION AND INSTABILITIES: STRESS WAVES IN A SLIDING CONTACT

S. Barbarin,[1] J.A.C. Martins,[2] and M. Raous[1]

[1]Laboratoire de Mécanique et d'Acoustique
13402 Marseille Cedex 20, France
[2]Instituto Superior Tecnico
1096 Lisboa Codex, Portugal

INTRODUCTION

In their accurate experimental works, Villechaise, Mouwakeh, Zeghloul (1992), have shown the occurrence of isolated fast stress waves in the contact area of a deformable block sliding on a rigid plane. Each wave is coupled with a jump in the total tangential force.

A numerical model using a variable friction coefficient gives a good description of the overall stresses in the solid but does not describe the repetition of the jumps (Raous and Topin, 1989).

With the intention of characterizing this phenomenon, we present here a first approach of stability for contact and friction by introducing a discrete energy criterion (Hill, 1958; Chateau and Nguyen, 1991; Martins et al., 1994) to characterize any possible initiation of a jump. For this purpose, we have to compute the smallest eigenvalue of a large symmetrized matrix constructed on the basis of the contact condition.

THE QUASISTATIC SOLUTION

We compute the quasistatic solution of an elastic problem with unilateral contact and Coulomb friction through an incremental formulation. The LEMKE mathematical programming method is used as solver (Raous et al., 1988). The contact forces and the isochromes are given for one step in Figure 1. Experimental data are the ones of Zeghloul (1992). The friction coefficient is $\mu = 1.1$. The progression of a single stress wave is slightly observed but the phenomenon is not repetitive as in the experiment. A variable friction coefficient (static and dynamic values) gives also this kind of results.

CRITERIA FOR DIVERGENCE DYNAMIC INSTABILITY

At relevant times along the above incremental process, dynamic stability may be analysed according to the propositions given next.

Let $\mathbf{M}$ be the symmetric positive definite mass matrix and $\mathbf{K}$ be the symmetric positive semi-definite stiffness matrix. Let also $\mathbf{U}$ and $\mathbf{R}$ be the equilibrium displacement and reaction vectors at some time τ. The subscripts d, f, z and s are respectively related to the nodes that are currently locked (prescribed displacement nodes and stick contact nodes), free (free nodes and currently separate contact nodes), in contact with zero reaction, or in contact with non-vanishing reaction on the friction cone (possible slip). The subscripts N and T denote normal and tangential components (for contact candidate nodes). A subscript i denotes an arbitrary node of some of the above sets. We denote by $K_{\dot{u}}$ the set of currently admissible (right) velocities v : $v_{di} = 0$; $v_{zNi} \leq 0$; $v_{sNi} = 0$ and $v_{sTi} = -|v_{sTi}|\ sign(R_{sTi})$. We denote by $K_{\dot{r}}(v)$ the set of currently admissible (right) reaction rates w for some given velocities v : $w_{fi} = 0$; $w_{zNi} \leq 0$; $w_{zNi}\, v_{zNi} = 0$; $|w_{zTi}| \leq -\mu\, w_{zNi}$ and $w_{zTi} = \mu\, w_{zNi}\, sign(v_{zTi})$, if $v_{zTi} \neq 0$;
$w_{sTi}\, sign(R_{sTi}) + \mu\, w_{sNi} \leq 0$ and $[w_{sTi}\, sign(R_{sTi}) + \mu\, w_{sNi}]\ |v_{sTi}| = 0$.

Proposition 1 :

If $\exists\lambda \geq 0$ and $\mathbf{V} \neq 0$ such that $\mathbf{V} \in K_{\dot{u}}$ and $(\lambda^2\mathbf{M} + \mathbf{K})\mathbf{V} \in K_{\dot{r}}(\mathbf{V})$ then

$$\mathbf{u}(t) = \mathbf{U} + \mathbf{V}\begin{cases} \exp[\lambda(t-\tau)]\ , & if\ \lambda > 0 \\ [1 + (t-\tau)]\ , & if\ \lambda = 0 \end{cases}$$

$$\mathbf{r}(t) = \mathbf{R} + (\lambda^2\mathbf{M} + \mathbf{K})\mathbf{V}\begin{cases} \exp[\lambda(t-\tau)]\ , & if\ \lambda > 0 \\ [1 + (t-\tau)]\ , & if\ \lambda = 0 \end{cases}$$

satisfy the dynamic equations $\mathbf{M\ddot{u}}(t) + \mathbf{Ku}(t) = \mathbf{f}(\tau) + \mathbf{r}(t)$
together with the classical unilateral frictional contact conditions at all $t \in [\tau, \tau+\varepsilon[$, with $\varepsilon > 0$ sufficiently small and with the initial conditions at $t = \tau$:

$$\mathbf{u}(\tau) = \mathbf{U} + \mathbf{V}\ ,\ \ \mathbf{r}(\tau) = \mathbf{R} + (\lambda^2\mathbf{M} + \mathbf{K})\mathbf{V}\ ,\ \ \dot{\mathbf{u}}(\tau) = \begin{cases} \lambda\mathbf{V}\ , & if\ \lambda > 0 \\ \mathbf{V}\ , & if\ \lambda = 0 \end{cases}.$$

for arbitrary, sufficiently small, $|\mathbf{V}|$. Hence the equilibrium state corresponding to $\mathbf{U}$ and $\mathbf{R}$ is dynamically unstable (a divergence instability).

With the definitions of the bilinear forms

$$a^*(\mathbf{v},\mathbf{v}) = \mathbf{Kv.v} + \sum_i\ [\mu\, sign(\mathbf{R}_{sTi})\,\mathbf{Kv}]_{sNi}\,\mathbf{v}_{sTi}$$

$$m^*(\mathbf{v},\mathbf{v}) = \mathbf{Mv.v} + \sum_i\ [\mu\, sign(\mathbf{R}_{sTi})\,\mathbf{Mv}]_{sNi}\,\mathbf{v}_{sTi}$$

for all $\mathbf{v} \in K_{\dot{u}}$, we also have the result :

Proposition 2 :

If $m^*(\mathbf{v},\mathbf{v}) \geq 0$ and $a^*(\mathbf{v},\mathbf{v}) > 0$ for all $\mathbf{v} \in K_{\dot{u}}$, then a dynamic instability by divergence of the type in Proposition 1 cannot occur.

In typical situations along the numerical incremental solution no 'z' nodes exist, i.e., no node currently in contact has a zero reaction. In these circumstances the assumptions of Proposition 1 are satisfied if for some $\lambda \geq 0$ we have :

$$(\lambda^2 \mathbf{M}^* + \mathbf{K}^*)\mathbf{V}^* = 0\ ,\ 0 \neq \mathbf{V}^* = \left\{ \begin{array}{c} \mathbf{V}_f \\ \mathbf{V}_{sT} \end{array} \right\}\ ,\ \mathbf{V}_{sTi} = -\ |\mathbf{V}_{sTi}|\ sign(\mathbf{R}_{sTi})\ ,$$

$$\text{where} \qquad \mathbf{K}^* = \left[\begin{array}{cc} K_{f,f} & K_{f,sT} \\ K^*_{sT,f} & K^*_{sT,sT} \end{array} \right] \quad , \quad \mathbf{M}^* = \left[\begin{array}{cc} M_{f,f} & M_{f,sT} \\ M^*_{sT,f} & M^*_{sT,sT} \end{array} \right]$$

$\mathbf{K}^*_{sTi,p} = \mathbf{K}_{sTi,p} + \mu\, sign(\mathbf{R}_{sTi})\, \mathbf{K}_{sNi,p}$, with $p = f$ or sT. With these definitions the bilinear forms of Proposition 2 may be written as follows :

$$a^*(\mathbf{v}, \mathbf{v}) = \mathbf{K}^*_S \mathbf{v}^*\ \mathbf{v}^* \qquad \text{where} \qquad \mathbf{K}^*_S = \tfrac{1}{2}\left(\mathbf{K}^* + \mathbf{K}^{*T}\right)$$

$$m^*(\mathbf{v}, \mathbf{v}) = \mathbf{M}^*_S \mathbf{v}^*.\mathbf{v}^* \qquad \text{where} \qquad \mathbf{M}^*_S = \tfrac{1}{2}\left(\mathbf{M}^* + \mathbf{M}^{*T}\right)$$

$$\text{with } \mathbf{v}^* = \left\{ \begin{array}{c} \mathbf{v}_f \\ \mathbf{v}_{sT} \end{array} \right\}\ ,\ \mathbf{v}_{sTi} = -\ |\mathbf{v}_{sTi}|\, sign(\mathbf{R}_{sTi})\ .$$

If, in addition, the mass matrix $\mathbf{M}$ is diagonal (hence $\mathbf{M}^*_S = \mathbf{M}^*$ diagonal and $m^*(\mathbf{v}, \mathbf{v}) > 0$ for the admissible $\mathbf{v}$) then the single condition $a^*(\mathbf{v}, \mathbf{v}) > 0$ is obtained.

In this case the stability criterion of Chateau and Nguyen (1991) would give the same condition $a^*(\mathbf{v}, \mathbf{v}) > 0$. This condition is also a sufficient condition for the impossibility of initiating instantaneous jumps of the type discussed in Martins et al. (1994).

NUMERICAL SOLUTION AND RESULTS

Hence we can apply 3 numerical methods which gave us the same results :

1) computation of the smallest eigenvalue of $\mathbf{K}^*_S$ using the Power Method one or two times : we first compute the largest eigenvalue in modulus λ_{max} of $\mathbf{K}^*_S$; if λ_{max} is negative we already have the smallest eigenvalue of $\mathbf{K}^*_S$ else we compute the largest eigenvalue in modulus of $(\mathbf{K}^*_S - \lambda_{max} I)$ corresponding to the smallest eigenvalue of $\mathbf{K}^*_S$;

2) minimization under constraint using SSOR method with projection;

3) use of Cholesky method to check the positive definite character of the matrix.

When using the first method, the eigenvector gives the initial direction of the possible jump (see figure 1).

Table 1. Results for two different values of the friction coefficient. STI, SLI, SEP represent respectively the number of stick, sliding or separate nodes.

$\mu = 0.6$					$\mu = 1.1$				
U_T	STI	SLI	SEP	Eigenv.	U_T	STI	SLI	SEP	Eigenv.
0.600	40	1	0	0.255	0.610	40	1	0	0.255
0.667	38	1	2	0.250	0.730	38	1	2	0.247
0.677	33	6	2	0.183	0.830	36	1	4	0.239
0.680	32	7	2	-0.031	0.900	35	2	4	-1.267
0.800	11	27	3	-1.863	1.000	33	3	5	-4.409
1.225	0	37	4	-2.444	5.000	0	33	8	-13.10

In Table 1, both the evolution of the contact condition and of the smallest eigenvalue of $\mathbf{K}^*_S$ are given when the prescribed tangential displacement increases (normal displacement is kept at $U^0_N = -0.36$mm).

A possible jump is obtained for $U_T = 0.68$mm when $\mu = 0.6$, and for $U_T = 0.9$mm when $\mu = 1.1$. The predicted value on forces is 40% lower than the one giving the first jump in the experiment ($\mu = 1.1$).

The directions of the possible jump are given by the eigenvectors (see figure 1). The shape is analogous to the stress wave of the experiment.

Various computations show that a negative eigenvalue of $\mathbf{K}_S^*$ does not occur for small friction ($\mu < 0.3$) and occurs very early when friction is large. In table 1, it can be observed ($\mu = 1.1$) that the negative eigenvalue is obtained very early (for $U_T = 0.9$mm, 35 of the 41 contact nodes are still stuck).

Figure 1. Contact condition, isochromes and eigenvector for $U_T = -1.25$mm ($\mu = 1.1$).

CONCLUSION

The results obtained by analysing the sign of $a^*(\mathbf{v}, \mathbf{v})$ predict a possible initiation of a jump earlier than the first jump observed in the experiment. The direction of the possible jump is in good agreement with the photoelasticity measurements of Villechaise and Zeghloul.

Since $a^*(\mathbf{v}, \mathbf{v}) \leq 0$ for some $0 \neq \mathbf{v} \in K_{\hat{u}}$ is just a necessary condition for a dynamic instability by divergence (Prop.2) , we now study numerically the sufficient conditions for divergence instabilities (Prop.1) and also the conditions for flutter instabilities.

So far, for the given loading program and in the absence of damping, divergence instability is never obtained and complex eigenvalues of $\mathbf{K}^*$ (flutter instability) with small real part appear a little earlier than the negative eigenvalues of $\mathbf{K}_S^*$. Damping is now under consideration. It will allow a clearer view of flutter instabilities.

Aknowledgements. This work is part of a bilateral project between France (C.N.R.S.) and Portugal (J.N.I.C.T.) .

REFERENCES

Chateau, X., and Nguyen, Q.S., 1991, Buckling of elastic structures in unilateral contact with or without friction, *Eur. J. Mech., A/Solids*,10(1), 71-89

Hill, R., 1958, A general theory of uniqueness and stability in elastic-plastic solids, *Journal of Mechanics Physics Solids*, 6, 236-249

Martins, J.A.C., Monteiro Marques, M.D.P., and Gastaldi, F., 1994, On an example of non-existence of solution to a quasistatic frictional contact problem, *Eur. J. Mech., A/Solids*, 13(1), 113-133

Raous, M., Chabrand, P., and Lebon F., 1988, Numerical methods for frictional contact problems and applications, *Journal de Mécanique Théo. Appl.*, special issue, supplement n°1 to vol.7

Raous, M., and Topin, S., 1989, Ondes de contraintes dans un contact glissant, 9ᵉ Congrès Français de Mécanique, Metz, 5-8 Sept.

Zeghloul, T., 1992, Etude des phénomènes d'adhérences et de glissements dans un contact entre solides: approche expérimentale et modélisation, *Thesis*, Poitiers, 13 Nov.

UNILATERAL CONTACT WITH BOUNCES

S. Durand

Laboratoire de Mécanique des Solides
Ecole Polytechnique
91128 Palaiseau Cedex
France

INTRODUCTION

The aim of the study is to take in interest in the dynamic behaviour of rigid body systems in unilateral contact with friction. The friction law is the law of Coulomb. Which are the parameters which determine the transition between a smooth motion and a motion with collisions ? In order to obtain some elements of answer, we study a simple system with two degrees of freedom. The system may be seen as a model of an elastic rod drawn on the ground. It results from our study that bounces occur if the friction coefficient is greater than a critical value.

DYNAMIC EQUATIONS

The displayed system (figure 1) consists of two bars of respective lenghts l_1 et l_2 without mass and of two springs located in points A et B. The mass m is concentrated in point C. A uniform translational velocity is applied in point A.

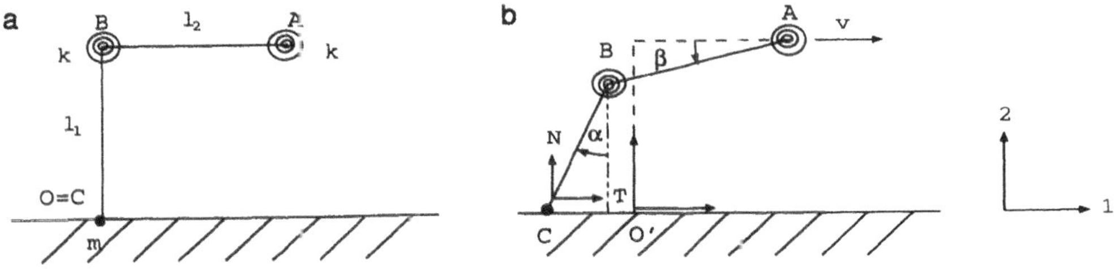

Figure 1. (a) initial state, (b) state at t>0

Contact Mechanics, Edited by M. Raous *et al.*
Plenum Press, New York, 1995

α is the angle between the axis Oy and the bar BC and β the angle between the axis Ox and the bar AB.
N and T represent respectively the normal and tangential components of the reaction at point C.
The equations of motion are :

$$ml_1l_2\sin(\beta-\alpha)\ddot{\beta}+ml_1^2\ddot{\alpha}+ml_1l_2\cos(\beta-\alpha)\dot{\beta}^2+k(\alpha-\beta) \quad (1)$$
$$+mgl_1\sin\alpha-Tl_1\cos\alpha-Nl_1\sin\alpha=0$$
$$ml_1l_2\sin(\beta-\alpha)\ddot{\alpha}+ml_2^2\ddot{\beta}-ml_1l_2\cos(\beta-\alpha)\dot{\alpha}^2+k(2\beta-\alpha) \quad (2)$$
$$-mgl_2\cos\beta-Tl_2\sin\beta+Nl_2\cos\beta=0$$

We assume that the stiffness of the springs is large enough for the angles α et β to remain small.
(u_1, u_2) is the displacement of point C in the moving frame in uniform translation with the velocity v. Linearising the equations :

$$u_1 = l_2 - l_2\cos\beta + l_1\sin\alpha$$
$$u_2 = l_1 - l_2\sin\beta - l_1\cos\alpha$$

we obtain :

$$u_1 = l_1\alpha$$
$$u_2 = -l_2\beta$$

The linearised equations of motion in terms of (u_1, u_2) are :

$$m\ddot{u}_1+\frac{k}{l_1^2}u_1+\frac{k}{l_1l_2}u_2-T=0 \quad (3)$$
$$m\ddot{u}_2+\frac{k}{l_1l_2}u_1+\frac{2k}{l_2^2}u_2+mg-N=0 \quad (4)$$

SYSTEM EVOLUTION

The system, initially at rest and in contact, is suddenly set in motion, the initial conditions are :

$$u_1(0)=0 \qquad \dot{u}_1(0)=-v$$

In the initial state, the normal force is not zero, two evolutions are possible a priori : positive sliding and sticking. If we assume that positive sliding occurs, the sliding velocity found is negative, sticking is thus the only possible evolution.

Sticking

$$\dot{u}_1(t) = -v$$
$$u_1(t) = -vt$$

Taking into account the contact condition $u_2 = 0$, we have :

$$T = -\frac{k}{l_1^2}vt$$
$$N = mg - \frac{k}{l_1l_2}vt$$

At the beginning of this phase, $|T| < \mu N$. The normal force decreases and the tangential force increases in absolute value, this phase of motion ends at $t = t_1$, when $T = \mu N$.

Sliding

Taking into account the Coulomb law $T = -\mu N$ and the contact condition $u_2 = 0$, the equation of motion is :

$$\ddot{u}_1 + \frac{k}{ml_1}\left(\frac{1}{l_1} + \frac{\mu}{l_2}\right) u_1 = -\mu g \tag{5}$$

Equation (5) admits a stationary solution :

$$u_1 = -\frac{\mu mg l_1^2 l_2}{k(l_2 + \mu l_1)}$$

We set :

$$\Omega^2 = \frac{k}{ml_1}\left(\frac{1}{l_1} + \frac{\mu}{l_2}\right)$$

The initial conditions of this phase are : $u_1(t_1) = -\frac{\mu g}{\Omega^2}$ et $\dot{u}_1(t_1) = -v$.

The solution of equation (5) is :

$$u_1(t) = -\frac{v}{\Omega}\sin\Omega(t - t_1) - \frac{\mu g}{\Omega^2}$$

$$N(t) = \frac{mgl_2}{l_2 + \mu l_1} - \frac{k}{l_1 l_2}\frac{v}{\Omega}\sin\Omega(t - t_1)$$

The normal force decreases, it becomes zero if the condition :

$$\frac{mgl_2}{l_2 + \mu l_1} - \frac{kv}{l_1 l_2 \Omega} < 0$$

is satisfied, which is equivalent to :

$$\mu > \mu_c = \frac{l_2}{l_1}\left(\frac{mg^2 l_2^2}{kv^2} - 1\right)$$

The system behaviour is studied successively for the three cases : $\mu < \mu_c$, $\mu > \mu_c$ and $\mu = \mu_c$.

- **If $\mu < \mu_c$:**

Sliding goes on. The normal force never vanishes, there is always sliding, and u_1 oscillates around the stationary solution (figure 2).

$$u_{1st} = -\frac{\mu g}{\Omega^2}$$

$$u_1(t) = -\frac{\mu g}{\Omega^2} - \frac{v}{\Omega}\sin\Omega(t - t_1)$$

- **Si $\mu > \mu_c$:**

The normal force vanishes at $t = t_2$. Sliding is not possible any more for $t > t_2$ as the normal force becomes negative. We thus suppose that separation occurs.

Separation

$$N = T = 0$$

The equations of motion are :

$$m\ddot{u}_1 + \frac{k}{l_1^2}u_1 + \frac{k}{l_1 l_2}u_2 = 0 \tag{6}$$

$$m\ddot{u}_2 + \frac{k}{l_1 l_2}u_1 + \frac{2k}{l_2^2}u_2 = -mg \tag{7}$$

The initial conditions of this phase are :

$$u_1(t_2) = -\frac{mgl_1l_2}{k}, \quad \dot{u}_1(t_2) = -\sqrt{v^2 - \frac{l_2^2 g^2}{l_1^2\Omega^2}}, \quad u_2(t_2) = 0 \quad \text{and} \quad \dot{u}_2(t_2) = 0$$

The solution of the system (6)(7) is :

$$\begin{aligned} u_1(t) &= \alpha_+ B_+ \cos(\omega_+(t-t_2)+\phi_+) + \alpha_- B_- \cos(\omega_-(t-t_2)+\phi_-) + \frac{mgl_1l_2}{k} \\ u_2(t) &= B_+ \cos(\omega_+(t-t_2)+\phi_+) + B_- \cos(\omega_-(t-t_2)+\phi_-) - \frac{mgl_2^2}{k} \end{aligned}$$

with :

$$\omega_-^2 = \frac{k}{2ml_1^2l_2^2}\left(l_2^2 + 2l_1^2 - \sqrt{l_2^4 + 4l_1^4}\right) \qquad \omega_+^2 = \frac{k}{2ml_1^2l_2^2}\left(l_2^2 + 2l_1^2 + \sqrt{l_2^4 + 4l_1^4}\right)$$

$$\alpha_+ = \frac{l_2^2 - 2l_1^2 + \sqrt{l_2^4 + 4l_1^4}}{2l_1l_2} \qquad \alpha_- = \frac{l_2^2 - 2l_1^2 - \sqrt{l_2^4 + 4l_1^4}}{2l_1l_2}$$

The study of u_2 in the neighbourhood of t_2 shows that the contact actually breaks :

$$u_2(t_2) = \dot{u}_2(t_2) = \ddot{u}_2(t_2) = 0 \quad \text{et} \quad u_2^{(3)}(t_2) > 0$$

This motion phase ends with the restoration of the contact $u_2 = 0$, the mass gets in contact with the ground with a negative normal velocity : a collision occurs. Equations of motion during the collision are :

$$\begin{aligned} m(\dot{u}_1^+ - \dot{u}_1^-) &= P_x \\ m(\dot{u}_2^+ - \dot{u}_2^-) &= P_y \end{aligned}$$

P_x and P_y are the contact impulses.
We assume that the impact is inelastic (the restitution coefficient e is zero).

$$\dot{u}_2^+ = 0$$

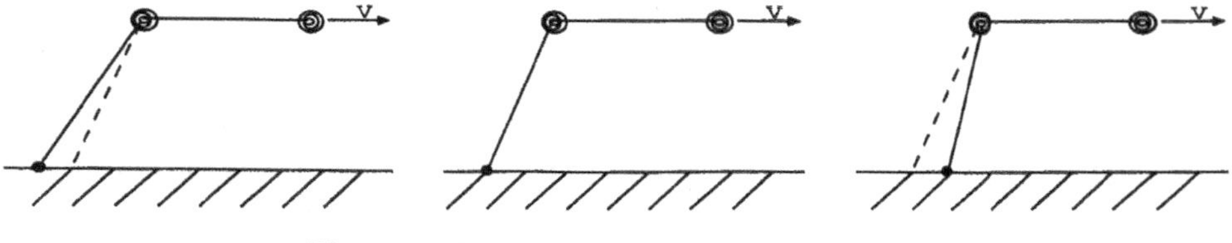

Figure 2. Motion of the system for $\mu < \mu_c$

At the beginning of the impact, the sliding velocity V_i is positive.

$$V_i = v + \dot{u}_1^-$$

Two evolutions are possible : the sliding velocity remains positive during the whole duration of the impact, or it vanishes during the impact and remains zero until the end of the impact.
If the first evolution occurs :

$$P_x = -\mu P_y$$
$$\dot{u}_1^+ = \mu \dot{u}_2^- + \dot{u}_1^-$$

The solution is valid if :

$$V_f > 0 \quad \Leftrightarrow \quad v + \dot{u}_1^- + \mu \dot{u}_2^- > 0$$

If the second evolution occurs, then :

$$\dot{u}_1^+ = -v$$

The solution is valid if :

$$| P_x | \leq u P_y \quad \Leftrightarrow \quad v + \dot{u}_1^- + \mu \dot{u}_2^- \leq 0$$

Results

We submitted above the solutions for the first phases of motion. The solutions for all possible evolutions are set up. The transition instants between the successive phases as well as the values of all variables at these instants are determined numerically, which allows us to follow the evolution of the successive bounces.
The value of μ_c which determines the coming out of bounces depends on the geometry and on the velocity at which the system is drawn. The greater is v, the more bounces occur for small friction coefficient. For a given value of v, the bounces amplitude increases with the friction coefficient. When the friction coefficient is such that sticking occurs during the impact, the motion is periodic. Figures 3, 4 et 5 show the evolution according to the friction coefficient. For larger μ_c values, a periodic motion occurs as soon as $\mu > \mu_c$ (figure 6).
The transition value μ_c depends on the initial conditions. Figure 7 corresponds to initial conditions $u_1(0) = 0, \quad \dot{u}_1(0) = 0$.
In the frame of inelastic impacts, the motion is periodic or shows a certain regularity, which is not true any more if we assume that impacts are elastic (figures 8 and 9).

- **If $\mu = \mu_c$:**

As in the case $\mu > \mu_c$, the normal force becomes zero at $t = t_2$. Separation may occur. But the study of u_2 in the neighbourhood of t_2 shows that the contact does not break.

$$u_2(t_2) = \dot{u}_2(t_2) = \ddot{u}_2(t_2) = u_2^{(3)}(t_2) = 0 \quad u_2^{(4)}(t_2) < 0$$

Sliding goes on, the evolution is the same as for $\mu < \mu_c$.

CONCLUSION

For given initial conditions, we have a unique dynamic behaviour. If the friction coefficient is greater than a critical value μ_c, the smooth motion turns towards an evolution with bounces. The motion with bounces seems stable as we establish numerically

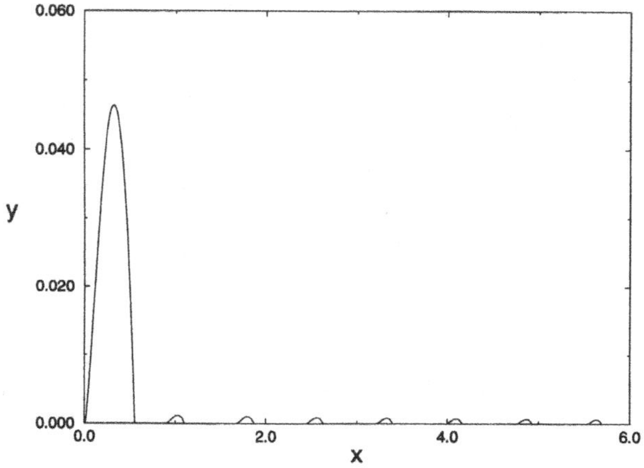

Figure 3. Trajectory of point C. $v = 1.5, \mu = 0.5$

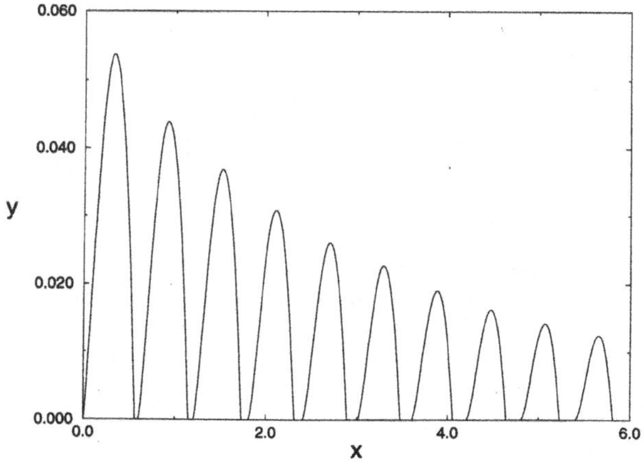

Figure 4. Trajectory of point C. $v = 1.5, \mu = 1.5$

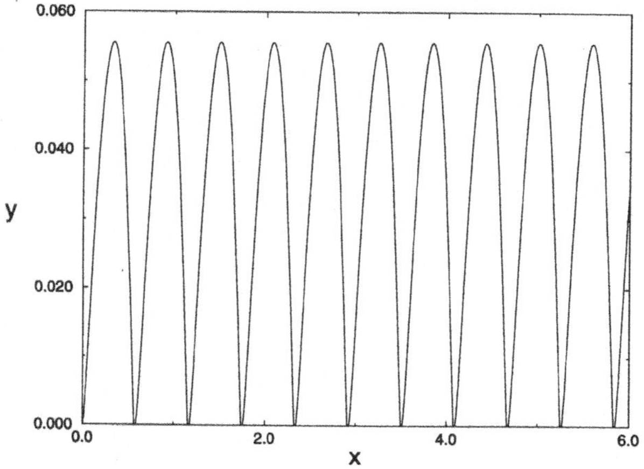

Figure 5. Trajectory of point C. $v = 1.5, \mu = 2$

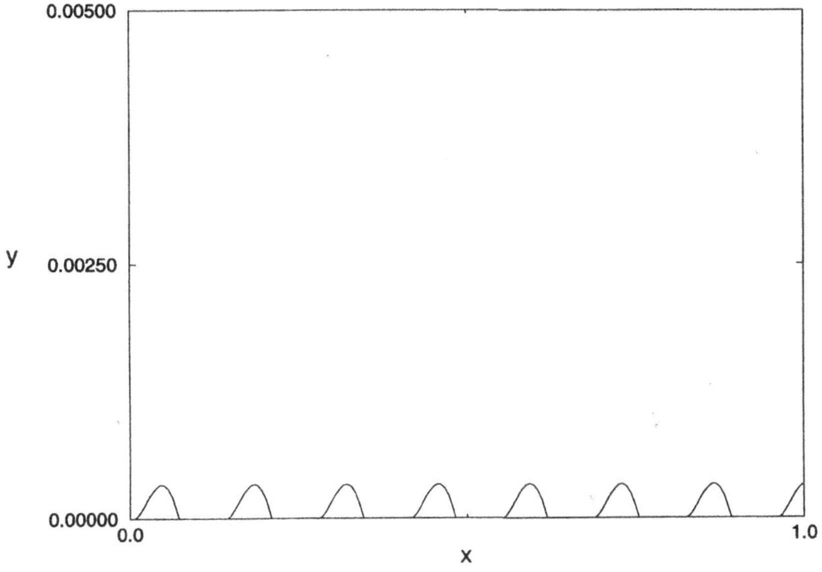

Figure 6. Trajectory of point C. $v = 0.5, \mu = 4$

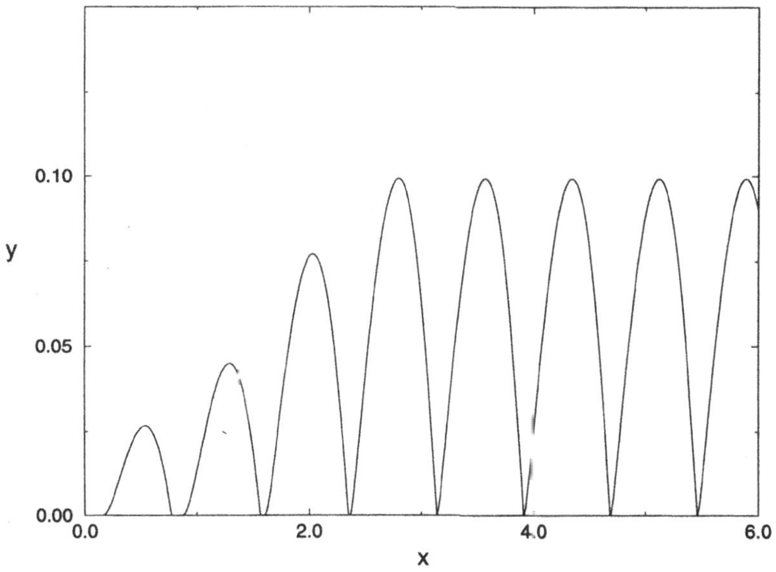

Figure 7. Others initial conditions. $v = 2, \mu = 2$

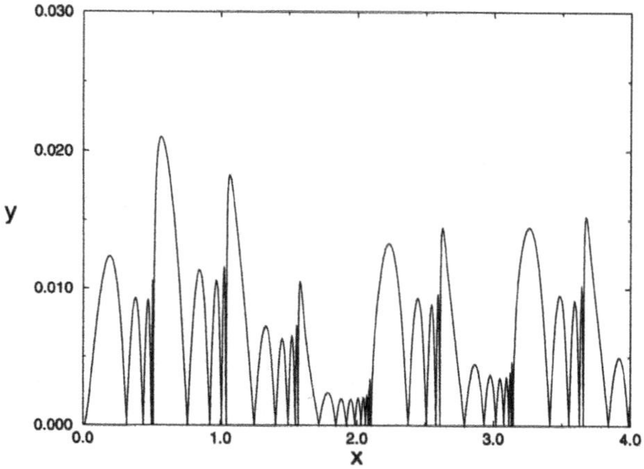

Figure 8. Elastic impact. $v = 1, \mu = 1$

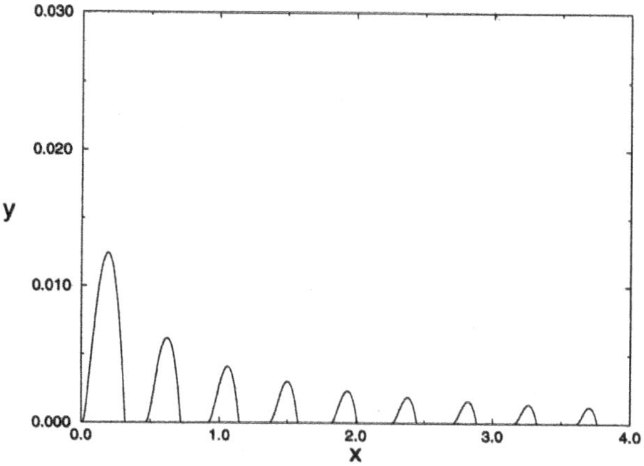

Figure 9. Inelastic impact. $v = 1, \mu = 1$

that the bounces amplitude does not increase indefinitely. If we apply the existing stability criteria to our example we do not obtain valid results. This is due to the fact that these are static criteria. The problem of a dynamic stability criterion remains open.

REFERENCES

Chateau, X., and Nguyen, Q.S., 1991, Buckling of elastic structures in unilateral contact with or without friction, *Eur. J. Mech. A/Solids*, **10**, n°1

Klarbring, A., 1990, Examples of non-uniqueness and non-existence of solutions to quasistatic contact problems with friction, *Ingenieur-Archiv 60*, 529-541

Martins, J.A.C., Monteiro Marques, M.D.P., and Gastaldi, F., 1994, On an example of non-existence of solution to a quasistatic frictional contact problem, *Eur. J. Mech. A/Solids*, **13**, n°1

SLIP DISPLACEMENT DEPENDENT FRICTION IN QUASI-STATIC ELASTICITY

Ioan R. Ionescu and Jean-Claude Paumier

Laboratoire de Modélisation et Calcul
Université Joseph Fourier, IMAG, BP 53
38041 Grenoble Cedex 9, France

INTRODUCTION

In this work we consider the static contact problem between an elastic body and a foundation. On the contact interface we use the Coulomb law in the special case of a prescribed normal pressure.

An important step in the understanding of the *stick-slip phenomenon* was done by Rabinowicz (1959) which pointed out that *the coefficient of friction* μ *varies, rather than be constant, with the tangential displacement, i.e.* $\mu = \mu(|u_T|)$. Stick-slip is then a result of the *slip weakening*, i.e. the fall of the friction force with slip.

The experimental and theoretical works on the mechanism of stick-slip involving slip dependent friction concerns only zero-dimensional problems (tangential motions with friction of a rigid block connected to a spring). As it is pointed out by Ionescu and Paumier (1994), for *the dynamical problem with slip rate dependent friction*, there exists a qualitative difference between the behaviour of the solutions in the one-dimensional and zero-dimensional cases with the same friction law.

It is the aim of this work to throw some light on problems of elasto-statics with slip dependent friction in N-dimensional cases, $1 \leq N \leq 3$. We refer to Ionescu-Paumier (to appear) for a complete description of assumptions and proofs.

THE GENERAL PROBLEM

Let $\Omega \subset \mathbf{R}^N$ be a bounded domain with boundary $\Gamma = \partial\Omega$ divided into three disjoint parts $\Gamma = \bar{\Gamma}_d \cup \bar{\Gamma}_c \cup \bar{\Gamma}_f$ with $meas(\Gamma_d) > 0$. The mechanical problem with friction consists in finding the displacement field $u : \Omega \longrightarrow \mathbf{R}^N$ such that

$$\sigma(u) = A\epsilon(u), \quad \text{div}\,\sigma(u) + r = 0 \quad \text{in} \quad \Omega, \tag{1}$$

$$u = d \quad \text{on} \quad \Gamma_d, \qquad \sigma(u)n = l \quad \text{on} \quad \Gamma_c, \tag{2}$$

Contact Mechanics, Edited by M. Raous *et al.*
Plenum Press, New York, 1995

$$\sigma(u)n \cdot n = -S \quad \text{on} \quad \Gamma_f, \tag{3}$$

$$\sigma_\tau(u) = -S\mu(|u_\tau|)u_\tau/|u_\tau| \quad \text{if} \quad u_\tau \neq 0 \text{ on } \Gamma_f, \tag{4}$$

$$|\sigma_\tau(u)| \leq \mu(0)S \quad \text{if} \quad u_\tau = 0 \text{ on } \Gamma_f, \tag{5}$$

where $\mathcal{A}$ is a symmetric and uniformly positively defined fourth order tensor, $\sigma(u)$ is the stress tensor, $\epsilon(u) = \frac{1}{2}(\nabla u + \nabla^T u)$ is the small strain tensor, $\sigma_\tau(u) = \sigma(u)n - (\sigma(u)n \cdot n)n$ is the tangential stress and $u_\tau = u - (u \cdot n)n$ is the tangential displacement. Here $\mu : \Gamma_f \times \mathbf{R}_+ \longrightarrow \mathbf{R}_+$ is the friction coefficient and r represents the given body forces, d the imposed displacement, l the load on Γ_c and $S \geq 0$ is the normal load which is imposed on Γ_f.

In order to simplify (1)-(5) let us consider $u^{stuck} : \Omega \longrightarrow \mathbf{R}^N$ the "stuck solution" of the following auxiliary elastic problem without friction:

$$\sigma(u^{stuck}) = \mathcal{A}\epsilon(u^{stuck}), \qquad \text{div}\, \sigma(u^{stuck}) + r = 0, \quad \text{in } \Omega,$$

$$u^{stuck} = d \quad on \quad \Gamma_d, \qquad \sigma(u^{stuck})n = l \quad \text{on} \quad \Gamma_c,$$

$$\sigma(u^{stuck})n \cdot n = -S, \qquad u_\tau^{stuck} = 0 \quad \text{on} \quad \Gamma_f.$$

Let us denote by $-f$ the tangential stress on Γ_f which corresponds to the stuck solution, i.e. $f := -\sigma_\tau(u^{stuck})$, and let $\bar{u} = u - u^{stuck}$. The homogenized problem with friction is: find the displacement $\bar{u} : \Omega \longrightarrow \mathbf{R}^N$ such that

$$\sigma(\bar{u}) = \mathcal{A}\epsilon(\bar{u}), \qquad \text{div}\, \sigma(\bar{u}) = 0, \quad \text{in } \Omega, \tag{6}$$

$$\bar{u} = 0 \quad \text{on} \quad \Gamma_d, \qquad \sigma(\bar{u})n = 0 \quad \text{on} \quad \Gamma_c, \tag{7}$$

$$\sigma(\bar{u})n \cdot n = 0 \quad \text{on} \quad \Gamma_f, \tag{8}$$

$$\sigma_\tau(\bar{u}) + \mu(|\bar{u}_\tau|)S\bar{u}_\tau/|\bar{u}_\tau| = f \quad \text{if} \quad \bar{u}_\tau \neq 0 \quad \text{on } \Gamma_f, \tag{9}$$

$$|\sigma_\tau(\bar{u}) - f| \leq \mu(0)S \quad \text{if} \quad \bar{u}_\tau = 0 \quad \text{on } \Gamma_f. \tag{10}$$

EXISTENCE AND UNIQUENESS RESULTS

In the following we assume that $f \in [L^p(\Gamma_f)]^N$ and $S \in L^\infty(\Gamma_f)$ with $S \geq 0$.

Let us denote by $V := \{v \in [H^1(\Omega)]^N / v = 0 \quad \text{on} \quad \Gamma_d\}$. It is a closed subspace of $[H^1(\Omega)]$ equipped with the inner product $< u, v >_V := \int_\Omega \mathcal{A}\epsilon(u) \cdot \epsilon(v)dx$ and the associated norm $\|v\|_V$. For $1 \leq p < 2(N-1)/(N-2)$ let us denote by

$$\gamma_\tau : V \to L_p := \{z \in [L^p(\Gamma_f)]^N / \; z(x) \cdot n(x) = 0 \quad a.e. \quad x \in \Gamma_f\},$$

the compact operator which associates to all $v \in V$ the tangential component $\gamma_\tau(v)$ of its trace on Γ_f, i.e. $\gamma_\tau(v) := v - (v \cdot n)n$ along Γ_f, $\forall v \in V$. Let $V_1 = \ker(\gamma_\tau)$ and let V_2 be the subspace of V orthogonal to V_1. Let $q > 2(N-1)/N$ and $z \in L_q$ and let us denote by $P_q z = \bar{u}$ the solution of (6)-(8) and $\sigma_\tau(\bar{u}) = z$ on Γ_f. Let $K : V_2 \longrightarrow V_2$ be the symmetric, (strictly) positive and compact operator $Kv := P_2\gamma_\tau(v)$, $\forall v \in V_2$.

Then there exists a positive and increasing sequence $(b_n)_{n \geq 0}$, with $b_n \to +\infty$, which represents the singular values of K. If we denote by W_n the finite dimensional subspace of eigenfunctions which corresponds to the singular value b_n then we have
$V_2 = W_0 \oplus W_1 \oplus \ldots \oplus W_n \oplus \ldots$

Let us denote by $j : V \times V \longrightarrow \mathbf{R}_+$ and $f : V \longrightarrow \mathbf{R}$ the functions:

$$j(u,v) = \int_{\Gamma_f} S\mu(|\gamma_\tau(u)|)|\gamma_\tau(v)|dx, \quad f(v) = \int_{\Gamma_f} f.\gamma_\tau(v)dx, \quad \forall u,v \in V.$$

Then the following quasi-variational inequality represents the *variational approach* of (6)-(10): find $\bar{u} \in V$, such that

$$< \bar{u}, \bar{u} - v >_V - j(\bar{u},\bar{u}) - j(\bar{u},v) \leq f(\bar{u} - v), \quad \forall v \in V. \tag{11}$$

Let us introduce the energy function $\mathcal{W} : V \longrightarrow \mathbf{R}$ given by

$$\mathcal{W}(v) = \frac{1}{2}\|v\|_V^2 + \int_{\Gamma_f} SH(|\gamma_\tau(v)|)dx - f(v), \quad \forall v \in V, \tag{12}$$

where H is the primitive $H(x,u) = \int_0^u \mu(x,s)ds$.

It can be shown that if $\bar{u} \in V$ is a local extremum for $\mathcal{W}$ then $\bar{u}$ is a solution of (11) and the following theorem gives sufficient conditions for a solution of (11) to be stable.

Theorem 1 *If $\bar{u} \in V$ is a solution of (11) and*

$$-ess \inf_{x\in\Gamma_f} [S(x)\partial_u\mu(x, |\bar{u}_\tau(x)|)] < b_0,$$

then $\bar{u}$ is an isolated local minimum for $\mathcal{W}$.

Now we give the existence and uniqueness results:

Theorem 2 *There exists at least a global minimum $\bar{u}$ for $\mathcal{W}$. Moreover $\bar{u}$ is a solution of (11).*

In the study of the uniqueness of the solution an important role is played by the parameter $\bar{b}$ measuring the slip weakening and given by

$$\bar{b} := -ess \inf_{x\in\Gamma_f} [S(x) \inf_{u\in\mathbf{R}_+} \partial_u\mu(x,u)]. \tag{13}$$

Theorem 3 *If $\bar{b} < b_0$, then (11) has a unique solution. Moreover this solution depends continuously on f, i.e. the application $f \longrightarrow \bar{u}$ is continuous from L_2 to V.*

THE ANTI-PLANE PROBLEM

The problem, which will be stated in this section was compiled as the simplest example able to emphasise the main difficulties of the general problem. It is sufficiently "simple" to permit some analytical computations and sufficiently "complicate" to point out some mathematical properties and some mechanical phenomena as non-uniqueness, stability, catastrophic behaviour, nonhomogeneous bifurcation and slip localisation.

Let us consider the shearing of an infinite linear elastic slab bounded by the planes $x_1 = l$, $x_1 = 0$, $x_2 = h$ and $x_2 = 0$ as in Figure 1. On $\Gamma_f = [0,l] \times \{h\} \times \mathbf{R}$ the slab is in contact with friction with a rigid body which pushes it with the constant normal force $\sigma_{22} = -S$ i.e. u, σ satisfy (3) − (5) along Γ_f.

Along $\Gamma_u = [0,l] \times \{0\} \times \mathbf{R}$ the displacement is imposed:

$$u_1 = 0, \quad u_2 = 0, \quad u_3 = B \quad \text{along} \quad \Gamma_u,$$

and on $\Gamma_c = \{0,l\} \times [0,h] \times \mathbf{R}$ the normal displacement and the tangential stress are vanishing:

$$u_1 = 0, \quad \sigma_{12} = \sigma_{13} = 0 \quad \text{along} \quad \Gamma_c.$$

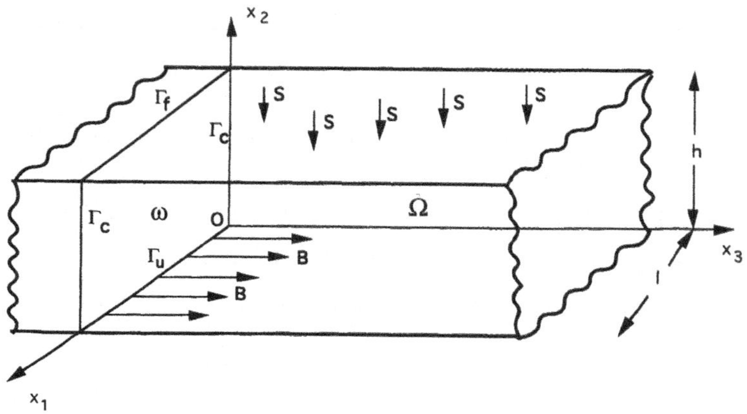

Figure 1 *Geometry of the anti-plane problem.*

Let us assume that the displacement field is 0 with respect to Ox_1, u_2 depends on x_2 alone and u_3 does not depend on x_3.

The equilibrum equation $div\ \sigma = 0$ becomes: $u_2''(x_2) = 0, \quad \Delta u_3(x_1, x_2) = 0.$

We get $u_2(x_2) = -\dfrac{S}{\lambda + 2G} x_2$ where $\lambda, G > 0$ are the Lamé coefficients. If we denote by $\omega :=]0, l[\times]0, h[$ and $w := u_3$ the problem (1)-(5) reads:

$$\Delta w(x_1, x_2) = 0 \quad \text{in} \quad \omega, \tag{14}$$

$$\partial_{x_1} w(l, x_2) = \partial_{x_1} w(0, x_2) = 0, \quad \forall\, x_2 \in]0, h[, \tag{15}$$

$$w(x_1, 0) = B, \quad \forall\, x_1 \in]0, l[, \tag{16}$$

$$G\partial_{x_2} w(x_1, h) = -\mu(|w(x_1, h)|)S\ sign\ (w(x_1, h)) \ \text{ if } \ w(x_1, h) \neq 0, \tag{17}$$

$$G\,|\partial_{x_2} w(x_1, h)| \leq \mu(0)S, \quad \text{if} \quad w(x_1, h) = 0. \tag{18}$$

HOMOGENEOUS SLIDING

Assume that μ, S and B do not depend on x. Now we will focus our attention on the *homogeneous sliding solutions* of this problem, i.e. we look for w^0 solution of (14)-(18) such that $\partial_{x_1} w^0(x_1, h) = 0$ for all $x_1 \in]0, l[$. If we denote by

$$U := w^0(x_1, h), \tag{19}$$

$$w^{stuck}(x_1, x_2) := B(1 - x_2/h), \tag{20}$$

$$\bar{w}^0(x_1, x_2) := Ux_2/h, \tag{21}$$

then from (14)-(16) we deduce $w^0 = w^{stuck} + \bar{w}^0$. We remark that the first singular value of the operateur K is $b_0 \; = \; G/h$.

Let us introduce the function $g : \mathbf{R}_+ \longrightarrow \mathbf{R}_+$ given by

$$g(u) := b_0 u + \mu(u)S, \qquad \forall u \in \mathbf{R}_+.$$

If we denote by $-t$ the tangential stress, which corresponds to the "stuck case", i.e. $t := -G\partial_{x_2} w^{stuck} = GB/h$, then (17)-(18) reads:

$$g(U) = t \quad \text{if} \quad U > 0, \tag{22}$$

$$t \in [-g(0), g(0)] \quad \text{if} \quad U = 0, \tag{23}$$

$$-g(-U) = t \quad \text{if} \quad U < 0. \tag{24}$$

Let us introduce the *homogeneous energy function*
$\mathcal{W}_0 : \mathbf{R} \longrightarrow \mathbf{R}$ given by $\mathcal{W}_0(U) := \mathcal{E}_0(U) + \mathcal{F}_0(U) + C$,
where $\mathcal{E}_0(U) := \int_0^h \sigma_{23}^o(U)\epsilon_{23}^o(U)dx_2 = \frac{1}{2}G(U-B)^2/h$ is the *potential elastic energy* and $\mathcal{F}_0(U) := \int_0^U S\mu(|u|)sign(u)du$ is the *work of the friction force* from the reference configuration to the current one.
If we denote by $H(u) = \int_0^u \mu(s)ds$ we deduce $\mathcal{F}_o(U) = SH(|U|)$.
Choosing the constant $C = -\frac{1}{2}GB^2/h$ we get

$$\mathcal{W}_0(U) = \frac{1}{2}b_0U^2 + SH(|U|) - tU.$$

Since $\mathcal{W}_0$ is not differentiable in $u = 0$ let us denote by
$\mathcal{W}_0'(0) := [\mathcal{W}_o'(0-), \mathcal{W}_o'(0+)] = [-g(0)-t, g(0)-t]$ (the derivative in the sense of Clark)
and
$\mathcal{W}_0'(u) := \{g(|u|)sign(u) - t\}$ for $u \neq 0$.
One can easely deduce that (22)-(24) is equivalent with $0 \in \mathcal{W}_0'(U)$, i.e. the solution of (22)-(24) is a stationary point for $\mathcal{W}_0$.

Here the value of the parameter $\bar{b}$ (see (13)) is $\bar{b} = -S\inf_{u\in\mathbf{R}_+}\mu'(u)$.

We will distinguish two cases.

The **first case**, called the *regular case*, corresponds to $\bar{b} \leq b_0$.
In this case g is an increasing function, $\mathcal{W}_0$ is a strictly convex function, the equations (22)-(24) (or $0 \in \mathcal{W}_0'(U)$) have a unique solution. This solution corresponds to a global minimum for $\mathcal{W}_0$ and depends continuously on the load t (i.e. on B).

The **second case** is called the *irregular case* and corresponds to $\bar{b} > b_0$.
In this case g is not an increasing function, equations (22)-(24) (or $0 \in \mathcal{W}_0'(U)$) *have not anymore an unique solution* and *the energy is not convex*. This means that a solution U of $0 \in \mathcal{W}_0'(U)$ may correspond to a local minimum or a local maximum for $\mathcal{W}_0$. But it is easy to prove that if U is a solution of (22)-(24) and $b_0 + S\mu'(|U|) > 0$, (i.e. $g'(U) > 0$) then U is a *local* (isolated) *minimum* for $\mathcal{W}_0$. Moreover there exists at least a solution of $0 \in \mathcal{W}_0'(U)$ which corresponds to a *global minimum* for $\mathcal{W}_0$.

In order to fix the ideas we will suppose for simplicity that g is increasing on $[0, \alpha_1]$ and on $[\alpha_2, +\infty)$ and it is deacreasing on $[\alpha_1, \alpha_2]$. If $U \geq 0$ from (22) we obtain the branch of homogeneous slip $U = g^{-1}(t)$ which is schematically represented in Figure 2.

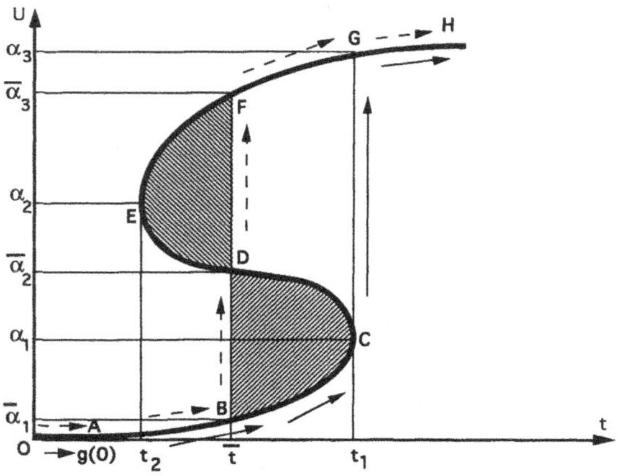

Figure 2 *The homogeneous slip $U = g^{-1}(t)$ versus the tangential load t. Solutions of the quasistatic loading process selected by the global minimum criterion (dotted arrows) and by the (perfect) delay criterion (solid arrows).*

The Maxwell line BF is constructed such $meas(\ BCD\ ) = meas(\ DEF\ )$, i.e. $\int_{\bar{\alpha}_1}^{\bar{\alpha}_3} g(s)ds = \bar{t}(\bar{\alpha}_3 - \bar{\alpha}_1)$ which means that $\mathcal{W}_0(\bar{\alpha}_1) = \mathcal{W}_0(\bar{\alpha}_3)$ for $t = \bar{t}$.

Let us analyse in the next the dependence of the homogeneous slip U on the load t when t is increasing, i.e. we imagine a *quasi-static loading process.* Notice that there exists *no (global) continuous solutions* $U : I \longrightarrow \mathbf{R}$ for I an open interval with $t_1, t_2 \in I$.

The solutions on the branch CDE corresponds to a local maximum for $\mathcal{W}_0$ (i.e. they are not stable) and the solutions on the branches $OABC$ and $EFGH$ corresponds to a local minimum for $\mathcal{W}_0$ (i.e. they are stable). Using the Maxwell line BF one can easely deduce that the solutions of the branches OAB and FGH corresponds to a global minimum for $\mathcal{W}_0$. For $t = \bar{t}$ we deal with two global minima $U = \bar{\alpha}_1$ and $U = \bar{\alpha}_3$.

From Figure 2 we can also remark that for $t \in [0, t_2[\cup]t_1, +\infty[$ the solution $U(t)$ is uniquely determined but for $t \in [t_2, t_1]$ two or three solutions are avaible. Let us give two criteria which select a solution between the stable ones.

The **first criterion**, called *the global minimum criterion*, selects as physical solution the global minimum for $\mathcal{W}_0$ at each t and it is shown in Figure 2 by dotted arrows. Using this criterion $U(t) = 0$ for $t \in [0, S\mu(0)]$ (no slip) and $U(t)$ is smoothly increasing for $t \in [S\mu(0), \bar{t}[$ up to $\bar{\alpha}_1$ (i.e. $U(\bar{t}-) = \bar{\alpha}_1$). For $t > \bar{t}$ the slip is suddenly increasing up to $\bar{\alpha}_3$ (i.e. $U(\bar{t}+) = \bar{\alpha}_3$) and then it follows the branch FGH.

The **second criterion**, called the *(perfect) delay convention*, comes from the catastrophe theory (see Poston and Stewart, 1978) and selects as physical solution the solution which jumps (has a discontinuity) as late as possible (i.e. when it has no other choice). Using this criterion $U(t) = 0$ (no slip) for $t \in [0, S\mu(0)]$ and $U(t)$ is smothly increasing for $t \in [S\mu(0), t_1]$ up to α_1 (i.e. $U(t_1-) = \alpha_1$). For $t > t_1$ the slip is suddenly increasing up to α_3 (i.e. $U(t_1+) = \alpha_3$) and follows the branch GH. The part of the branch of solutions selected by this criterion is shown in Figure 2 by solids arrows.

If we denote by $\bar{w} := w - w^{stuck}$ (given by (20)) then (14)-(18) becomes:

$$\Delta \bar{w} = 0 \quad \text{in} \quad \omega, \tag{25}$$

$$\partial_{x_1}\bar{w}(l, x_2) = \partial_{x_1}\bar{w}(0, x_2) = \bar{w}(x_1, 0) = 0 \quad \forall\, x_1, x_2 \in]0, l[\times]0, h[, \tag{26}$$

$$G\partial_{x_2}\bar{w}(x_1, h) + \mu(|\bar{w}(x_1, h)|)S sign(\bar{w}(x_1, h)) = t, \quad \text{if} \quad \bar{w}(x_1, h) \neq 0, \tag{27}$$

$$|G\partial_{x_2}\bar{w}(x_1, h) - t| \leq \mu(0)S \quad \text{if} \quad \bar{w}(x_1, h) = 0. \tag{28}$$

We know that $b_0 = G/h$ and, after some algebra, we get the other singular values of the operateur K:

$$b_n = \frac{Gn\pi}{l} cth\left(\frac{n\pi h}{l}\right) \quad \forall n \geq 1. \tag{29}$$

The corresponding eigenfunctions are

$$\phi_0(x_1, x_2) = x_2/h,$$

$$\phi_n(x_1, x_2) = C_n sh\left(\frac{n\pi}{l}x_2\right)\cos\left(\frac{n\pi}{l}x_1\right) \quad \forall n \geq 1,$$

where C_n are choosen such that $\int_0^l \varphi_n^2(x_1)dx_1 = 1$ with $\varphi_n(x_1) := \phi_n(x_1, h)$.

We consider the space $V = \{v \in H^1(\omega)/v(x_1, 0) = 0 \text{ a.e. } x_1 \in (0, l)\}$ and let $\mathcal{W} : V \longrightarrow \mathbf{R}$ be the energy function

$$\mathcal{W}(v) = \frac{G}{2}\int_\omega |\nabla v|^2 + \int_0^l H(|v(x_1, h)|)dx_1 - t\int_0^l v(x_1, h)dx_1 \tag{30}$$

which corresponds to (12) in the general case.

We can now give a result on stability and uniqueness.
Let U be a solution of (22)-(24) corresponding by (21) to $\bar{w}^0$ a solution of (25)-(28). Then we have that:

i) if $\bar{b} < b_0$ holds *then* $\bar{w}^0$ *is the unique solution* of (25)-(28) in V,
ii) if $b_0 + S\mu(|U|) > 0$ holds *then* $\bar{w}^0$ *corresponds to an (isolated) local minimum for* $\mathcal{W}$ in V,
iii) if U is a global minimum for $\mathcal{W}_0$ then $\bar{w}^0$ *is a global minimum for* $\mathcal{W}$ in V.

Using the Liapounov-Schmidt method we get the following results on the branches of solutions and bifurcation.

Theorem 4 *Let us suppose in the next that* $\mu \in C^2$, $\mu(0) = \mu''(0) = 0$.
Let (t_0, U_0) *a solution* (t, U) *of (22)-(24).*
In a neighborhood of $(t_0, U_0\phi_0) \in \mathbf{R} \times V$ *we have:*
i) There exists a branch $(t, \bar{w}^0) = (t, U\phi_0)$ *of solutions for (25)-(28) with* $U(t_0) = U_0$ *which is the unique branch of homogeneous sliding solutions,*
ii) If $b_n + S\mu'(|U_0|) \neq 0$ *for all* $n \geq 1$ *then* $(t, \bar{w}^0)$ *is the unique branch of solutions for (25)-(28) in* V,
iii) If $b_j + S\mu'(|U_0|) = 0$ *with* $j \geq 1$ *and* $\mu''(|U_0|) \neq 0$ *then* $(t_0, U_0\phi_0)$ *is a cusp bifurcation point, where the branch of homogeneous sliding solutions* $(t, \bar{w}^0)$ *intersects tranversally a branch of nonhomogeneous sliding solutions* $(t, \bar{w}^j)$ *(see Figure 3).*

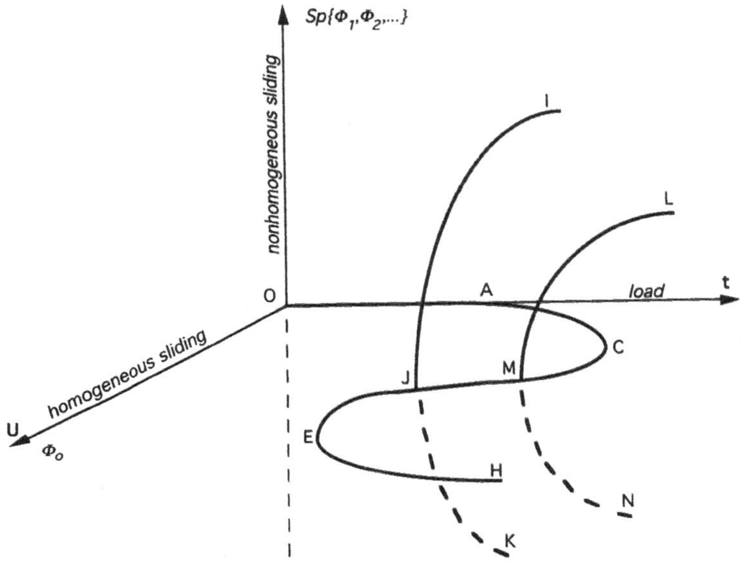

Figure 3 *Two cusp bifurcation points (J and M), where the branch of homogeneous sliding (OACMJEH) intersects transversally two nonhomogeneous sliding branches (IJK and LMN).*

Remark: The above result on nonhomogeneous bifurcation branches intersecting transversally an other branch of homogeneous sliding solutions can be extended to a more general case (see I., P., to appear).

CONCLUSION

In a quasi-static process, when the load t is increasing, there exist equilibrum positions (the limit point C for instance) where a *typical catastrophic event* is present. If we consider not only the homogeneous solution but also nonhomogeneous sliding solutions, the perfect delay criterion does not merely work as in Figure 2. With a quasistatic analysis it is difficult to predict the new equilibrum position. Only the *stability analysis of the homogeneous sliding in the associated dynamical problem* can give some answers to this question. The fact that some branches of nonhomogeneous sliding equilibrum intersect the homogeneous sliding branch is a good reason to think that the *slip localisation* phenomenon can be captured in the model and can be described by this last type of analysis.

REFERENCES

Ionescu I.R., Paumier J.-C., 1994, On the Contact Problem with slip Rate Dependent Friction in Elastodynamics, European J. of Mech., A/Solids, **13**, n° 4, 555-568

Ionescu I.R., Paumier J.-C., to appear, On the Contact Problem with slip Dependent Friction in Elastostatics

Poston T., Stewart I., 1978, Catastrophe Theory and its Applications, Pitman

Rabinowicz E., 1959, A Study of Stick-Slip Process, in Friction and Wear (R. Davies editor), Elviser, London, 149-161

MODELIZATION BY FINITE ELEMENTS OF CUTTING AND PUNCHING PROBLEMS USING AN ALGORITHM BASED ON THE IMPLICIT STANDARD MATERIALS, APPLIED TO THE FRICTION

L. Bousshine[1] and G. de Saxcé[2]

[1]National Higher School of Electricity and Mechanics, Casablanca, Morocco
[2]Mechanics of Materials and Structures, Polytechnic Faculty of Mons, Belgium

SUMMARY

A new class of materials called Implicit Standard Materials is proposed. It allows to generalize Fenchel's inequality, and then to recover normality law for non standard behaviour, in particular for unilateral contact with dry friction and granular materials. In terms of finite element method, an algorithm based on the non linear mathematical programming is used to solve every step of the incremental problem. The numerical applications show the convergence of the method. The significative sensitivity of the solution to the non associativity of the soil flow rule can be quotted.

THE IMPLICIT STANDARD MATERIALS

For this family of materials, we postulate the existence of a potential b(x,y) such that for any y (resp. x), the partial function b(.,y) (resp. b (x,.)) is convex. b is said a **bipotential** if the following inequality is satisfied :

$$\forall\ (x',y'),\ b(x',y') \geq x'.y' \tag{1}$$

A couple (x,y) is said extremal if the equality is achieved in (1) :

$$b(x,y) = x.y \tag{2}$$

Hence, x and y are related by subdifferential mappings :

$$y \in \partial_x b(x,y),\ x \in \partial_y b(x,y) \tag{3}$$

These relations define a (multivalued) constitutive law if we consider that the physical behaviour corresponds to the extremal couples [1].

Contact Mechanics, Edited by M. Raous *et al.*
Plenum Press, New York, 1995

Let Ω_1 and Ω_2 two bodies initially in contact at a point, $\dot{u}$ the relative velocity at this point of Ω_1 with respect to Ω_2 and t the reaction acting onto Ω_1 from Ω_2. It can be proved the unilateral contact with Coulomb's dry friction is represented by the following bipotential (1992b) :

$$b_c(-\dot{u},t) = \Psi_{R_-}(-\dot{u}_n) + \Psi_{K_\mu}(t) + \mu t_n \|-\dot{u}\| \tag{4}$$

where occurs the indicatory function of Coulomb's cone :

$$K_\mu = \{ (t_n,t_t) \text{ such that } \leq \mu t_n \} \tag{5}$$

Hence, the unilateral contact law with dry friction can be written using (3b) applied to (4) :

$$-(\dot{u}_n + \mu \|\dot{u}_t\|, \dot{u}_t) \in \partial\Psi_{K_\mu}(t) \tag{6}$$

A similar situation occurs with the granular material stresses belong to the following set (fig.1).

$$K_\sigma = \{(s_m, s) \text{ such that } \|s\| \leq r(c - s_m \tan\varphi)\} \tag{7}$$

where s_m is the hydrostatic pressure, s the stress deviator, c is the cohesion, φ is the friction angle.

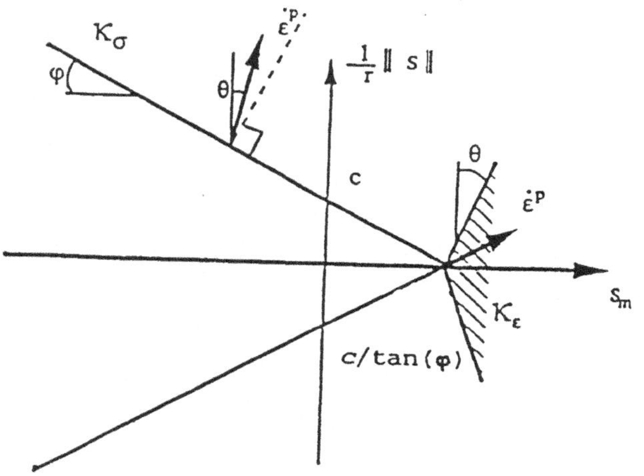

Figure 1. Rudnicki-Rice constitutive law

For a realistic material, the flow rule is generally non-associated and caracterized by a plastic dilatancy angle θ within the range from 0 to φ . the following bipotential is introduced:

$$b_p(\dot{\varepsilon}^p, \sigma) = \Psi_{K_\varepsilon}(\dot{\varepsilon}^p) + \frac{c\,\dot{e}^p_m}{\tan\varphi} + \Psi_{K_\sigma}(\sigma) - r(\tan\varphi - \tan\theta)\left(\frac{c}{\tan\varphi} - s_m\right)\|\dot{e}^p\| \quad (8)$$

where $\dot{e}^p_m$ is the trace of the plastic strain rate tensor, $\dot{e}^p$ the plastic strain rate deviator and σ the stress tensor; then, the plastic strain rate belongs to the following set of admissible plastic strain rates :

$$K_\varepsilon = \left\{(\dot{e}^p_m, \dot{e}^p) \text{ such that } \dot{e}^p_m \geq r\tan\vartheta\|\dot{e}^p\|\right\} \quad (9)$$

Hence, a new formulation of Rudnicki-Rice constitutive law [5] is obtained applying (3b) to (8) :

$$(\dot{e}^p_m + r(\tan\varphi - \tan\theta)\|\dot{e}^p\|, \dot{e}^p) \in \partial\Psi_{K_\sigma}(\sigma) \quad (10)$$

ELASTOPLASTIC EVOLUTION AND TIME INTEGRATION SCHEME

Assuming the strain decomposition, if the strain history $\varepsilon(t)$ is imposed, the couple $(\sigma(t), \varepsilon^p(t))$ is a solution of the following multivalued differential equation system of the first order :

$$\begin{aligned} S\dot{\sigma} + \dot{\varepsilon}^p &= \dot{\varepsilon}(t), \\ \dot{\varepsilon}^p &\in \partial_\sigma b_p(\dot{\varepsilon}^p, \sigma) \end{aligned} \quad (11)$$

Similarly, for the friction contact, one has to solve :

$$\begin{aligned} S_c\dot{t} + \dot{u}^f &= \dot{u}(t), \\ \dot{u}^f &\in \partial_t b_c(-\dot{u}^f, t) \end{aligned} \quad (12)$$

where S_C is a fictitious contact compliance matrix introduced for the numerical regularization of the non smooth law. In the following, the index 1 is relative to the end of the step, the symbol Δ to the increment and τ is the time. The implicit scheme :

$$\begin{aligned} \Delta\varepsilon &= S\Delta\sigma + \Delta\tau\,\dot{\varepsilon}^p_1, \\ \Delta u &= S_c\Delta t + \Delta\tau\,\dot{u}^f_1 \end{aligned} \quad (13)$$

leads, after some computations, to the following incremental law of implicit normality :

$$\Delta\sigma = \frac{\partial \Delta b_p}{\partial \Delta\varepsilon}(\Delta\varepsilon, \Delta\sigma) \; ,$$
$$\Delta t = \frac{\partial \Delta b_c}{\partial(-\Delta u)}(-\Delta u, \Delta t) \tag{14}$$

VARIATIONAL PRINCIPLES AND NUMERICAL ALGORITHM

Let Ω be a body of boundary S, in possible contact on the part S_2 of S, subjected during a time step to imposed surface tractions Δp on the part S_1, and imposed displacements Δv on the remaining part $S_0 = S - S_1 - S_2$. As proposed first in de Saxcé (1991-92a), the so-called **bifunctional** is introduced

$$\Delta\beta\,(\Delta u, \Delta\sigma) = \int_\Omega \Delta b_p\,(\Delta\varepsilon(\Delta u), \Delta\sigma)\, d\Omega + \int_{S_2} \Delta b_c\,(-\Delta u, \Delta t(\Delta\sigma))\, dS - \int_{S_1} \Delta p . \Delta u\, dS - \int_{S_0} \Delta t(\Delta\sigma) . \Delta v\, dS \tag{15}$$

in order to extend the classical calculus of variation. Indeed, let the couple $(\Delta u, \Delta\sigma)$ be an exact solution of the corresponding boundary value problem. Then, it can be proved that this couple is simultaneously solution of the two following variational principles Ladevèze (1985) :

$$\inf \left\{ \Delta\beta\,(\Delta u^k, \Delta\sigma) : \Delta u^k \; K.A. \right\},$$
$$\inf \left\{ \Delta\beta\,(\Delta u, \Delta\sigma^s) : \Delta\sigma^s \; S.A. \right\} \tag{16}$$

Introducing the usual nodal displacement increment vector ΔU, the nodal force increment vector ΔF, the displacement field shape function matrix N and the strain field shape function matrix B, the finite element method leads to the following discretized bifunctional :

$$\Delta\beta\,(\Delta U, \Delta\sigma) = \int_\Omega \Delta b_p\,(B\Delta U, \Delta\sigma) + \int_{S_2} \Delta b_c\,(-N\Delta U, \Delta t)\, dS - \Delta F^T \Delta U \tag{17}$$

Combining the structural equilibrium equations resulting from the minimum condition (16a) with the incremental laws (14), it can be seen that the solution of the boundary value problem must verify the following equation system :

$$\Delta\sigma_r = \frac{\partial \Delta b_p}{\partial \Delta\varepsilon}(B\Delta U, \Delta\sigma) - \Delta\sigma = 0 \quad (14)$$

$$\Delta t_r = \frac{\partial \Delta b_c}{\partial(-\Delta u)}(-N\Delta U, \Delta t) - \Delta t = 0 \quad (15) \qquad (18)$$

$$\Delta F_r = \int_\Omega B^T \Delta\sigma \, d\Omega - \int_{S_2} N^T \Delta t dS - \Delta F = 0$$

In principle, the equations (18a) and (18b) should be satisfied anywhere. In practical implementation, the integrals are numerically computed by Gauss integration sheme. Hence, the local equations are only considered at Gauss points.

The bipotential theory may be qualified as a **constructive** method in the sense that it suggests pertinent ideas to find numerical algorithms. For instance, the following iterative method was tested with success. At each iteration, there are two stages:

- a **global stage** for which $\Delta\sigma$ and Δt are fixed and a new approximation of ΔU is obtained by solving the minimization problem (16a) by a non linear Mathematical Programming software;

- a **local stage** for which ΔU is fixed and a new approximation of $\Delta\sigma$ and Δt is computed by applying the fixed point method to the equations (18a) and (18b).

In fact, it can be noted the similarity of this formulation with one of the key ideas of the large time increment or LATIN method proposed by Ladeveze (1985). Using the terminology of this last approach, the couple (ΔU ,$\Delta\sigma$), solution of the boundary value problem, is at the intersection of the non lineare subspace Γ defined by the constitutive law (18a et 18b) and the affin subspace defined by Eq. (18c) (Fig. 2).

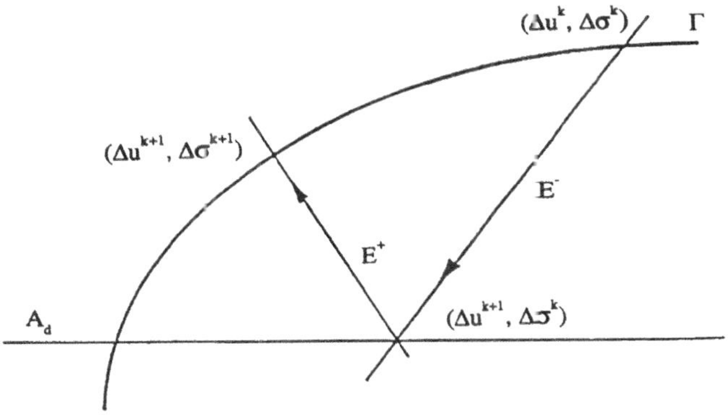

Figure 2. The Latin method

The solving of (18a-18b) corresponds to the local stage, associated with an upwards search direction E^+, and the solving of (18c) to the globale stage, with a downwards search direction E^-; For every step, the solving procedure is based on an iterative method.

DISCUSSION

The numerical testing shown that the iterative procedure is **convergent** in all the numerous applications considered. The main numerical difficulty is to condition the optimization problem by a suitable scaling of the unknowns ΔU and the objective function $\Delta\beta$ de Saxcé (1992c). Hence, the minimization problem is reduced to an equivalent form :

$$Inf\{ \ \Delta\Phi(\Delta U,\Delta\sigma) = \int_{\Omega} \Delta b_p(B\Delta U,\Delta\sigma)d\Omega + \int_{S_2} \Delta b_c(-N\Delta U,\Delta t)dS - \Delta F^T\Delta U; \quad \Delta U = \Delta\bar{U} \ on\ S_0 \ \} \tag{19}$$

The problem is solved by the non linear mathematical programing software MINOS developped by Murthag and Sanders of Standford University, for the scaled quantities :

$$\Delta\bar{U} = \frac{\Delta U}{U_R} \ , \ \Delta\Phi = \frac{\Delta\Phi}{\Phi_R} \tag{20}$$

For displacement controled problems, the reference displacement U_R is estimated from the imposed displacement on S_0. Let D be the physical space dimension (for plane strain problems, D = 2). at the first step elastic, by usual scale arguments, the following estimator of the total energy can be taken :

$$\Phi_R = E_R\, U_R^2\, X_R^{D-2} \tag{21}$$

where E_R is a reference elastic modulus and X_R is a reference length of the structure. For the subsequent steps, in order to take into account of the reduction of the energy resulting from the plastic yielding, a plastification ratio is introduced de Saxcé (1992c). the estimator

$$\Phi_R = E_R\, U_R^2\, X_R^{D-2}(1 - \tau_p) \tag{22}$$

is computed from the values at the previous step.
For load controled problems, U_R and Φ_R are estimate as :

$$U_R = \frac{P_R}{E_R X_R^{D-2}(1-\tau_p)} \ , \ \Phi_R = U_R P_R \tag{23}$$

where P_R is a reference imposed force.

Concerning the local stage, an alternative method consists in using an **augmented lagrangian** and Usawa iteration de Saxcé (1991-92a). The existence of the bipotential allows to use a direct projection onto Coulomb's cone. Hence, in the local stage, the fixed point method is applied to the following equation, equivalent to (6) :

$$t = proj((t_n - \rho(\dot{u}_n + \mu\|\dot{u}_t\|), t_t - \rho\dot{u}_t), K_\mu) \qquad (24)$$

In the point of view of simplicity and computation time, there is improvement with respect to the usual formulation with two projections, the first one for the unilateral contact, and the second one for the sliding law.

Another idea is to replace in the global stage the optimization procedure by a classical **Newton's iteration**. Because of the existence of the bifunctional, the tangent operator is symmetric and positive definite. Once again, significative reduction of the computation time may be obtained, with respect to usual method with non symmetric and ill-conditionned tangent operators de Saxcé (1994).

Another relevant question is the comparizon Newton's iteration **versus** mathematical programming. In authors opinion, the first algorithm is very fast but the convergence uncertain. On the other hand, the mathematical programming is a very reliable tool, even for large steps. The principal drawback is nowaday the computation time. A good compromise is to combine the two kind of steps.

In a theoretical point of view, another relevant feature of the bipotential approach is the easy generalization of the well-known bound theorems of the limit analysis to the Implicit Standard Materials, including friction contact de Saxcé (1993).

APPLICATIONS

Punching

Let us consider a rectangular piece, fixed on the lower side, traction free on the lateral sides, and subjected to uniform imposed displacement on the upper one, due to the action of a rigid punch. The material is elastic perfectly plastic with Young's modulus E=21000 ksi, Poisson's ratio ν=0,3, Von Mises criterion and the yield stress σ_0 =40 ksi. The problem was computed for two values of the friction coefficient, μ =0,05 and μ =0,15. The distribution of the normal and the tangential reactions are shown at fig. 3 and 4. For μ =0,05, a significative sticking region is observed.

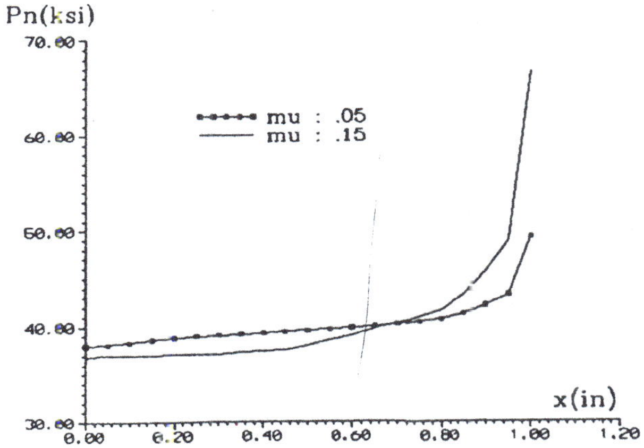

Figure 3. Normal reaction distribution on OA.

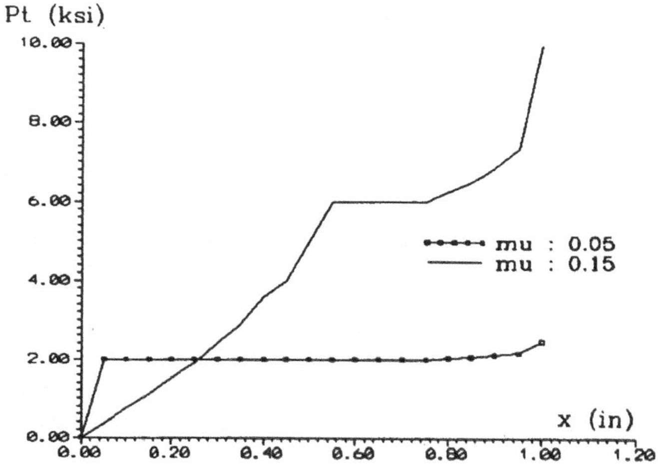

Figure 4. Tangential reaction distribution on OA.

Cutting

In this section are presenting some results of the numerical simulation of the cutting process of granular materials. From the industrial point of view, this kind of computation is of great interest in application such as oil drilling. After simplification of the realistic problem, the solid of fig 5 in plane strain condition is considered. The displacement at a significative distance of the cutter are fixed. The cutter is assumed rigid and subjected to an horizontal uniform displacement. The material elastic-perfectly plastic, governed by Rudnicki-Rice constitutive law with the following material parameters : E = 50000 MPa , $\nu = 0.33$, cohesion stress c = 30 MPa and variable friction $\varphi = 40°,30°,20°,10°,5°$ and plastic dilatancy angle $\theta = 40°,20°,10°,0°$. After meshing raffinement analysis, a mesh with 81 finite elements and 332 d.o.f. is considered.

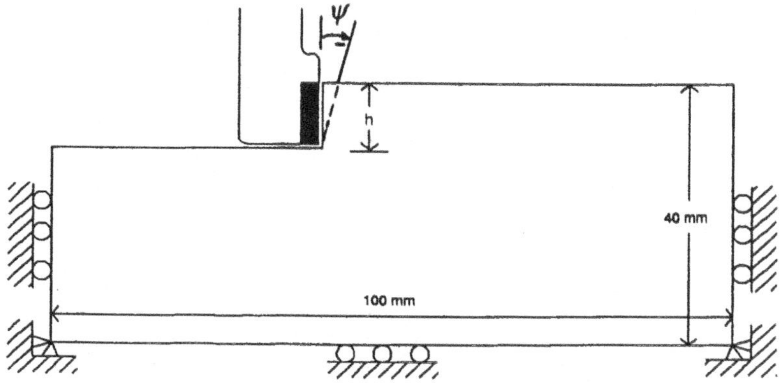

Figure 5. Modelization of the cutting process.

A sensitivity analysis of the load-displacement curves and the limit load with respect to geometrical and material parameters is performed. In the following, the load is the horizontal force acting on the cutter. The limit load is of practical interest because corresponding to the collapse mechanism representing the chip formation. First of all, it was observed that the limit load is an affin function of the cutting depth with a slop of 0.85 N/mm. The sensitivity of the loading curve to the cutting angle ψ is very weak as shown at Fig. 6.

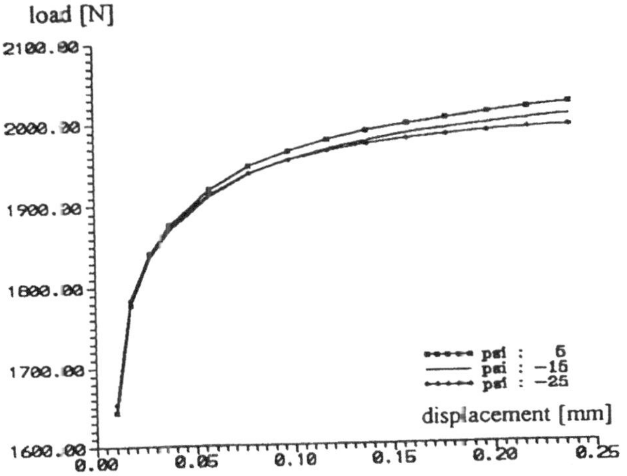

Figure 6. sensitivity of the loading curve to the cutting angle.

The influence of the loading curve to the internal friction angle is much more significative as given at fig. 7. The sensitivity of the limit load to φ is more less parabolic.

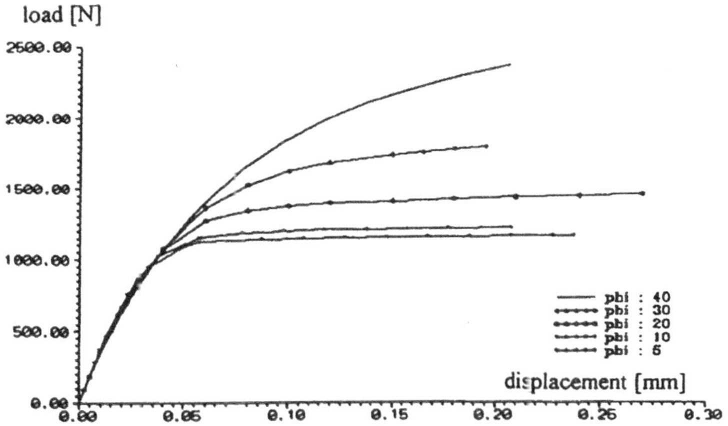

Figure 7. load-displacement curves

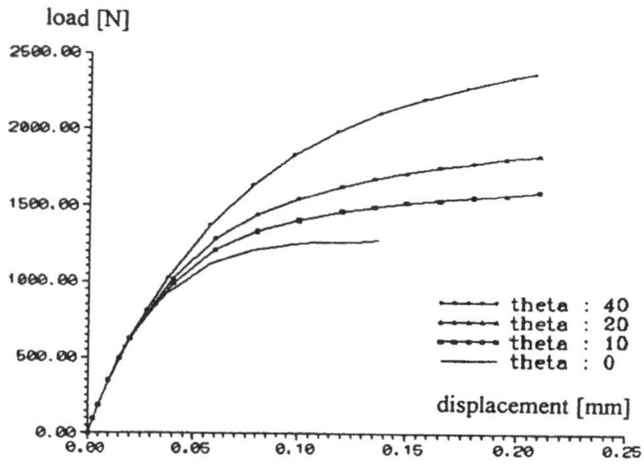

Figure 8. sensitivity of the loading curve to the plastic dilatancy angle.

Loading curves corresponding to various values of the plastic dilatancy angle φ , ploted at Fig. 8 show the great sensitivity of the results with respect to this parameter. This suggests to the experimentators the interest to determine the non associativity characteristics of the granular material constitutive laws.

REFERENCES

Ladeveze P., 1985, Sur une famille d'algorithmes en mécanique des structures, Comptes-rendu, Académie des Sciences-Paris, 300. Serie II, n°2, 41-44, 1985.

De Saxcé G., 1992a, The variational inequations for the problem of unilateral contact with friction, Proc. 2ème Congrès National Belge de Mécanique Théorique et Appliquée.

De Saxcé G., Feng Z.Q., 1991, New Inequation and Functional for Contact with Friction: the Implicit Standard material Approach. Mech. of Structures and Machines., Vol. 19, N°3.

De Saxcé G., 1992b, Une généralisation de l'inégalité de Fenchel et ses applications aux lois constitutives, C.R. Acad. Sci. Paris, t. 314, Série II.

Rudnicki J. and Rice J.R., 1975, Conditions for the localization of deformation in pressure-sensitive dilatant materials, J. Mech. Phys. Solid., Vol. 23, pp. 371-390.

De Saxcé G., Bousshine L., 1993, On the extension of limit analysis théorèms to the non associated flow rules in soils and to the contact with Coulomb's friction, Proc. XI, Polish Conference on Computer Methods Mechanics. Vol. 2. pp. 815-822, Kilce (Poland), mai 11-14.

Murtagh B.A., Sanders M.A., Minos 5.1 User's guide. Technical Rapport Sol 83 - 2OR.

De Saxcé G., Berga A., Bousshine L., 1992a, The Implicit Standard Material Approach for Non Associated Plasticity in Soil Mechanics, Proc. Int. Cong. on Num. Meth. in Eng. and Appl. Sciences, Vol. 1, pp. 585-594, Conception (Chile), 16-20 november.

De Saxcé G., Berga A, 1994, Elastoplastique Finite Element Analysis of Soil Problems with Implicit Standard Material Constitutive Laws, Revur Européenne des éléments finis, Vol. 3, N° 2.

SOME COMPUTATIONAL ASPECTS OF STRUCTURAL DYNAMICS PROBLEMS WITH FRICTIONAL CONTACT

Maciej Wronski[1] and Michel Jean[2]

[1]Division de la Modélisation Numérique en Mécanique
Université de Technologie de Compiègne, France
[2]Laboratoire de Mécanique et Génie Civil
Université de Montpellier II, France

INTRODUCTION

Contact and friction phenomena are common and very important in many fields of mechanical engineering. Recently, a significant research effort has been made to develop effective computational methods for finite element analysis of nonlinear problems in structural dynamics. This work may contribute to the discussion about different aspects of the implicit contact dynamics analysis. We begin with a brief presentation of the standard Newmark and Newton-Raphson approach. We propose several modifications to these schemes, presenting a set of first-order integration algorithms, and analyzing their numerical characteristics and utility in the numerical treatment of contact.

CLASSICAL MODELS

The most common method for dynamic implicit analysis in structural mechanics is the Newmark method. It is based on the following assumptions concerning the relation between displacement, velocity and acceleration:

$$\begin{aligned} X^{t+\Delta t} &= X^t + \Delta t\,\dot{X}^t + \Delta t^2\,[(\tfrac{1}{2} - \alpha)\ddot{X}^t + \alpha\ddot{X}^{t+\Delta t}] \\ \dot{X}^{t+\Delta t} &= \dot{X}^t + \Delta t\,[(1-\beta)\ddot{X}^t + \beta\ddot{X}^{t+\Delta t}] \end{aligned} \tag{1}$$

together with the dynamics equation:

$$M(X)\ddot{X} = F(X, \dot{X}, t) + R_c \tag{2}$$

Contact Mechanics, Edited by M. Raous *et al.*
Plenum Press, New York, 1995

The parameters α et β determine the stability and precision of the algorithm. Using these approximations, we transform the nonlinear dynamics equations to an incremental form

$$\hat{K}^k \Delta X = F_{ext}^{t+\Delta t} + F_{int}^k + F_{acc}^k + R_c^{k+1} \tag{3}$$

where $\hat{K}^k = K^k + \frac{\beta}{\alpha \Delta t} C^k + \frac{1}{\alpha \Delta t^2} M^k$ is effective stiffness,

$F_{acc}^k = -\frac{1}{\alpha \Delta t^2} M^k \left[\left(X^k - X^t\right) - \Delta t \dot{X}^t - \Delta t^2 \left(\frac{1}{2} - \alpha\right) \ddot{X}^t\right]$ are inertia forces,

and F_{int}, F_{ext}, R_c are internal, external and contact forces, respectively.

This set of simultaneous equations is usually solved for the displacement as a primary unknown by the Newton-Raphson method. The velocities and accelerations are then consecutively updated. In the implicit formulation, contact reactions and contact surfaces are also unknown; to complete the mathematical formulation, Signorini's condition and Coulomb friction law are assumed. Contact forces R_c are iteratively computed within an inner loop, using a penalty, flexibility, Lagrangian, or mathematical programming approach. In this work we have used the flexibility method (Jean and Touzot, 1988; Wronski, 1994), which can be easily applied to nonlinear 2D and 3D contact problems with arbitrary geometry and arbitrary material behaviour law. In this method, the problem is reduced to contact degrees of freedom, and contact forces are computed successively at each node using a relaxation method.

The Newmark method is unconditionally stable for $\beta \geq \frac{1}{2}$ et $\alpha \geq \frac{1}{4}\left(\frac{1}{2} + \beta\right)^2$. In standard applications without contact, the values corresponding to trapezoidal rule ($\beta = \frac{1}{2}$, $\alpha = \frac{1}{4}$) are commonly used. It is a second order scheme, which ensures a good stability and regularity of the solution. In contact problems however, higher order approximation does not necessarily mean better accuracy, and may even be superfluous. In the moments of sudden change of contact conditions (impact, release of contact), the velocity and acceleration are not continuous, and excessive regularity constraints may lead to serious errors. For the same reason, multi-step formulae are inadequate for contact problems. In Newmark method, these undesirable effects may be avoided by a particular choice of approximation parameters. For parameters $\beta = \alpha = \frac{1}{2}$, most commonly used (Chaudhary, 1986; Bathe, 1988), current acceleration does not depend on its previous value (but it still has to be computed and stored to update the velocity). Other values are also used, for example $\beta = 0.7$, $\alpha = 0.4$, where numerical dissipation is introduced to compensate sudden changes of velocity during impact charges (Wriggers et al., 1990).

FIRST ORDER INTEGRATION ALGORITHMS

A simple and effective solution to integrate dynamics equations with contact is to employ first order approximations. They do not introduce any regularity constraints on the kinematic variables, avoiding velocity estimation errors in the instants of a discontinuity of contact conditions. In this paper we describe three algorithms, based on the model proposed by Jean (to be published), and a more general weighted residual approach of Zienkiewicz et al. (1980, 1984).

Algorithm I

Let's write the dynamic equation in the form

$$M(X) d\dot{X} = F(X, \dot{X}, t) dt + R_c dt \tag{4}$$

Suppose that the mass matrix vary slowly with time; we can integrate (4) between consecutive time configurations t and $t + \Delta t$, using following approximations:

$$\int M(X)\, d\dot{X} = M(X^{t+\Delta t})(\dot{X}^{t+\Delta t} - \dot{X}^t) \tag{5}$$

$$\int_t^{t+\Delta t} F(t)\, dt = \Delta t \left[(1-\xi)F^t + \xi F^{t+\Delta t}\right] \tag{6}$$

$$\int_t^{t+\Delta t} R_c\, dt = \Delta t\, R_c^{t+\Delta t} \tag{7}$$

and

$$X^{t+\Delta t} - X^t = \Delta t\,[(1-\alpha)\dot{X}^t + \alpha \dot{X}^{t+\Delta t}] \tag{8}$$

A standard approximation of internal force derivative gives

$$\begin{aligned} F^{k+1} &\cong F^k + \frac{\partial F}{\partial X}(X^{k+1} - X^k) + \frac{\partial F}{\partial \dot{X}}(\dot{X}^{k+1} - \dot{X}^k) \\ &= F_{int}^k - K^k \Delta X - C^k \Delta \dot{X} \end{aligned} \tag{9}$$

We obtain then the recursive form of (4), expressed in terms of displacement

$$\hat{K}^k \Delta X = \hat{F}^k + \hat{F}_{acc}^k + R_c^{k+1} \tag{10}$$

$$X^{k+1} = X^k + \Delta X \tag{11}$$

where the final step configuration $t + \Delta t$ is approximated by current iteration $(k+1)$, and

$$\hat{K}^k = \frac{1}{\alpha(\Delta t)^2} M^k + \frac{\xi}{\alpha \Delta t} C^k + \xi K^k \tag{12}$$

$$F_{acc}^k = -\frac{1}{\alpha(\Delta t)^2} M^k \left[(X^k - X^t) - \Delta t \dot{X}^t\right] \tag{13}$$

$$\hat{F}^k = (1-\xi)\left(F_{int}^t + F_{ext}^t\right) + \xi\left(F_{int}^k + F_{ext}^{t+\Delta t}\right) \tag{14}$$

At the end of every time step, the velocity is updated

$$\dot{X}^{t+\Delta t} = \left(1 - \frac{1}{\alpha}\right)\dot{X}^t + \frac{1}{\alpha \Delta t}\left(X^{t+\Delta t} - X^t\right) \tag{15}$$

Algorithm II

The viscous damping matrix C is introduced in (9) as the derivative of velocity dependent terms in the constitutive equations, and assembled from elementary matrices. If we wish to itroduce damping effects, the Rayleigh damping may be also assumed: $C = \alpha_1 K_T + \alpha_2 M$. In this case, the dissipation forces may be directly introduced to the dynamics equation as $M\ddot{X} + C\dot{X} = F + R_c$. Supposing a slow variation in time of C matrix, we integrate this equation between t and $t + \Delta t$

$$M^k(\dot{X}^{t+\Delta t} - \dot{X}^t) + C^k(X^{t+\Delta t} - X^t) = \int_t^{t+\Delta t} [F(t) + R_c]\, dt \tag{16}$$

Using basic assumptions (5) – (7), we obtain another formulation for inertia forces and effective stiffness

$$\hat{K}^k = \frac{1}{\alpha(\Delta t)^2} M^k + \frac{1}{\Delta t} C^k + \xi K^k \tag{17}$$

$$F_{acc}^k = -\frac{1}{\alpha(\Delta t)^2}\left[\left(M^k + \alpha \Delta t C\right)\left(X^k - X^t\right) - \Delta t M^k \dot{X}^t\right] \tag{18}$$

Algorithm III

In the precedent algorithms, we have used the same approximations for displacement and velocity. It is possible to generalize this approach by introducing two different approximation parameters (α for the displacement, θ for the velocity), as proposed by Zienkiewicz, Wood and Taylor (1980). Equation (16) is then replaced by

$$\begin{aligned} M(\dot{X}^{t+\Delta t} - \dot{X}^t) + \Delta t C[(1-\theta)\dot{X}^t + \theta \dot{X}^{t+\Delta t}] \quad + & \\ \Delta t K[(1-\theta)X^t + \theta X^{t+\Delta t}] \quad = & \quad \int_t^{t+\Delta t} [F(t) + R_c]\, dt \end{aligned} \qquad (19)$$

It integrates to

$$M^k(\dot{X}^{t+\Delta t} - \dot{X}^t) + C^k[(1-\theta)X^t) + \theta X^{t+\Delta t}] = \int_t^{t+\Delta t} [F(t) + R_c]\, dt \qquad (20)$$

Finally, we obtain

$$\hat{K}^k = \frac{1}{\alpha(\Delta t)^2} M + \frac{\theta}{\alpha \Delta t} C + \xi K^k \qquad (21)$$

$$F^k_{acc} = -\frac{1}{\alpha(\Delta t)^2} M \left[(X^k - X^t) - \Delta t \dot{X}^t\right] \qquad (22)$$

$$-\frac{\theta}{\alpha \Delta t} C \left[(X^k - X^t) - \Delta t(1 - \frac{\alpha}{\theta})\dot{X}^t\right] \qquad (23)$$

If viscous damping is negligible, all three versions of the algorithm presented above are strictly identical. The only differences depend on the formulation used for dissipation forces. In linear case, the algorithm is unconditionally stable for parameters $\xi \geq \frac{1}{2}$, $\alpha \geq \frac{1}{2}$, $\theta \geq \alpha$, and reaches the maximal precision for $\alpha = \frac{1}{2}$.

Compared to the Newmark scheme, it better adapts to the non-smooth contact conditions. The kinematic variables are not supposed to be regular, and the acceleration terms need not be updated to continue the iterative resolution process. Another important advantage is the facility of parametric control over artificial (algorithmic) dissipation. For $\alpha > \frac{1}{2}$ and $\xi > \frac{1}{2}$ the damping effect increases, especially for higher frequencies. This may be useful in the modelling of dynamic processes, where the suppression of any spurious participation of higher modes may be desirable. Newmark method does not offer this type of numerical damping control. The non-dissipative trapezoidal rule is not adequate for contact applications; otherwise, the method possesses a numerical dissipation, but the choice of parameter values is very limited when contact is considered.

PSEUDO-INVERSE METHOD

The standard approach to solve a nonlinear set of equilibrium or dynamics equations, is the Newton-Raphson method. The unknown displacement vector is computed iteratively

$$X^{k+1} = X^k + \hat{K}^{-1}\left(\hat{F}^k + F^k_{acc} + R^{k+1}_c\right) \qquad (24)$$

where effective matrix $\hat{K}$ and force vectors $\hat{F}$, F_{acc}, are functions of displacement and internal state variables. To obtain a good convergence rate in nonlinear problems, the $\hat{K}$ matrix is updated every time step or iteration. Assembling and inverting $\hat{K}$

(usually by Gauss type solvers) become very costly in applications involving a large number of elements. For this reason, explicit time integration methods are commonly used for large scale, fast dynamics problems (as crash simulations), where they are very effective. However, their conditional stability requires very small time increments, even for standard, "slow" structural dynamics applications. To accelerate the resolution process, still preserving the implicit formulation, we propose to introduce a *pseudo-inverse* approximation $\tilde{K}^{-1}$ of the effective stiffness (Wronski, 1994). Equation solvers need not be used in this case, and the solution is directly obtained by a simple matrix-vector multiplication. This method is particularly interesting when coupled with the flexibilty method for contact computation. In this case, contact flexibility matrix is also directly calculated as a part of $\tilde{K}^{-1}$, reduced to contact degrees of freedom.

To introduce the pseudo-inverse approximation, we suppose that the mass matrix is diagonal and the viscous damping is negligible. The effective stiffness matrix $\hat{K}$ reduces to

$$\hat{K} = \frac{1}{\alpha(\Delta t)^2} M \left(I + \alpha\xi(\Delta t)^2 M^{-1} K \right) \tag{25}$$

Introducing definitions

$$A = \frac{1}{\alpha(\Delta t)^2} M, \qquad B = -\xi A^{-1} K \tag{26}$$

we can transform (25) to the following form:

$$\hat{K} = A(I - B), \qquad \hat{K}^{-1} = (I - B)^{-1} A^{-1} \tag{27}$$

The inversion of the diagonal mass matrix is trivial. Let's now consider the term $(1 - B)^{-1} = -\xi A^{-1} K$. It can be proved for an arbitrary matrix B, that the series $I + B + B^2 + \dots$ converge to $(I - B)^{-1}$, if its spectral radius is less than unity

$$\rho(B) < 1 \Rightarrow (I - B)^{-1} = \lim_{N \to \infty} (I + \sum_{n=1}^{N} B^n) \tag{28}$$

If we neglect higher order terms, we obtain the following approximation:

$$(I - B)^{-1} \cong I + B \tag{29}$$

The spectral radius of B matrix, equal to its maximal eigenvalue, has the upper bound set by

$$\rho(B) = max_i |\lambda_i(B)| \leq inf(\|B\|) \tag{30}$$

where $\|B\|$ is an arbitrary matrix norm. For the norm defined by the maximal sum of column elements $\sum_{j=1}^{n} |b_{ij}|$, we obtain

$$\|B\| = max_i \left(\sum_{j=1}^{n} |b_{ij}| \right) < 1 \tag{31}$$

If we introduce (26) into (31), the validity condition of the pseudo-inverse approximation (30) becomes

$$max_i \left((\Delta t)^2 \sum_{j=1}^{n} \left| \frac{k_{ij}}{m_{ii}} \right| \right) < 1 \tag{32}$$

which is true for all integration methods presented above, if $\alpha \leq 1$ and $\xi \leq 1$. We can then replace the exact inverse of $\hat{K}$ in equation (24) by

$$\tilde{K}^{-1} = \alpha \Delta t^2 M^{-1} \left(I - \alpha \xi \Delta t^2 K M^{-1} \right) \cong \hat{K}^{-1} \tag{33}$$

The quality of this approximation depends only on the significance of inertia effects, which are determined by the relation between the elements of tangent stiffness and mass matrices, and on the time step size. However, the condition (32) is necessary, but not sufficient to guarantee the convergence of the iterative procedure. In practice, the time step size often needs to be significantly diminished, even below the theoretical limit.

From the functional point of view, we can place this method between explicit and implicit schemes. The obvious advantage of this approach is that the solution can be obtained without actually inversing the stiffness matrix, as in an explicit scheme. The time step, limited by the criterion of the pseudo-inverse approximation, does not depend on stability conditions. In the numerical analysis, this criterion may also be used for the automatic control of time increments, ensuring a steady convergence rate and the same approximation error.

This method may be very effective for large scale applications, concerning fast dynamic processes with contact (especially when treated by the flexibility method), where inertia forces are dominant over the material stiffness. In such applications, $\tilde{K}^{-1}$ approaches to the inverse of the diagonal mass matrix, and may be easily computed.

NUMERICAL EXAMPLES

To illustrate the behaviour of the presented algorithms, we consider a numerical example of an elasto-plastic ball, bouncing against a flat elastic surface (figure 1).

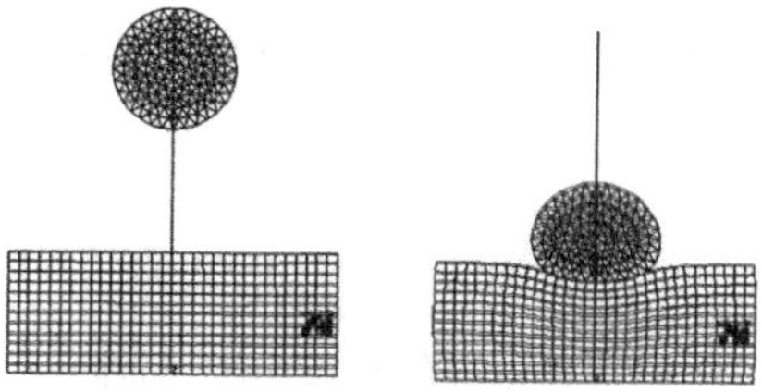

Figure 1. Bouncing ball simulation

The model includes large deformations, and a nonlinear material behaviour of the ball. The unilateral contact between the block surface and the ball is treated by the flexibility method. No viscous damping is assumed. The following figures show the comparative results, obtained with different methods and their integration parameters. The Newmark trapezoidal rule has a troublesome oscillatory behaviour, and do not converge in this example, even for very small time increments. Good results have been obtained for parameters $\alpha = 0.4$, $\beta = 0.7$, and for $\alpha = \beta = \frac{1}{2}$, although some residual oscillations were still observed in this case (figure 2).

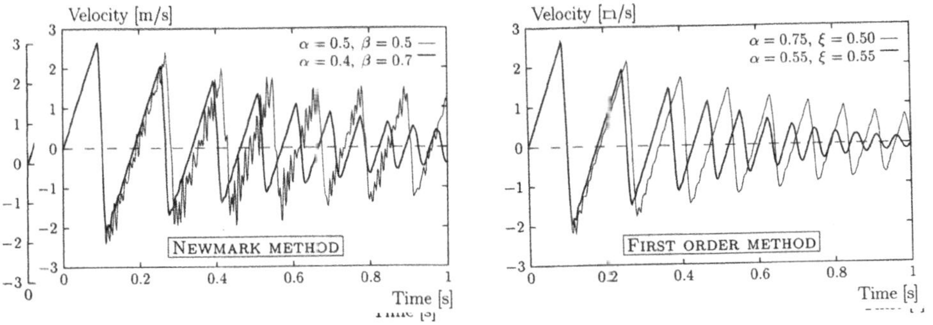

Figure 2. Ball velocity. Newmark and first order algorithms

The first order scheme gives more regular solution, especially if numerical dissipation ($\alpha = 0.75$) is introduced, as shown on the same figure. It has also a better stability in the function of the time step size. The effect of the algorithmic dissipation is most pronounced in higher modes, which can be well observed in the surface oscillations (figures 3 and 4). Side effects of the numerical damping are changes of both jump amplitude and frequency, but they are not very significant. A bigger amplitude error is made if long time increments are used, because of an underestimation of the impact energy. Finally, the pseudo-inverse approach was also tested, and gave exactly the same results as standard Newton-Raphson algorithm. Yet, to satisfy the numerical criterion of the inverse approximation of this method, the time step had to be considerably diminished, and total computation time actually increased.

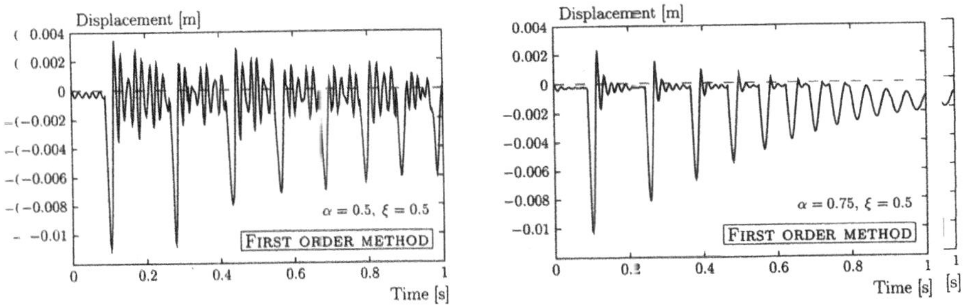

Figure 3. Elastic surface oscillations. Numerical damping effect

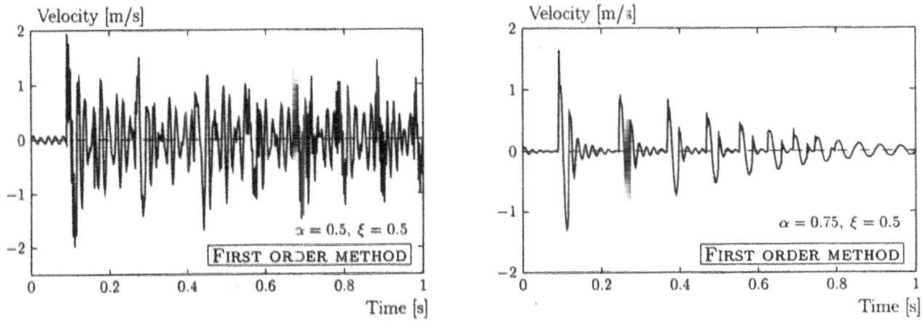

Figure 4. Numerical damping of high frequency elastic surface vibrations

CONCLUDING REMARKS

We have proposed in this paper a simple first-order integration scheme for the numerical integration of dynamics equations. The numerical simulations show its usefulness in contact applications. The algorithm gives a stable solution, even in the moments of sudden velocity changes during impacts. The artificial dissipation is not necessary for the convergence of the algorithm. When it is used, it allows damping of high frequencies not under consideration, and may be easily controlled by the modification of integration coefficients.

We have also presented a pseudo-inverse modification of standard Newton-Raphson approach. Although its effectiveness in "slow" dynamics problems is to be proved, it may be useful in large scale applications with many contact degrees of freedom. Continuing research on further improvements of this method, we also investigate a possibility of its application to the dynamic regularization of highly nonlinear, ill-conditioned quasistatic problems with contact.

REFERENCES

Bathe K.J., Mijailovich S., 1988, Finite element analysis of frictional contact problems, *Journal Mec. Theor. Appl.*, vol. 7, pp. 31–45. Special issue, supplement N° 1.

Chaudhary A., Bathe K.J., 1986, A solution method for static and dynamic analysis of three-dimensional contact problems with friction, *Comp. Struct.*, vol. 24, no. 6, pp. 855–873.

Jean M., Frictional contact in collections of rigid or deformable bodies: Numerical simulations of geometrical motions, to be published in: Mechanics of Geomaterial Interfaces, Elsevier Science Publishers.

Jean M., Touzot G., 1992, Implementation of unilateral contact and dry friction in computer codes dealing with large deformation problems. *Journal Mec. Theor. Appl.*, vol. 7, pp. 145–160. Special issue, supplement N° 1.

Wriggers P., Vu Van T., and Stein E., 1990, Finite element formulmation of large deformation impact-contact problems with friction, *Comp. Struct.*, vol. 37, no. 3, pp. 319–331.

Wronski M., 1994, Couplage du Contact et du Frottement avec la Mécanique Non Linéaire des Solides en Grandes Déformations, PhD thesis, Université de Technologie de Compiègne.

Zienkiewicz O. et al., 1984, A unified set of single step algorithms. Part I: general formulation and applications, *Int. J. Numer. Methods Eng.*, vol. 20, pp. 1529–1552.

Zienkiewicz O., Wood W.L. and Taylor R.L., 1980, An alternative single-step algorithm for dynamic problems, *Earthq Engng Struct. Dyn.*, vol. 8, no. 1, pp. 31–40.

CONTINUUM MECHANICS MODELLING OF LARGE DEFORMATION CONTACT WITH FRICTION

Alain Curnier,[1] Qi-Chang He,[1] and Anders Klarbring[2]

[1] Département de Génie Mécanique,
Ecole Polytechnique Fédérale de Lausanne,
CH-1015 Lausanne

[2] Department of Mechanical Engineering,
Linköping Institute of Technology,
S-58183 Linköping

INTRODUCTION

Material forming, vehicle crash and human joints are but a few typical situations where *large displacement contact mechanics* plays a crucial role. Although classical computational methods combined with special constrained optimisation techniques have been successfully developed by many authors for solving such contact problems [see e.g. Kikuchi and Oden 1985], the underlying *continuum mechanics formulation* is lagging behind, probably due to its complexity. Most of the work is restricted to unilateral contacts without friction [Ciarlet and Necas, 1985; Ciarlet 1988; Curnier et al., 1992] and a tentative theory with friction is just emerging [Klarbring et al., 1991; Laursen and Simo, 1991, 1993; Klarbring, 1994].

The purpose of this paper is to propose a formulation of unilateral contact between two deformable continuous solids undergoing large displacements with friction at the interface, in the line of the two last quoted references. A characteristic of the model is that kinematic and static contact quantities (relative velocity and stress vector) and governing equations (action-reaction and moment balance) are defined not only when the two bodies are effectively in contact but also when they are slightly separated or penetrated. This enables treating the unilateral contact conditions as a tribological law rather than as boundary conditions, which is advantageous for implementing numerical solution procedures. Another feature of the formulation is that it is entirely developed in intrinsic form without resorting to surface coordinates. Yet, the formulation remains limited to slow kinematics and dynamics (i.e. inertia forces are neglected and impact discontinuities disregarded) and thus particularly suited for rate independent friction problems.

Contact Mechanics, Edited by M. Raous *et al.*
Plenum Press, New York, 1995

The article outline is classical in so far as contact kinematics, statics, energetics and tribology are successively developed. For *kinematics*, the model relies on the projection of a particle located on the boundary of one body on the potential contact surface of the other. Provided a well identified condition is fulfilled, this projection establishes a bijection between the neighbouring potential contact surfaces. Then, a measure of the relative velocity, which is objective even when the two particles are distant, is proposed. Regarding *statics*, the principle of action and reaction is stated in terms of the first Piola-Kirchhoff stress vector and extended for two particles which are not necessarily in contact but put in bijection by the above projection. The balance of moments turns out to play a non trivial role when a separation or a penetration occurs. Using the principle of energy conservation and the established bijection, the contact stress vector is then put in *energetic* duality with the relative contact velocity defined earlier. Finally, *tribological laws* including the unilateral contact law (Signorini conditions) as well as rate independent friction laws (based on Coulomb's isotropic criterion and some anisotropic extensions) are formulated.

As required by an intrinsic formulation, the direct notation is used throughout: vectors are denoted by small bold letters ($\mathbf{v}$) and tensors by capital bold letters ($\mathbf{T}$). Often the point at which a function is evaluated is omitted as a shorthand which entails that functions are denoted by the same letter as their values in spite of the resulting ambiguities ($\mathbf{y} = \mathbf{y}(\mathbf{x},t)$). The spatial form of a field is indicated by an italic $f = \mathit{f}(\mathbf{y},t) = \mathit{f}[\mathbf{y}(\mathbf{x},t),t] = f(\mathbf{x},t)$.

CONTACT KINEMATICS

In this section, the geometric and kinematic quantities found suitable for measuring the relative motion of two deformable bodies in eventual contact are defined. To begin, basic elements of solid kinematics are reviewed with particular emphasis on surface description.

Solid surface kinematics

A deformable solid is identified with the domain Ω of $\mathbf{R}^3$ it occupies in some reference configuration. Accordingly, a particle is identified by its reference position vector $\mathbf{x} \in \Omega$. Denoting by $\mathbf{y}(\mathbf{x},t)$ the current position vector of the particle $\mathbf{x}$ at time t, the motion $\mathbf{y}$ of body Ω is a smooth invertible mapping noted

$$\mathbf{y} = \mathbf{y}(\mathbf{x},t)\ ,\quad \mathbf{y} : \ \Omega \times [0,T] \rightarrow \mathbf{R}^3.$$

The velocity and the transformation gradient are defined by

$$\dot{\mathbf{y}} = \dot{\mathbf{y}}(\mathbf{x},t) = \partial_t \mathbf{y}(\mathbf{x},t)\ ,\quad \mathbf{F} = \mathbf{F}(\mathbf{x},t) = \nabla_{\mathbf{x}}\, \mathbf{y}(\mathbf{x},t) \quad (0 < J(\mathbf{x},t) = \det\mathbf{F} < \infty).$$

The boundary $\partial\Omega$ of Ω is assumed to be divided into three disjoint parts, a part where the motion is prescribed, another where the traction will be specified and the ***reference potential contact surface*** $\Gamma \subseteq \partial\Omega$. The ***deformed potential contact surface*** Γ_t is the image of the reference one Γ by the (invertible) motion (Fig.1)

$$\Gamma_t = \mathbf{y}(\Gamma,t) = \{\ \mathbf{y} \in \mathbf{R}^3 \mid \mathbf{y} = \mathbf{y}(\mathbf{x},t)\ ,\ \mathbf{x} \in \Gamma\ \}.$$

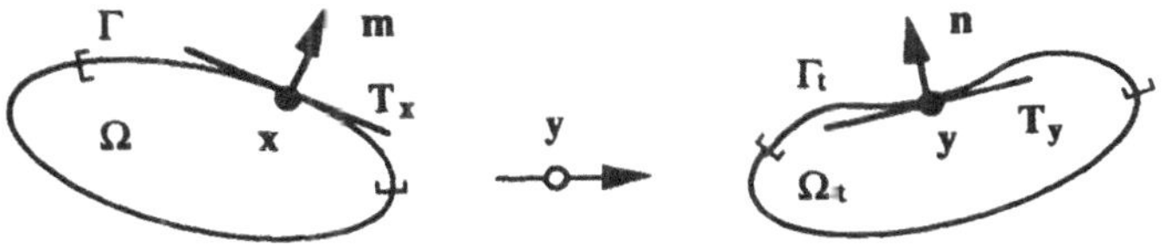

Fig. 1. Solid surface kinematics.

In continuum mechanics, the surface $\partial\Omega$ is usually characterised by its outward unit normal alone since only its orientation is relevant to the theory. In the coming contact mechanics formulation, both the orientation and the curvature of the contact surface Γ are needed, which presumes that it is orientable and regular. Accordingly, the reference and deformed contact surfaces Γ and Γ_t are characterised by their ***outward unit normal*** vectors $\mathbf{m}$ and $\mathbf{n}$ as well as their 3D ***gradient*** tensors $\mathbf{M}$ and $\mathbf{N}$ at each point $\mathbf{x}$ and $\mathbf{y} = \mathbf{y}(\mathbf{x},t)$, respectively:

$$\mathbf{m} = \mathbf{m}(\mathbf{x}) \perp \Gamma\,, \quad \|\mathbf{m}\| = 1\,, \qquad \mathbf{n} = \boldsymbol{n}(\mathbf{y},t) = \boldsymbol{n}[\mathbf{y}(\mathbf{x},t),t] = \mathbf{n}(\mathbf{x},t) \perp \Gamma_t\,, \quad \|\mathbf{n}\| = 1,$$

$$\mathbf{M} = \mathbf{M}(\mathbf{x}) = \nabla_{\mathbf{x}}\,\mathbf{m}(\mathbf{x})\,, \qquad \mathbf{N} = \boldsymbol{N}(\mathbf{y},t) = \nabla_{\mathbf{y}}\,\boldsymbol{n}(\mathbf{y},t)\,, \quad (\nabla_{\mathbf{x}}\mathbf{n} = \nabla_{\mathbf{y}}\,\boldsymbol{n}\,\mathbf{F}).$$

The planes orthogonal to $\mathbf{m}(\mathbf{x})$ and $\boldsymbol{n}(\mathbf{y},t)$ are called the *tangent planes* to Γ at $\mathbf{x}$ and to Γ_t at $\mathbf{y}$, respectively, and are denoted

$$\mathrm{T}_{\mathbf{x}} = \{\mathbf{v} \in \mathbf{R}^3 \mid \mathbf{v} \cdot \mathbf{m}(\mathbf{x}) = 0\}\,, \qquad \mathrm{T}_{\mathbf{y}} = \{\mathbf{w} \in \mathbf{R}^3 \mid \mathbf{w} \cdot \boldsymbol{n}(\mathbf{y},t) = 0\}.$$

Since $\mathbf{n}$ and $\mathbf{N}$ are essential for the formulation, it is important to indicate how they can be calculated. Denoting by "$\wedge$" the cross product of two vectors, an element of reference surface $\mathrm{d}\mathbf{s} = \mathrm{d}\mathbf{x} \wedge \mathrm{d}\mathbf{x}' = \mathbf{m}\,\mathrm{dA}$ with $\mathrm{d}\mathbf{x}, \mathrm{d}\mathbf{x}' \in \mathrm{T}_{\mathbf{x}}$, its area $\mathrm{dA} = \|\mathrm{d}\mathbf{s}\|$ and its unit orientation $\mathbf{m}$ are transformed in the deformed configuration into $\mathrm{d}\mathbf{s}_t = \mathrm{d}\mathbf{y} \wedge \mathrm{d}\mathbf{y}' = \mathbf{n}\,\mathrm{dA}_t$, $\mathrm{dA}_t = \|\mathrm{d}\mathbf{s}_t\|$ and $\mathbf{n}$, respectively, by means of the transformation co-gradient $\mathbf{F}^*$ ($= \mathrm{J}\mathbf{F}^{-\mathrm{T}}$ since $\exists\,\mathbf{F}^{-1}$) as

$$\mathrm{d}\mathbf{s}_t[\mathbf{y}(\mathbf{x},t),t] = \mathbf{F}^*(\mathbf{x},t)\,\mathrm{d}\mathbf{s}(\mathbf{x})\,, \quad \boldsymbol{n}[\mathbf{y}(\mathbf{x},t),t] = \frac{\mathbf{F}^*(\mathbf{x},t)\,\mathbf{m}(\mathbf{x})}{\|\mathbf{F}^*(\mathbf{x},t)\,\mathbf{m}(\mathbf{x})\|}\,, \tag{1}$$

$$\mathrm{dA}_t[\mathbf{y}(\mathbf{x},t),t] = \mathrm{j}(\mathbf{x},t)\,\mathrm{dA}(\mathbf{x})\,, \quad 0 < \mathrm{j}(\mathbf{x},t) = \|\mathbf{F}^*(\mathbf{x},t)\,\mathbf{m}(\mathbf{x})\| < \infty\,. \tag{2}$$

The calculation of the normal gradient $\mathbf{N}$ is more complicated [e.g. Hansson and Klarbring, 1990]. Whether the reference surface Γ is described in implicit or in explicit (parametric) form, a non-unit normal $\tilde{\mathbf{m}}$ and its 3D gradient $\tilde{\mathbf{M}}$ can easily be found. A non-unit normal $\tilde{\mathbf{n}}$ to the deformed surface Γ_t and its gradient can then be calculated by the chain rule as

$$\tilde{\mathbf{n}} = \tilde{\boldsymbol{n}}(\mathbf{y},t) = \mathbf{F}^*\,\tilde{\mathbf{m}}\,, \qquad \tilde{\mathbf{N}} = \nabla_{\mathbf{y}}\,\tilde{\boldsymbol{n}}(\mathbf{y},t) = \mathbf{F}^*[\tilde{\mathbf{M}} - (\nabla_{\mathbf{x}}^{t}\,\mathbf{F}^{*-1})\tilde{\boldsymbol{n}}\,]\mathbf{F}^{*\mathrm{T}}\,,$$

where the transpose of the 3rd order tensor $\in = \nabla_{\mathbf{x}}\mathbf{F}^{*-1}$ is defined as $\mathbf{x} \cdot (\in \mathbf{X}) = [\in^{t}\mathbf{x}] : \mathbf{X}$. The normal $\tilde{\mathbf{n}}$ can always be normalised to unity, after what its gradient can be derived as

$$\mathbf{N} = \boldsymbol{N}(\mathbf{y},t) = [\mathbf{I} - \mathbf{n} \otimes \mathbf{n}]\frac{\tilde{\mathbf{N}}}{\|\tilde{\mathbf{n}}\|} = [\mathbf{I} - \mathbf{n} \otimes \mathbf{n}]\frac{\mathbf{F}^*[\tilde{\mathbf{M}} - (\nabla_{\mathbf{x}}^{t}\,\mathbf{F}^{*-1})(\mathbf{F}^*\,\tilde{\mathbf{m}})]\mathbf{F}^{*\mathrm{T}}}{\|\mathbf{F}^*\,\tilde{\mathbf{m}}\|}\,,$$

where "$\otimes$" represents the tensor product of two vectors defined by $[\mathbf{a} \otimes \mathbf{b}]\mathbf{x} = (\mathbf{b} \cdot \mathbf{x})\mathbf{a}$, $\forall\mathbf{x}$.

Contact geometry

Consider now ***two*** bodies Ω and Ω' subject to finite deformations and bound to contact each other (Fig. 2). For describing their relative motion, it is found convenient to take one of

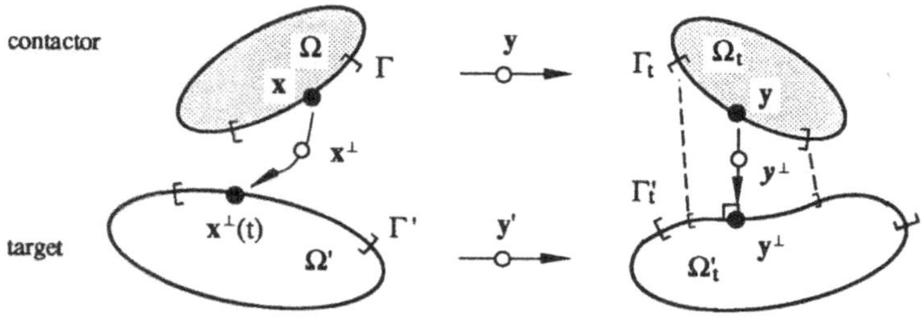

Fig. 2. Contact kinematics.

them, say Ω', for reference and to describe the approach of the other, thus Ω, with respect to it. The relative body Ω is called the *contactor* and the reference one Ω' the *target*. Accordingly, the potential contact surface $\Gamma \subset \partial\Omega$ is called the ***contact surface*** and the corresponding one $\Gamma' \subset \partial\Omega'$ the ***target surface***. Of course, all quantities introduced so far on Ω are reproduced on Ω' with a prime. The union $\Omega \cup \Omega'$ is called the *contact system*. The key idea of the following geometric formulation is to establish a *bijection* between Γ and a portion $\Gamma^{\perp}$ of Γ' based on a (unique, regular) *projection* of Γ_t on Γ'_t at each time t, i.e. even when Ω_t on Ω'_t are separated or have penetrated one another.

For measuring the normal approach of the particle $\mathbf{x} \in \Gamma$ located at $\mathbf{y}(\mathbf{x},t) \in \Gamma_t$ with respect to the target surface Γ' currently deformed into $\Gamma'_t = \mathbf{y}'(\Gamma',t)$ (Fig.3), it is natural to look on Γ' for the ***"proximal particle"*** $\mathbf{x}^{\perp} \in \Gamma'$, which is currently located at the ***closest point*** $\mathbf{y}^{\perp} = \mathbf{y}'(\mathbf{x}^{\perp},t) \in \Gamma'_t$ to $\mathbf{y}(\mathbf{x},t)$, defined by

$$\mathbf{x}^{\perp} = \underset{\mathbf{x}' \in \Gamma'}{\operatorname{argmin}} \frac{1}{2} \| \mathbf{y}(\mathbf{x},t) - \mathbf{y}'(\mathbf{x}',t) \|^2 .$$

The necessary condition for the (halved squared) distance separating $\mathbf{y}$ from $\mathbf{y}'$ to have a local minimum (or maximum) at $\mathbf{y}^{\perp}$ is $[\mathbf{y}(\mathbf{x},t) - \mathbf{y}'(\mathbf{x}^{\perp},t)] \cdot \mathbf{F}'(\mathbf{x}^{\perp},t)\,d\mathbf{x}' = 0$, $\forall d\mathbf{x}' \in \mathbf{T}_{\mathbf{x}^{\perp}}$. Since $d\mathbf{x}'$ belongs to the tangent plane $\mathbf{T}_{\mathbf{x}^{\perp}}$ to Γ' at $\mathbf{x}^{\perp}$, $d\mathbf{y}' = \mathbf{F}'(\mathbf{x}^{\perp},t)\,d\mathbf{x}'$ belongs to its image $\mathbf{T}_{\mathbf{y}^{\perp}}$ at $\mathbf{y}^{\perp}$ and consequently the closest point is also ***an orthogonal projection*** of $\mathbf{y}(\mathbf{x},t)$ on Γ'_t, thus the notations $\mathbf{y}^{\perp}$ and $\mathbf{x}^{\perp}$. It follows that the proximal particle $\mathbf{x}^{\perp}$ is equivalently characterised by the necessary condition

$$[\mathbf{I} - \mathbf{n}'(\mathbf{x}^{\perp},t) \otimes \mathbf{n}'(\mathbf{x}^{\perp},t)]\,[\mathbf{y}(\mathbf{x},t) - \mathbf{y}'(\mathbf{x}^{\perp},t)] = \mathbf{0} ,$$

where $\mathbf{n}^{\perp} = \mathbf{n}'(\mathbf{x}^{\perp},t) = \boldsymbol{n}'[\mathbf{y}'(\mathbf{x}^{\perp},t),t]$ is the deformed target normal at $\mathbf{y}^{\perp}$. Provided it is unique (a provision which is discussed later), the solution of this implicit equation can be written $\mathbf{x}^{\perp} = \mathbf{x}^{\perp}(\mathbf{x},t)$. The vector separating the particle $\mathbf{x} \in \Gamma$ at $\mathbf{y}(\mathbf{x},t) \in \Gamma_t$ from the proximal particle $\mathbf{x}^{\perp} \in \Gamma'$ at $\mathbf{y}'(\mathbf{x}^{\perp},t) \in \Gamma'_t$ at time t is called the ***normal gap vector*** $\mathbf{g}$ and its (only) component along the target normal, the ***signed contact distance*** g_n

$$\mathbf{g} = \mathbf{g}(\mathbf{x},t) = \mathbf{y}(\mathbf{x},t) - \mathbf{y}'[\mathbf{x}^{\perp}(\mathbf{x},t),t] = g_n(\mathbf{x},t)\, \mathbf{n}'[\mathbf{x}^{\perp}(\mathbf{x},t),t] , \tag{3}$$

$$g_n = g_n(\mathbf{x},t) = \{\mathbf{y}(\mathbf{x},t) - \mathbf{y}'[\mathbf{x}^{\perp}(\mathbf{x},t),t]\} \cdot \mathbf{n}'[\mathbf{x}^{\perp}(\mathbf{x},t),t] , \quad (\mathbf{g}_T = \mathbf{g} - g_n\, \mathbf{n}^{\perp} = \mathbf{0}) .$$

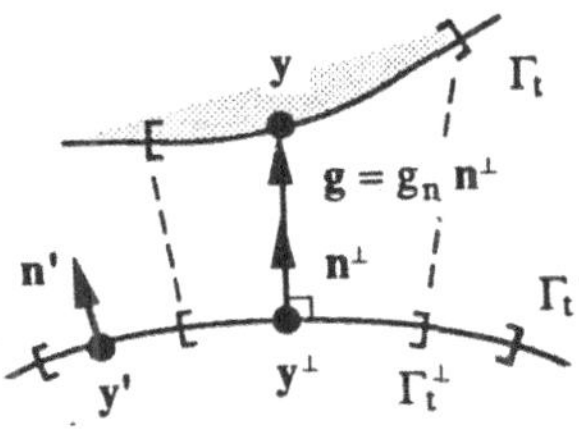

Fig. 3. Projection and gap vector.

It is emphasised that the gap vector and the signed distance remain well defined even if the two bodies penetrate each other, a situation which is not precluded at this stage. In fact, the signed distance defines the *geometric contact status* of the particle **x** :

$$g_n > 0 \leftrightarrow gap\,, \qquad g_n = 0 \leftrightarrow contact\,, \qquad g_n < 0 \leftrightarrow penetration\,.$$

In order to assess the (local) uniqueness and regularity of the projection $\mathbf{y}^{\perp} = y^{\perp}(\mathbf{y},t)$, it is expedient to find its spatial gradient $\mathbf{P}^{\perp} = \nabla_y\, y^{\perp}(\mathbf{y},t)$ presuming it exists, and then look if its determinant remains finite and positive, respectively. More exactly, since $\mathbf{P}^{\perp}$ maps any contactor line element into the target tangent plane $\mathbf{T}_{y^{\perp}}$, it is the restriction $\mathbf{P}_T^{\perp}$ of $\mathbf{P}^{\perp}$ to the contactor tangent plane $\mathbf{T}_y$ which is relevant because it controls the elongation of *tangent* line elements during the projection as $dy^{\perp} = \mathbf{P}_T^{\perp}(y,t)dy$. Applying the chain rule to the spatial characterisation of $y^{\perp}$, $\mathbf{P}^{\perp}(y,t)$ and $\mathbf{P}_T^{\perp}(y,t)$ can be found to be characterised by

$$[\mathbf{I} - \boldsymbol{n}^{\perp} \otimes \boldsymbol{n}^{\perp} + g_n \boldsymbol{N}^{\perp}]\, \mathbf{P}^{\perp} = [\mathbf{I} - \boldsymbol{n}^{\perp} \otimes \boldsymbol{n}^{\perp}]\ ,\quad \mathbf{P}_T^{\perp} = \mathbf{P}^{\perp}\,[\mathbf{I} - \boldsymbol{n} \otimes \boldsymbol{n}]\ ,\quad \mathbf{P}_T^{\perp} : \mathbf{T}_y \rightarrow \mathbf{T}_{y^{\perp}}\,.$$

Although $\mathbf{P}_T^{\perp}$ is of rank ≤ 2, its cofactor $\mathbf{P}_T^{\perp *}$ is a well defined rank ≤ 1 tensor through $(\mathbf{P}_T^{\perp}\, dy) \wedge (\mathbf{P}_T^{\perp}\, dz) = \mathbf{P}_T^{\perp *}\,(dy \wedge dz)$. Accordingly, an element $ds_t = \boldsymbol{n}\, dA_t$ of Γ_t is projected on Γ_t' into $ds_t^{\perp} = \boldsymbol{n}^{\perp} dA_t^{\perp}$ as in (1)-(2) with $\mathbf{F}^*$ and j replaced by $\mathbf{P}_T^{\perp *}$ and $j^{\perp} = \| \mathbf{P}_T^{\perp *} \boldsymbol{n} \|$, respectively, and, by exploiting the above characterisation of $\mathbf{P}_T^{\perp}$, it can be shown that

$$j^{\perp}(y,t) = \frac{dA_t^{\perp}}{dA_t} = \frac{-\,\boldsymbol{n}(y,t) \cdot \boldsymbol{n}'(y^{\perp},t)}{|\,1 - (\kappa_1^{\perp} + \kappa_2^{\perp})\, g_n + \kappa_1^{\perp} \kappa_2^{\perp}\, g_n^2\,|}\ , \tag{4}$$

where $\kappa_1^{\perp}$, $\kappa_2^{\perp}$ are the principal curvatures of Γ_t' at $y^{\perp}$. Now, the projection $\mathbf{y}^{\perp}$ is *locally unique* only if $j^{\perp}(y,t) < \infty \Leftrightarrow g_n \neq 1/\kappa_1^{\perp}$ or $1/\kappa_2^{\perp}$, i.e. only if y is not a curvature center. It is globally unique and can effectively be written $\mathbf{y}^{\perp} = y^{\perp}(\mathbf{y},t)$ only if $\mathbf{y}$ is not a focal point (an equidistant intersection of normals) of Γ_t' (Fig. 4). Further, $\mathbf{y}^{\perp}$ is locally the *closest* (i.e. the optimum is a local minimum) if and only if there is no focal point of Γ_t' between $\mathbf{y}$ (included) and $\mathbf{y}^{\perp}$ along the normal $\mathbf{n}^{\perp}$ [Thorpe, 1979] or, more restrictively and globally, if $|\,g_n\,| < \rho_{min}(\mathbf{y}^{\perp}) = 1/\max(|\kappa_1^{\perp}|, |\kappa_2^{\perp}|)$, $\forall \mathbf{y}^{\perp}$, i.e. if the distance separating $\mathbf{y}$ from $\mathbf{y}^{\perp}$ is smaller than the minimum radius of curvature of Γ_t' at any $\mathbf{y}'$. Conversely, the projection $\mathbf{P}_T^{\perp}$ stays of rank 2 if $j^{\perp}(y,t) \neq 0 \Leftrightarrow \mathbf{n} \cdot \mathbf{n}^{\perp} \neq 0$ and *positive semi-definite* if $j^{\perp}(y,t) > 0 \Leftrightarrow \mathbf{n} \cdot \mathbf{n}^{\perp} < 0$, i.e. if the outward normals to the contactor and target deformed surfaces are opposite (Fig. 4). In case of contact ($g_n = 0$ and $\mathbf{n} = -\,\mathbf{n}^{\perp}$), $j^{\perp} = 1$, and in case of a plane target ($\kappa_1^{\perp} = \kappa_2^{\perp} = 0$) $j^{\perp} = -\mathbf{n} \cdot \mathbf{n}^{\perp}$, as expected. The opposite normal condition is also crucial in case of *self contact* as a criterion for splitting $\partial\Omega$ into Γ and Γ'. Finally, provided a condition $0 < j^{\perp}(y,t) < M$ for all $\mathbf{y} \in \Gamma_t$ is satisfied, the projection $\mathbf{y}^{\perp}$ establishes, at each instant t, a bijection $\mathbf{y}^{\perp}$ between the deformed contact surface Γ_t and a subset $\Gamma_t^{\perp}$ of the deformed target surface Γ_t', and by composition with the contactor motion and target inverse motion, a ***bijection*** $\mathbf{x}^{\perp}$ between

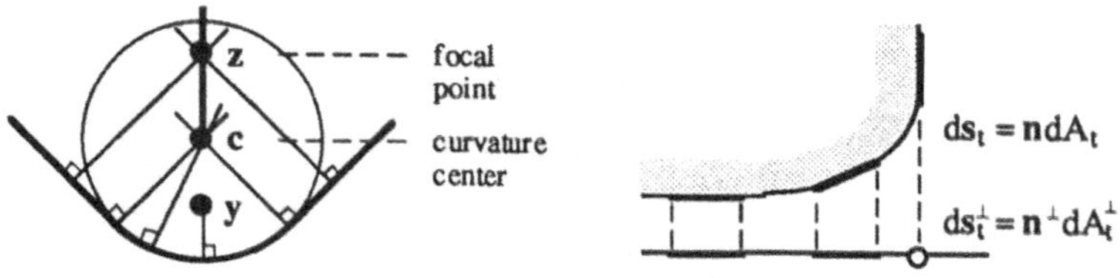

Fig. 4. Projection uniqueness and regularity.

the reference contact surface Γ and a subset $\Gamma^{\perp}$ of the reference target surface Γ' (Fig. 2)

$$\mathbf{x}^{\perp} = \mathbf{x}^{\perp}(\mathbf{x},t) = \mathbf{y}'^{-1}\{\mathbf{y}^{\perp}[\mathbf{y}(\mathbf{x},t)],t\}\ , \quad \mathbf{x}^{\perp} = \mathbf{y}'^{-1} \circ \mathbf{y}^{\perp} \circ \mathbf{y} : \Gamma \leftrightarrow \Gamma^{\perp}\ .$$

It follows that an element $ds = \mathbf{m}\, dA$ of Γ is transformed on Γ' into $ds^{\perp} = \mathbf{m}^{\perp} dA^{\perp}$ as in (1)-(2) again but with $\mathbf{F}^*$ and j replaced by

$$\mathbf{B}^*(\mathbf{x},t) = \mathbf{F}'^{*-1}[\mathbf{x}^{\perp}(\mathbf{x},t),t]\ \mathbf{P}_T^{\perp *}[\mathbf{y}(\mathbf{x},t),t]\ \mathbf{F}^*(\mathbf{x},t)\ ,$$

$$0 < j^{\perp}(\mathbf{x},t) = \| \mathbf{B}^*(\mathbf{x},t)\, \mathbf{m}(\mathbf{x}) \| = \frac{j(\mathbf{x},t)\, j^{\perp}[\mathbf{y}(\mathbf{x},t),t]}{j'[\mathbf{x}^{\perp}(\mathbf{x},t),t]} < \infty\ .$$

Contact kinematics

By construction, the gap vector is a measure of the normal approach of the contactor particles from the target surface. A measure of the *tangential slip* is hard to exhibit because the impact time t^i and location $\mathbf{x}^i = \mathbf{x}^{\perp}(\mathbf{x},t^i)$ of the particle $\mathbf{x}$ on the target surface Γ' are ignored and because the later is curved. Nevertheless a measure of the *slip rate* can be proposed. To begin, the *gap vector rate* can be derived from its definitions (3) to be

$$\dot{\mathbf{g}} = \dot{\mathbf{g}}(\mathbf{x},t) = \dot{\mathbf{y}}(\mathbf{x},t) - \dot{\mathbf{y}}'[\mathbf{x}^{\perp}(\mathbf{x},t),t] = \dot{g}_n(\mathbf{x},t)\ \mathbf{n}'[\mathbf{x}^{\perp}(\mathbf{x},t),t] + g_n(\mathbf{x},t)\ \dot{\mathbf{n}}'[\mathbf{x}^{\perp}(\mathbf{x},t),t]\ ,$$

where $\dot{\mathbf{y}}^{\perp}$ and $\dot{\mathbf{n}}^{\perp}$ denote the total time derivatives of $\mathbf{y}^{\perp}$ and $\mathbf{n}^{\perp}$ defined by

$$\dot{\mathbf{y}}'[\mathbf{x}^{\perp}(\mathbf{x},t),t] = \partial_t \mathbf{y}'[\mathbf{x}^{\perp}(\mathbf{x},t),t] + \mathbf{F}'[\mathbf{x}^{\perp}(\mathbf{x},t),t]\ \dot{\mathbf{x}}^{\perp}(\mathbf{x},t)\ ,$$

$$\dot{\mathbf{n}}'[\mathbf{x}^{\perp}(\mathbf{x},t),t] = \partial_t \mathbf{n}'[\mathbf{x}^{\perp}(\mathbf{x},t),t] + \mathbf{N}'[\mathbf{x}^{\perp}(\mathbf{x},t),t]\ \mathbf{F}'[\mathbf{x}^{\perp}(\mathbf{x},t),t]\ \dot{\mathbf{x}}^{\perp}(\mathbf{x},t)\ .$$

Noting that $\dot{\mathbf{n}}^{\perp}$, $\partial_t \mathbf{n}^{\perp}$, $\mathbf{F}^{\perp}\dot{\mathbf{x}}^{\perp}$ and $\mathbf{N}^{\perp}\mathbf{F}^{\perp}\dot{\mathbf{x}}^{\perp}$ are all separately orthogonal to $\mathbf{n}^{\perp}$ (because $\mathbf{n}'$ is a unit vector, $\mathbf{F}'\dot{\mathbf{x}}'$ a tangent vector at $\mathbf{y}'$ and $\mathbf{N}'$ a map from the tangent space $\mathbf{R}^3_{\mathbf{y}'}$ into the tangent plane $\mathbf{T}_{\mathbf{y}'}$) and projecting the gap vector rate expressions along the normal $\mathbf{n}^{\perp}$ and on the tangent plane $\mathbf{T}_{\mathbf{y}^{\perp}}$, one obtains the *normal* and the *tangential gap rates*

$$\dot{\mathbf{g}}_n = \dot{\mathbf{g}}_n(\mathbf{x},t) = \{\mathbf{n}'[\mathbf{x}^{\perp}(\mathbf{x},t),t] \otimes \mathbf{n}'[\mathbf{x}^{\perp}(\mathbf{x},t),t]\}\ \dot{\mathbf{g}}(\mathbf{x},t) = \dot{g}_n(\mathbf{x},t)\ \mathbf{n}'[\mathbf{x}^{\perp}(\mathbf{x},t),t]\ ,$$

$$\dot{\mathbf{g}}_T = \dot{\mathbf{g}}_T(\mathbf{x},t) = \{\mathbf{I} - \mathbf{n}'[\mathbf{x}^{\perp}(\mathbf{x},t),t] \otimes \mathbf{n}'[\mathbf{x}^{\perp}(\mathbf{x},t),t]\}\ \dot{\mathbf{g}}(\mathbf{x},t) = g_n(\mathbf{x},t)\ \dot{\mathbf{n}}'[\mathbf{x}^{\perp}(\mathbf{x},t),t]\ .$$

Note that $\dot{\mathbf{g}}_T$ is not a slip rate measure since it vanishes when contact occurs, as anticipated. The *relative velocity* of the particle $\mathbf{x} \in \Gamma$ located at $\mathbf{y}(\mathbf{x},t) \in \Gamma_t$ with respect to the proximal particle $\mathbf{x}^{\perp} \in \Gamma'$ located at $\mathbf{y}'(\mathbf{x}^{\perp},t) \in \Gamma'_t$ at time t is usually defined as the difference

$$\dot{\mathbf{d}} = \dot{\mathbf{d}}(\mathbf{x},\mathbf{x}^{\perp},t) = \partial_t \mathbf{y}(\mathbf{x},t) - \partial_t \mathbf{y}'(\mathbf{x}^{\perp},t) = \dot{\mathbf{y}}(\mathbf{x},t) - \partial_t \mathbf{y}'(\mathbf{x}^{\perp},t)\ .$$

This velocity is not objective however since, in a change of reference frame defined by $\mathbf{y}^* = \mathbf{R}(t)\mathbf{y} + \mathbf{b}(t)$ (where $\mathbf{R}$ is a rotation and $\mathbf{b}$ a translation), $\dot{\mathbf{y}}^* = \mathbf{R}\dot{\mathbf{y}} + \dot{\mathbf{R}}\mathbf{y} + \dot{\mathbf{b}}$ with similar relationships for $\mathbf{y}'$ and $\partial_t \mathbf{y}'$, and consequently $\dot{\mathbf{d}}^* = \mathbf{R}\dot{\mathbf{d}} + \dot{\mathbf{R}}\mathbf{g} \neq \mathbf{R}\dot{\mathbf{d}}$. Thus $\dot{\mathbf{d}}$ is objective if and only if $g_n = 0$, i.e. when contact persists. One way to make it objective is to cancel the parasitic effect of the gap lever arm by subtracting its velocity. More specifically, using the gap vector rate relationships, the ***contact velocity*** is defined as

$$\mathring{\mathbf{g}} = \mathring{\mathbf{g}}(\mathbf{x},t) = \dot{\mathbf{y}}(\mathbf{x},t) - \partial_t \mathbf{y}'[\mathbf{x}^\perp(\mathbf{x},t),t] - g_n(\mathbf{x},t)\, \dot{\mathbf{n}}'[\mathbf{x}^\perp(\mathbf{x},t),t]$$

$$= \dot{g}_n(\mathbf{x},t)\, \mathbf{n}'[\mathbf{x}^\perp(\mathbf{x},t),t] + \mathbf{F}'[\mathbf{x}^\perp(\mathbf{x},t),t]\, \dot{\mathbf{x}}^\perp(\mathbf{x},t)\ ,$$

where the circle "°" superposed on $\mathbf{g}$ aims at pointing out that $\mathring{\mathbf{g}}$ has the dimension of a velocity but is not the time derivative of $\mathbf{g}$ (or any other function). When contact persists ($g_n = 0$), $\mathring{\mathbf{g}}$ coincides with $\dot{\mathbf{d}}$ and, when a separation or a penetration occurs ($g_n \neq 0$), it remains objective since then $\mathbf{n}^{\perp *} = \mathbf{R}\mathbf{n}^\perp$, $\dot{\mathbf{n}}^{\perp *} = \mathbf{R}\dot{\mathbf{n}}^\perp + \dot{\mathbf{R}}\mathbf{n}^\perp$ and it is clear that $\mathring{\mathbf{g}}^* = \mathbf{R}(\dot{\mathbf{d}} - g_n\dot{\mathbf{n}}^\perp) + \dot{\mathbf{R}}(\mathbf{g} - g_n\mathbf{n}^\perp) = \mathbf{R}\mathring{\mathbf{g}}$.

Projecting $\mathring{\mathbf{g}}$ along the normal and on the tangent plane to the deformed target surface at $\mathbf{y}^\perp$, the ***normal gap rate*** is recovered and the ***tangential slip rate*** uncovered

$$\mathring{g}_n = \mathring{g}_n(\mathbf{x},t) = \{\mathbf{n}'[\mathbf{x}^\perp(\mathbf{x},t),t] \otimes \mathbf{n}'[\mathbf{x}^\perp(\mathbf{x},t),t]\}\ \mathring{\mathbf{g}}(\mathbf{x},t) = \dot{g}_n(\mathbf{x},t)\ \mathbf{n}'[\mathbf{x}^\perp(\mathbf{x},t),t] = \dot{g}_n\ ,$$

$$\mathring{\mathbf{g}}_T = \mathring{\mathbf{g}}_T(\mathbf{x},t) = \{\mathbf{I} - \mathbf{n}'[\mathbf{x}^\perp(\mathbf{x},t),t] \otimes \mathbf{n}'[\mathbf{x}^\perp(\mathbf{x},t),t]\}\ \mathring{\mathbf{g}}(\mathbf{x},t) = \mathbf{F}'[\mathbf{x}^\perp(\mathbf{x},t),t]\ \dot{\mathbf{x}}^\perp(\mathbf{x},t)\ .$$

Therefore, the *normal approach rate* $\mathring{g}_n$ is equal to the signed distance rate $\dot{g}_n$ whereas the *tangential slip rate* $\mathring{\mathbf{g}}_T$ is equal to the proximal particle velocity $\dot{\mathbf{x}}^\perp$ on Γ' convected on Γ'_t by $\mathbf{F}^\perp$, which keeps it tangent. This result is sensible since $\dot{\mathbf{x}}^\perp$ represents a sound material measure of the slip rate (invariant in a change of reference frame) just as the arc described by $\mathbf{x}^\perp(t)$ on Γ' would be a good measure of the slip.

Remark. The contact velocity $\mathring{\mathbf{g}}$ introduced in this study slightly differs from the one introduced by [Klarbring, 1994] and denoted $\mathbf{r}$. The difference reduces in subtracting from $\dot{\mathbf{d}}$ the total normal rate instead of the partial one or equivalently in taking for the tangential rate the proximal particle velocity on the target surface instead of taking it on the level surface passing through the contactor point $\mathbf{y}$

$$\mathring{\mathbf{g}} = \dot{\mathbf{d}} - g_n\dot{\mathbf{n}}^\perp = \dot{g}_n\, \mathbf{n}^\perp + \mathbf{F}^\perp\dot{\mathbf{x}}^\perp \neq \mathbf{r} = \dot{\mathbf{d}} - g_n\partial_t\, \mathbf{n}^\perp = \dot{g}_n\, \mathbf{n}^\perp + [\mathbf{I} + g_n\mathbf{N}^\perp]\mathbf{F}^\perp\dot{\mathbf{x}}^\perp\ .$$

Both are objective. They coincide when contact occurs and slightly differ otherwise.

CONTACT STATICS

In this section, the (quasi-) static quantities found necessary for modelling the interaction of two deformable solids in eventual contact are introduced and the governing principles of action and reaction and moment balance are extended to distant particles. To begin, basic elements of solid statics are reviewed with particular attention paid to the assumptions relevant for contacts.

Solid surface statics

In mechanics, interactions between two bodies are modelled either by at-a-distance volume forces or by contact surface forces, according as the bodies are distant or contiguous.

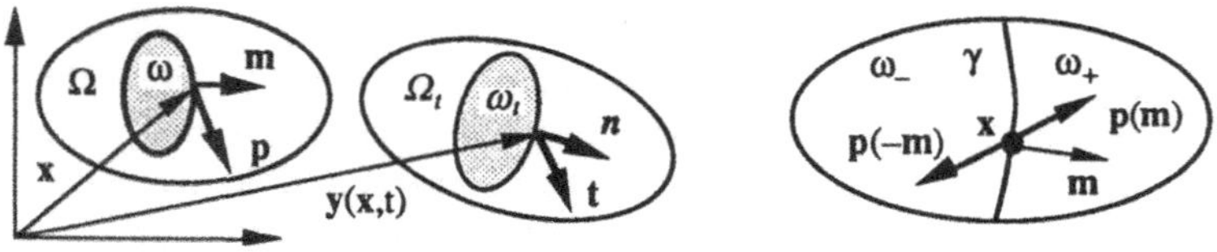

Fig. 5. Nominal stress vector and classical action-reaction principle.

For a continuous deformable solid, an at-a-distance force **k** can be defined as the volume integral of a distribution $\boldsymbol{\kappa}$ of force per unit of reference volume called the nominal force density and a contact force **q** as the surface integral of a distribution **p** of force per unit of reference area called the *first Piola-Kirchhoff or **nominal stress vector***. It is next assumed that these external forces induce in the body Ω internal cohesion forces analogous to contact forces and it is postulated that nominal stress vectors identical to **p** act on the boundary ∂ω of any part $\omega \subseteq \Omega$ (Euler-Cauchy localisation principle). It is further postulated that the stress vector **p(x,t)** acting on ∂ω depends only on the orientation of ∂ω, i.e. on the unit normal **m(x)** (Cauchy's postulate). Cohesion and contact forces are then unified by introducing the normal **m** in the contact stress as well: **p** = **p[x,t,m(x)]** (Fig. 5). Finally, neglecting inertia forces (hereby excluding impacts) for simplicity, the sum of the volume and surface forces acting in and on ω are postulated to be in equilibrium, together with their moments. In summary, the (quasi-) statics of a deformable solid rests on the two principles of *equilibrium of forces and moments*

$$\exists \boldsymbol{\kappa} : \omega \times [0,T] \rightarrow \mathbf{R}^3 ,\ \exists \mathbf{p} : \omega \times [0,T] \times S_x \rightarrow \mathbf{R}^3 \mid \forall \omega \subseteq \Omega ,\ \forall t \in [0,T]$$

$$\mathbf{k}(\omega,t) + \mathbf{q}(\partial\omega,t) = \int_{\omega} \boldsymbol{\kappa}(\mathbf{x},t)\, dV + \int_{\partial\omega} \mathbf{p}[\mathbf{x},t,\mathbf{m}(\mathbf{x})]\, dA = \mathbf{0} , \tag{5}$$

$$\hat{\mathbf{k}}(\omega,t) + \hat{\mathbf{q}}(\partial\omega,t) = \int_{\omega} \mathbf{y}(\mathbf{x},t) \wedge \boldsymbol{\kappa}(\mathbf{x},t)\, dV + \int_{\partial\omega} \mathbf{y}(\mathbf{x},t) \wedge \mathbf{p}[\mathbf{x},t,\mathbf{m}(\mathbf{x})]\, dA = \mathbf{0} , \tag{6}$$

where S_x denotes the unit sphere. Under sufficient continuity, the balance of forces implies the existence of the nominal stress tensor (Cauchy's theorem), after what, using the divergence and localisation theorems, the two principles take the *local forms*

$$\exists \mathbf{P} : \omega \times [0,T] \rightarrow \mathbf{R}^3 \mid \mathbf{p}[\mathbf{x},t,\mathbf{m}(\mathbf{x})] = \mathbf{P}(\mathbf{x},t)\, \mathbf{m}(\mathbf{x})$$

$$-\operatorname{Div} \mathbf{P}(\mathbf{x},t) = \boldsymbol{\kappa}(\mathbf{x},t) ,\quad \forall \mathbf{x} \in \Omega ,\ \forall t \in [0,T] , \tag{7}$$

$$\mathbf{P}(\mathbf{x},t)\, \mathbf{F}^T(\mathbf{x},t) = \mathbf{F}(\mathbf{x},t)\, \mathbf{P}^T(\mathbf{x},t) ,\quad \forall \mathbf{x} \in \Omega ,\ \forall t \in [0,T] . \tag{8}$$

The balance of forces also implies the principle of action and reaction across an interface γ cutting ω in two adjacent parts ω_- and ω_+ in its classical global and local forms (Fig. 5)

$$\mathbf{q}(\gamma,t) + \mathbf{q}(-\gamma,t) = \int_{\gamma} \mathbf{p}[\mathbf{x},t,\mathbf{m}(\mathbf{x})]\, dA + \int_{\gamma} \mathbf{p}[\mathbf{x},t,-\mathbf{m}(\mathbf{x})]\, dA = \mathbf{0} ,\ \forall \gamma ,$$

$$\mathbf{p}[\mathbf{x},t,\mathbf{m}(\mathbf{x})] = -\,\mathbf{p}[\mathbf{x},t,-\mathbf{m}(\mathbf{x})] ,\quad \forall \mathbf{x} \in \gamma ,\ \forall t \in [0,T] . \tag{9}$$

The proof relies essentially on the fact that the surface of ω_- in contact with ω_+ is γ with normal **m** whereas the surface of ω_+ in contact with ω_- is −γ, i.e. the *same* surface γ with opposite normal $-\mathbf{m}$. Moreover the proof can be carried out either in the reference configuration or in the deformed one since matter is continuous across the interface at all times by motion continuity and invertibility.

Contact statics

Consider now the contact system $\Omega \cup \Omega'$ previously defined. The potential contact and target surfaces Γ and Γ' are usually separated in the reference configuration and remain so in the deformed configuration at this stage of the formulation. Consequently the action and reaction principle cannot be derived from force equilibrium in the classical way. The key idea of the following static formulation is to extend the classical principle of action and reaction to distant particles by postulating that contact forces act at a distance (!) between closest points or proximal particles (Fig.6). More specifically, it is postulated that for any part $\omega \cup \omega'$ of the contact system $\Omega \cup \Omega'$ such that $\mathbf{x}^{\perp}(\omega \cap \Gamma, t) \subset (\omega' \cap \Gamma')$, the mutual contact interaction of ω and ω' is fully defined by that of $\gamma = \omega \cap \Gamma$ on $\gamma^{\perp} = \mathbf{x}^{\perp}(\gamma, t) \subset \gamma' = \omega' \cap \Gamma'$.

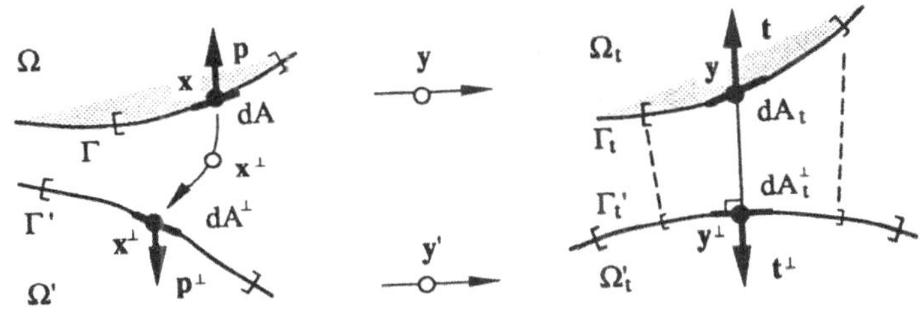

Fig. 6. Action-reaction principle generalisation.

Thus, in addition to the force and moment equilibrium (5) and (6) for any part $\omega \subseteq \Omega$ and any part of $\omega' \subseteq \Omega'$, the following ***generalised principle of action and reaction or contact force and moment balance*** is postulated

$$\forall \gamma \subseteq \Gamma, \ \forall \gamma' \subseteq \Gamma' \mid \gamma^{\perp} = \mathbf{x}^{\perp}(\gamma,t) \subset \gamma' \ \text{and} \ \forall t \in [0,T],$$

$$\mathbf{q}(\gamma,t) + \mathbf{q}(\gamma^{\perp},t) = \int_{\gamma} \mathbf{p}[\mathbf{x},t,\mathbf{m}(\mathbf{x})]\, dA + \int_{\gamma^{\perp}} \mathbf{p}'[\mathbf{x}^{\perp},t,\mathbf{m}'(\mathbf{x}^{\perp})]\, dA' = \mathbf{0},$$

$$\mathbf{q}(\gamma' \setminus \gamma^{\perp},t) = \int_{\gamma' \setminus \gamma^{\perp}} \mathbf{p}'[\mathbf{x}',t,\mathbf{m}'(\mathbf{x}')]\, dA' = \mathbf{0},$$

$$\hat{\mathbf{q}}(\gamma,t) + \hat{\mathbf{q}}(\gamma^{\perp},t) = \int_{\gamma} \mathbf{y}(\mathbf{x},t) \wedge \mathbf{p}[\mathbf{x},t,\mathbf{m}(\mathbf{x})]\, dA + \int_{\gamma^{\perp}} \mathbf{y}'(\mathbf{x}^{\perp},t) \wedge \mathbf{p}'[\mathbf{x}^{\perp},t,\mathbf{m}'(\mathbf{x}^{\perp})]\, dA' = \mathbf{0}.$$

Contact force balance completed by moment balance are sometimes called the strong law of action and reaction. Using the bijection $\mathbf{x}^{\perp}$, the integrals over $\gamma^{\perp}$ can be transferred over the contact surface by carrying out the change of variable $\mathbf{x}^{\perp} = \mathbf{x}^{\perp}(\mathbf{x},t)$ with $dA^{\perp} = j^{\perp}(\mathbf{x},t)\, dA$

$$\mathbf{q}(\gamma,t) + \mathbf{q}[\mathbf{x}^{\perp}(\gamma,t),t] = \int_{\gamma} \{\mathbf{p}[\mathbf{x},t,\mathbf{m}(\mathbf{x})] + j^{\perp}(\mathbf{x},t)\, \mathbf{p}'[\mathbf{x}^{\perp}(\mathbf{x},t),t,\mathbf{m}'(\mathbf{x}^{\perp}(\mathbf{x},t))]\}\, dA = \mathbf{0},$$

$$\hat{\mathbf{q}}(\gamma,t) + \hat{\mathbf{q}}[\mathbf{x}^{\perp}(\gamma,t),t] = \int_{\gamma} \{\mathbf{y}(\mathbf{x},t) \wedge \mathbf{p}[\mathbf{x},t,\mathbf{m}(\mathbf{x})]$$
$$+ \mathbf{y}'[\mathbf{x}^{\perp}(\mathbf{x},t),t] \wedge j^{\perp}(\mathbf{x},t)\mathbf{p}'[\mathbf{x}^{\perp}(\mathbf{x},t),t,\mathbf{m}'(\mathbf{x}^{\perp}(\mathbf{x},t))]\}\, dA = \mathbf{0}.$$

Assuming sufficient continuity, the force and moment equilibrium take the local forms

$$\mathbf{p}[\mathbf{x},t,\mathbf{m}(\mathbf{x})] + j^{\perp}(\mathbf{x},t)\ \mathbf{p}'[\mathbf{x}^{\perp}(\mathbf{x},t),t,\mathbf{m}'(\mathbf{x}^{\perp}(\mathbf{x},t))] = \mathbf{0}\ ,\quad \forall \mathbf{x} \in \Gamma\ ,\ \forall t \in [0,T]\ , \tag{10}$$

$$\mathbf{p}'[\mathbf{x}',t,\mathbf{m}'(\mathbf{x}')] = \mathbf{0}\ ,\quad \forall \mathbf{x}' \in \Gamma'\backslash\Gamma^{\perp}\ ,\ \forall t \in [0,T]\ , \tag{11}$$

$$\mathbf{y}(\mathbf{x},t) \wedge \mathbf{p}[\mathbf{x},t,\mathbf{m}(\mathbf{x})] + \mathbf{y}'[\mathbf{x}^{\perp}(\mathbf{x},t),t] \wedge j^{\perp}(\mathbf{x},t)\mathbf{p}'[\mathbf{x}^{\perp}(\mathbf{x},t),t,\mathbf{m}'(\mathbf{x}^{\perp}(\mathbf{x},t))] = \mathbf{0}\ .$$

Substituting the force balance $j^{\perp}\mathbf{p}^{\perp} = -\mathbf{p}$ in the moment one and recognising the gap vector in the relative lever arm $\mathbf{y} - \mathbf{y}^{\perp} = \mathbf{g} = g_n\, \mathbf{n}^{\perp}$, the moment equilibrium further simplifies into

$$\mathbf{g}(\mathbf{x},t) \wedge \mathbf{p}[\mathbf{x},t,\mathbf{m}(\mathbf{x})] = \mathbf{0}\ ,\quad \forall \mathbf{x} \in \Gamma\ ,\ \forall t \in [0,T]\ . \tag{12}$$

Hence balance of contact forces implies that the nominal stress vectors must be collinear and opposite as in the continuous case (9) but not equal in magnitude in contrast with (9), since (10) involves area weighting factors due to deformation and projection (recall that $j^{\perp} = j\, \jmath^{\perp}/j'$ in general and $j^{\perp} = j/j'$ when contact occurs). Balance of moments (12) is trivially satisfied if there is contact ($g_n = 0$) but requires that the stress vectors be collinear with the gap vector (i.e. orthogonal to the deformed target) if a gap or a penetration exists ($g_n \neq 0$)

$$\mathbf{p} = -j^{\perp}\mathbf{p}^{\perp} = \alpha\, g_n\, \mathbf{n}^{\perp}\ \text{ if }\ g_n \neq 0\ .$$

Hence balance of moments plays a non trivial role when the proximal particles are distant.

To match the choice of the contactor for reference and the definition of the gap vector and contact velocity on the contact surface, the ***contact stress vector*** is defined as the one exercised by the target on the contactor, i.e. as $\mathbf{p}$ (rather than $\mathbf{p}^{\perp}$). This contact stress can be decomposed into a normal *contact pressure* and a tangential *contact shear* as

$$\mathbf{p}[\mathbf{x},t,\mathbf{m}(\mathbf{x})] = p_n[\mathbf{x},t,\mathbf{m}(\mathbf{x})]\ \mathbf{n}^{\perp}(\mathbf{x},t) + \mathbf{p}_T[\mathbf{x},t,\mathbf{m}(\mathbf{x})]\ ,\quad p_n[\mathbf{x},t,\mathbf{m}] = \mathbf{p}[\mathbf{x},t,\mathbf{m}] \cdot \mathbf{n}^{\perp}(\mathbf{x},t)\ ,$$

where $\mathbf{n}^{\perp}(\mathbf{x},t) = \mathbf{n}'[\mathbf{x}^{\perp}(\mathbf{x},t),t]$ is the current target normal at $\mathbf{y}^{\perp}$. This decomposition of the stress acting on the contact surface along and across the normal to the target surface is somewhat artificial. It is a consequence of the unsymmetric treatment of the contactor and target through the projection. It is emphasised that tension is not precluded at this stage. The contact pressure defines the *static contact status* of the particle $\mathbf{x}$

$$p_n > 0 \leftrightarrow \textit{tension}\ ,\quad p_n = 0 \leftrightarrow \textit{neutral}\ ,\quad p_n < 0 \leftrightarrow \textit{compression}\ .$$

Remark. The action and reaction principle $\mathbf{p} = -j^{\perp}\mathbf{p}^{\perp}$ introduced in this study differs from the one introduced by [Curnier et al., 1992] by the projection ratio $\jmath^{\perp}$ included in $j^{\perp} = j\, \jmath^{\perp}/j'$. In our previous study, this projection ratio $\jmath^{\perp}$ was overlooked. When expressed in terms of Cauchy stress vectors $j(\mathbf{x},t)\ \mathbf{t}(\mathbf{y},t) = \mathbf{p}(\mathbf{x},t)$ the principle takes the form $\mathbf{t}(\mathbf{y}) + \jmath^{\perp}\, \mathbf{t}'(\mathbf{y}^{\perp}) = \mathbf{0}$ or $\mathbf{t}(\mathbf{y})dA_t + \mathbf{t}'(\mathbf{y}^{\perp})dA_t^{\perp} = \mathbf{0}$ and not $\mathbf{t}(\mathbf{y}) + \mathbf{t}'(\mathbf{y}^{\perp}) = \mathbf{0}$ as misstated. This latter form is correct in contact only.

CONTACT ENERGETICS

In this section, contact static quantities are put in energetic duality with the contact kinematic variables by means of the principle of energy conservation. To begin this principle is recalled for a deformable solid.

Solid energy conservation

Neglecting kinetic energy, the principle of energy conservation for a deformable solid stipulates that the internal stress power must be equal to the external power developed by the at-a-distance and contact forces

$\forall \omega \subseteq \Omega\ ,\ \forall t \in [0,T]\ ,\ P_i(\omega,t) = P_e(\omega,t) + \partial P_e(\partial\omega,t)$ or explicitly:

$$\int_{\omega} \mathbf{P}(\mathbf{x},t) : \dot{\mathbf{F}}(\mathbf{x},t)\, dV = \int_{\omega} \boldsymbol{\kappa}(\mathbf{x},t) \cdot \dot{\mathbf{y}}(\mathbf{x},t)\, dV + \int_{\partial\omega} \mathbf{p}[\mathbf{x},t,\mathbf{m}(\mathbf{x})] \cdot \dot{\mathbf{y}}(\mathbf{x},t)\, dA \qquad (13)$$

where " : " denotes the scalar product for second order tensors and $\dot{\mathbf{F}}$ the transformation gradient rate. Invariance of this principle in an arbitrary change of frame of reference implies the principle of force equilibrium (5) (translation) and moment equilibrium (6) (rotation).

Contact energy conservation

For the contact system $\Omega \cup \Omega'$, the *principle of energy conservation* takes the form

$$\forall(\omega \cup \omega') \subseteq (\Omega \cup \Omega')\ |\ \mathbf{x}^{\perp}(\omega \cap \Gamma,t) \subset (\omega' \cap \Gamma')\ ,\ \forall t \in [0,T]\ ,$$

$$P_i(\omega,t) + P_i'(\omega',t) = P_e(\omega,t) + P_e'(\omega',t) + \partial P_e(\partial\omega,t) + \partial P_e'(\partial\omega',t)\ ,$$

where ω or ω' can be empty and P_i', P_e' and $\partial P_e'$ are defined by adding primes in (13). Decomposing the boundary of ω and ω' into their contact and complementary parts along $\partial\omega = (\partial\omega\backslash\gamma) \cup \gamma$ and $\partial\omega' = (\partial\omega'\backslash\gamma') \cup \gamma'$ yields the equivalent right hand side

$$P_e(\omega,t) + P_e'(\omega',t) + \partial P_e(\partial\omega\backslash\gamma,t) + \partial P_e(\partial\omega'\backslash\gamma',t) + \partial P_e(\gamma,t) + \partial P_e'(\gamma',t)\ ,$$

where the sum of the last two terms represent the ***contact stress power*** with explicit form

$$P_c(\gamma,\gamma',t) = \int_{\gamma} \mathbf{p}[\mathbf{x},t,\mathbf{m}(\mathbf{x})] \cdot \dot{\mathbf{y}}(\mathbf{x},t)\, dA + \int_{\gamma'} \mathbf{p}'[\mathbf{x}',t,\mathbf{m}'(\mathbf{x}')] \cdot \dot{\mathbf{y}}'(\mathbf{x}',t)\ dA'.$$

Decomposing the target surface γ' into $\gamma' = (\gamma'\backslash\gamma^{\perp}) \cup \gamma^{\perp}$ and using the bijection $\mathbf{x}^{\perp}$, the integral over the target part $\gamma^{\perp} = \mathbf{x}^{\perp}(\gamma,t) \subset \gamma'$ can be transferred over the contact surface by carrying out the change of variable $\mathbf{x}^{\perp} = \mathbf{x}^{\perp}(\mathbf{x},t)$ with $dA^{\perp} = j^{\perp}(\mathbf{x},t)\, dA$

$$P_c(\gamma,t) = \int_{\gamma} \{\mathbf{p}[\mathbf{x},t,\mathbf{m}(\mathbf{x})] \cdot \dot{\mathbf{y}}(\mathbf{x},t) + j^{\perp}(\mathbf{x},t)\ \mathbf{p}'[\mathbf{x}^{\perp}(\mathbf{x},t),t,\mathbf{m}'(\mathbf{x}^{\perp}(\mathbf{x},t))]$$
$$\cdot\, \partial_t \mathbf{y}'[\mathbf{x}^{\perp}(\mathbf{x},t),t]\}\ dA + \int_{\gamma'\backslash\gamma^{\perp}} \mathbf{p}'[\mathbf{x}',t,\mathbf{m}'(\mathbf{x}')] \cdot \dot{\mathbf{y}}'(\mathbf{x}',t)\ dA'.$$

Using the local action-reaction principle $j^{\perp}\mathbf{p}^{\perp} = -\,\mathbf{p}$ and $\mathbf{p}' = \mathbf{0}$ on $\gamma'\backslash\gamma^{\perp}$, the contact velocity definition $\mathring{\mathbf{g}} = \dot{\mathbf{y}} - \partial_t\, \mathbf{y}^{\perp} - g_n\dot{\mathbf{n}}^{\perp}$, the moment principle $\mathbf{p} = \alpha\, g_n\, \mathbf{n}^{\perp}$ if $g_n \neq 0$ and remembering that $\mathbf{n}^{\perp} \cdot \dot{\mathbf{n}}^{\perp} = 0$ successively yields

$$P_c(\gamma,t) = \int_{\gamma} \mathbf{p}[\mathbf{x},t,\mathbf{m}(\mathbf{x})] \cdot \{\dot{\mathbf{y}}(\mathbf{x},t) - \partial_t\, \mathbf{y}'[\mathbf{x}^{\perp}(\mathbf{x},t),t]\}\ dA$$
$$= \int_{\gamma} \mathbf{p}[\mathbf{x},t,\mathbf{m}(\mathbf{x})] \cdot [\ \mathring{\mathbf{g}}(\mathbf{x},t) + g_n(\mathbf{x},t)\ \dot{\mathbf{n}}'[\mathbf{x}^{\perp}(\mathbf{x},t),t]]\ dA$$

$$P_c(\gamma,t) = \int_\gamma \mathbf{p}[\mathbf{x},t,\mathbf{m}(\mathbf{x})] \cdot \mathring{\mathbf{g}}(\mathbf{x},t)\, dA \; .$$

This last equation shows that the nominal contact stress $\mathbf{p}$ is effectively conjugate to the relative contact velocity $\mathring{\mathbf{g}}$. Both measures are objective.

TRIBOLOGICAL LAWS

Just as rheology is subdivided into elasticity, viscosity, plasticity and damage for a solid material with infinitely many possible laws in each category, *tribology* can be subdivided into adhesion, lubrication, friction and wear for a pair of solid materials and the number of possible laws is also enormous. In this last section, a normal unilateral contact law and a tangential threshold friction law are formulated to complete and illustrate the formulation on the material behaviour side. To begin an elastic law is mentioned for the solids.

Solid constitutive laws

To fix ideas, a hyperelastic law is selected for modelling the material behaviour of the contactor and the target, although other constitutive laws could be considered. More specifically, the nominal stress is assumed to derive from a stored elastic energy function

$$\mathbf{P}(\mathbf{x},t) = \frac{\partial W}{\partial \mathbf{F}}[\mathbf{x},\mathbf{F}(\mathbf{x},t)] \; . \tag{14}$$

A safe way to get a nominal law which is objective and satisfy moment equilibrium a priori is to start from a material law between the Cauchy-Green metric tensor $\mathbf{C} = \mathbf{F}^T\mathbf{F}$ and the conjugate second Piola-Kirchhoff symmetric stress tensor $\mathbf{S}$ and to deduce $\mathbf{P}$ from $\mathbf{P} = \mathbf{FS}$.

Unilateral contact law

Classically, a unilateral contact law is characterised by a geometric condition of no-penetration, a static condition of no-tension and a mechanical complementarity condition [Signorini, 1959]. In the present context of two bodies undergoing finite deformations, these three Signorini conditions can be written in terms of the signed contact distance and the nominal pressure to form the *nominal unilateral contact law*

$$g_n \geq 0 \;, \quad p_n \leq 0 \;, \quad g_n p_n = 0 \;, \tag{15}$$

where the particle and time arguments $(\mathbf{x},t)$ are omitted for clarity. This no-adhesion law excludes geometric penetration as well as static tension so that the possible *contact status* for the particle $\mathbf{x}$ are (Fig. 7):

$g_n > 0$ and $p_n = 0 \leftrightarrow$ *gap no-pressure* ; $g_n = 0$ and $p_n \leq 0 \leftrightarrow$ *contact with compression* .

By means of convex analysis, the multivalued law $p_n = p_n[g_n]$ and its inverse $g_n = g_n[p_n]$ can be shown [Moreau, 1973, 1979] to derive in the sense of subgradients from two conjugate non differentiable convex potentials, namely the indicator function $I_{\mathbf{R}^+}$ of the positive half-line $\mathbf{R}^+$ and its Legendre-Fenchel transform, the indicator function $I_{\mathbf{R}^-}$ of the polar negative half-line $\mathbf{R}^-$. Hence the unilateral contact law (15) can be condensed in either

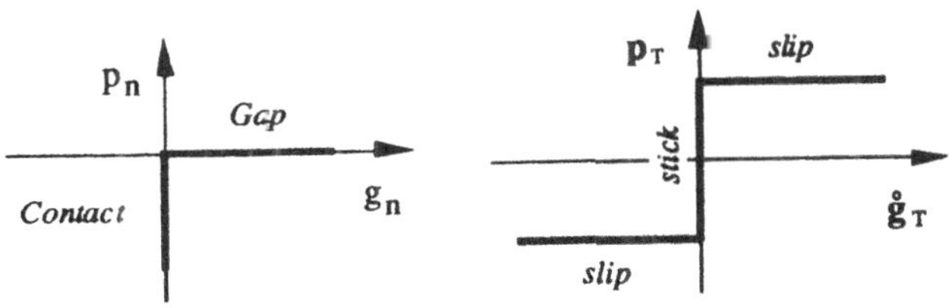

Fig. 7. Normal unilateral contact law and tangential threshold friction law.

one of two inverse inclusions

$$p_n \in \partial I_{\mathbf{R}^+}(g_n) \quad \text{or} \quad g_n \in \partial I_{\mathbf{R}^-}(p_n) \; , \tag{16}$$

where $\partial I_{\mathbf{R}^+}$ denotes the subdifferential of $I_{\mathbf{R}^+}$. The law (16) could be called *hyperunilateral* or *hypernonadhesive* by analogy with the hyperelastic law (14).

Threshold friction law

Classically, a (rate independent) dry friction law is characterised by a kinematic slip rule, a static friction criterion and a mechanical complementarity condition. Taking Coulomb's law of isotropic friction as an illustration, these three relations can be written in terms of the contact velocity and the contact shear to form the *nominal threshold friction law*

$$\mathring{g}_T = \| \mathring{g}_T \| \frac{\mathbf{p}_T}{\| \mathbf{p}_T \|} \; , \quad \| \mathbf{p}_T \| + \mu p_n \le 0 \; , \quad \| \mathring{g}_T \| (\| \mathbf{p}_T \| + \mu p_n) = 0 \; , \tag{17}$$

where $\mu \ge 0$ is the friction coefficient and $p_n \le 0$ is momentarily assumed to be *constant*. This tribological law excludes slip below the threshold $-\mu p_n$ so that the possible *friction status* of the particle **x** are (Fig. 7) :

$\| \mathring{g}_T \| = 0$ and $\| \mathbf{p}_T \| \le -\mu p_n \;\leftrightarrow\;$ *stick and shear* ;

$\| \mathring{g}_T \| \ne 0$ and $\| \mathbf{p}_T \| = -\mu p_n \;\leftrightarrow\;$ *slip and friction* .

The slip rule being normal (associated) to the criterion (the circle of radius $-\mu p_n$), it satisfies the normality rule. Consequently, using the same tools as for the unilateral law, the friction law (17) can be condensed into

$$\mathbf{p}_T \in \partial I^*_{C(p_n)}(\mathring{g}_T) = -\mu p_n \, \partial \| \mathring{g}_T \| \quad \text{or} \quad \mathring{g}_T \in \partial I_{C(p_n)}(\mathbf{p}_T) \; , \tag{18}$$

$$C(p_n) = \{ \mathbf{p}_T \mid \| \mathbf{p}_T \| \le -\mu p_n \;\; (p_n \text{ given}) \} \; ,$$

where $C(p_n)$ denotes the convex disc of radius $-\mu p_n$. More complicated friction behaviours can be modelled by using other convex sets [He & Curnier, 1993]. For instance, orthotropic steady friction can be accommodated by taking an elliptical disc instead of a circular one.

Frictional contact law

In addition to the usual restrictions of constitutive theory (causality, objectivity, entropy inequality ...) [Telega, 1988; Zmitrowicz, 1993], a frictional contact law must comply with the new requirement imposed by the balance of moments that the friction shear must be zero when a gap exists. Moreover, a realistic friction law must comply with the widely verified experimental law of friction of Amontons-Coulomb, which stipulates that the friction force is proportional to the load and thus the shear proportional to the pressure. An objective law of frictional contact which fulfils both requirements can be constructed by combining the normal unilateral contact law of Signorini (15) with the tangential friction law of Coulomb (17) with *variable* pressure. This Signorini-Coulomb law can equivalently be written in the form (16) *and* (18), provided $\mathbf{C}(p_n)$ is still understood as the disc section of Coulomb's cone $\mathbf{C} = \{\mathbf{p} \mid \| p_T \| \leq -\mu p_n, p_n \leq 0\}$ at the depth p_n (and not as the cone itself). The complete frictional contact law is not associated to Coulomb's cone [Michalowski and Mroz, 1978].

Remark. If a univalued frictional contact law with normal and tangential compliances (penalties) is preferred to the multivalued law (15)-(17), then the balance of moments requires using the same compliance coefficients in the normal and tangential directions in order for the contact stress to remain collinear with the gap vector (which is no longer orthogonal to the target).

CONCLUSION

A continuum formulation of unilateral contact problems between two deformable solids undergoing finite deformations with friction at the interface has been proposed. The condition $0 < (4) < M$ implicitly defines the class of addressed problems. The resulting boundary value problem consists in solving the force equilibrium (7) in $\Omega \cup \Omega'$ and (10)-(11) on $\Gamma \cup \Gamma'$, together with two constitutive laws such as (14) for the solids and a tribological law such as (15)-(17) at the interface (satisfying moment balances (8) and (12) a priori) and with proper motion and traction boundary conditions on $\partial\Omega_y \cup \partial\Omega'_y$ and $\partial\Omega_p \cup \partial\Omega'_p$. Such problems are successfully solved by a combination of numerical methods.

REFERENCES

Ciarlet P.G. and Necas J., 1985, Unilateral problems in nonlinear, three-dimensional elasticity, *Arch. Rational Mech. Anal.*, 87, 319-338.

Ciarlet P.G., 1988, *Mathematical Elasticity. Vol I: Three-dimensional Elasticity*, North-Holland, Amsterdam.

Curnier A., He Q.-C and Telega J.J., 1992, Formulation of unilateral contact between two elastic bodies undergoing finite deformations, *C. R. Acad. Sc. Paris*, t. 314, série II, 1-6.

Hansson E. and Klarbring A., 1990, Rigid contact modelled by CAD surface, *Eng. Comput.*, 7, 344-348.

He Q.-C and Curnier A., 1993, Anisotropic dry friction between two orthotropic surfaces undergoing large displacements, *Eur. J. Mech., A/Solids*, 12(5), 631-666.

Kikuchi N. and Oden J.T., 1988, *Contact Problems in Elasticity: a study of variational inequalities and finite element methods*, SIAM, Philadelphia.

Klarbring A., Mikelic A. and Shillor M., 1991, The rigid punch problem with friction, *Int. J. Engng. Sci.*, 29 (6), 751-768.

Klarbring A., 1994, Large displacement frictional contact: a continuum framework for finite element discretisation, submitted. Also in: *Comp. Solid Mech.*, COMETT lecture notes, EPFL, Lausanne.

Laursen T.A. and Simo J.C., 1991, Formulation and regularisation of frictional contact problems for Lagrangian finite element computations, *Complas III.*, 36, 395-407.

Laursen T.A. and Simo J.C., 1993, A continuum-based finite element formulation for the implicit solution of multibody, large deformation frictional contact problems, *Int. J. Num. Meth. Eng.*, 36, 3451-3485.

Michalowski R. and Mroz Z., 1978, Associated and non-associated sliding rules in contact friction problems, *Arch. Mech.*, 30, 259-276.

Moreau J.J., 1973, On unilateral constraints, friction and plasticity, in: *New Variational Techniques in Mathematical Physics*, G. Capriz and G. Stampacchia ed., CIME, II, Ed. Cremonese, Roma, 175-322.

Moreau J.J., 1979, Application of convex analysis to some problems of dry friction, in: *Trends of Pure Mathematics to Mechanics. Vol. II*, H. Zorski ed., Pitman, 263-280.

Signorini A., 1959, Questioni di elasticità non linearizzata, *Rend. Mat.*, 18, 95-139.

Telega J.J., 1988, Topics on unilateral contact problems of elasticity and inelasticity, in: *CISM: Nonsmooth Mechanics and Applications*, J.J. Moreau and P.D. Panagiotopoulos ed., Springer,

Thorpe J. A., 1978, *Elementary Topics in Differential Geometry*, Springer-Verlag, Berlin.

Zmitrowicz A., 1993, Constitutive modelling of anisotropic phenomena of friction wear and frictional heat, report 381/1342, IMP, PAN, Gdansk, 1-234.

NUMERICAL METHODS FOR DEALING WITH FRICTIONAL CONTACT PROBLEMS IN FINITE ELASTOPLASTIC DEFORMATION

Patrick Chabrand and Frédéric Dubois

Laboratoire de Mécanique et d'Acoustique
31, Chemin Joseph Aiguier
13402 Marseille Cedex 20

INTRODUCTION

Modelling metal forming processes involves studying strongly non-linear problems. The non-linearities arise from the finite elastic-plastic deformation and from the unilateral conditions under consideration. Our aim in the present study was to compare some numerical methods for dealing with discretized frictional contact problems in the context of metal forming processes.

The possible applications include modelling the effects of blankholders depending on whether or not they are fitted with a drawbead. In the forming of sheet metal parts, the quality of the final products in terms of the absence of wrinkles and necking depends on the restraining forces developed by the blankholder. In the present study we carried out a numerical analysis of the friction under the blankholder. For this purpose we investigated the influence of the main parameters (shape of the tools, friction, clamping pressure, speeds) on the restraining forces. Drawbead simulators have been designed by Renault with the view to comparing experimental and numerical datas. To deal with this problem, a 2-D analysis can be performed in the case of plane strain or in the axisymmetrical case, where only the region beneath the blankholder and the die is taken into account. Both types of blankholder (those including drawbeads and otherwise) are considered. In the case of the plane blankholder (one without a drawbead), an elastic model with small deformation is appropriate whereas in that of the blankholder with a drawbead, a finite elastic-plastic analysis has to be performed. This leads to comparing the numerical methods taking both linear and non-linear behaviour.

One of the main difficulties arising with the numerical problem is how to link up the strongly non-linear equations relating to large deformations with the inequations relating to contact and friction. Since the radii of the drawbead and the die shoulders are very small, a large number of elements are required, and due to the complex

Contact Mechanics, Edited by M. Raous *et al.*
Plenum Press, New York, 1995

forms of all the tools which have both small and large curvatures, the number of candidate nodes in the contact can be large. There is therefore a crucial need for efficient methods of dealing with this particular class of frictional contact problems. Generally speaking solving industrial contact problems requires accurate and efficient methods. The efficiency of the method obviously depends on the size of the problem and on the number of saturated contact constraints. The performances of these methods also depend strongly on the contact kinematics (small or large slip), the constitutive law (linear or non-linear) and on whether the static or dynamic hypothesis is adopted.

In the first part of the paper, we describe the model used to characterize the finite deformation arising from the presence of the drawbead. The elastic-plastic formulation incorporates an isotropic hardening law and is based on hyperelastic regularization. A multiplicative decomposition of the deformation gradient is carried out. The model is numerically implementated by means of an elastic predictor/plastic corrector algorithm. The corrector step is carried out using a radial return algorithm.

In the second part, the frictional contact problem and the numerical methods implemented in our finite element code are presented. These methods belong to the classes most frequently used to solve contact with friction problems, such as penalty, augmented Lagrangian, linear complementarity, minimization with projection (conjugate gradient, relaxation, etc.) methods.

Lastly some applications are described in the third part, where for the sake of comparison, these methods are used to solve various exemples. The first example involves a fairly simple test in linear elasticity with small deformation. In the second example we test the validity and the performances of the methods in the case of finite elasto-plasticity during the stretching of a sheet with a hemispherical punch. The last example involves simulating a blankholder fitted with a drawbead.

FINITE STRAIN ELASTO-PLASTIC MODELLING

Formulation

The kinematic description of finite strain plasticity used in this study is based on the local multiplicative decomposition of the deformation gradient F into its elastic F^e and plastic F^p counterparts, as proposed by (Lee, 1969):

$$F = F^e . F^p \tag{1}$$

where F^e defines the local, stress-free unloaded configuration. The isotropic elastic behaviour of the material is determined by the free energy function :

$$\Psi\left(b^e, \bar{\epsilon}^p\right) = \Psi^e\left(b^e\right) + \Psi^p\left(\bar{\epsilon}^p\right) \tag{2}$$

where the elastic potential Ψ^e is an isotropic function of the left Cauchy Green strain tensor b^e and where $\bar{\epsilon}^p$ denotes the isotropic hardening variable and Ψ^p the hardening energy. The Kirchhoff strain tensor is then given by the classical constitutive equation (Sidoroff, 1982):

$$\tau = 2\rho_0 b^e \frac{\partial \Psi^e}{\partial b^e} \tag{3}$$

The plastic response of the material is described using the classical Von-Mises yield condition. The associative flow rule determined by the principle of maximum plastic dissipation is written as:

$$d^p = \lambda\sqrt{\frac{3}{2}}\frac{dev\tau}{|dev\tau|} \tag{4}$$

where d^p is the plastic strain rate, λ is the plastic multiplier and $dev\tau$ denotes the deviatoric part of the Kirchhoff stress. The yield function associated with (4) is defined as :

$$\sqrt{\frac{3}{2}}|dev\tau| - Y\left(\bar{d^p}\right) = 0 \tag{5}$$

where $\bar{d^p}$ is the equivalent plastic strain and $Y\left(\bar{d^p}\right)$ the isotropic hardening law.

From the computational point of view, the above constitutive equations can be integrated using an elastic predictor/plastic corrector algorithm (Dubois, 1994). Assume that at time t_n with the configuration Ω_n of the body, the state variables (F_n^e, F_n^p, τ_n) are known. Let Δu_{n+1} be the (known) incremental displacement between the configurations Ω_n and Ω_{n+1}. Use the notion of operator splitting, the total deformation gradient at time t_{n+1} can be computed. We make then an elastic prediction, assuming that the plastic variables remain frozen ($F_{n+1}^p = F_n^p$). The trial elastic part of the deformation gradient as well as the elastic trial Kirchhoff stress can be then computed. If the stress lies outside the elastic region a plastic correction step is then carried out using a classical radial return algorithm (Simo and Taylor, 1985).

Finite element modelling

In the displacement based finite element method, the discretized form of the equilibrium equations are used to calculate an estimated incremental displacement. A modified Newton-Raphson method is used to deal with the non-linear equations arising from the constitutive equations and the large deformations. The algorithm is organized as follows: loading is given as a sequence of loading steps. At a given loading step, an iterative process is then performed to solve the set of non-linear equations written on successive intermediate configurations. We denote Ω_n the reference configuration at the end of loading step n. Configuration Ω_{n+1} is computed iteratively from Ω_n (which is known from the previous loading step) by iterations i of the Newton-Raphson method with corresponding intermediate configurations Ω_{n+1}^i.
Let u_n be the nodal displacement vector at the end of loading step n. We denote Δu_{n+1}^i the incremetal displacement between configurations Ω_n and Ω_{n+1}^i and du^i the displacement between Ω_{n+1}^i and Ω_{n+1}^{i+1}. Hence

$$\Delta u_{n+1}^{i+1} = \Delta u_{n+1}^i + du^i \tag{6}$$

$$u_{n+1}^{i+1} = u_n + \Delta u_{n+1}^{i+1} \tag{7}$$

The equilibrium equation of the configuration Ω_{n+1}^i is:

$$\left\{Res^i\right\} = \left\{F^{int}\left(u_{n+1}\right)^i\right\} - \left\{F_{n+1}^{ext}\right\} - \left\{R^i\right\} = 0 \tag{8}$$

where $\{Res^i\}$ denotes the equilibrium residual vector obtained by assembling the internal and external forces. $\left\{F^{int}\left(u_{n+1}\right)^i\right\}$ is the discrete load vector corresponding to the internal stresses, $\left\{F_{n+1}^{ext}\right\}$ is the discrete load vector corresponding to the external forces excluding the contact forces. $\{R^i\}$ is the unknown vector of the contact forces. A linearized form of (8) is given by :

$$[K_T]\{du\}^{i+1} = -\left\{F^{int}(u_{n+1})^i\right\} + \left\{F^{ext}_{n+1}\right\} + \left\{R^{i+1}\right\} \tag{9}$$

where K_T is the tangent stiffness matrix.

FRICTIONAL CONTACT

We restrict this part of the study to the contact between a deformable body and a rigid obstacle. Let n be the outward normal unit vector to the obstacle. We use the following decomposition into the normal and tangential components of the displacements (u) and the contact force vector (R):

$$u_N = u.n, \quad u_T = u - u_N n, \quad R = R_N n + R_T \tag{10}$$

Let Δg_N be the gap function at the start of the loading step. The unilateral contact conditions can then be written in terms of the relative displacement (Δw) as follows:

$$\Delta w_N = \Delta u_N + \Delta g_N \quad \Delta w_N \geq 0, \quad R_N \geq 0, \quad \Delta w_N.R_N = 0 \tag{11}$$

As a contact condition for the tangential direction we take Coulomb's law of friction, written below in an incremental form, in which Δg_T denotes the tangential incremental displacement of the rigid body and Δw_T the relative tangential displacement:

$$\Delta w_T = \Delta u_T - \Delta g_T \qquad \|R_T\| \leq \mu R_N \begin{cases} \|R_T\| < \mu R_N \Rightarrow \Delta w_T = 0 \\ \|R_T\| = \mu R_N \Rightarrow \Delta w_T = -\alpha R_T, \alpha \geq 0 \end{cases} \tag{12}$$

The numerical methods available for dealing with the discretized frictional contact problem set in equations (11) and (12) are the penalty/Lagrangian methods, the minimization under constraints of a quadratic functional (Gauss-Seidel methods with projection) methods, and Lemke's method, which can be used to solve a linear complementarity problem associated with the contact and friction conditions.

Augmented Lagrangian formulation by Simo and Laursen

The method used in this study is that described by Simo and Laursen (1992). Formulating Coulomb's friction law by analogy with the theory of plasticity leads to the following definition of the slip surface:

$$f_s = \|R_T\| - \mu R_N = 0 \tag{13}$$

If $f_s < 0$, no slip occurs (stick contact) and when $f_s = 0$, slip is impending. The condition f_s ¿ 0 is not admissible. The use of these notations makes it possible to write the Coulomb friction law as the following Kuhn-Tucker conditions:

$$f_s \leq 0, \quad \Delta w_T = -\Delta\xi \frac{\partial f_s}{\partial R_T}, \quad \Delta\xi \geq 0, \quad \Delta\xi.f_s = 0 \tag{14}$$

Additively decomposing R_N and R_T into their penalty and Lagrange multiplier parts yields:

$$R_N = <\lambda_N - \epsilon_N w_N> \tag{15}$$

$$\Delta R_T = \Delta\lambda_T - \epsilon_T\left(\Delta w_T + \Delta\xi\frac{\partial f_s}{\partial R_T}\right) \tag{16}$$

Introducing the unknown Lagrange terms into the problem leads to an additive loop (indicated below by superscript k) in which the Lagrange multipliers are updated. At each of these iterations, the Newton-Raphson method (iteration indicated below by superscript i) is used to determine the equilibrium state. The normal component (R_N) of the contact forces is computed by :

$$^kR^i_{N_{n+1}} =< {}^k\lambda_{N_{n+1}} - \epsilon_N\left({}^k\Delta w^i_{N_{n+1}}\right) > \tag{17}$$

When a backward Euler scheme associated with a radial return algorithm is used to determine the tangential component (R_T) of the contact forces, a trial value of (R_T) is first calculated assuming a sticking state:

$$^kR^{i(trial)}_{T_{n+1}} = R_{T_n} + {}^k\Delta\lambda_T - \epsilon_T\left({}^k\Delta w^i_{T_{n+1}}\right) \tag{18}$$

The corrected value of (R_T) is then evaluated using a radial return algorithm, which is a straightforward extension of those used in plasticity, (see Giannakopoulos, 1990 and Wriggers and al., 1990):

$$^kR^i_{T_{n+1}} = {}^kR^{i(trial)}_{T_{n+1}} - \Delta\xi\frac{^kR^{i(trial)}_{T_{n+1}}}{\|{}^kR^{i(trial)}_{T_{n+1}}\|} \tag{19}$$

where: $\Delta\xi = 0 \;\; if \;\; f^{trial}_{s_{n+1}} \leq 0 \quad and \quad \Delta\xi = \frac{f^{trial}_{s_{n+1}}}{\epsilon_T} \;\; if \;\; f^{trial}_{s_{n+1}} > 0$
After convergence of the Newton-Raphson iterations, the updated forms of the Lagrange multipliers are then given by the following expressions:

$$^{k+1}\lambda_{N_{n+1}} =< {}^k\lambda_{N_{n+1}} - \epsilon_N\left({}^k\Delta w_{N_{n+1}}\right) > \tag{20}$$

$$^{k+1}\Delta\lambda_{T_{n+1}} = {}^k\Delta\lambda_{T_{n+1}} - \epsilon_T\left(\Delta w^k_T + \Delta\xi\frac{^kR^{(trial)}_{T_{n+1}}}{\|{}^kR^{(trial)}_{T_{n+1}}\|}\right) \tag{21}$$

Minimization with projection method

Here we use a straightforward extension of the method presented by (Raous et al., 1988) for dealing with elasticity with infinitesimal deformations. In the contactless case, (R^{i+1} vanishes in (9)) at each equilibrium iteration. As K_T is a symmetric definite positive matrix, it can easily be proved that the problem can be set as the following minimization one, where for the sake of simplicity, index n is not used:

$$\left\{\begin{array}{l} \text{Find the vector } du^{i+1} \text{ in } R^{2no} \text{ (no being the number of mesh nodes)} \\ \text{such that :} J\left(du^{i+1}\right) \leq J\left(v\right)\forall v \in R^{2no} \text{ with :} \\ J(v) = \frac{1}{2}v^tK_Tv - v^tF, \quad with \quad F = -\left\{F^{int}\left(u_{n+1}\right)^i\right\} + \left\{F^{ext}_{n+1}\right\} \end{array}\right. \tag{22}$$

As shown in (Glowinski et al., 1976) the frictionless contact problem can be written as a minimization under constraints problem, in which the constraints are the discretized form of the unilateral conditions and where the unknown vector du^{i+1} is sought in convex K defined by: K= { v $\in R^{2no}$ such that $\Delta w^{i+1}_{N_j} = \Delta w^i_{N_j} + v_{N_j} \geq 0, \forall j \in$ I} where I is the set of contact nodes. This problem is then solved using a Gauss-Seidel with projection method. In the case with friction, a fixed point algorithm on the sliding threshold leads, as shown by Duvaut (1980), to solving a sequence of minimized constraint problems where the constraints affect only the normal components of the contact nodes, as in frictionless contact. These problems include an additive term for the tangential components of the contact nodes corresponding to the friction, which is updated at each fixed point iteration as shown below:

$$\begin{cases} \textbf{Find b fixed point of the application: } s^{i+1} \rightarrow \mu R_N\left(du^{i+1}(s)\right) \\ \text{with } du^{i+1} \text{ such that: } Q\left(du^{i+1}\right) \leq Q(v) \quad \forall v \in K \\ Q(v) = \frac{1}{2} v^t K_T v - v^t\left(F + \epsilon_T b^{i+1}\right) \end{cases} \tag{23}$$

where w_N and w_T are the normal and tangential components of the vector w defined in (11) and (12).

ϵ_T concerns the friction and is defined as follows:
$$\begin{cases} \epsilon_T = 1 & if\ \Delta w_T(v) < 0 \\ \epsilon_T = -1 & if\ \Delta w_T(v) > 0 \\ \epsilon_T = -1 & otherwise \end{cases}$$

This method is a very robust one, which can be used to solve a large class of contact problems in the context of infinitesimal elastic deformation. In the context of finite elastoplasticity, it requires lengthy computations and can therefore only be used with an acceleration process such as Aitken's one (Raous et al., 1994).

Linear complementarity problem

For dealing with the three-dimensional case, Klarbring and Bjorkman (1988) have introduced a piecewise linear friction law, approximating Coulomb's friction law. This discretization procedure then makes it possible to write the friction relations as complementarity conditions and then to set the problem as a linear complementarity one. In the present study, we shall restrict ourselves to a two-dimensional analysis. In this situation Klarbring and Bjorkman's approach leads to introducing two new variables λ and ϕ which define the boundary of the Coulomb's cone. The Kuhn Tucker conditions for the frictional Coulomb problem can then be written in the following form:

$$\begin{cases} R_T \in C(R_N) = \{P_T, \quad \phi_m(P_T, R_N) \geq 0, m = 1, 2\} \\ \phi_1(P_T, R_N) = -P_T + \mu R_N \\ \phi_2(P_T, R_N) = P_T + \mu R_N \\ \Delta w_T = -\sum_{m=1}^{2} \lambda_m \frac{\partial \phi_m}{\partial R_T} \\ \lambda_m \geq 0, \quad \phi_m \geq 0, \quad \lambda_m \phi_m = 0, \quad for \quad m = 1, 2 \end{cases} \tag{24}$$

Thanks to a condensation procedure, two connected systems can be written. The first one deals only with the normal and tangential components of the contact nodes which, upon introducing variables λ and ϕ, are all constrained by complementarity conditions. This linear complementarity problem with a 2NC by 2NC square singular matrix (NC being the number of contact nodes) can then be straighforwardly solved using a pivot algorithm such as Lemke's method. The second system deals with the nodes which are not involved in the contact. This is a non constrained problem in which the only unknown vector is the displacement one. Its solution obviously depends on the solution of the previous system, and can be obtained using a more classical algorithm.

NUMERICAL EXAMPLES

In this section, the algorithms presented above are tested and compared on two simple frictional contact problems. The methods have been implemented in the finite element code developed by Renault and several University partners (Jameux and Tathi, 1995). The first example is that of the indentation of an elasto-plastic block by a rigid cylinder. The second exemple is that of the stretching of a sheet with a hemispherical punch. Lastly, we perform a numerical study on a blankholder fitted with a drawbead.

Indentation test

Here an elastoplastic block is indented into a rigid cylinder . The cylinder has a diameter of 60mm and the block measures 40mm by 80mm. The block is discretized using the four-node elasto-plastic element (Q4/P0) described in (Dubois, 1994) and has the following material properties : E = 2.10^5MPa, $\nu = 0.3$. The hardening law is given by : $\sigma_0 = 593(0.00384 + \bar{\epsilon}_p)^{0.202}$. The coefficient of friction between the die and the workpiece is μ =0.2. The mesh and the deformed geometry are shown in figure 1.

As the results obtained with all three methods are very similar, comparisons were made between the CPU times, which are given in Table 1.

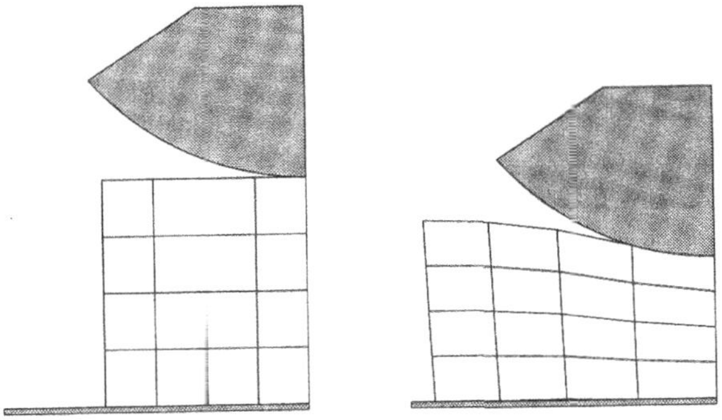

Figure 1 . Undeformed and deformed geometry of indentation

Table 1 . Computational times

Method	Gauss-Seidel	Lemke	Augmented Lagrangian
CPU Times	15"3	8"4	17"5

Stretching of a sheet with a hemispherical punch

The second example is a benchmark test invented by Wagoner et al. (1990) to provide a simple test for comparing the available analytical tools for sheet metal analysis. As our aim here is to make further comparisons between the numerical methods developed for dealing with the frictional problem, the mechanical results will not be presented. A complete description of the problem can be found in Chabrand et al. (1994), where our results are compared with those of the other users of this benchmark test. The stretching test, which yielded the geometry and material data given in figure 2 is a study on the axisymmetric situation. The mesh used is that proposed by the authors of the benchmark test: it comprises 15 nodes in the lengthwise and 3 in the breadthwise direction. The sheet was meshed using 28 Q4/P0 elements. The computations were carried out using both frictionless and frictional (μ = 0.15 and μ = 0.3) interfaces. With the Lagrangian method, $\epsilon_T = 10^4$ and the maximum number of updatings was 3. Table 2 gives the computational times obtained with each of the methods in those three situations.

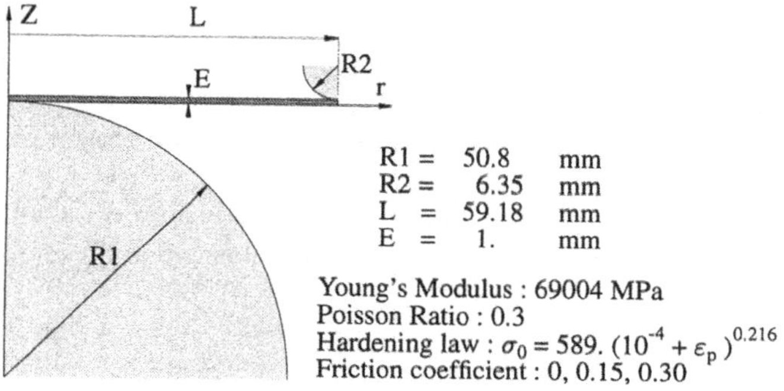

Figure 2 . Stretching of a sheet with a hemispherical punch : geometry and material data

Table 2 . Computational times

Method	Gauss-Seidel	Lemke	Augmented Lagrangian
$\mu = 0.$	5'01"	1'41"	1'38"
$\mu = 0.15$	6'16"	1'47"	3'48"
$\mu = 0.3$	7'07"	1'50"	4'23"

Drawbead test

The tool geometry and properties of the material are given in figure 3. The sheet was modelled with two layers of Q4/P0 elements.

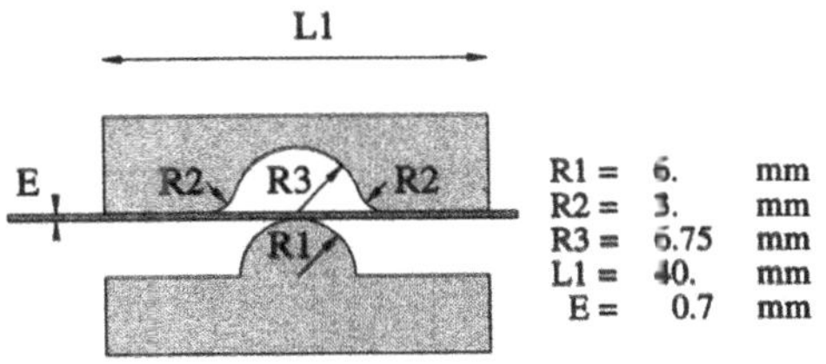

Young's Modulus : 0.1909 10^6 MPa
Poisson Ratio : 0.3
Hardening law : $\sigma_0 = 673.6\ (0.010095 + \varepsilon_p)^{0.176}$
Friction coefficient on flat part: 0.17
Friction coefficient on rounded part : 0.0 to 0.17

Figure 3 . Drawbead test : geometry and material data.

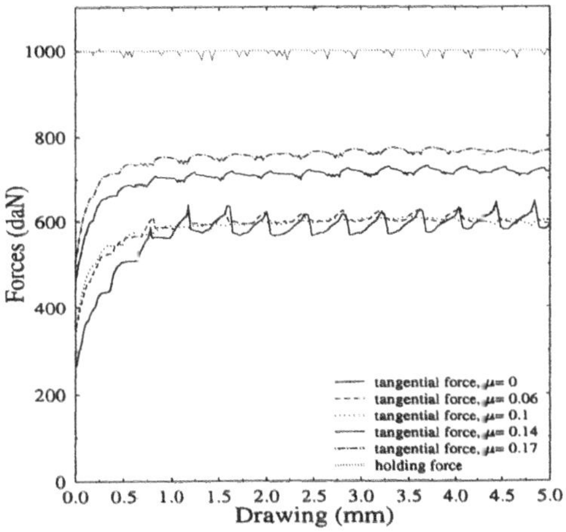

Figure 4 . Restraining forces versus drawing.

The analysis was carried out under the plane strain assumption. During the locking phase the blankholder moves up until the prescribed holding force is reached. At the end of the locking, a tangential displacement is prescribed on the right hand side of the sheet. During this step a specific algorithm that we developed (Chabrand and Dubois, 1995), is used to drive the blankholder keeping a constant holding force. The resistance of the strip to be moved through the drawbead is caused by friction and bending and unbending. Our aim was to carry out a fine analysis of this phenomenon by studing the contact state and the hardening, particularly in the thickness of the sheet. A fairly good agreement was found to exist between our numerical results and the experimental measurements. Due to the presence of the drawbead and the rounded part of the tool, the number of contact nodes was very low and Lemke's method turned out to be the most efficient. The stationary solution is reached after pulling out the sheet 1.5mm. Figure 4 gives the restraining forces as a function of the drawing. The results shown in figure 4 illustrate the influence of various friction coefficients between the rounded parts between the tools and the sheet metal and the need for a fine analysis to be able to analyse various friction situations (such as the lubricating boundary, mixed and hydrodynamic regimes) depending on the part of the tools with which the contact occurs.

CONCLUSION

No general conclusion can obviously be drawn from the examples dealt with here. However, Lemke's algorithm turned out to be a suitable method for dealing with these two-dimensional frictional contact problems where the number of contact nodes was not too large. In fact, both the displacements and the contact forces can be straightforwardly computed in this way without performing any additive iterations. Furthermore, there is no need to use a penalty parameter as in the augmented Lagrangian and penalty methods.

REFERENCES

Chabrand P., Dubois F, Gelin J.C., 1994, Modelling drawbeads in sheet metal forming, submitted for publication.

Chabrand P., Dubois F., 1995, A study on the friction in the deep drawing context, NUMIFORM'95, The 5th International Conference on Numerical methods in industrial forming processes, Cornell University, Ithaca, New York, June 18-21, 1995.

Dubois, F., 1994, Contact et frottement en grandes déformations élastoplastiques. Application à l'emboutissage, Thesis, Université d'Aix-Marseille II, décembre 1994

Duvaut, G., Equilibre d'un solide élastique avec contact unilatéral et frottement de Coulomb, C.R. Acad. Sc., Paris 290 A, pp 263-265

Giannakopoulos, A.E., 1989, The return mapping method for the integration of friction con stitutive relations, *Computers & Structures*, vol. 32, pp 157-167.

Glowinski, R., Lions, J.L., Trémolières, R., 1976, Analyse numérique des inéquations variationnelles, Dunod, Paris.

Jameux, J.P., Tathi, B., 1995, A joint project for the numerical simulation of 3D sheet metal forming processes with quasi-static and dynamic approaches, NUMIFORM'95, The 5th International Conference on Numerical methods in industrial forming processes, Cornell University, Ithaca, New-York, June 18-21, 1995.

Lee, E.H., 1969, Elastic plastic deformation at finite strain, *ASME Trans. J. Appl. Mech.*, 36, 1-6.

Lee, J.K., Wagoner, R.H., Nakamachi, E., 1990, A benchmark test for sheet forming analysis, Report N0. ERC/NSM-S-90-22, The Ohio state university, july 1990.

Raous, M., Chabrand, P., and Lebon, F., 1988, Numerical methods for frictional contact problems and applications, *App. Th. Mech. Jour.*, Supp. to vol 7, 111:128.

Raous, M., and Barbarin, S., 1992, Conjugate gradient for frictional contact, *in*: "Contact Mechanics", Presses Polytechniques et Universitaires Romandes .

Raous, M., Barbosa, H., and Latil, J.C., 1994, Aitken acceleration of Gauss Seidel with projection algorithm for unilateral contact problems with friction, submitted for publication to *Comm. Appl. Numer. Meth.*

Sidoroff, F., 1982, Incremental constitutive equation for large strain elasto plasticity, Int. J. Engng Sci, Vol 20, N01, pp 19-26.

Simo J.C., Taylor R.L., 1985, Consistent tangent operators for rate-independent elastoplasticity, *Comp. Meth. in Appl. Mech. and Eng.*, vol.48, pp. 101-118.

Simo J.C., Laursen T. A., 1992, An augmented lagrangian treatment of contact problems involving friction, *Computers & Structures*, vol. 42, pp. 97-116.

Wriggers, P., Vu Van, T., Stein, E., 1990, Finite element formulation of large deformation impact-compact problems with friction, *Computers & Structures*, vol. 37, pp.319-331.

COUPLED TREATMENT OF FRICTIONAL CONTACT PROBLEMS IN A REDUCED SYSTEM AND APPLICATIONS IN LARGE DEFORMATION CONTEXT

Zhi-Qiang Feng and Mathieu Domaszewski

Group MMS/GM, Institut Polytechnique de Sévenans
90010 Belfort, France

INTRODUCTION

In many metal forming processes and other industrial applications such as joint problems, two or more bodies come into contact with friction and they may undergo large deformation. In such processes nonlinearities arise in geometry, constitutive relation as well as variable contact and friction conditions. Since these nonlinearities make it hard to solve such problems, in particular the contact nonlinearities are non-smooth, a reliable and efficient solution algorithm is necessary. In the literature, large deformation contact problems with friction have almost exclusively been treated by penalty methods (Hallquist, 1983; ANSYS, 1993). Much information points to the fact that the penalty methods have some drawbacks concerning numerical stability and precision, in particular for friction simulation. By means of a theory called ISM (Implicit Standard Materials), De Saxcé and Feng (1991) propose an augmented Lagrangian formulation in which, the unilateral contact and the friction are coupled. The frictional contact problem is treated by solving a reduced system. The aim of the present paper is to constitute a brief outline and an extension of previous research in large deformation context and to present some computational examples of nonlinear analysis of contact problems.

LOCAL CONTACT MODELLING

For notational convience, a two-body contact problem is considered. We assume that the contact with friction may occur between some points $\mathbf{x}_A$ and $\mathbf{x}_C$ of two bodies A and C. Consequently, the contact and friction laws can be written only in terms of relative displacements $\mathbf{w} = \mathbf{u}_A - \mathbf{u}_C$ and of contact reactions $\mathbf{r} = \mathbf{r}_A = -\mathbf{r}_C$. The unilateral contact conditions for each contact point may be stated in a complementarity form:

$$w_n + g \geq 0; \quad r_n \geq 0 \text{ and } (w_n + g)\, r_n = 0 \tag{1}$$

Coulomb's dry friction criterion for each contact point is:

Contact Mechanics, Edited by M. Raous *et al.*
Plenum Press, New York, 1995

$$\|\mathbf{r}_t\| < \mu r_n \qquad \text{if } \|\mathbf{w}_t\| = 0 \tag{2a}$$

$$\mathbf{r}_t = -\mu r_n \frac{\mathbf{w}_t}{\|\mathbf{w}_t\|} \qquad \text{if } \|\mathbf{w}_t\| \neq 0 \tag{2b}$$

where n and t denote respectively the normal and tangential direction to the contact surface, μ is the friction coefficient, g is the initial contact gap and $\|\cdot\|$ denotes the euclidian norm.
Let κ_μ be the convex friction set defined by

$$\kappa_\mu = \{ (\mathbf{r}_t, r_n) \text{ such that } \|\mathbf{r}_t\| \leq \mu r_n \} \tag{3}$$

With the usual notation of an indicator function, the contact with friction can be represented by the following bipotential (De Saxcé and Feng, 1991) :

$$b(-w_n, -\mathbf{w}_t, r_n, \mathbf{r}_t) = \mu r_n \|\mathbf{w}_t\| + \Psi_{\mathbf{R}^+}(w_n + g) + \Psi_{\kappa\mu}(\mathbf{r}_t, r_n) \tag{4}$$

Then, the frictional contact constitutive laws can be written in a sub-differential form:

$$r_n \in -\partial\Psi_{\mathbf{R}^+}(w_n + g), \qquad \mathbf{r}_t \in -\mu r_n\, \partial_{\mathbf{w}_t}(\|\mathbf{w}_t\|) \tag{5}$$

From convex analysis, the above contact and friction laws are equivalent to the variational inequality:
Find $(\mathbf{r}_t, r_n) \in \kappa_\mu$ such that

$$\forall\, (\mathbf{r}_t{}^*, r_n{}^*) \in \kappa_\mu\,; \qquad (w_n + g - \mu\|\mathbf{w}_t\|)\,(r_n{}^* - r_n) + \mathbf{w}_t \cdot (\mathbf{r}_t{}^* - \mathbf{r}_t) \leq 0 \tag{6}$$

where $$-\mathbf{w}_t = \lambda \frac{\mathbf{r}_t}{\|\mathbf{r}_t\|} \quad \text{for some } \lambda \geq 0$$

In order to avoid nondifferentiable potentials that occur in nonlinear mechanics, such as in contact problems, it is convenient to use the augmented Lagrangian method. This method, applied to the variational inequality, leads to the following implicit equation:

$$\mathbf{r} = proj\,((r_n + \rho_n(w_n + g - \mu\|\mathbf{w}_t\|),\, \mathbf{r}_t + \rho_t \mathbf{w}_t),\, \kappa_\mu) \tag{7}$$

where ρ_n and ρ_t are real positive numbers, which can be chosen according to the eigenvalue of the contact flexibility matrix.

For numerical solution of implicit equation (7), Uzawa's algorithm can be used, which leads to an iterative procedure involving one predictor-corrector step:

- predictor
$$\begin{cases} \tau_n^{i+1} = r_n^i + \rho_n^i \left(r_n^i + g - \mu\|\mathbf{w}_t^i\|\right) \\ \tau_t^{i+1} = \mathbf{r}_t^i + \rho_t^i\, \mathbf{w}_t^i \end{cases} \tag{8}$$

- corrector
$$\left(r_n^{i+1}, \mathbf{r}_t^{i+1}\right) = proj\left(\left(\tau_n^{i+1}, \tau_t^{i+1}\right), \kappa_\mu\right) \tag{9}$$

It is important to note that, as compared to classical penalty methods, this approach allows to obtain very precise and stable results. This is valid for both 2-D and 3-D contact problems with Coulomb friction. An explicit expression of Eq.(9) is given by Klarbring (1992).

GLOBAL NUMERICAL ALGORITHM

Generally, nonlinear mechanical behaviors of solid media are represented by an equilibrium equation:

$$\mathbf{Res(x)} = -\ \mathbf{Fint(x)} + \mathbf{Fext(x)} + \mathbf{Reac(x)} = \mathbf{0} \tag{10}$$

where **Res(x)** denotes the residual vector, **Fint(x)** is the internal forces vector and **Fext(x)** external forces vector. **Reac(x)** is the contact reactions vector. This equation is strongly nonlinear, because of finite strains and large displacements of solid. Besides, the constitutive law of contact with friction is usually represented by inequalities and the contact potential is even nondifferentiable. Instead of solving this equation in consideration of all nonlinearities at the same time, our idea is to separate the nonlinearities to overcome the complexity of calculation and to improve the numerical stabilities. A step by step Newton-Raphson algorithm is often used to solve the nonlinear equations. A linearized form of Eq.(10) is:

$$\begin{cases} \mathbf{K\ dx} = \mathbf{Res}^* + \mathbf{Reac} \\ \mathbf{x} = \mathbf{x} + \mathbf{dx} \end{cases} \tag{11}$$

where **K** is the tangent stiffness matrix, $\mathbf{Res}^* = -\ \mathbf{Fint(x)} + \mathbf{Fext(x)}$. As **dx** and **Reac** are both unknown, Eq.(11) cannot be directly solved. First, vector **Reac** is determined by Eqs.(8,9) in a reduced system which only concerns the contact nodes. Then, vector **dx** can be computed in the whole structure, by means of contact reactions as external loading. Unlike the penalty or Lagrangian multiplier methods, our method neither changes the global stiffness matrix nor increases the number of degrees-of-freedom. So it is very easy to implement the contact and friction problems in an existing general purpose finite element code by this method. Besides, the contact conditions are accurately satisfied and the mathematical programming or iterative procedure is very simple and elegant in this model.

NUMERICAL EXAMPLES

Many application examples, academic or industrial, have been carried out by means of the present method. Due to space limitation, we present briefly some numerical examples concerning metal forming and joint problems. The first example (Figure 1) simulates a rubber shock absorber. The second example (Figure 2) simulates an upsetting test. Figure 3 shows a three dimensional contact problem.

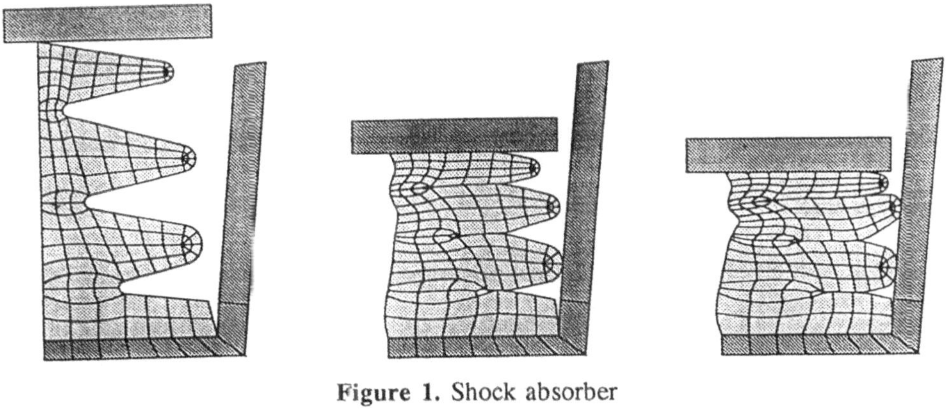

Figure 1. Shock absorber

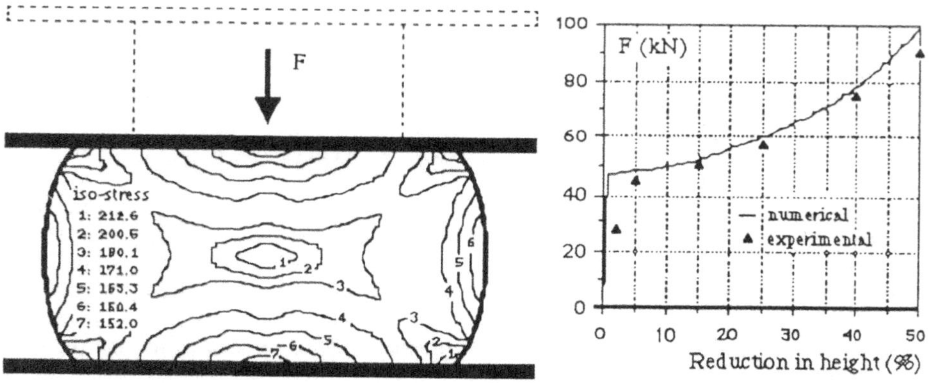

Figure 2. Upsetting test

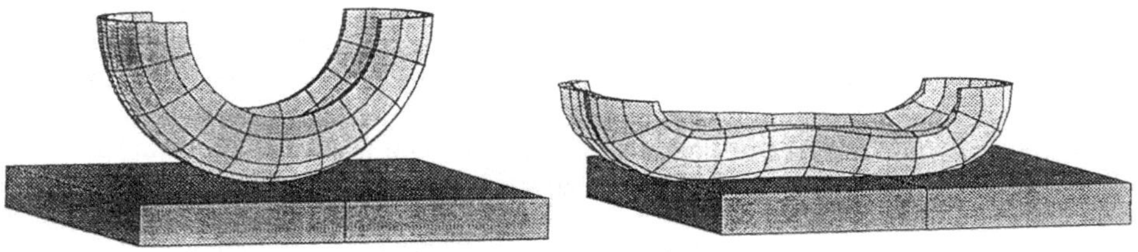

Figure 3. Three dimensional contact problem

CONCLUSIONS

The main purpose of this paper is to present briefly a frictional contact formulation in a large deformation context. The numerical results demonstrate that the proposed algorithms, for the local analysis of frictional contact problems and for the global solution of equilibrium equation, are capable of handling a wide range of engineering applications. These include situations in which:

- the contact occurs between several deformable bodies or between deformable bodies and rigid bodies in two or three dimension problems, the constitutive law of contact bodies being linear or nonlinear with finite strain and large displacements;
- the contact may be sticking or sliding, with or without Coulomb friction;
- the repeated "contact, non contact" and "sticking, sliding" are permitted;
- the relative displacement between contact bodies can be large;
- the contact surface can have small curvature radius, even sharp corners.

REFERENCES

ANSYS, 1993, "ANSYS User's Manual for Revision 5.0, Vol. IV - Theory", Swanson Analysis Systems.

De Saxcé, G. and Feng, Z.Q., 1991, New inequality and functional for contact with friction: The implicit standard material approach, *Mech. Struct. & Mach.*, **19**, 301:325.

Hallquist, J.O., 1983, "NIKE2D: A Vectorized, Implicit, Finite Deformation, Finite Element Code for Analyzing the Static and Dynamic Response of 2-D Solids", LLDL.

Klarbring, A., 1992, Mathematical programming and augmented Lagrangian methods for frictional contact problems, in: "1st Contact Mechanics International Symposium", A. Curnier, ed., PPUR, 369:390.

GEOMETRIC PROPERTIES OF UNILATERAL CONTACT CONSTRAINTS

Jean H. Heegaard

Division of Applied Mechanics
Mechanical Engineering Department
Stanford University
Stanford, CA 94305-4040

INTRODUCTION

A majority of available solutions to the unilateral contact problem implicitly consider small or moderate slip between a set of deformable bodies. Such contact problems are often solved using the Finite Element Method, in which the contacting surfaces are discretized into contact elements typically consisting of a striker node interacting with a target node or a linear facet. The most recent formulations treat the contact problem by using continuum-based finite element formulations, *i.e.* checking contact conditions and integrating contact forces at Gauss points instead of considering them at nodal points (Laursen and Simo, 1993; Wriggers and Imhof, 1993). However, these new approaches still include piecewise linear geometries to describe the contacting bodies boundary, introducing kinematics discontinuities when large slip occur.

In the present contribution, we explore some geometric properties of 3D surfaces in order to express unilateral contact constraints occurring during large slip over *smooth contact regions.* We start by constructing a 3D node-to-surface contact element. The projection of the striker point on any type of smooth 3D surface and its first variation are derived. The signed distance is further characterized by one uniqueness and two orthogonality conditions. The respective constraint equations are derived, and expressions for their first and second variations are calculated using intrinsic properties of smooth surfaces related to their first and second *fundamental forms.* The first variation typically conveys nonlinear expressions for the contact forces, which are then consistently linearized using the second variation

The resulting nonsmooth optimization problem with unilateral constraints is solved using *augmented Lagrangian* multipliers, which at equilibrium hold the contact forces. The main features of this method are to lead to well conditioned problems and to provide

Contact Mechanics, Edited by M. Raous *et al.*
Plenum Press, New York, 1995

exact solutions with respect to contact, in contrast to penalty methods. Hence, these elements rigorously enforce the Signorini conditions (i.e. non penetrability, no tension, and complementarity between gap and contact forces). A *generalized Newton method* proposed by Alart and Curnier (1991) conveying local quadratic rate of convergence is used to solve the nonlinear and nonsmooth equilibrium equations.

The resulting contact element was implemented in the contact problem oriented Finite Element code TACT (Curnier, 1985; Heegaard and Curnier, 1994). A comprehensive 3D model of the human patella (knee-cap) sliding over the femur during knee flexion illustrates how the present contact formulation could be applied in biomechanical engineering practice.

3D LARGE SLIP CONTACT ON A RIGID OBSTACLE

For simplicity we consider only two contacting bodies, denoted B^1 and B^2 in their reference configuration and arbitrarily called *striker* and *target* respectively. Furthermore, it is assumed that the target body is fixed up to a rigid body motion, so that only its reference configuration B^2 need to be considered. In order to characterize existence and uniqueness conditions for the gap vector, geometric properties of smooth surfaces in 3D Euclidian space $\mathbf{R}^3$ are first recalled. In previous studies (*e.g.* Parisch, 1989), such smooth curves were approximated by a set of bilinear patches, joined together at corner nodes. More generally, a surface $\mathcal{S}$ is represented in parametric form by a mapping (Figure 1)

$$\mathbf{y} = \mathbf{y}(\boldsymbol{\xi}) : \mathcal{B} \subset \mathbf{R}^2 \rightarrow \mathcal{S} \subset \mathbf{R}^3$$

defined over a simply connected bounded domain $\mathcal{B}$ of $\mathbf{R}^2$ (typically the unit square).

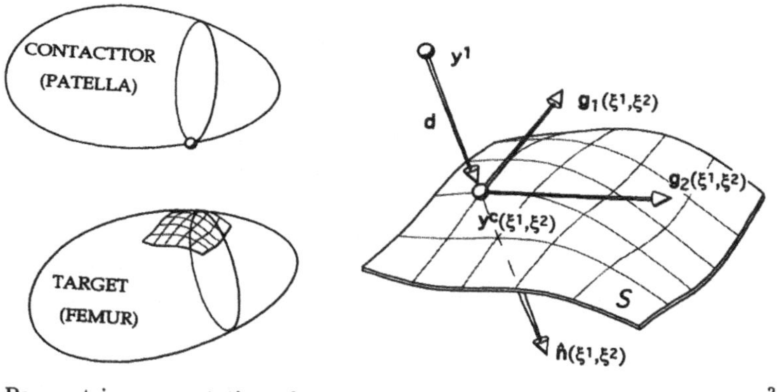

Figure 1. Parametric representation of a surface $\mathcal{S}$ viewed as a mapping from $\mathcal{B} \subset \mathbf{R}^2$ into $\mathbf{R}^3$.

Furthermore, smoothness of $\mathcal{S}$ requires that $\mathbf{y}$ be at least continously differentiable ($\mathbf{y} \in C^1$) so that the Jacobian matrix of the mapping exists and is of maximal rank 2:

$$\mathbf{J} = \nabla_{\boldsymbol{\xi}} \mathbf{y} = \frac{\partial y_i}{\partial \xi^\alpha} \qquad (i = 1,3 \quad \alpha = 1,2) \tag{1}$$

The columns of $\mathbf{J}$ represent the tangent vectors to the surface along the directions ξ^α, and the full rank condition ensures their linear independence.

A local tangent coordinate frame $\mathcal{T} = (\mathbf{g}_1, \mathbf{g}_2, \hat{\mathbf{n}})$ associated to $\mathbf{y}$ is introduced next, where

$$\mathbf{g}_\alpha = \mathbf{g}_\alpha(\boldsymbol{\xi}) = \frac{\partial \mathbf{y}(\boldsymbol{\xi})}{\partial \xi^\alpha} = \mathbf{y}_{,\alpha} \qquad \alpha = 1,2 \tag{2}$$

are the tangent vectors at $\mathbf{y}(\boldsymbol{\xi})$, spanning the tangent plane to $\mathcal{S}$ at that point and

$$\hat{\mathbf{n}} = \hat{\mathbf{n}}(\boldsymbol{\xi}) = \frac{\mathbf{g}_1 \wedge \mathbf{g}_2}{\| \mathbf{g}_1 \wedge \mathbf{g}_2 \|} \tag{3}$$

denotes the unit normal to the surface. The parameters orientation is chosen so as to let $\hat{\mathbf{n}}$ be *inward* to the target body B^2. Having a complete description of the target surface at hand, the gap vector $\mathbf{d}$ between a striker point $\mathbf{y}^1 \in B^1$ and $\mathcal{S}$ can be expressed as follow

$$\mathbf{d} = \mathbf{d}(\mathbf{y}^1) = (\mathbf{y}^c - \mathbf{y}^1) \tag{4}$$

where $\mathbf{y}^c = \mathbf{y}(\boldsymbol{\xi}^c)$ stands for the projection of $\mathbf{y}^1$ on $\mathcal{S}$ satisfying

$$\boldsymbol{\xi}^c = \boldsymbol{\xi}^c(\mathbf{y}^1) = \underset{\boldsymbol{\xi} \in B^2}{\operatorname{argmin}} \| \mathbf{y}(\boldsymbol{\xi}) - \mathbf{y}^1 \| \tag{5}$$

The signed contact distance d_n is defined in the usual way

$$d_n = d_n(\mathbf{y}^1) = \mathbf{d} \cdot \hat{\mathbf{n}} \tag{6}$$

The critical point $\boldsymbol{\xi}^c$ satisfying (5) is obtained by solving the following minimization problem

$$\min_{\boldsymbol{\xi} \in B^2} \frac{1}{2} \left(\mathbf{y}(\boldsymbol{\xi}) - \mathbf{y}^1 \right) \cdot \left(\mathbf{y}(\boldsymbol{\xi}) - \mathbf{y}^1 \right) \tag{7}$$

whose minimum is characterized by

$$\begin{cases} \mathbf{g}_1 \cdot \mathbf{d} = 0 & (a) \\ \mathbf{g}_2 \cdot \mathbf{d} = 0 & (b) \\ \boldsymbol{\eta}^{\mathrm{T}} \mathbf{B} \boldsymbol{\eta} > 0 \ \forall \ \boldsymbol{\eta} \in \mathbb{R}^2 & (c) \end{cases} \tag{8}$$

where $\mathbf{B}$ is the Hessian of the objective function introduced in (7) evaluated at $\boldsymbol{\xi}^c$. Equations (8–a) and (8–b) respectively indicate that the gap vector $\mathbf{d}$ is directed along $\hat{\mathbf{n}}$ (*i.e.* $\mathbf{d} = d_n \hat{\mathbf{n}}$) and that a local minimum is indeed obtained. A simple expression of $\mathbf{B}$ can be found by using quantities related to the *first* and *second fundamental forms* of $\mathcal{S}$. To this end the covariant *metric tensor* $g_{\alpha\beta}$ associated to $\mathcal{T}$ (first fundamental form)

$$g_{\alpha\beta} = \mathbf{g}_\alpha \cdot \mathbf{g}_\beta \tag{9}$$

and the *curvature tensor* associated to the second fundamental form of $\mathcal{S}$ (see *e.g.* Flugge, 1972)

$$b_{\alpha\beta} = \mathbf{g}_{\alpha,\beta} \cdot \hat{\mathbf{n}} \tag{10}$$

are introduced. For further use, the *conjugacy property* of the metric tensor is recalled *i.e.*

$$g_{\alpha\gamma}\, g^{\gamma\beta} = \mathbf{g}_\alpha \cdot \mathbf{g}^\beta = \delta_\alpha^\beta \qquad (i.e. \ \ g_{\alpha\beta} = (g^{\alpha\beta})^{-1})$$

where $g^{\alpha\beta}$ are the contravariant components of the metric tensor. The corresponding contravariant base vectors are defined by

$$\mathbf{g}^\alpha = g^{\alpha\beta}\, \mathbf{g}_\beta = g_{\alpha\beta}^{-1}\, \mathbf{g}_\beta \tag{11}$$

The Hessian $\mathbf{B}$ can then be expressed as

$$\mathbf{B} = B_{\alpha\beta}[\mathbf{g}^\alpha \otimes \mathbf{g}^\beta] = (d_n\, b_{\alpha\beta} + g_{\alpha\beta})[\mathbf{g}^\alpha \otimes \mathbf{g}^\beta] \quad (\alpha, \beta = 1, 2) \tag{12}$$

The *normal curvature vector* $\mathbf{k}_n$ at a point $\mathbf{y}$ on $\mathcal{S}$, along the tangent direction $\mathbf{t} = (t^1, t^2)$ is defined as:

$$\mathbf{k}_n = \kappa_n \hat{\mathbf{n}}$$

where κ_n is the (normal) curvature of the intersection curve between $\mathcal{S}$ and the plane spanned by $\mathbf{t}$ and $\hat{\mathbf{n}}(\mathbf{y})$. It is known that κ_n can be expressed as the ratio between the second and the first fundamental forms of $\mathcal{S}$ (see *e.g.* Faux and Pratt, 1987)

$$\kappa_n = \frac{1}{\rho_n} = \frac{b_{\alpha\beta}\, dt^\alpha\, dt^\beta}{g_{\alpha\beta}\, dt^\alpha\, dt^\beta}$$

where ρ_n is the usual radius of curvature.

There are two *principal directions* of normal curvature $\mathbf{t}^*_\alpha$, $(\alpha = 1, 2)$ which are those directions of the tangents $\mathbf{t}$ for which κ_n takes extreme values (the *principal normal curvature* κ_1 and κ_2). The *focal points* $\mathbf{f}^\alpha$, $(\alpha = 1, 2)$ of $\mathcal{S}$ at $\mathbf{y}$ are defined as (see *e.g.* Thorpe, 1979):

$$\mathbf{f}^\alpha = \mathbf{y} + \kappa_\alpha \hat{\mathbf{n}}$$

The *Gaussian curvature* K and the *mean curvature* H are respectively defined as

$$\mathrm{K} = \kappa_1 \kappa_2$$
$$\mathrm{H} = \frac{1}{2}(\kappa_1 + \kappa_2)$$

and are *intrinsic* to the surface $\mathcal{S}$ in the sense that they are invariants (*i.e.* independent of the parameterization of $\mathcal{S}$). It can be shown (*e.g.* Thorpe, 1979), that under a parametrization of $\mathcal{S}$, such that its tangent base vectors $\mathbf{g}^*_\alpha$ are along the principal directions of curvature $\mathbf{t}^*_\alpha$, the off-diagonal terms $g_{\alpha\beta}$ and $b_{\alpha\beta}$ vanish. Furthermore, the normal principal curvature κ_α become (Farin, 1990)

$$\kappa_\alpha = \frac{\mathbf{g}^*_{\alpha,\xi^\alpha} \cdot \hat{\mathbf{n}}}{\| \mathbf{g}^*_\alpha \|^2}$$

In this particular case, the positive-definiteness of $\mathbf{B}$ required as a minima condition in (8–c) can be interpreted, using (12), as

$$\begin{cases} d_n > -\dfrac{1}{\kappa_\alpha} & \{\alpha \mid \kappa_\alpha \geq 0\} \quad \alpha \in \{1, 2\} \\[2ex] d_n < -\dfrac{1}{\kappa_\alpha} & \{\alpha \mid \kappa_\alpha < 0\} \quad \alpha \in \{1, 2\} \end{cases} \tag{13}$$

Convexity of the surface as seen from a striker $\mathbf{y}^1$ outside of the target body B^2 is characterized by the signs of κ_1 and κ_2, as summarized in Figure 2. Hence, when the projection $\mathbf{y}^c$ of $\mathbf{y}^1$ is an elliptic or a parabolic point, d_n will be minimum if $\mathbf{y}^1$ is on the half line bounded by the *closest focal point* $\mathbf{f}^\alpha$ of $\mathcal{S}$ at $\mathbf{y}^c$ and oriented along $(\mathbf{y}^c - \mathbf{f}^\alpha)$. In the case of a saddle point, d_n will be a minimum only if $\mathbf{y}^1$ lies within the segment spanned by both focal points $\mathbf{f}^\alpha$ (in that case they lie on each side of $\mathcal{S}$). These conditions can thus be used efficiently to determine existence and uniqueness of the projection $\mathbf{y}^c$ of $\mathbf{y}^1$.

$\kappa_1\ \kappa_2 > 0$	elliptic	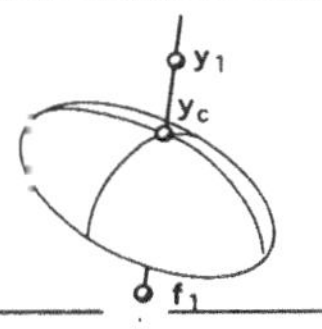
$\kappa_1\ \kappa_2 = 0$	parabolic	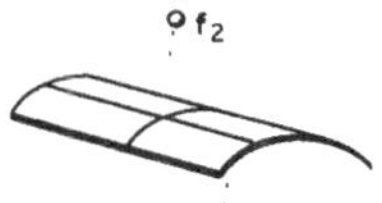
$\kappa_1\ \kappa_2 < 0$	hyperbolic (saddle point)	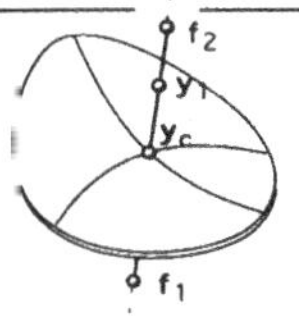

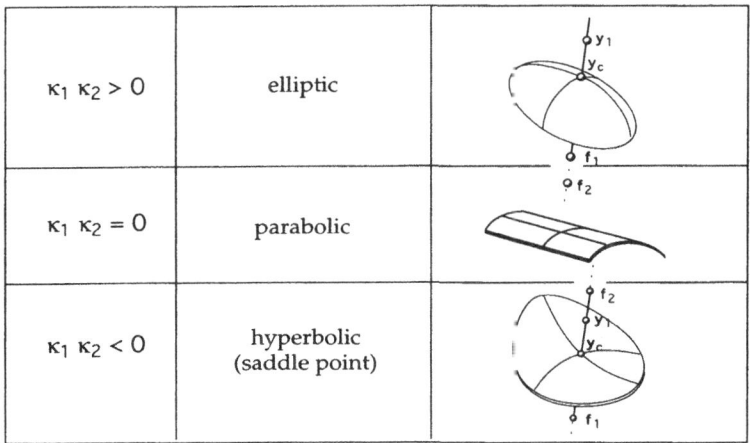

Figure 2. A point on $\mathcal{S}$ is either *elliptic*, *parabolic* or *hyperbolic* depending on the signs of κ_1 and κ_2

Equilibrium of the contacting bodies is expressed using the principle of virtual work, which further requires differentiation of the signed gap distance d_n. We first express a striker's actual position $\mathbf{y}^1$ in terms of its displacement vector $\mathbf{u}^1$ by $\mathbf{y}^1 = \mathbf{x}^1 + \mathbf{u}^1$, where $\mathbf{x}^1$ represents the striker's reference position. Assuming then that the striker's projection $\mathbf{y}^c$ exists, is unique, and is known, the first variation $d\,d_n$ of d_n is easily derived by differentiating (6)

$$d\,d_n = \nabla_{\mathbf{u}^1}\, d_n\, d\mathbf{u}^1 = (\nabla_{\mathbf{u}^1}\, \mathbf{y}^c\, d\mathbf{u}^1 - d\mathbf{u}^1)\cdot\hat{\mathbf{n}} + \mathbf{d}\cdot(\nabla_{\mathbf{u}^1}\hat{\mathbf{n}}\, d\mathbf{u}^1)$$

Recalling that $\hat{\mathbf{n}}\cdot\hat{\mathbf{n}} = 1$, so that $\hat{\mathbf{n}}_{,\alpha}$ is perpendicular to $\hat{\mathbf{n}}$, the last expression simplifies into

$$d\,d_n = (d\,\mathbf{y}^c - d\,\mathbf{u}^1)\cdot\hat{\mathbf{n}}$$

The first variation of the projection $\mathbf{y}^c$ being given by

$$d\,\mathbf{y}^c = \nabla_{\mathbf{u}^1}\, \mathbf{y}^c\big(\xi^1(\mathbf{y}^1),\xi^2(\mathbf{y}^1)\big)d\mathbf{u}^1 = \mathbf{g}_1\, d\xi^1 + \mathbf{g}_2\, d\xi^2 \tag{14}$$

confirms that $d\,\mathbf{y}^c$ is orthogonal to $\hat{\mathbf{n}}$ so that $d\,d_n$ simply becomes

$$d\,d_n = -\hat{\mathbf{n}}\cdot d\mathbf{u}_1 \tag{15}$$

This last equation reflects the now well established characteristic of a contact distance variation which is normal to the target surface $\mathcal{S}$. According to the principle of virtual work this variation characterizes the balance of contact forces (Curnier and Alart, 1988). Using a mixed form extension of this principle Alart and Curnier, 1991, the corresponding *augmented Lagrangian* contact force operator $\mathbf{f} \in \mathbb{R}^3$ takes the form (Heegaard and Curnier, 1993):

$$\mathbf{f} = \begin{cases} \begin{pmatrix} -(f_n + rd_n)\,\hat{\mathbf{n}} \\ d_n \end{pmatrix} & \text{if } (f_n + rd_n) \le 0 \quad \text{(contact)} \\ \begin{pmatrix} \mathbf{0} \\ -f_n/r \end{pmatrix} & \text{if } (f_n + rd_n) > 0 \quad \text{(gap)} \end{cases} \tag{16}$$

where r is a regularization parameter, and f_n is the Lagrange multiplier holding the unknown contact force magnitude. This contact element contribution combined with standard finite elements leads to a *mixed formulation* of the contact problem (typical of Lagrangian based methods) in the sense that both primal and dual quantities are independent variables. It follows that the determination of the element status (*i.e.* contact or gap) is based on a linear combination $f_n + rd_n$ of the kinematic variable $d_n(\mathbf{u}^1)$ *and* the static variable f_n.

Consistent linearization of the nonlinear contact force operator $\mathbf{f}$ requires the computation of the first variation of $\hat{\mathbf{n}}$ along $d\mathbf{u}^1$. This variation has already been calculated in the particular case where the ξ^α are lines of curvature (Hansson and Klarbring, 1990) *i.e.* when $\mathbf{g}_\alpha$ is a principal direction of normal curvature. Here this result is generalized for any parametrization ξ^α of $\mathcal{S}$. The first variation of $\hat{\mathbf{n}}$ can be written as

$$d\,\hat{\mathbf{n}} = \nabla_{\mathbf{u}^1}\,\hat{\mathbf{n}} d\mathbf{u}_1 = \hat{\mathbf{n}}_{,\alpha}\; d\xi^\alpha \tag{17}$$

The first term in the righthand side of (17), $\hat{\mathbf{n}}_{,\alpha}$, lies in the tangent plane spanned by $\mathbf{g}_1$ and $\mathbf{g}_2$ so that it can be written as a linear combination of these base vectors

$$\hat{\mathbf{n}}_{,\alpha} = c_\alpha^\gamma\; \mathbf{g}_\gamma$$

The coefficients c_α^γ are determined by taking the scalar product $\mathbf{g}_\sigma$ on both sides of the last expression *i.e.*

$$\hat{\mathbf{n}}_{,\alpha} \cdot \mathbf{g}_\sigma = c_\alpha^\gamma\; \mathbf{g}_\gamma \cdot \mathbf{g}_\sigma = c_\alpha^\gamma\; g_{\gamma\sigma}$$

Furthermore, from the identity $\mathbf{g}_\alpha \cdot \hat{\mathbf{n}} = 0$ it can be seen that

$$\hat{\mathbf{n}}_{,\beta} \cdot \mathbf{g}_\alpha = -\hat{\mathbf{n}} \cdot \mathbf{g}_{\alpha,\beta} = -b_{\alpha\beta}$$

and by using the conjugacy property of the metric tensor (11)

$$\hat{\mathbf{n}}_{,\alpha} \cdot \mathbf{g}_\sigma\; g^{\sigma\gamma} = -b_{\alpha\sigma}\; g^{\sigma\gamma} = c_\alpha^\lambda\; g_{\lambda\sigma}\; g^{\sigma\gamma} = c_\alpha^\gamma$$

the components of $\hat{\mathbf{n}}_{,\alpha}$ in the tangent reference frame become

$$c_\alpha^\gamma = -g^{\sigma\gamma}\; b_{\alpha\sigma}$$

Thus, the following expression for the *Weingarten map* $\hat{\mathbf{n}}_{,\alpha}$ is obtained

$$\hat{\mathbf{n}}_{,\alpha} = -g^{\sigma\beta}\; b_{\alpha\sigma}\; \mathbf{g}_\beta = b_{\alpha\sigma}\; \mathbf{g}^\sigma \tag{18}$$

which is a measure (up to sign) of the rate of change of $\hat{\mathbf{n}}$ (*i.e.* the turning of $\hat{\mathbf{n}}$ since $\hat{\mathbf{n}}$ is normed) along the coordinate curve ξ^α.

The variation of $d\xi^\alpha$ in (17) must still be expressed in terms of $d\mathbf{u}^1$. This is achieved by taking the total differential of (8–a) and (8–b)

$$d\,\mathbf{g}_\alpha \cdot (\mathbf{y}^c - \mathbf{y}^1) + \mathbf{g}_\alpha \cdot d\,(\mathbf{y}^c - \mathbf{y}^1) = 0\;, \qquad \alpha = 1,2 \tag{19}$$

Furthermore, according to (2) $d\,\mathbf{g}_\alpha$ is given by$\mathbf{g}_{\alpha,\beta}\; d\xi^\beta$, so that by using (12) and (14), $d\boldsymbol{\xi}$ is obtained by solving the system

$$\mathbf{B}d\boldsymbol{\xi} = \mathbf{J}^T\; d\mathbf{u}^1$$

where it is recalled that $\mathbf{J}$ is the Jacobian matrix of the mapping $\mathbf{y}(\boldsymbol{\xi})$. It follows that $d\boldsymbol{\xi}$ expresses in terms of $d\mathbf{u}^1$ as

$$\begin{aligned} d\xi^1 &= \frac{1}{h}(B_{22}\,\mathbf{g}_1 - B_{12}\,\mathbf{g}_2) \cdot d\mathbf{u}^1 \\ d\xi^2 &= \frac{1}{h}(B_{11}\,\mathbf{g}_2 - B_{12}\,\mathbf{g}_1) \cdot d\mathbf{u}^1 \end{aligned} \tag{20}$$

where h is the determinant of $\mathbf{B}$.

Substituting $\hat{\mathbf{n}}_{,\alpha}$ and $d\xi^\alpha$ in (17) by their expression found in (18) and (20) the first variation of $\hat{\mathbf{n}}$ becomes

$$d\,\hat{\mathbf{n}} = \Big(-g^{\sigma\beta}\, b_{1\sigma}\, \mathbf{g}_\beta \frac{1}{h}\left(B_{22}\,\mathbf{g}_1 - B_{12}\,\mathbf{g}_2\right) - g^{\sigma\beta}\, b_{2\sigma}\, \mathbf{g}_\beta \frac{1}{h}\left(B_{11}\,\mathbf{g}_2 - B_{12}\,\mathbf{g}_1\right)\Big) \cdot d\mathbf{u}^1$$

Replacing the covariant base vectors $\mathbf{g}_\beta$ by their contravariant duals $\mathbf{g}^\beta$ and by using the fact that for any 3 vectors $\mathbf{u}, \mathbf{v}, \mathbf{w}$ of $\mathbb{R}^n$, $(\mathbf{v}\cdot\mathbf{w})\mathbf{u} = [\mathbf{u}\otimes\mathbf{v}]\mathbf{w} = [\mathbf{u}\otimes\mathbf{w}]\mathbf{v}$ the last equation becomes

$$\begin{aligned} d\,\hat{\mathbf{n}} = -\frac{1}{h}\Big(& b_{11}\left(B_{22}\,[\mathbf{g}^1\otimes\mathbf{g}_1] - B_{12}\,[\mathbf{g}^1\otimes\mathbf{g}_2]\right) + \\ & b_{22}\left(B_{11}\,[\mathbf{g}^2\otimes\mathbf{g}_2] - B_{12}\,[\mathbf{g}^2\otimes\mathbf{g}_1]\right) + \\ & b_{12}\big(B_{11}\,[\mathbf{g}^1\otimes\mathbf{g}_2] + B_{22}\,[\mathbf{g}^2\otimes\mathbf{g}_1] - \\ & \mathrm{B}_{12}\left([\mathbf{g}^1\otimes\mathbf{g}_1] + [\mathbf{g}^2\otimes\mathbf{g}_2]\right)\big)\Big)\, d\mathbf{u}^1 \end{aligned} \qquad (21)$$

In this expression, the gradient of $\hat{\mathbf{n}}$ is expressed in a *mixed base* of covariant and contravariant base vectors, leading to a loss of symmetry of the associated matrix (unless the base vectors form an orthonormal basis). An otherwise symmetric tensor in $[\mathbf{g}_\alpha\otimes\mathbf{g}_\beta]$ or $[\mathbf{g}^\alpha\otimes\mathbf{g}^\beta]$ looses its symmetry when expressed in $[\mathbf{g}^\alpha\otimes\mathbf{g}_\beta]$ since generally, $\mathbf{g}^\alpha$ and $\mathbf{g}_\alpha$ have not the same length and directions. To recover symmetry in (21) $d_{\mathbf{u}}\,\hat{\mathbf{n}}$ can be expressed in either covariant or contravariant basis. From a computational point of view, simpler expressions are obtained when expressing $d\,\hat{\mathbf{n}}$ in the contravariant basis $[\mathbf{g}^\alpha\otimes\mathbf{g}^\beta]$. This is done by *raising* the covariant indices of the $\mathbf{g}_\alpha$, *i.e.* expressing the $\mathbf{g}_\alpha$'s in the contravariant basis $\mathbf{g}^\alpha$

$$\mathbf{g}_\alpha = g_{\alpha\beta}\,\mathbf{g}^\beta$$

After rearranging terms and simplifying, the following symmetric expression is obtained for the first variation of $\hat{\mathbf{n}}$ in terms of $d\mathbf{u}_1$

$$\begin{aligned} d\,\hat{\mathbf{n}} = -\frac{1}{h}\Big(& (d_n g_{11} b + g b_{11})\,[\mathbf{g}^1\otimes\mathbf{g}^1] + \\ & (d_n g_{22} b + g b_{22})\,[\mathbf{g}^2\otimes\mathbf{g}^2] + \\ & (d_n g_{12} b + g b_{12})\left([\mathbf{g}^1\otimes\mathbf{g}^2] + [\mathbf{g}^2\otimes\mathbf{g}^1]\right)\Big) d\mathbf{u}^1 \end{aligned}$$

or in a more concise form

$$d\,\hat{\mathbf{n}} = -\frac{1}{h}\,(d_n\, g_{\alpha\beta}\, b + g\, b_{\alpha\beta})\,[\mathbf{g}^\alpha\otimes\mathbf{g}^\beta]\, d\mathbf{u}^1 \qquad (22)$$

where g and b are the determinants of $g_{\alpha\beta}$ and $b_{\alpha\beta}$ respectively.

The local 4×4 Jacobian matrix $\nabla_{\mathbf{u}}\,\mathbf{f}$ of the augmented Lagrangian contact force $\mathbf{f}$ (16) becomes

$$\nabla_{\mathbf{u}}\,\mathbf{f} = \begin{cases} -\nabla_{\mathbf{u}}\left[(f_n + r d_n)\hat{\mathbf{n}}\right] & \text{if } (f_n + r d_n) \le 0 \quad \text{(contact)} \\ \begin{bmatrix} \mathbf{0} & \mathbf{0} \\ \mathbf{0}^{\mathrm{T}} & -1/r \end{bmatrix} & \text{if } (f_n + r d_n) > 0 \quad \text{(gap)} \end{cases}$$

where $\mathbf{u}$ is now defined by $\mathbf{u} = (\mathbf{u}^1, f_n)^{\mathrm{T}} \in \mathbb{R}^4$. Hence a more explicit form for this Jacobian is

$$\nabla_{\mathbf{u}} \mathbf{f} = \begin{cases} \begin{bmatrix} \mathbf{M} & -\hat{\mathbf{n}} \\ -\hat{\mathbf{n}}^{\mathrm{T}} & 0 \end{bmatrix} & \text{if } (f_n + r d_n) \leq 0 \quad \text{(contact)} \\ \\ \begin{bmatrix} \mathbf{0} & \mathbf{0} \\ \mathbf{0}^{\mathrm{T}} & -1/r \end{bmatrix} & \text{if } (f_n + r d_n) > 0 \quad \text{(gap)} \end{cases} \tag{23}$$

where $\mathbf{M}$ and $\mathbf{0}$ are 3×3 matrices. The expression of $\mathbf{M}$ is deduced from (16), (15) and (22) *i.e.*

$$\mathbf{M} = r\,[\hat{\mathbf{n}} \otimes \hat{\mathbf{n}}] + \frac{(f_n + r d_n)}{h}\,(d_n\, g_{\alpha\beta}\, b + g\, b_{\alpha\beta})\,[\mathbf{g}^\alpha \otimes \mathbf{g}^\beta] \tag{24}$$

The assembly of these 4×4 local matrices produces the global generalized contact Jacobian $\bar{\mathbf{J}}$ of all the contact elements force contribution. The closed form expressions obtained for $d\, d_n$ and $d\,\hat{\mathbf{n}}$ allow thus to handle contact problems were the rigid target surface ∂B^2 is described by any (smooth) mapping from $\mathbb{R}^2$ into $\mathbb{R}^3$.

NUMERICAL APPLICATIONS

The development of Total Knee Replacement prostheses (TKR) requires knowledge about the knee joint's biomechanics. Although much attention has been paid to the tibio-femoral joint biomechanics, much less is known about the patello-femoral joint (knee-cap). The present example focuses on the analysis of patellar 3D kinematics and quasi-static analysis by constructing a 3D finite element model of the patello-femoral joint (Fig. 3). The boundary conditions includes prescribed trajectories of the tibial

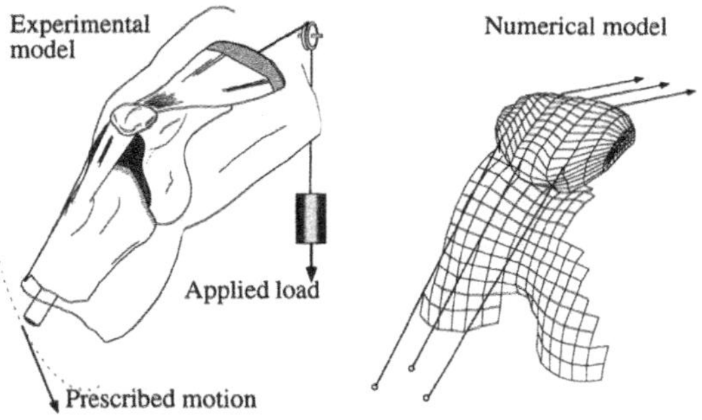

Figure 3. Schematic representation of the patello-femoral 3D model (left) and its corresponding FE mesh (right)

insertions of the patellar tendon and prescribed forces (representing the quadriceps muscle pulling forces) acting on the patella. These conditions where obtained from a preliminary experiment using Roentgen stereo-photogrammetric analysis (RSA) (Heegaard *et al.*, 1994). The 3D geometry and structure of the patella were obtained from CT-scans reconstruction. Patello-femoral contact was modeled by a set of large slip contact elements. The strikers were represented by the nodes on the external surface of the patellar cartilage layer (100 nodes). The rigid surface geometry of the femur was accurately measured using a stereophotogrammetric curve reconstruction (SCR) system (Meijer *et al.*, 1989) and was fitted by a set of Hermite bicubic patches. The quasi-rigid body kinematics of the patella, was expressed as a set of 3 Eulerian rotations and 3 translations. The kinematics of the patella during knee flexion was measured ex-

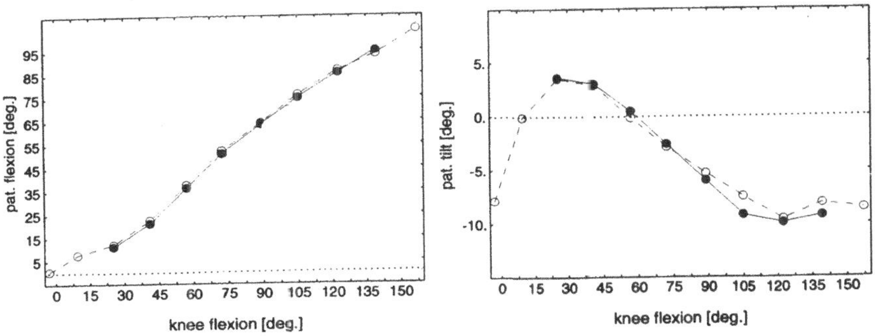

Figure 4. Two rotations of the patellar 3D kinematics as a function of knee flexion (left: flexion, right: tilt, as defined in (Heegaard *et al.* 1994).

perimentally, and some results are plotted on Figure 4, using dashed lines and hollow symbols. Numerical results, obtained from the 3D model are plotted using solid lines and plain symbols.

The excellent accuracy obtained with this model to predict 3D patellar motions during knee flexion could be accounted for precise geometry and boundary conditions specifications and for a rigorous treatment of the contact conditions. Furthermore, the large slip contact capacities of the 3D element were well illustrated in this example. Finally, unlike rigid body models, the present model could also compute the stresses occurring in the patella during knee flexion.

CONCLUSION

The fundamental geometric properties characterizing unilateral 3D large slip contact between a deformable body and a rigid smooth obstacle were derived. This derivation illustrated the three classical stages of unilateral finite element implementation, namely the computation of the gap distance, the associated contact force and the corresponding tangent matrix. The gap distance, contact force and tangent matrix were derived in explicit form in the case of contact with a parametrized obstacle. This formalism has been applied to target surfaces parametrized as bi-cubic Hermite polynomial patches leading to piecewise smooth surfaces (Heegaard and Curnier, 1994). The resulting contact element exhibited a quadratic rate of convergence. Its formulation and implementation have been validated on benchmark problems, and its capabilities were illustrated in a biomechanics study of the human knee, involving contact between bodies described by complex shapes. Further development should include friction and the consideration of both contacting bodies being deformable.

REFERENCES

Alart, P., and Curnier, A. (1991). A mixed formulation for frictional contact problems prone to Newton like methods. *Comp. Meth. Appl Mech. Engng.*, **9**, 353–375.

Curnier, A. (1985). TACT : A Contact Analysis Program. *Pages 97–116 of:* Maceri, F., and Del Piero, G. (eds), *Unilateral Problems in Structural Analysis - 2, CISM 304*. Wien: Springer-Verlag.

Curnier, A., and Alart, P. (1988). A generalized Newton method for contact problems with friction. *Méch. Théo. Appl.*, **Suppl.[1]:7**, 67–82.

Farin, G. (1990). *Curves and Surfaces for Computer Aided Design*. Boston: Academic Press.

Faux, I. D., and Pratt, M. J. (1987). *Computational geometry for design and manufacture*. Chichester: Horwood.

Flugge, W. (1972). *Tensor Analysis and Continuum Mechanics*. Berlin: Springer-Verlag.

Hansson, E., and Klarbring, A. (1990). Rigid Contact modelled by CAD surface. *Eng. Comput.*, **7**, 334–348.

Heegaard, J. H., and Curnier, A. (1993). An augmented Lagrangian method for discrete large slip contact problems. *Int. J. Num. Meth. Engng.*, **36**, 569–593.

Heegaard, J. H., and Curnier, A. (1994). Geometric Properties of 2D and 3D Unilateral Large Slip Contact Operators. *submitted to Comp. Meth. Appl. Mech. Engng.*

Heegaard, J.H., Leyvraz, P.F., van Kampen, A., Rakotomanana, L., Rubin, P.J., and Blankevoort, L. (1994). Influence of Soft Structures on Patellar 3D Tracking. *Clin. Orthop.*, **299**, 235–243.

Laursen, T.A., and Simo, J.C. (1993). A continuum-based finite element formulation for the implicit solution of multibody, large deformation frictional contact problems. *Int. J. Num. Meth. Engng.*, **36**, 3451–3485.

Meijer, R.C., Huiskes, R., and Kauer, J.M. (1989). A stereophotogrammetric method for measurements of ligament structure. *J Biomechanics*, **22**, 177–184.

Parisch, H. (1989). A consistent tangent stiffness matrix for three-dimensional non-linear contact analysis. *Int. J. Num. Meth. Engng.*, **28**, 1803–1812.

Thorpe, J.A. (1979). *Elementary Topics in Differential Geometry*. New York: Springer-Verlag.

Wriggers, P., and Imhof, M. (1993). On the treatment of nonlinear unilateral contact problems. *Arch. Appl. Mech.*, **63**, 116–129.

AN ADAPTIVE FINITE ELEMENT TECHNIQUE FOR NONLINEAR CONTACT PROBLEMS

P. Wriggers and O. Scherf

Institut für Mechanik
Technische Hochschule Darmstadt
Hochschulstr. 1
64289 Darmstadt
Germany

INTRODUCTION

Contact problems in engineering are often associated with very complex geometries. For such problems only numerical methods yield the required solutions of the associated mathematical model. Here the finite element method has been proven to be a flexible tool, especially when applied to nonlinear problems. Since numerical methods yield approximate solutions it is necessary to control the errors inherited in the method. During the last ten years research activities have been focused on adaptive techniques providing automatically a numerical modell which is accurate and reliable.

In this paper we develop an adaptive method for contact problems which ensures successive improvement of the numerical solution via an iterative solution procedure to refine the finite element mesh. This method will be formulated for frictionless contact problems for the case of linear and nonlinear elastic bodies.

The objective in adaptive techniques is to obtain a mesh which is optimal in the sense that the computational costs involved are minimal under the constraint that the error in the finite element solution is beyond a certain limit. Since the computational effort can be linked to the number of unknown of the finite element mesh the task is to find a mesh with minimum unknowns or nodes for a given error tolerance. Due to the fact that the dependence of the number of unknowns on the finite element error is very implicit this nonlinear optimization problem has to be solved iteratively. Adaptive methods rely on error indicators and error estimators which can be computed *a priori* or *a posteriori*. For an overview over different techniques, see e.g. Johnson (1987) and references therein. Based on the error distribution a new partially refined mesh can then be constructed which yields a better approximate solution. To obtain an optimal mesh in the sense of cost effectiveness it is desirable to design the mesh such that the error contributions of the elements are equidistributed over the mesh. During the last years a growing number of papers has been devoted to this topic and applied to problems of solid and fluid mechanics, see e.g. Zienkiewicz and Taylor (1988), Zienkiewicz et al.

(1988), Peraire et al. (1987). The methods rely on error estimators which have been developed so far in different versions. The estimators which are most frequently used in the literature are residual based error estimators, see e.g. Babuska and Rheinboldt (1978) or Johnson and Hansbo (1992), or error estimators which use superconvergence properties, see e.g. Zienkiewicz and Zhu (1987).

For frictionless contact problems *a priori* error estimators have been derived for linear elastic bodies, see e.g. Kikuchi and Oden (1988), Hlavacek et al. (1988). An adaptive method for problems with unilateral constraints has been developed by Lee et al. (1991) who treated as an example a free surface flow problem. In our paper we use the results derived by Wriggers et al. (1994) for the contact of two elastic bodies to derive an adaptive method for problems with large elastic strains. In Wriggers et al. (1994) a residual based error estimator has been developed following an approach persued by Johnson (1991) for unilateral membrane problems.

The paper is organized as follows. First we state the kinematical constraint conditions associated with contact. Then the boundary value problem for linear elastic contact problems is summarized. The boundary value problem for large elastic deformation is developed next together with its incremental form. The discretization of both formulations is given and error estimators for linear elasticity are stated. Finally an algorithm for an adaptive method is developed which relies on the error estimates of the linear theory applied to the incremental nonlinear problem. The h–adaptive algorithm is applied to an test example with known analytical solution and to a problem exhibiting large elastic deformations.

FORMULATION OF THE CONTACT PROBLEM

Let us consider two elastic bodies $\mathcal{B}^\alpha, \alpha = 1, 2$, each of them occupying the bounded domain $\Omega^\alpha \subset R^2$. The boundary Γ^α of a body $\mathcal{B}^\alpha$ can in general be splitted into three parts: Γ^α_σ with prescribed surface loads, Γ^α_u with prescribed displacements and Γ^α_c where the two bodies $\mathcal{B}^1$ and $\mathcal{B}^2$ come into contact.

Next the contact conditions, the strong and weak form of the associated elasticity problem and the penalty regularization of the contact constraints will be stated.

Contact kinematics

Assume that two bodies come into contact. In that case we have to find the minimum distance of a point on the surface of one body with respect to the other one. The associated mathematical formulation for this case can be found in e.g. Laursen and Simo (1991), Wriggers and Miehe (1994). It yields the non–penetration condition in terms of the coordinates $\mathbf{x}^\alpha = \mathbf{X}^\alpha + \mathbf{u}^\alpha$ of the current configuration of the bodies $\mathcal{B}^\alpha$

$$[\mathbf{x}^1 - \hat{\mathbf{x}}^2(\bar{\boldsymbol{\xi}})] \cdot \mathbf{n}_o \geq 0 \,. \tag{1}$$

Here $\bar{\boldsymbol{\xi}}$ denote the surface coordinates of body $\mathcal{B}^2$ and $\mathbf{n}_o$ the surface normal with respect to the current configuration. The point $\hat{\mathbf{x}}^2$ is found from the minimal distance problem which associates to every point $\mathbf{x}^1$ on Γ^1 a point $\bar{\mathbf{x}}^2$ on Γ^2 via

$$\|\mathbf{x}^1 - \hat{\mathbf{x}}^2(\bar{\boldsymbol{\xi}})\| = \min_{\mathbf{x}^2 \subseteq \Gamma^2} \|\mathbf{x}^1 - \mathbf{x}^2\| , \tag{2}$$

see e.g. Wriggers and Miehe (1994). This defines also the normal vector in the contact interface as

$$\mathbf{n}_o = \frac{\mathbf{x}^1 - \hat{\mathbf{x}}^2}{\|\mathbf{x}^1 - \hat{\mathbf{x}}^2\|} . \tag{3}$$

In case of small deformations we can use the displacement field instead of the deformation itself. The associated linearization then leads to the non–penetration condition

$$(\mathbf{u}^1 - \hat{\mathbf{u}}^2) \cdot \mathbf{n}_o + g \geq 0 , \tag{4}$$

where the inital gap g between the two bodies is given by $g = (\mathbf{X}^1 - \hat{\mathbf{X}}^2) \cdot \mathbf{n}_o$.

In view of the penalty formulation which will be applied to solve the contact problems we introduce a penetration function as follows:

$$g_n^- = \begin{cases} \| \mathbf{x}^1 - \hat{\mathbf{x}}^2(\bar{\boldsymbol{\xi}}) \| & \text{if } [\mathbf{x}^1 - \hat{\mathbf{x}}^2(\bar{\boldsymbol{\xi}})] \cdot \mathbf{n}_o < 0 \\ 0 & \text{otherwise} \end{cases} \tag{5}$$

and similar for the linear case

$$\mathbf{u}_n^- = \begin{cases} (\mathbf{u}^2 - \bar{\mathbf{u}}^1) \cdot \mathbf{n}_o + g & \text{if } (\mathbf{u}^2 - \bar{\mathbf{u}}^1) \cdot \mathbf{n}_o + g < 0 \\ 0 & \text{otherwise} . \end{cases} \tag{6}$$

Functions (5) and (6) indicate when a body has penetrated another and show in which parts of Γ^α the constraint equations preventing penetration have to be activated. Thus (5) or (6) can be used to determine the contact area $\Gamma_c^\alpha \subseteq \Gamma^\alpha$.

Remark: In the case of contact between a rigid surface and a deformable body the above equations also hold. Then we set $\bar{\mathbf{u}}^2 \equiv \mathbf{0}$ and $\mathbf{n}_o$ represents the normal of the rigid body.

The boundary value problem in elasticity

In the previous section we distinguished between the deformations $\mathbf{x}^\alpha$ to define the penetration function. This is no longer necessary since the subsequent equations are valid for every point $\mathbf{X} \in \mathcal{B}^\alpha$. Therefore we omit the index α in the following for convenience knowing that in case of different constitutive equations for $\mathcal{B}^1$ and $\mathcal{B}^2$ we have to make a distinction.

We first summarize the equations associated with linear elasticity. The relation between strains $\boldsymbol{\varepsilon}$ and displacements $\mathbf{u}$ is given by

$$\boldsymbol{\varepsilon}(\mathbf{u}) = \frac{1}{2}(\nabla \mathbf{u} + \nabla^T \mathbf{u}) . \tag{7}$$

Local equilibrium yields

$$\operatorname{div} \boldsymbol{\sigma} + \hat{\mathbf{b}} = \mathbf{0} , \tag{8}$$

where $\hat{\mathbf{b}}$ denotes the body forces. The evaluation of the constitutive tensor $\boldsymbol{C}$ at the undeformed state leads to Hooke's law which relates stresses $\boldsymbol{\sigma}$ and strains linearily:

$$\boldsymbol{\sigma} = \boldsymbol{C}_0 : \boldsymbol{\varepsilon} \,, \tag{9}$$

with $\boldsymbol{C}_0$ as the standard fourth order material tensor. Boundary conditions are

$$\mathbf{u} = \hat{\mathbf{u}} \text{ on } \Gamma_u \,, \qquad \boldsymbol{\sigma}\mathbf{n} = \hat{\mathbf{t}} \text{ on } \Gamma_\sigma \,, \tag{10}$$

where $\mathbf{n}$ denotes the outward unit normal vector of the considered area. Inequality (4) completes the formulation when contact constraints are present.

These equations can be recast in a weak or variational formulation as follows:

$$a(\mathbf{u}, \mathbf{v} - \mathbf{u}) \geq f(\mathbf{v} - \mathbf{u}) \,, \tag{11}$$

with

$$a(\mathbf{u}, \mathbf{w}) = \int_\Omega \boldsymbol{\varepsilon}(\mathbf{u}) : \boldsymbol{C}_0 : \boldsymbol{\varepsilon}(\mathbf{w}) \, d\Omega \,,$$

$$f(\mathbf{w}) = \int_\Omega \hat{\mathbf{b}} \cdot \mathbf{w} \, d\Omega + \int_{\Gamma_\sigma} \hat{\mathbf{t}} \cdot \mathbf{w} \, d\Gamma \tag{12}$$

and $\Omega = \cup_\alpha \Omega^\alpha$. The problem is now to find $\mathbf{u} \in \mathbf{K}$ such that (12) is fulfilled for all $\mathbf{v} \in \mathbf{K}$ with

$$\mathbf{K} = \{\mathbf{v} \in \mathbf{V} \mid (\mathbf{v}^2 - \bar{\mathbf{v}}^1) \cdot \mathbf{n}_o + g \geq 0 \text{ on } \Gamma_c\} \tag{13}$$

and

$$\mathbf{V} = \{\mathbf{v} \in [H^1(\Omega)]^2 \mid \mathbf{v} = \mathbf{0} \text{ on } \Gamma_u\} \,. \tag{14}$$

Due to the inequality constraint on the displacement field this problem is nonlinear even for the linear elastic case. The solution of the contact problem is obtained using the penalty method, see e.g. Luenberger (1989). This technique replaces the variational inequality by an unconstraint problem with regard to the contact constraint (4) as follows: Find $\mathbf{u}_\varepsilon \in \mathbf{V}$ such that

$$a(\mathbf{u}_\varepsilon, \mathbf{v}) + c^-(\mathbf{u}_\varepsilon, \mathbf{v}) = f(\mathbf{v}) \quad \forall \, \mathbf{v} \in \mathbf{V} \,, \tag{15}$$

where $\mathbf{V}$, $a(\mathbf{u}_\varepsilon, \mathbf{v})$ and $f(\mathbf{v})$ are defined as above and

$$c^-(\mathbf{u}_\varepsilon, \mathbf{v}) = \int_{\Gamma_c} \varepsilon \, u_{\varepsilon n}^- \, v_n \, d\Gamma \,. \tag{16}$$

$u_{\varepsilon n}^-$ has already been defined in (6) and $v_n = \mathbf{v} \cdot \mathbf{n}_o$. The penalty parameter ε is a positive constant. It can be shown, see e.g. Kikuchi and Oden (1988) or Carstensen et al., that the solution of the regularized problem will converge to the solution of the original contact problem as ε tends to infinity.

Now the equations for finite elasticity problems are summarized. These equations can be formulated with respect to the current or the reference configuration. Since there is no preference for the choice of the configuration which is used to formulate the equations we choose here that one which is related to the minimal numerical effort. For finite elasticity this is the current configuration $\varphi(\mathcal{B})$.

The kinematical relations yield as strain measure the left Cauchy Green tensor $\mathbf{b}$

$$\mathbf{b} = \mathbf{F}\,\mathbf{F}^T\,, \tag{17}$$

where $\mathbf{F} = \operatorname{Grad}\mathbf{x}$ is the deformation gradient. Further we recall the local equilibrium equation:

$$\operatorname{div}\boldsymbol{\sigma} + \hat{\mathbf{b}} = \mathbf{0} \tag{18}$$

with the Cauchy $\boldsymbol{\sigma}$ and the body force $\hat{\mathbf{b}}$. The stress tensor $\boldsymbol{\sigma}$ is related to the strains via the following equation

$$\boldsymbol{\sigma} = 2\,\frac{\partial W}{\partial \mathbf{b}}\,\mathbf{b}\,, \tag{19}$$

where W is the stored strain energy function of the elastic material. To complete the standard boundary value problem of a body undergoing finite elastic deformations we have to formulate the boundary conditions for the tractions $\hat{\mathbf{t}} = \mathbf{F}\,\mathbf{S}\,\mathbf{N}$ on Γ_σ and for the deformation $\hat{\mathbf{u}} = \mathbf{u}$ on Γ_u. Here $\mathbf{N}$ denotes the surface normal defined in the reference configuration and $\mathbf{S}$ the second Piola–Kirchhoff stress tensor.

In the case of unilateral contact the space of deformations is restricted by the constraint (1). For a thorough treatment of this situation for two elastic bodies fulfilling the conditions of polyconvexity, see Curnier et al. (1992). These authors also established the existence of the solution of the associated boundary value problem when formulated as a minimization problem. Another proof of the existence of the solution for the finite deformation contact problem can be found in Ciarlet (1988) or Kikuchi and Oden (1988) where the latter restricted themselves to the contact between a deformable body and a rigid obstacle using a different definition of the gap function.

We now state the minimization problem which is the point of departure for the subsequent discretizations using finite elements. Let Π denote the functional of the unconstraint problem

$$\Pi(\mathbf{u}) = \int_{\varphi(\Omega)} W(\mathbf{b})\,d\Omega - \int_{\varphi(\Omega)} \hat{\mathbf{b}}\cdot\mathbf{u}\,d\Omega - \int_{\varphi(\Gamma_\sigma)} \hat{\mathbf{t}}\cdot\mathbf{u}\,d\Gamma \Longrightarrow MIN \tag{20}$$

which has to be minimized with respect to the constraint (1) leading to a non-convex minimization problem. Now we introduce a penalty functional to approximately fulfill the constraint condition (1). For this purpose we use the penetration function (5). This leads to

$$\Pi_\varepsilon(\mathbf{u}) = \Pi(\mathbf{u}) + \frac{1}{2}\int_{\varphi(\Gamma_c)} \varepsilon\,(g_n^-)^2\,d\Gamma \Longrightarrow MIN\,. \tag{21}$$

The solution of this minimization problem is now obtained by using the directional derivative to arrive at the weak form of equilibrium $f_u(\mathbf{v}) = -D\,\Pi_\varepsilon(\mathbf{u})\cdot\mathbf{v}$. Note that for a formulation in the current configuration we have to use the so-called *pull back* and *push forward* operations to compute the directional derivative which can then also be viewed in the light of a Lie derivative

$$f_u(\mathbf{v}) = -\int_{\varphi(\Omega)} \boldsymbol{\sigma} : \delta\mathbf{e}\,d\Omega + \int_{\varphi(\Omega)} \hat{\mathbf{b}}\cdot\mathbf{v}\,d\Omega + \int_{\varphi(\Gamma_\sigma)} \hat{\mathbf{t}}\cdot\mathbf{v}\,d\Gamma - \int_{\varphi(\Gamma_c)} \varepsilon\,g_n^-\,\delta g_n^-\,d\Gamma = 0 \tag{22}$$

where the variations $\delta\mathbf{e}$ and δg_n^- have to be computed as follows:

$$\begin{aligned} \delta\mathbf{e} &= \frac{1}{2}(\nabla_x\mathbf{v} + \nabla_x^T\mathbf{v})\,, \\ \delta g_n^- &= [\,\mathbf{v}^1 - \hat{\mathbf{v}}^2(\bar{\boldsymbol{\xi}})\,]\cdot\mathbf{n}^o\,. \end{aligned} \tag{23}$$

The test function $\mathbf{v}$ has to fulfill the boundary condition $\mathbf{v} = \mathbf{0}$ on Γ_u. Note, that even since $\delta\mathbf{e}$ looks like (7) it is a function of $\mathbf{u}$ which implicitly enters through the computation of ∇ with respect to the current coordinates x_i. For the derivation of the δg_n^- see e.g. Wriggers and Miehe (1994). Equation $f_u(\mathbf{v}) = 0$ is nonlinear in the displacement field $\mathbf{u}$, denoted by the index u, and thus an iteration procedure has to be used for its solution.

To establish error estimators we need a theoretical background which so far is only known for contact problems in linear elasticity, see e.g. Kikuchi and Oden (1988), Lee et al. (1991) or Wriggers et al. (1994). Thus next we will derive incremental equations of (22) to see whether they have the same structure as the linear equations and thus results for the linear theory also apply to the nonlinear problem. In general we will try to use the information of the incremental equations to adapt the finite element mesh which represent a tangent to the solution curve at a given deformation $\mathbf{u}$.

Using again the directional derivative to compute the incremental equations $a_u(\Delta\mathbf{u}, \mathbf{v}) + c^-(\Delta\mathbf{u}, \mathbf{v}) = D\,[\,D\Pi_\varepsilon(\mathbf{u}) \cdot \mathbf{v}\,] \cdot \Delta\mathbf{u}$ we arrive at

$$
\begin{aligned}
a_u(\Delta\mathbf{u}, \mathbf{v}) &= \int_{\varphi(\Omega)} \{\, \delta\mathbf{e} : \boldsymbol{C}_u : \Delta\mathbf{e} + \boldsymbol{\sigma} : \nabla_x \Delta\mathbf{u} \nabla_x \mathbf{v} \,\}\, d\Omega\,, \\
c_u^-(\Delta\mathbf{u}, \mathbf{v}) &= \int_{\varphi(\Gamma_c)} \{\, \varepsilon\, \delta g_n^-\, \Delta g_n^- + \varepsilon\, \Delta\delta g_n^- \,\}\, d\Gamma\,.
\end{aligned} \tag{24}
$$

Here $\boldsymbol{C}_u$ denotes the incremental constitutive tensor which depends on the deformation. The second term represents the influence of the stress state reached at the deformation $\mathbf{u}$. $c_u^-(\mathbf{v}, \Delta\mathbf{u})$ is the result of the linearization of the penalty term. Here the last term is associated with the change of the normal $\mathbf{n}_o$, as can be seen from (23). Note that both operators are symmetric with respect ot the test function $\mathbf{v}$ and the displacement increment $\Delta\mathbf{u}$: $a_u(\Delta\mathbf{u}, \mathbf{v}) = a_u(\mathbf{v}, \Delta\mathbf{u})$ and $c_u^-(\mathbf{v}, \Delta\mathbf{u}) = c_u^-(\Delta\mathbf{u}, \mathbf{v})$. Furthermore these operators are linear in $\mathbf{v}$ and $\Delta\mathbf{u}$. Thus they have the same structure as the operators associated with the linear elastic formulation. However both depend nonlinearily on $\mathbf{u}$ which means that e.g. the ellipticity of $a_u(\mathbf{v}, \Delta\mathbf{u})$ and $c_u^-(\mathbf{v}, \Delta\mathbf{u})$, which is needed in the derivation of the error estimator, has to be checked for every deformation state $\mathbf{u}$ during the incremental solution of the nonlinear problem. We note however that the notion of ellipticity is in this case only of local consequence and thus cannot be applied for an estimation in a global sense, see e.g. Marsden and Hughes (1983).

DISCRETIZATION

To discretize as well the linear as the nonlinear problem defined above we divide the domain Ω^α occupied by the bodies $\mathcal{B}^\alpha$ into non–overlapping finite elements T of diameter h_T and introduce a standard finite element space

$$
\mathbf{V}_h = \{\mathbf{v} \in \mathbf{V} \mid \mathbf{v} \in C(\Omega),\, \mathbf{v}|_T \in [P(T)]^2,\, \forall\, T\}\,, \tag{25}
$$

where $P(T)$ is a space of polynomials of degree p_T on T and p_T is a positive integer. The discrete finite element problem yields now in the nonlinear case: find $\mathbf{u}_h \in \mathbf{V}_h$ such that

$$f_{u_h}(\mathbf{v}) = \sum_{\alpha=1}^{2}\left[- \int_{\varphi(\Omega^\alpha)} \boldsymbol{\sigma}_h : \delta \mathbf{e}_h \, d\Omega + \int_{\varphi(\Omega^\alpha)} \hat{\mathbf{b}} \cdot \mathbf{v} \, d\Omega + \int_{\varphi(\Gamma_\sigma^\alpha)} \hat{\mathbf{t}} \cdot \mathbf{v} \, d\Gamma \right] - \int_{\varphi(\Gamma_c)} \varepsilon \, g_{nh}^- \, \delta g_{nh}^- \, d\Gamma = 0 \qquad \forall \, \mathbf{v} \in \mathbf{V}_h \, . \tag{26}$$

To solve this nonlinear equation Newton's method is applied which leads to the following iterative scheme for the displacement field at the incremental step $i+1$:

$$\begin{aligned} a_{u_h^i}(\Delta \mathbf{u}_h^{i+1}, \mathbf{v}) + c_{u_h^i}^-(\Delta \mathbf{u}_h^{i+1}, \mathbf{v}) &= f_{u_h^i}(\mathbf{v}) \, , \\ \mathbf{u}_h^{i+1} &= \mathbf{u}_h^i + \Delta \mathbf{u}_h^{i+1} \, . \end{aligned} \tag{27}$$

In case of the linear theory we have to find the displacement field $\mathbf{u}_h \in \mathbf{V}_h$ such that

$$a(\mathbf{u}_h, \mathbf{v}) + c^-(\mathbf{u}_h, \mathbf{v}) = f(\mathbf{v}) \quad \forall \, \mathbf{v} \in \mathbf{V}_h \, . \tag{28}$$

ERROR ESTIMATOR

Let $\mathbf{u}_\varepsilon$ and $\mathbf{u}_h$ denote the exact penalty solution of (15) or (22) and the discrete FEM–solution of (26) or (28), respectively. With

$$\mathbf{e} = \mathbf{u}_\varepsilon - \mathbf{u}_h \tag{29}$$

we define an error measure as follows:

$$r = \alpha \|\nabla \mathbf{e}\|_{L2(\Omega)}^2 + c^-(\mathbf{e}, \mathbf{e}) \, , \qquad \alpha > 0 \, . \tag{30}$$

The error consists of two parts: an error in the strain components and an error resulting from the contact constraint. This equation is used in the linear and the nonlinear case where in the latter α depends on the displacement state which represents an equilibrium configuration. Now we like to recall the results for the linear elastic contact problem which have been derived by Wriggers et al. (1994). There the convergence of the solution due to penalty regularization to the solution of the variational inequality (11) was shown for an elliptic operator $a(\mathbf{u}, \mathbf{u})$. Furthermore the interpolation error was estimated leading to

$$r \leq C \, [E(h, \mathbf{u}_h, \hat{\mathbf{b}})]^2 = C \sum_T [E_T(h_T, \mathbf{u}_h, \hat{\mathbf{b}}_T)]^2 \, , \tag{31}$$

with the constant C being independent on h. E_T can be computed for each element in the finite element mesh as follows

$$\boxed{\begin{aligned} E_T^2 = \; & h_T^2 \int_T |\operatorname{div} \boldsymbol{\sigma}_h + \hat{\mathbf{b}}|^2 d\Omega \; + \; h_T \int_{\partial T \cup \Omega} 1/2 \, |[\mathbf{t}_h]|^2 d\Gamma \; + \\ & h_T \int_{\partial T \cup \Gamma_\sigma} |\hat{\mathbf{t}} - \mathbf{t}_h|^2 \, d\Gamma \; + \; h_T \int_{\partial T \cup \Gamma_c} |\varepsilon \, u_{hn}^- \, \mathbf{n}_o - \mathbf{t}_h|^2 \, d\Gamma \end{aligned}} \tag{32}$$

Inequality (31) yields an upper bound for the error which is bounded by the deviation of the discrete solution from equilibrium and the element size. The first and the third term of the right hand side contribute to the error bound if the local equilibrium and the traction boundary condition, respectively, are violated. Local equilibrium requires that $[\mathbf{t}_h] = \mathbf{0}$ which is associated with the second term where $[\mathbf{t}_h]$ describes the jumps of the tractions over the interface. Further the term $-\varepsilon\, u_{\bar{h}n}^{-} \mathbf{n}_o$ can be interpreted as the contact pressure on Γ_c. Therefore the fourth term corresponds to fulfillment of local equilibrium in the contact interface. In the nonlinear case all terms of (32) have to be computed on the deformed element T.

The error estimator described above yields a measure between the exact penalty solution and its finite element approximation. What really is needed is the error between the exact solution of (11) and the approximate finite element solution. So far there do not exist computable error bounds for unilateral problems in elasticity. But we can make use of a result derived by Kikuchi and Oden (1988) to change the penalty parameter in such a way that an optimal convergence rate of the method is achieved. To this purpose we state the result of Kikuchi and Oden (1988) which was derived for a perturbed Lagrangian formulation of the contact problem

$$\| \mathbf{u} - \mathbf{u}_{\varepsilon h} \|_1 + | \sigma - \sigma_{\varepsilon h} |^* \leq C_3\, h + C_4\, \varepsilon^{-1}\, h^{-1/2} \tag{33}$$

where σ denotes the contact pressure. From this equation it is clear that an optimal convergence rate can be obtained for $\varepsilon \approx h^{-3/2}$. According to this relation we develop now the following update at iteration $k+1$ for the penalty parameter in the contact interface

$$\varepsilon_{k+1} = \varepsilon_0 \left(\frac{h_{k+1}}{h_0} \right)^{-\frac{3}{2}} , \tag{34}$$

where ε_0 and h_0 are the starting values at the beginning of the adaptive iteration.

MESH ADAPTION

To state an h–adaptive method which bases on the error estimator (32) we define

$$\begin{aligned} a_T &= \int_T | \operatorname{div} \boldsymbol{\sigma}_h + \hat{\mathbf{b}} |^2 d\Omega \,, \\ b_T &= \int_{\partial T \cup \Omega} 1/2 \, | [\mathbf{t}_h] |^2 d\Gamma + \int_{\partial T \cup \Gamma_\sigma} | \hat{\mathbf{t}} - \mathbf{t}_h |^2 \, d\Gamma + \int_{\partial T \cup \Gamma_c} | \varepsilon\, u_{\bar{h}n}^{-}\, \mathbf{n}_o - \mathbf{t}_h |^2 \, d\Gamma \,, \end{aligned} \tag{35}$$

such that

$$E_T^2 = a_T\, h_T^2 + b_T\, h_T \,. \tag{36}$$

Now we can use (31) to establish an error bound as follows

$$r \leq C\, E^2 = C \sum_T E_T^2 \leq TOL_0 \,, \tag{37}$$

which can be used to terminate the adaptive iteration. Provided that the actual mesh is an optimal mesh, i.e. that the total error E is equally distributed between elements, we can write

$$E^2 = \sum_T E_T^2 = N\, E_T^2 \,, \tag{38}$$

with N being the number of finite elements T in the actual mesh. Based on this assumption we obtain the local criterion to check whether a specific element has to be refined or not:

$$a_T\, h_T^2 + b_T\, h_T \leq \frac{TOL_1}{N}\,, \tag{39}$$

which has to be checked for all elements. In case (39) is violated the mesh has to be refined.

Now we can state the overall algorithm of our h–adaptive method for contact problems. The algorithm consists of the following steps:

1. Set initial values: $l = 0$, $\lambda_0 = 0$, $\Delta\lambda$
2. Generation of start mesh: M_l
3. Loop over load increments : $\lambda_{l+1} = \lambda_l + \Delta\lambda$
   - Iteration loop to solve nonlinear problem
   - IF $\lambda_{l+1} < \lambda_{max}$ THEN
     - Set; $l = l + 1$
     - Mesh optimization

       Compute $E_T^2 = h_T^2\, a_T + h_T\, b_T$

       $E_T^2 > TOL_1\,/\,N \Longrightarrow$ element refinement
     - Generate new mesh M_l

       Delaunay triangularization

       Smoothing
     - Interpolate displacement field on new mesh
   - ELSE
     - IF $E^2 \leq TOL_1$ THEN

       STOP
     - ELSE

       Compute $E_T^2 = h_T^2\, a_T + h_T\, b_T$

       $E_T^2 > TOL_1\,/\,N \Longrightarrow$ element refinement

       Set: $l = l + 1$, generate mesh M_l

       Solve nonlinear problem for mesh M_l
     - ENDIF
   - ENDIF

The mesh is defined via a parametric surface description of the boundaries. All loads, boundary constraints and contact constraints are defined with respect to these surfaces. A Delaunay triangularization is then used to create the successive meshes during the adaptive process, for the associated algorithm see Sloan (1987).

NUMERICAL EXAMPLES

In this section we apply the methodology derived in the sections above to a linear and a nonlinear contact problem to show the performance of the adaptive process. All algorithms and finite elements have been implemented in the Finite Element Analysis Program (FEAP), developed by R. L. Taylor, see Zienkiewicz and Taylor (1988).

Hertzian contact problem

The first example is the well known Hertzian problem of an elastic cylinder pressed by a point load against a rigid foundation, see Fig. 1a.

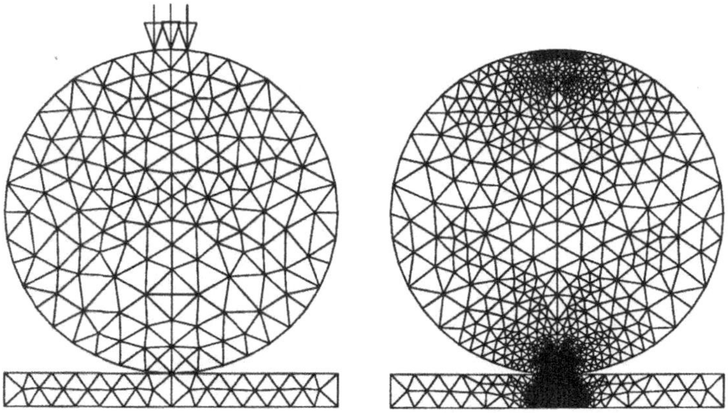

Fig. 1 Hertzian contact: a) System b) Refined mesh

Plane strain conditions are assumed. With respect to the adaptive algorithm the single load is applied to the discrete model as a surface load to avoid singularities in the structure. Throughout the computations we use linear triangular elements. Due to symmetry only one half of the system is discretized. The first mesh and the mesh after 4 refinements are given in Fig. 1a and Fig. 1b, respectively. In Tab. 1 the presented error estimator is compared with the estimator according to Zienkiewicz and Zhu (1992), Babuska and Miller (1987) and the analytical solution.

Tab. 1. Efficiency of different error estimators

	Error estimator for contact	Error estimator Babuska/Miller	Error estimator Zienkiewicz/Zhu	Analytical solution
Mesh number	7	6	6	
Nodes	2035	2666	2658	
Contact pressure	495	495	494	495

We see that the derived error estimator leads to a final mesh with less nodes and thus is more efficient when contact is present.

As a second example we consider a block which is pushed into a almost rigid tool, see Fig. 2a. The block consists of a compressible Neo–Hookean material with a shear modulus being a hundred times smaller than that of the tool material.

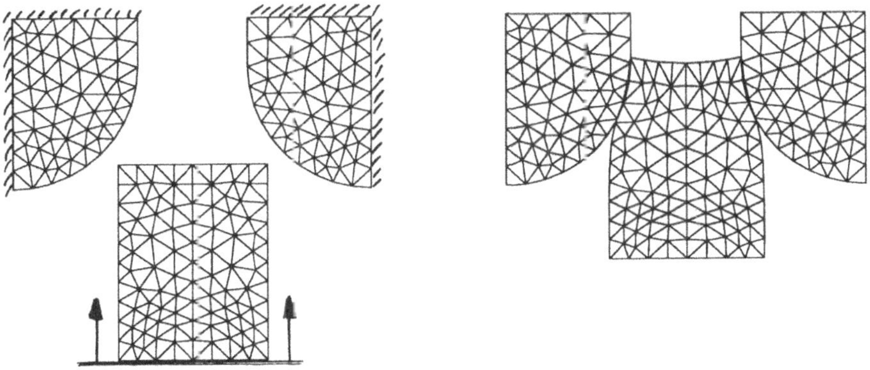

Fig. 2 Large deformation: a) System b) Deformed mesh

The load is applied in one increment which yields the deformed configuration shown in Fig. 2b for the initial mesh. After that, the load is kept constant during the refinement stages. Fig. 3a shows the error distribution of the second mesh and Fig. 3b the resulting next mesh refinement.

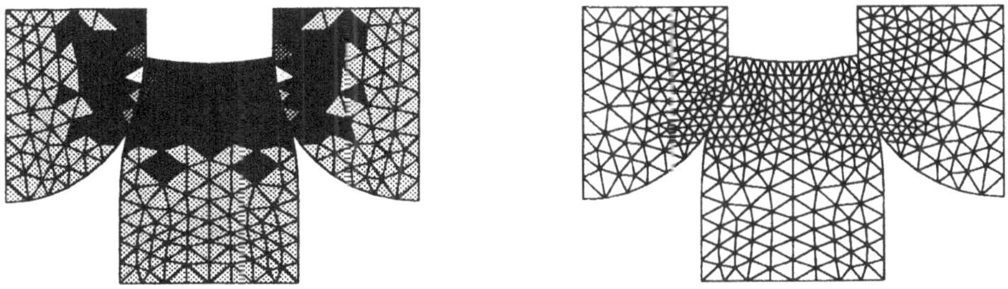

Fig. 3 Adaptive method: a) Error distribution b) Mesh refinement

We observe that the error estimator also works for finite elasticity problems when we exclude stability problems.

REFERENCES

Babuska, I. and Miller, A., 1987, A feedback finite element method with a posteriori error estimation: Part I. The finite element method and some basic properties of the a posteriori error estimation, *Comp. Meth. Appl. Mech. Engrg.*, **61**, 1–40.

Babuska, I. and Rheinboldt, W., 1978, Error estimates for adaptive finite element computations, *J. Num. Analysis*, **15**, 736–754.

Carstensen, C., Scherf, O. and Wriggers, P., A posteriori estimate and adaptive mesh refinement for contact of elastic bodies, submitted to *Comput. Meth. Appl. Mech. Engrg.*

Ciarlet, P., 1988, "Mathematical Elasticity, I: 3–D Elasticity", North Holland, Amsterdam.

Curnier, A., Qi-Chang, H. and Telega, J. J., 1992, Formulation of unilateral contact between two elastic bodies undergoing finite deformation, *C. R. Acad. Sci*, t. 314, Serie II, 1–6.

Hlavacek, I., Haslinger, J., Necas, J. and Lovisek, J., 1988, "Solution of Variational Inequalities in Mechanics", Springer, New York.

Johnson, C., 1987, "Numerical Solutions of Partial Differential Equations by the Finite Element Method", Cambridge Press, New York.

Johnson, C., 1991, Adaptive finite element methods for the obstacle problem, Technical Report, Chalmers University of Technology, Göteborg.

Johnson, C., and Hansbo, P., 1992, Adaptive finite element methods in computational mechanics, *Comput. Meth. Appl. Mech. Engrg.*, **101**, 143–181.

Kikuchi, N., and Oden, J. T., 1988, "Contact Problems in Elasticity: A Study of Variational Inequalities and Finite Element Methods", SIAM, Philadelphia.

Laursen, T. A. and Simo, J. C., 1991, On the formulation and numerical treatment of finite deformation frictional contact problems, *in*: "Computational Methods in Nonlinear Mechanics", P. Wriggers, W. Wagner, eds., Springer, Berlin.

Lee, C. Y., Oden, J. T. and Ainsworth, M., 1991, Local a posteriori error estimates and numerical results for contact problems and problems of flow trough porous media, *in*: "Nonlinear Computational Mechanics",671–689, P. Wriggers, W. Wagner, eds., Springer, Berlin.

Luenberger, D. G., 1989, "Linear and Nonlinear Programming", Addison Wesley, Reading.

Marsden, J. E., Hughes, T. J. R., 1983, "Mathematical Foundations of Elasticity", Prentice Hall, Englewood Cliffs.

Peraire, J., Vahdati, M., Morgan, K. and Zienkiewicz, O. C., 1987, Adaptive meshing for compressible flow computations, *J. Comp. Phys.*, **72**, 449–466.

Sloan, S. W., 1987, A fast algorithm for constructing delaunay triangularization in the plane, *Adv. Eng. Software*, **9**, 34–55.

Wriggers, P. and Miehe, C., 1994, Contact constraints within coupled thermomechanical analysis – A finite element model, *Comput. Meth. Appl. Mech. Engrg.*, **113**, 301–319.

Wriggers, P., Scherf, O., Carstensen, C., 1994, Adaptive techniques for the contact of elastic bodies, *in*: "Recent Developments in Finite Element Analysis", T. J. R. Hughes, E. Onate, O. C. Zienkiewicz, eds., CIMNE, Barcelona.

Zienkiewicz, O. C., Liu, Y. C. and Huang, G. C., 1988, Error estimation and adaptivity in flow formulation for forming problems, *Int. J. Num. Meth. Engrg.*, **25**, 23–42.

Zienkiewicz, O. C., and Taylor R. L., 1988, "The Finite Element Method, 4th ed., Vol. I", McGraw Hill, London.

Zienkiewicz, O. C., and Zhu, J. Z., 1987, A simple error estimator and adaptive procedure for practical engineering analysis, *Int. J. Num. Meth. Engrg.*, **24**, 337–357.

Zienkiewicz, O. C. and Zhu, J. Z., 1992, The superconvergent patch recovery and a posteriori error estimates. Part 1: The recovery technique, *Int. J. Num. Meth. Engrg.*, **33**, 1331–1364.

NUMERICAL ANALYSIS OF PERTURBED CONTACT PROBLEMS

Hachmi Ben Dhia

LMSSM, URA 850
Ecole Centrale de Paris
Grande Voie des Vignes
92295 Chatenay Malabry Cedex
France

INTRODUCTION

For the numerical simulations of contact problems involving the action of a rigid tool on a deformable structure, the analytical descriptions of the surface of the rigid body which are those actually used in these simulations, are generally approximations of the desired shape. These approximations or perturbations of the data introduce additional sources of numerical errors. Many different reasons, such as CAD descriptions, meshing of the surface of the rigid body, industrial elaboration of this body or numerical integrations of the contact laws, lead to the analysis of perturbed contact problems. This remark has motivated the work reported in this paper.

First, we establish error estimate results that control the deviation of the perturbed contact problem's solution from the solution of the exact one by the deviation of the perturbed shape from the exact one. To prove these results, our basic idea consists in using the penalty method. These theoretical results, reported in section 3, are carried out for two model contact problems. The first one models the action of a rigid obstacle on an elastic membrane and is a classical Signorini problem. The second one models the action of a rigid tool on the surface of a thin linear homogenous Khirchhoff-Love plate. These model problems, the associated perturbed ones and their variational formulations are defined in section 2.

Then, by using the established error estimate results, we suggest, in the last section, some mesh-controlled perturbations of a given contact problem in order to handle by the finite element method ultimate contact situations. These are typically situations where the unknown contact area includes parts of the surface of the rigid body where the radius of curvature are sufficiently small, when compared to the mesh size (we assume implicitly here that low-order finite elements are used). We notice that for such cases, and by proceeding in a classical manner, a kind of contact "locking" of the numerical solution may occur. To handle such contact problems numerically, sufficiently refined meshes are required. But, even when these refined meshes are generated by adaptive techniques, the solution of the global problem may require a lot of memory and CPU time resources, especially when implicit schemes are used. Here, we show that, for mesh-dependent ultimate contact situations encountered in the simulations of deep drawing processes, our mesh-controlled perturbation strategy gives an effective method of avoiding such locking, even for coarse meshes. Moreover, for a given mesh and a given contact algorithm, significant CPU time savings are obtained, without deterioration in quality of the global numerical solution. These illustrations are reported in the final section.

Contact Mechanics, Edited by M. Raous *et al.*
Plenum Press, New York, 1995

MODELS OF PERTURBED CONTACT PROBLEMS

The first model considered here is a Signorini-like unilateral contact problem which consists in finding a scalar deflection field u and a contact scalar load field R satisfying the following system:

$$-\Delta u = R \text{ , in } \omega \tag{1}$$

$$u = 0, \text{ on } \partial\omega \tag{2}$$

$$u \leq \psi, \text{ in } \omega \tag{3}$$

$$R \geq 0, \text{ in } \omega \tag{4}$$

$$R\,(\,u - \psi\,) = 0, \text{ in } \omega \tag{5}$$

where Δ is the bidimensional Laplacian, ω is a bounded regular domain of R^2 occupied by an elastic membrane and $\partial\omega$ is the boundary of this domain. The surface of the obstacle is assumed to be described by a function ψ defined from ω into R. In the sequel, we will refer to (1)-(5) as the exact or idealised contact Signorini problem.

Now, we associate to this exact contact problem the following perturbed one obtained by modifying the shape of the obstacle's surface. The latter consists in finding a perturbed deflection field, denoted by u_p, and a perturbed contact reaction field, denoted by R_p, which are solutions of the following system:

$$-\Delta u_p = R_p \text{ , in } \omega \tag{6}$$

$$u_p = 0, \text{ on } \partial\omega \tag{7}$$

$$u_p \leq \psi_p, \text{ in } \omega \tag{8}$$

$$R_p \geq 0, \text{ in } \omega \tag{9}$$

$$R_p\,(\,u_p - \psi_p\,) = 0, \text{ in } \omega \tag{10}$$

where ψ_p is another given function defined from ω into R.

A second model which will be studied in the sequel is the unilateral contact problem between a linear elastic Kirchhoff-Love clamped plate and a rigid obstacle. This problem consists in finding a vector valued displacement field $\mathbf{u} = (u_1, u_2, u_3)$, defined on the middle surface of the plate and a scalar valued field of normal reactions, denoted by R_n, such that:

$$-\mathbf{B}\mathbf{u} = R_n\,\mathbf{n}, \text{ in } \omega \tag{11}$$

$$u_i = 0, \text{ on } \partial\omega, \text{ for } i = 1, 3 \tag{12}$$

$$\frac{\partial u_3}{\partial \nu} = 0, \text{ on } \partial\omega \tag{13}$$

$$\mathbf{u}\,.\,\mathbf{n} \leq g, \text{ in } \omega \tag{14}$$

$$R_n \geq 0, \text{ in } (\omega) \tag{15}$$

$$R_n\,(x)(\,\mathbf{u}(x)\,.\,\mathbf{n}(T(x)) - g(x)\,) = 0, \text{ a.e. } x \text{ in } \omega \tag{16}$$

where B is the Kirchhoff-Love operator (see e.g. Destuynder, 1986), **n** is the outward unit normal to the surface of the obstacle , T(x) is the so-called target point (assumed here to be unique) of point x, ν is the outward unit normal to the boundary $\partial\omega$ of the plate's middle surface and g is the so-called gap function. To this problem, we associate the following perturbed one which consists in finding a vector valued displacement field $\mathbf{u}_p$ defined on the middle surface of the plate and a scalar valued field of normal reactions, denoted by R_{np}, such that:

$$-\mathbf{B}\mathbf{u}_p = R_{np}\,\mathbf{n}, \text{ in } \omega \tag{17}$$

$$u_{ip} = 0, \text{ on } \partial\omega, \text{ for } i = 1, 3 \tag{18}$$

$$\frac{\partial u_{3p}}{\partial \nu} = 0, \text{ on } \partial\omega \tag{19}$$

$$\mathbf{u}_p \,.\, \mathbf{n} \leq g+p, \text{ in } \omega \tag{20}$$

$$R_{np} \geq 0, \text{ in } \omega \tag{21}$$

$$R_{np}(x)(\,\mathbf{u}_p(x)\,.\,\mathbf{n}(T(x)) - g(x) - p(x)\,) = 0, \text{ a.e. } x \text{ in } \omega \tag{22}$$

where p is a perturbation function, modifying the gap function g and it can be noticed that the function p modifies the shape of the real obstacle (O) along the normal vector **n**, to generate a modified obstacle (O_p), as shown in figure 1.

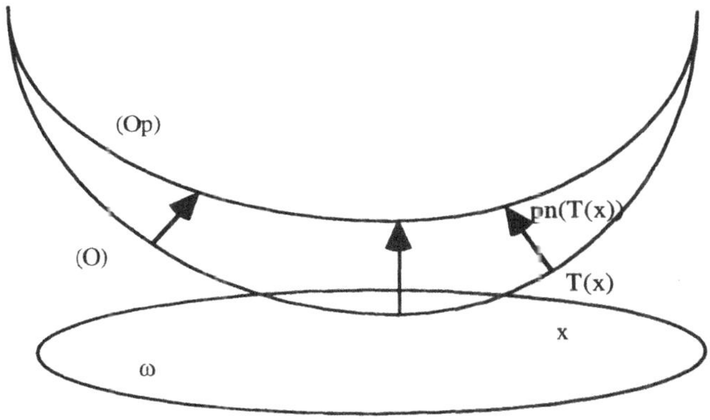

Figure 1. A contact perturbation along the normal vector to the rigid obstacle's surface

Primal variational formulations of the Signorini problems

The variational formulations of the Signorini problem and the associated perturbed problem, defined by (1)-(5) and (6)-(10), respectively, can be obtained in a classical manner. The first one reads (Duvaut and Lions, 1972; Kikuchi and Oden,1988):

$$\text{Find } u \in K;\ \forall\, v \in K,\ a\,(u,v-u) \geq 0, \tag{23}$$

where:

$$a\,(u,v) = \int_\omega \nabla u . \nabla v \; d\omega, \tag{24}$$

$$K = \{v \in V;\ v \leq \psi, \text{ in } \omega\} \tag{25}$$

and V is the following Sobolev space (Adams, 1975):

$$V = H_0^1\,(\omega), \tag{26}$$

The second one reads:

$$\text{Find } u_p \in K_p;\ \forall\ v \in K_p,\ a\ (u_p, v-u_p) \geq 0, \tag{27}$$

where a(.,.) is defined by (24) and where:

$$K_p = \{v \in V;\ v \leq \psi_p, \text{ in } \omega\} \tag{28}$$

It is well-known (Glowinski et al., 1981) that, under the following regularity hypotheses:

$$\psi \in L^2(\omega) \text{ and } \psi(\partial\omega) \geq 0, \tag{29}$$

$$\psi_p \in L^2(\omega) \text{ and } \psi_p(\partial\omega) \geq 0, \tag{30}$$

each one of the previous variational inequality problems admits a unique solution. Note that the hypotheses $\psi(\partial\omega) \geq 0$ and $\psi_p(\partial\omega) \geq 0$ are mechanical compatibility conditions which ensure that the convex sets K and K_p are not empty.

Primal Variational Formulations of the Plate Contact Problems

The variational formulations of the plate contact problems defined by (11)-(16) and (17)-(22), respectively, can also be derived in a rather classical manner. The first one reads:

$$\text{Find } \mathbf{u} \in \mathbf{K};\ \forall\ \mathbf{v} \in \mathbf{K},\ b(\mathbf{u}, \mathbf{v} - \mathbf{u}) \geq 0 \tag{31}$$

where:

$$b\ (\mathbf{u},\mathbf{v}) = \int_\omega D_{\alpha\beta\lambda\mu}\ \partial_{\lambda\mu}u_3\ \partial_{\alpha\beta}v_3\ d\omega\ +\ \int_\omega E_{\alpha\beta\lambda\mu}\ \gamma_{\lambda\mu}(\mathbf{u})\ \gamma_{\alpha\beta}(\mathbf{v})\ d\omega \tag{32}$$

$$\mathbf{K} = \{\mathbf{v} \in \mathbf{V};\ \mathbf{v}.\mathrm{n} \leq g, \text{ in } \omega\} \tag{33}$$

In (32), $D_{\alpha\beta\lambda\mu}$ and $E_{\alpha\beta\lambda\mu}$ are the second order deflection and the membrane rigidity tensors satisfying the classical symmetry and ellipticity properties (see e.g. Destuynder, 1986) and γ denotes the linearized strain tensor. In (33), $\mathbf{V}$ is the following Sobolev space (Adams, 1975):

$$\mathbf{V} = (H_0^1\ (\omega))\ \mathrm{x}\ H_0^2\ (\omega) \tag{34}$$

The second one reads:

$$\text{Find } \mathbf{u}_p \in \mathbf{K}_p;\ \forall\ \mathbf{v} \in \mathbf{K}_p,\ b\ (\mathbf{u}_p, \mathbf{v} - \mathbf{u}_p) \geq 0, \tag{35}$$

where b(.,.) is defined by (32) and where:

$$\mathbf{K}_p = \{\mathbf{v} \in \mathbf{V};\ \mathbf{v}.\mathbf{n} \leq g + p, \text{ in } \omega\} \tag{36}$$

It is easy to check that, assuming standard regularity hypotheses of the data (ω, D and E) and assuming the following regularity hypotheses on the gap functions:

$$g \in L^2(\omega) \text{ and } g(\partial\omega) \geq 0, \tag{37}$$

$$p \in L^2(\omega) \text{ and } (g+p)(\partial\omega) \geq 0, \tag{38}$$

each one of the problems defined by (31)-(34) and (35), (32), (36), (34), respectively, admits a unique solution.

The basic question of this paper is the stability of the contact solutions of the two models with respect to the perturbations of the shape of the idealised obstacles. This question is addressed in the following section.

CONTINUITY RESULTS WITH RESPECT TO THE PERTURBATIONS

Comparing the solutions of the exact contact problems, defined in the previous section, and these of their associated perturbed ones can not be accomplished directly since there is no trivial inclusion between the functional sets K and K_p. Nevertheless, we will show that this issue may be addressed, using the penalty method.

Analysis of the membrane perturbed contact problems

The basic result of this paper is the following:

Theorem 1: If the domain ω is sufficiently regular and if the obstacle functions ψ and ψ_p satisfy (29) and (30), respectively, with the following additional hypothesis:

$$\psi - \psi_p \in V \text{ (defined by (26))} \tag{39}$$

then we have:

$$N_V(u - u_p) \leq C\, N_V(\psi - \psi_p) \tag{40}$$

where C is a strictly positive constant, independent on ψ and ψ_p and where N_V denotes the natural norm of the Sobolev space V.

An immediate application of this result is the following one. Let ω be a polygonal domain, partitioned into regular triangles K and denote by $\mathcal{T}_h$ the resulting mesh and h the mesh size. Let V_h be the standard finite element space generated by continuous, piecewise linear functions satisfying the boundary condition and, finally, let Π_h denote the Lagrangian interpolation operator defined from $H^2(\omega)$ into V_h. Then we have the following practical result:

Corollary 1 : If the domain ω is polygonal and if ψ satisfies (29) with the following additional hypothesis:

$$\psi \in V \cap H^2(\omega) \tag{41}$$

then, by taking in (28) $\psi_p = \Pi_h \psi$, we have:

$$N_V(u - u_p) \leq C\, h\, |\psi|_{2,\omega} \tag{42}$$

where C is a strictly positive constant, independent on h and where $|*|_{2,\omega}$ is the classical seminorm 2, defined on $H^2(\omega)$.

Proof : the estimation result (42) is a straightforward consequence of theorem 1 and classical results in the theory of interpolation in the Sobolev spaces (Ciarlet, 1978).

To prove theorem 1, the basic idea consists in introducing the following approximations of problems (23)-(26) and (27), (24), (28), based on the penalty method:

$$\text{Find } u^\varepsilon \in V;\ \forall\, v \in V,\ a(u^\varepsilon,v) + \frac{1}{\varepsilon}\int_\omega (u^\varepsilon-\psi)^+ v\, d\omega = 0 \tag{43}$$

$$\text{Find } u_p^\varepsilon \in V;\ \forall\, v \in V,\ a(u_p^\varepsilon,v) + \frac{1}{\varepsilon}\int_\omega (u_p^\varepsilon-\psi_p)^+ v\, d\omega = 0 \tag{44}$$

where a(.;.) is defined by (24), ε is a strictly positive parameter and $(*)^+$ denotes the positive part of function (*).

It is well-known (Glowinski et al., 1981) that each one of the penalised contact problems (43) and (44) admits a unique solution in V, defined by (26). Moreover, we have the following stability result:

Theorem 2 : Under the hypotheses of theorem 1, the deviation of the solution of the perturbed penalised problem (44) from the solution of the penalised contact problem (43) is bounded by the deviation of the perturbed obstacle function from the real one. More precisely, we have:

$$N_V(u^\varepsilon - u_p^\varepsilon) \leq C\, N_V(\psi - \psi_p) \tag{45}$$

where C is a strictly positive constant, independent on ε, ψ and ψ_p.

Proof : Because of the hypothesis (39), one can choose in (43) and (44) $v = (u^\varepsilon - u^\varepsilon_p) - (\psi - \psi_p)$ and derive two equations. By subtracting them, the following one is obtained:

$$a(u^\varepsilon - u_p^\varepsilon, u^\varepsilon - u_p^\varepsilon - \psi + \psi_p) + \frac{1}{\varepsilon}\int_\omega \{(u^\varepsilon - \psi)^+ - (u_p^\varepsilon - \psi_p)^+\}\{u^\varepsilon - u_p^\varepsilon - \psi + \psi_p\}\, d\omega = 0 \qquad (46)$$

By using the monotony of the real function which associate $(x)^+$ to x, it is deduced from (46) that:

$$a(u^\varepsilon - u_p^\varepsilon, u^\varepsilon - u_p^\varepsilon - \psi + \psi_p) \leq 0 \qquad (47)$$

Using inequality (47), it may be easily checked that:

$$a(u^\varepsilon - u_p^\varepsilon, u^\varepsilon - u_p^\varepsilon) \leq a(u^\varepsilon - u_p^\varepsilon, \psi - \psi_p) \qquad (48)$$

The proof of theorem 2 is then achieved by using (48) and the ellipticity and continuity of the bilinear form a(.,.).

The result of theorem 1 and particularly the independence of constant C appearing in (45) on the penalty parameter is the first basic tool used to prove theorem 1. The second one is a quite classical result which compares the solution of a linear contact problem to the solution of the associated penalised one. The following result is available (Kikuchi and Oden, 1988; Glowinski et al., 1981):

Proposition 1 : The family of solutions of the penalised contact problems defined by (43) (respectively (44)) converges strongly in V to the solution of the problem (23)-(25) (respectively (27),(28)) when the penalty parameter ε goes to zero.

With these two results, one can now easily establish the proof of theorem 1

Proof of Theorem 1 : By using the triangular inequality of a norm twice, we have:

$$N_V(u - u_p) \leq N_V(u - u^\varepsilon) + N_V(u^\varepsilon - u_p^\varepsilon) + N_V(u_p^\varepsilon - u_p)$$

The desired result is an immediate consequence of this last inequality, theorem 2 and proposition 1.

Remark 1 : To ensure that the hypothesis (39) is fulfilled, it is, for example, sufficient to ask that both of the two obstacle functions ψ and ψ_p be in the space $H^1(\omega)$ with equal traces on the clamped part of the domain ω. This last hypothesis does not introduce any practical restriction since the behaviour of the obstacle in a neighbourhood of the clamped part of the boundary has no practical interest assuming, however, as in (29), that it is mechanically compatible with the Dirichlet boundary condition.

In some practical situations, the hypothesis (39) may not be satisfied. For example, when CAD tools are used to approximate an ideal shape of a given obstacle, it may happen that the approximate or perturbed shape is not sufficiently regular. For clarity, let us assume that ψ_p is just in the space $L^2(\omega)$. In this case, the technique developed to prove theorem 1 does not apply. Nevertheless, one can still prove the following estimation result.

Theorem 3 : If the domain ω is sufficiently regular and if the obstacle functions ψ and ψ_p satisfy (29) and (30), respectively, then, we have for sufficiently small penalty parameter:

$$N_V(u^\varepsilon - u_p^\varepsilon) \leq \frac{C}{\varepsilon} N_{L^2}(\psi - \psi_p)$$

Proof : By choosing in (43) and (44) $v = (u^\varepsilon - u^\varepsilon_p)$ and subtracting the resulting two equations, we obtain :

$$a(u^\varepsilon - u_p^\varepsilon, u^\varepsilon - u_p^\varepsilon) + \frac{1}{\varepsilon}\int_\omega \{(u^\varepsilon - \psi)^+ - (u_p^\varepsilon - \psi_p)^+\}\{u^\varepsilon - u_p^\varepsilon\}\, d\omega = 0 \qquad (49)$$

Then, by using the monotony of the real function which associates $(x)^+$ to x, the ellipticity and continuity of the bilinear for a(.,.) and a Young's inequality, the proof of theorem 3 may easily be achieved.

Analysis of the plate perturbed contact problems

As in the previous subsection, the basic result is the following one:

Theorem 4 : Assuming standard regularity hypotheses of the data (ω, D, E) and, in addition to (37) and (38), the following regularity of both of the perturbation function p and the unit normal to the obstacle's surface:

$$p\,\mathbf{n} \in \mathbf{V} \tag{50}$$

we have the following estimation result:

$$N_V(\mathbf{u}-\mathbf{u}_p) \leq C\, N_V(p\mathbf{n})$$

where C is a strictly positive constant independent on p.

The proof of this result follows in essence the proof of theorem 1 and is left to the reader.

Remark 2 : The same kind of estimation result can be obtained for Mindlin-Reissner plates by asking less regularity for the vector valued field (p **n**) than in (50). Typically, one should only ask that p be in V defined by (26) and **n** be in the Sobolev space $(W^{1,\infty}(\omega))^3$. This kind of perturbation is used for the numerical simulation reported in the following section.

NUMERICAL APPLICATION

Numerical simulations of general contact problems are often carried out using the finite element method. This method allows the calculation of numerical solutions which are approximations of the solutions of the continuous contact problems. In certain cases, the error can be estimated. It depends basically on the mesh size of the triangulation. For certain industrial applications such as the sheet metal forming process, a given mesh may fail to handle ultimate contact situations leading to a kind of mesh-dependent "contact locking phenomenon" (see Figure 2).

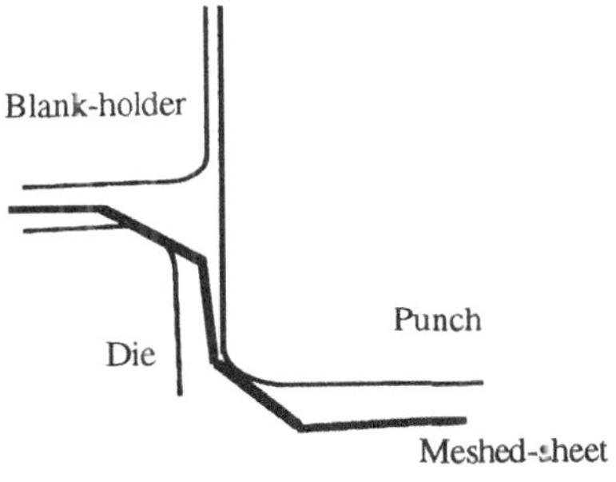

Figure 2. A mesh-dependent ultimate contact situation in the simulation of deep drawing processes

This phenomenon occurs when tools (the punch, the blank-holder and the die) force the meshed sheet to fit areas of their surfaces having radius of curvature significantly smaller than the size of the elements coming in contact with them. These situations can of course be handled by using adaptive techniques. Note that the results proved in the previous section may constitute a first step for building a numerical contact estimator. But, even with these techniques, a large number of elements may be required leading, especially when implicit methods are used to solve these problems, to unbearable requirements of memory and CPU time resources. This is the fundamental reason why explicit schemes are preferred to implicit ones to solve numerically such kind of complex processes.

By taking into account the theoretical results proved in the previous section, we suggest here as an alternative remedy to use mesh-dependent perturbations to handle these ultimate contact situations. The size of the perturbations has to be directly related to the size of the mesh in order to introduce the same order of error as that coming from the finite element method.

We have tested this numerical strategy to calculate the following industrial example of deep drawing of a body car piece, reported in figure 3.

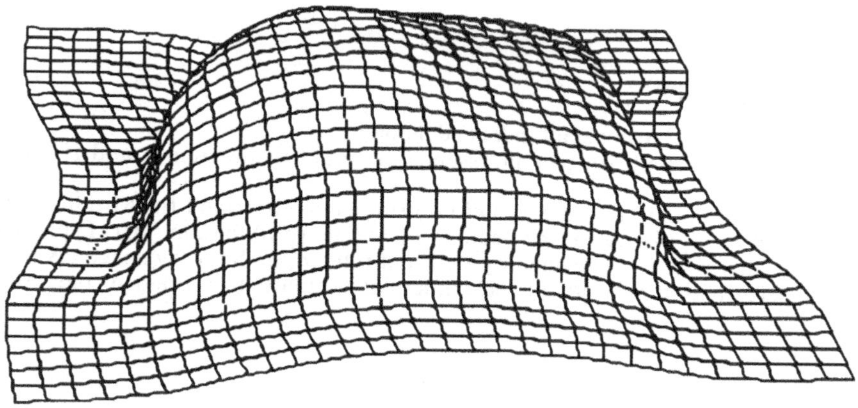

Figure 3. A deep drawing process calculated by a perturbed contact technique

For this test, only 900 quadrangular shell elements were used. Without perturbation of the shape of the punch, we were not able to carry out the simulation up to the end of the process. A kind of contact locking was observed, leading to the divergence of the global algorithm. By using a perturbation of the shape of the punch along its normal vector (as shown in Figure 1), we were able to reach the end of the process. Moreover, significant CPU time savings have been obtained during the steps where the non-perturbed contact simulation was converging and without loss of accuracy. This example shows that this technique is indeed effective, not in terms of giving more accurate results, but for obtaining results of simulations of contact problems which, otherwise, can not be solved by using coarse meshes.

Acknowledgements

This research has been partially supported by both of the French automotive groups PSA and RNUR. This support is gratefully acknowledged. We would also like to acknowledge the collaboration of D. Durville to the calculation of the numerical example.

REFERENCES

Adams, R.A., 1975, "Sobolev Spaces", Academic Press, New York.

Ciarlet, P.G., 1978 "The Finite Element Method for Elliptic Problems", North-Holland, Amsterdam, New York.

Destuynder, P., 1986, "Une Théorie Asymptotique des Plaques Minces en Elasticité Linéaire", Masson, Paris, New York.

Duvaut, G. and Lions, J.L., 1972, "Les Inéquations en Mécanique et en Physique", Dunod, Paris.

Kikuchi, N. and Oden, J.T., 1988, "Contact Problems in Elasticity", SIAM, Philadelphia.

Glowinski, R., Lions, J.L. and Trémolières, R., 1981, "Numerical Analysis of Variational Inequalities", North-Holland, Amsterdam, New York.

FRICTION AND PRECONDITIONERS

Frédéric Lebon, Pierre Alart and Philippe Doudet

Laboratoire de Mécanique et Génie Civil
Université Montpellier 2
Pl. E. Bataillon, 34095 Montpellier Cedex 5
France

INTRODUCTION

Many structural analysis problems are concerned with friction contact phenomena. These problems are difficult to formulate and even more to solve because they are governed by multivalued tribological laws and some numerical resolutions can lead to unsymmetric operators. This last disadvantage becomes crucial for very large problems involving three dimensional discretization and time evolution. This paper shows how to use a simple mixed formulation together with an efficient preconditioned generalized conjugate gradient algorithm coupled with a multilevel interpolation technique when dealing with frictional contact problems.

The augmented Lagrangian approach given by Alart and Curnier (1991) yields non linear and non-differentiable systems the unknowns of which are node displacements and Lagrangian multipliers identified with contact forces. The tangent matrix of the system being *non-symmetric, non-positive definite, ill-conditioned and with zeros on the diagonal*, generalized conjugate gradient methods introduced by Sonneveld, Wesseling and de Zeeuw are used. Appropriate preconditioners are necessary. In this paper, two kinds are presented : *incomplete factorization* and *two-grid factorization*. These solvers are performed in the context of non-linear multigrid methods.

In section 1, we introduce the mixed formulation for frictional contact problem. Section 2 is devoted to conjugate gradient methods, to incomplete factorization and numerical aspects of multilevel preconditioner. Finally, in section 3, efficiency is discussed and comparison between the two preconditioners is done.

MECHANICAL AND MATHEMATICAL SUMMARY

Contact and friction

Following a previous paper based on an augmented Lagrangian approach (Alart and Curnier, 1991), the equilibrium of a discretized elastic body in frictional contact with an obstacle is given by the system of equations (u is the displacement and λ the contact force) :

$$\left\{ \begin{array}{ll} \boldsymbol{F}_{int}(\boldsymbol{u}) - \boldsymbol{F}_{ext} + \boldsymbol{F}(\boldsymbol{u},\lambda) & = 0 \\ -\frac{1}{r}\left(\lambda - \boldsymbol{F}(\boldsymbol{u},\lambda) \right) & = 0 \end{array} \right. \qquad (1)$$

Notations are precised in Alart et al. (1994) in this volume.

Contact Mechanics, Edited by M. Raous *et al.*
Plenum Press, New York, 1995

Generalized Newton Method (GNM)

To solve the equation (1), we use a simultaneous treatment of both variables by Newton's method. Firstly it is useful to resolve the system of equations (1) in two parts depending on the two variables u and λ, a differentiable part G and a non differentiable one $\mathcal{F}$,

$$G(x) + \mathcal{F}(x) = 0 \tag{2}$$

$$G(x) = G(u,\lambda) = \begin{bmatrix} F_{int}(u) - F_{ext} \\ -\frac{1}{r}\lambda \end{bmatrix}, \quad \mathcal{F}(x) = \mathcal{F}(u,\lambda) = \begin{bmatrix} F(u,\lambda) \\ +\frac{1}{r}F(u,\lambda) \end{bmatrix}$$

An extension of Newton method to non-differentiable but continuous equations such as (2) consists in: $x^{i+1} = x^i - (\mathbf{K}_i + \mathbf{J}_i)^{-1} (G(x^i) + \mathcal{F}(x^i)), \mathbf{K}_i = \partial G(x^i) , \mathbf{J}_i \in \partial\mathcal{F}(x^i)$ (3) where $\partial\mathcal{F}(x^i)$ is the generalized Jacobian of $\mathcal{F}$ at x^i . The convergence of the GNM is discussed by Alart and Curnier (1991).

Multilevel interpolation

As proposed in former paper (Alart and Lebon, 1993), the Newton method is initialized using an interpolation from a coarser mesh. The displacements and the multipliers are interpolated by two different operators. To summarize, the non-linear problem is solved on the coarse mesh using the GNM. The solution is interpolated on the fine grid and becomes the initial value for the GNM on the fine mesh. This method is very efficient to accelerate the Newton method. Usually, the difficulty for the Newton method is to find the contact status on the contact surface. When the status are found, convergence is quadratic and even immediate in 2D because of the piecewise linear property of the operator. Multilevel interpolation gives a very good approximation of these status on the fine grid. Then, the convergence of the method is improved.

CONJUGATE GRADIENT SQUARRED METHOD (CGS) AND MULTIGRID

For large and sparse matrices A, it is attractive to use *generalized conjugate gradient method* to solve the equation (3). Preconditioning techniques are necessary to accelerate iterative methods. Different methods have been proposed. Traditionally, these methods are based on Jacobi, block Jacobi, SOR iterations, incomplete factorizations or element-by-element constructions. For our kind of matrix, numerical experiments have shown the efficiency of incomplete LU factorization (ILU/CGS) which has been introduced as preconditioner by Meijerink and van der Vorst (1977). If C denotes the preconditioner matrix, the preconditioned conjugate gradient squarred method is summarized in table 1.

Fahrat and Sobh (1989) have introduced a coarse/fine preconditioner (CF/CGS) for very ill-conditioned finite element problems. In this paper, we generalize this algorithm to our non-symmetric systems. The basic idea of the method is to assume that A is obtained through the refinement of a coarser mesh. The resulting matrix on the coarse mesh is denoted $\mathbf{A}_{cc*}$. The matrix $\mathbf{A}$ is splitted into four parts according to the mesh level, index c for coarse level and f for fine one. Then the matrix $\mathbf{A}$ is written in the following manner:

$$\mathbf{A} = \begin{bmatrix} \mathbf{A}_{ff} \mathbf{A}_{fc} \\ \mathbf{A}_{cf} \mathbf{A}_{cc} \end{bmatrix} = \begin{bmatrix} \mathbf{A}_{ff} & \mathbf{0} \\ \mathbf{A}_{cf} & \mathbf{A}_{cc} - \mathbf{A}_{cf}\mathbf{A}_{ff}^{-1}\mathbf{A}_{fc} \end{bmatrix} \begin{bmatrix} \mathbf{I} & \mathbf{A}_{ff}^{-1}\mathbf{A}_{fc} \\ \mathbf{0} & \mathbf{I} \end{bmatrix}. \tag{4}$$

The part $\mathbf{A}_{cc} - \mathbf{A}_{cf}\mathbf{A}_{ff}^{-1}\mathbf{A}_{fc}$ of this matrix corresponds to the condensation of the fine nodes. As proposed by Fahrat and Sobh (1989), this matrix may be approximated by the "coarse matrix" $\mathbf{A}_{cc*}$. As the matrices $\mathbf{A}_{cc*}$ and $\mathbf{A}_{cc} - \mathbf{A}_{cf}\mathbf{A}_{ff}^{-1}\mathbf{A}_{fc}$ coincide for one bar element, we postulate a *coarse/fine preconditioner* for the conjugate gradient squarred method:

$$\mathbf{C} = \begin{bmatrix} \mathbf{A}_{ff} & \mathbf{0} \\ \mathbf{A}_{cf} & \mathbf{A}_{cc*} \end{bmatrix} \begin{bmatrix} \mathbf{I} & \mathbf{A}_{ff}^{-1}\mathbf{A}_{fc} \\ \mathbf{0} & \mathbf{I} \end{bmatrix}. \tag{5}$$

Table 1. Preconditioned Congugate Gradient Squarred method.

Initialization

$x^0 \in \mathrm{R}^n$

$r^0 = b - \mathrm{A}x^0$

$z^0 = \mathrm{C}^{-1}r^0$

$p^0 = r^0$ and $q^0 = r^0$

Iterations

For k=0,1,...

$\boxed{y^k = \mathrm{C}^{-1}\mathrm{A}q^k}$

$\alpha^k = (z^0, z^k)/(z^0, y^k)$

$x^{k+1} = x^k + \alpha^k(p^k + p^k - \alpha^k y^k)$ and $r^{k+1} = r^k - \alpha^k \mathrm{A}(p^k + p^k - \alpha^k y^k)$

$\boxed{z^{k+1} = \mathrm{C}^{-1}r^{k+1}}$

$\beta^{k+1} = (z^0, z^{k+1})/(z^0, z^k)$

$p^{k+1} = z^{k+1} + \beta^{k+1}(p^k - \alpha^k y^k)$ and $q^{k+1} = p^k + \beta^{k+1}(p^k - \alpha^k y^k + \beta^{k+1} q^k)$

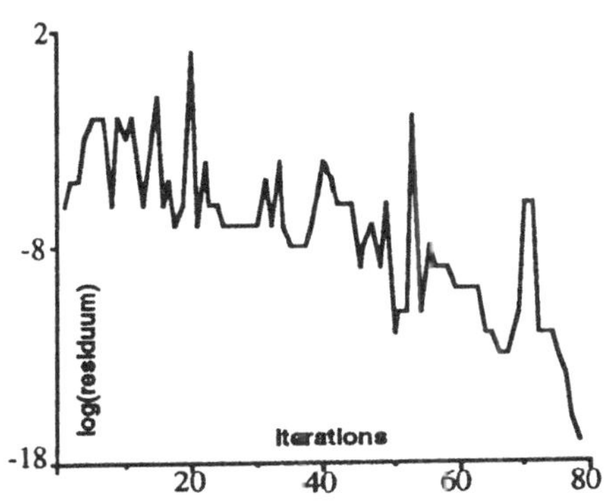

Figure 1a ILU/CGS (cahotic behavior)

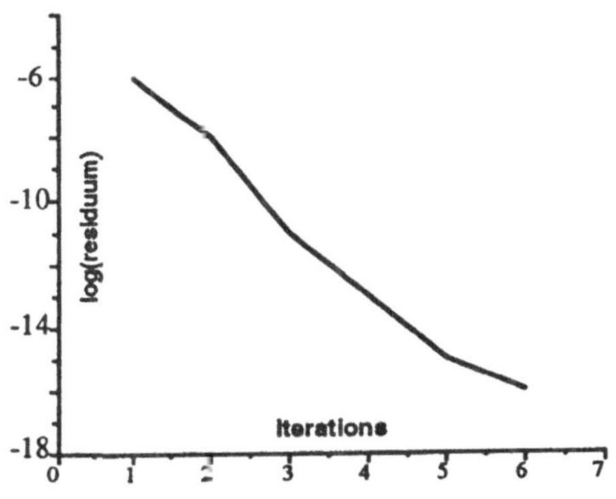

Figure 1b CF/CGS (superconvergence)

Figure 1. CGS behavior for ILU and CF preconditioners

At each iteration of the preconditioned conjugate gradient squarred, we have to solve twice such a system $z^k = \mathbf{C}^{-1} r^k$. For the coarse/fine preconditioner this linear system is written:

$$\begin{cases} \mathbf{A}_{ff} t_f = r_f \\ \mathbf{A}_{cc*} t_c = r_c - \mathbf{A}_{cf} t_f \\ \mathbf{A}_{ff} z_f = r_f - \mathbf{A}_{fc} z_c \end{cases} \tag{6}$$

We have to solve sequentially three sub-systems and to multiply twice a matrix by a vector. The matrix $\mathbf{A}_{cc*}$ is already factorized due to previous coarse mesh solution (multilevel interpolation). The coarse/fine preconditioner demands to solve two sub-systems with matrix $\mathbf{A}_{ff}$. It is possible to compute the exact solution using conjugate gradient squarred method without a preconditioner or to define the matrix **P** equals $(\mathbf{L} + \mathbf{D}^*)\mathbf{D}^{*-1}(\mathbf{U} + \mathbf{D}^*)$. **L** and **U** are the lower and the upper part of the matrix $\mathbf{A}_{ff}$. $\mathbf{D}^*$ is a generalized diagonal of $\mathbf{A}_{ff}$ because of the eventual zeros on the true diagonal **D** of $\mathbf{A}_{ff}$. The matrix **P** could be used either as a preconditioner of $\mathbf{A}_{ff}$ or as an approximation of $\mathbf{A}_{ff}$. To try to improve this technique, a relaxation parameter denoted ω $(0 < \omega < 2)$ is introduced in the definition of the matrix **P** (SOR/CGS) (see Alart and Lebon, 1994).

It is possible to give a multigrid interpretation of the preconditioner : the two first equalities in equation (6) correspond to a coarse grid correction and the third equality corresponds to a from-coarse-to-fine process. The CF/CGS algorithm can be viewed as a multigrid iteration accelerated by the conjugate gradient squarred method.

NUMERICAL RESULTS

An important remark for ILU/CGS method is the fluctuations of the residuum along the iterative process (figure 1a). These fluctuations are due to the ill-preconditioned matrices and to the acceleration technique. If the genelarized conjugate gradient method converges in a bad manner then the conjugate gradient method emphasizes strongly the variation of the convergence. The acceleration is an amplifier of the conjugate gradient defects. A second problem is that we have to determine an optimal level. The efficiency is reduced by the computer time necessary to the factorization, if the level increases. The computations are less efficient for high values of the level due to the cost of the factorization. It is necessary to obtain a good compromise between the reduction of the spectral radius of the method and the difficulties to compute an incomplete factorization of high level. Moreover, the number of ILU/CGS iterations does not depend very much on the Newton iteration, and the optimal level is the same for each Newton iteration (Alart and Lebon, 1994).

For the CF/CGS method, the superconvergence is obtained (figure 1b). The number of conjugate gradient squarred iterations is very small and independant of the number of degrees of freedom. We find a well-known result on more classical problems, such as symmetric, linear or non-linear problems. The difficulties are due to the solution of the linear system on the fine/fine grid. At each CF/CGS iteration it is necessary to solve two systems involving the fine/fine matrix. The fine/fine matrix is very better conditioned than the matrix on the fine grid because it corresponds to the high frequencies of the error. So the range of the eigenvalues is less important (Doudet, 1993).

For problems with global refinement, ILU/CGS method is the most efficient even if an optimal parameter is needed (Alart and Lebon, 1994). For local refinement, CF/CGS method is the most efficient method without finding of an optimal parameter (Alart and Lebon, 1994). On the contrary, SOR/CGS method has the advantage to be naturally implemented on a parallel computer (Alart, Lebon and de Moura, 1994).

REFERENCES

Alart, P., and Curnier, A., 1991, A mixed formulation for frictional contact problems prone to Newton like solution methods, *Comp. Meth. in Appl. Mech. and Eng.*, 92, 353:375.

Alart, P., and Lebon, F., 1993, Multigrid method applied to mixed formulation for frictional contact problem, Contact Mechanics Computational Techniques, Computational Mechanics Publications, Southampton, 227:234.
Alart, P., and Lebon, F., 1994, Solution of frictional contact problems using performing preconditioners, submitted.
Alart, P., Lebon, F., and de Moura, C.A., 1994, Frictional contact : augmented lagrangian, multigrid and parallelism, in preparation.
Alart, P., Lebon, F., Quittau, F., and Rey, K., 1994, Frictional contact problem in elastostatics : revisiting the uniqueness condition, this volume.
Doudet, P., 1993, Frottement et préconditionnement, D.E.A. report, Montpellier University.
Farhat, C., and Sobh N., 1989, A coarse/fine preconditioner for very ill-conditioned finite element problems, *Int. J. for Num. Meth. in Eng.*, 28, 1715-1723.
Meijerink, J.A., and van der Vorst, H.K., 1977, An iterative solution method for linear systems of which the coefficient matrix is a symmetric M-matrix, Math. Comp., 31, 148-162.
Sonnenveld, P., Wesseling, P., and de Zeeuw, P.M., 1985, Multigrid and conjugate gradient methods as convergence acceleration techniques, Multigrid Methods for Integral and Differential Equations, Clarendon Press, Oxford, 117-167.

DECOMPOSITION AND DESCENT METHODS FOR THE SOLUTION OF THE CONTACT PROBLEM BETWEEN RIGID AND DEFORMABLE SOLIDS

B. Radi,[1] and J.C. Gelin[2]

[1] Centre de Recherche de Mathématique de Bordeaux, 351 cours de la Libération, 33405 Talences
[2] Laboratoire de Mécanique Appliquée, 16 route de Gray, 25030 Besançon cedex

INTRODUCTION

The most common variational formulation of the frictional contact between a rigid solid and a deformable one leads generally to fixed points method for numerical solution procedures[1].

In this paper, we are interested in the unilateral and bilateral contact problems with friction, the nonlocal friction law is adopted.

The bilateral contact problem between two deformable solids is treated using decomposition methods. The main advantage of such methods is that it allow to treat the contact problem in each solid separetely.

In the framework of finite element approximation, numerical methods are proposed based on a condensation technique in the contact region for the solution of the contact problem with friction law with the given sliding limit.

The validation tests of these numerical methods are given.

A BILATERAL CONTACT BETWEEN DEFORMABLE SOLIDS

Problem Statement

Under external forces $t^k (k = 1, 2)$, two deformable solids S_1 and S_2 are assumed to have a contact zone. Each solid S_1 and S_2 have respectively a fixed part of the boundary (where the displacement is imposed to be zero) denoted by Γ_{u_1} and Γ_{u_2} as figure 1 shows.

Contact Mechanics, Edited by M. Raous *et al.*
Plenum Press, New York, 1995

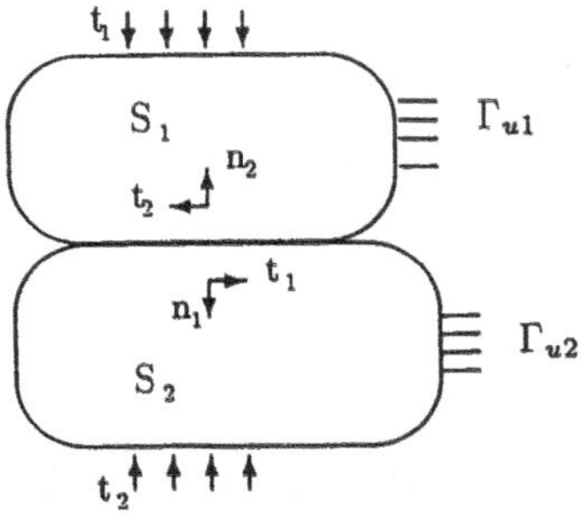

Figure 1. Contact between two deformable solids

Supposing a nonlocal friction law (Duvaut, 1980), the equations governing this contact problem, with small strains, are:

$$(P_f)\quad\left[\begin{array}{l}\left\{\begin{array}{rcll}\sigma_{ij,j}(u)+f_i &=& 0 & \\ \sigma_{ij}(u) &=& a_{ijml}\epsilon_{ml}(u) & \text{in } \Omega=S_1\cup S_2\\ \epsilon_{ij} &=& 1/2(u_{i,j}+u_{j,i}) & \end{array}\right.\\ \\ u_i = 0 \qquad\qquad\qquad\qquad \text{on } \Gamma_u=\Gamma_{u_1}\cup\Gamma_{u_2}\\ \sigma_{ij}.n_j = t_i \qquad\qquad\qquad\quad \text{on } \Gamma_t=\Gamma_{t_1}\cup\Gamma_{t_2}\\ \\ \left.\begin{array}{l} u_n^2-u_n^1 = 0 \Rightarrow \sigma_n=\sigma_n^1=-\sigma_n^2\\ \sigma_t=\sigma_t^1=\sigma_t^2\\ \\ \left\{\begin{array}{lcll}\|\sigma_t\| &\le& \nu_f|S(\sigma_n)| & \\ \|\sigma_t\| &<& \nu_f|S(\sigma_n)| & \Rightarrow\ u_t^1+u_t^2=0\\ \|\sigma_t\| &=& \nu_f|S(\sigma_n)| & \Rightarrow\ \exists\alpha\ge 0\ \ u_t^1+u_t^2=-\alpha\sigma_t\end{array}\right.\end{array}\right\}\ \text{on } \Gamma_c\end{array}\right.$$

S_1 and S_2 are placed in the open sets Ω_1 and Ω_2 of $I\!R^n$ $(n=1,2)$, and Γ_c is the contact zone between S_1 and S_2.

Variational Formulation

We introduce the functional space V_k defined as:

$$V_k \;=\; \{v|v=(v_i), v_i\in H^1(\Omega_k), v_i=0 \quad \text{on } \Gamma_{u_k}\} \tag{1}$$

where V_k stands for the set of all kinematically admissible displacement fields and let:

$$K \;=\; \{(v^1,v^2)\in V_1\times V_2/v_n^1=v_n^2 \text{ on } \Gamma_c\} \tag{2}$$

be the set of all the kinematically admissible displacement fields compatible with the bilateral contact condition. The set K is a non-empty subspace of $V_1\times V_2$.

The standard functionals introduced by Duvaut and Lions (1972) write as:

$$a_k(u^k,v^k)=\int_{\Omega_k}\epsilon_{ij}^k(u^k)a_{ijml}^k\epsilon_{ml}^k(v^k)d\Omega \tag{3}$$

and

$$L_k(v^k) \;=\; \int_{\Omega_k} f_i^k.v_i^k d\Omega+\int_{\Gamma_{tk}} t_i^k.v_i^k d\gamma \tag{4}$$

Supposing that $(u^k)_{k=1,2}$ are sufficiently regular, and if $(u^k)_{k=1,2}$ and $(\sigma^k)_{k=1,2}$ are the solutions to the contact problem (P_f), the following quasi-variational inequality (QVI) (P_1) have to be satisfied:

$$(P_1)\begin{cases} (u^1, u^2) \in K \\ \sum_{k=1}^{2} a_k(u^k, v^k - u^k) + \int_{\Gamma_c} \nu_f |S(\sigma_n)|(\|v_t^1 + v_t^2\| - \|u_t^1 + u_t^2\|)d\gamma \ \geq \ \sum_{k=1}^{2} L_k(v^k - u^k) \\ \forall (v^1, v^2) \in K \end{cases}$$

In order to solve the above QVI, the fixed point method is often used (Duvaut, 1980) that requires the solution of the intermediate problem (P_I) at each iteration:

$$(P_I)\begin{cases} \text{find } (u^1, u^2) \in K \\ \sum_{k=1}^{2} a_k(u^k, v^k - u^k) + \int_{\Gamma_c} g(\|v_t^1 + v_t^2\| - \|u_t^1 + u_t^2\|)d\gamma \ \geq \ \sum_{k=1}^{2} L_k(v^k - u^k) \\ \forall (v^1, v^2) \in K \end{cases}$$

where g is a positive function of $L^2(\Gamma_c)$.

We are interested to solve the intermediate problem (P_I). After the regularization of the euclidian norm $\|.\|$ and the dualization of the constraint $v_n^1 = v_n^2$, the problem (P_I) is equivalent to search a saddle point of a Lagrangian. This saddle point verifies the Euler equations that can write as a nonlinear system. By taking $v^1 \equiv 0$ and $v^2 \equiv 0$ successively in this nonlinear system, the system (5) is obtained (Radi and al., 1995):

$$\begin{cases} a_1(u^1, v^1) + \int_{\Gamma_c} g\phi_\epsilon(\|u_t^1\|)u_t^1.v_t^1 d\gamma + \int_{\Gamma_c} \lambda v_n^1 d\gamma \ = \ L_1(v^1) \quad \forall v^1 \in V_1 \\ a_2(u^2, v^2) + \int_{\Gamma_c} g\phi_\epsilon(\|u_t^2\|)u_t^2.v_t^2 d\gamma - \int_{\Gamma_c} \lambda v_n^2 d\gamma \ = \ L_2(v^2) \quad \forall v^2 \in V_2 \\ \int_{\Gamma_c} \mu(u_n^1 - u_n^2)d\gamma \ = \ 0 \quad \forall \mu \in L^2(\Gamma_c) \end{cases} \tag{5}$$

where

$$\phi_\epsilon(\|u_t^k\|) = \begin{cases} 1/\|u_t^k\| & \text{if } \|u_t^k\| > \epsilon \\ 1/\epsilon & \text{if } \|u_t^k\| \leq \epsilon \end{cases} \tag{6}$$

λ and μ are the Lagrange multipliers.

Two important remarks can be made (i) it is possible to decompose the problem into subproblems that can be treated separetely (ii) it is possible to find the rigid-deformable contact for each solid without any variable changing (Pinto and al., 1987).

Proposed Method

Using a finite element approximation, the nonlinear system (7) have to be solved:

$$\begin{cases} [A_1 + B(u^1)]u^1 + C_1\lambda \ = \ f^1 \\ A_2 + B(u^2)]u^2 - C_2\lambda \ = \ f^2 \\ C_1^t u^1 - C_2^t u^2 \ = \ 0 \end{cases} \tag{7}$$

where:
* A_k is the stiffness matrix of the domain Ω_k,
* f^k represents the body forces of the domain Ω_k,
* C_k is the interaction matrix between Ω_k and Γ_c,
* $B(u^k)$ is a rectangular matrix.

The proposed method is based on the substitution-condensation technique:

1. At the first iteration, $(u^{1(0)}, u^{2(0)})$ is given. This vector is the result of the contact problem without friction.
2. At the $(n+1)$th iteration, the displacement fields $(u^{1(n)}, u^{2(n)})$ are known.
3. The solution of the following system is calculated by eliminating the displacement fields u^1 and u^2 and using the conjugate gradient method to solve resulting condensed system:

$$\sum_{k=1}^{2} C_k^t[A_k + B(u^{k(n)})]^{-1}C_k\lambda \;=\; \sum_{k=1}^{2} C_k^t[A_k + B(u^{k(n)})]^{-1}f^k \tag{8}$$

(the condensed matrices being not computed explicitly (Radi and al. 1994)).

Let $(u^{1(n+1)}, u^{2(n+1)}, \lambda^{n+1})$ be the solution of the system.

4. The following test is done:

if $\|u^{k(n+1)} - u^{k(n)}\|/\|u^{k(n)}\| <$ TOL (TOL is given)

then $(u^{1(n+1)}, u^{2(n+1)})$ is the solution and λ^{n+1} corresponds to (Radi, 1992):

$$\begin{cases} \sigma_n^1(I) & = & \lambda^{n+1}(I) \\ \sigma_n^2(I) & = & -\lambda^{n+1}(I) \end{cases} \tag{9}$$

the friction stress is computed using the nonlinear nonlocal law proposed by Kikuchi and Oden (1988),

otherwise, replace $u^{1(n)}$ by $u^{1(n+1)}$ and $u^{2(n)}$ by $u^{2(n+1)}$ in the system (7) and repeat.

This algorithm is essentially based on the solution of the condensed system (8) whose the dimension is lower than the global dimension of the problem. This strategy gives an advantage to the use of the Lagrange multiplier in comparison with the disadvantage noted by Simo and Laursen (1992).

UNILATERAL CONTACT BETWEEN RIGID AND DEFORMABLE SOLIDS

The unilateral condition is written as follows:

$$u_n \;\leq\; 0,\; \sigma_n \;\leq\; 0 \text{ and } u_n.\sigma_n \;=\; 0 \text{ on } \Gamma_c \tag{10}$$

Let:

$$M = \{\mu \in H^{-1/2}(\Gamma_c)/\; \mu \geq 0 \text{ on } \Gamma_c\} \tag{11}$$

be the set of the Lagrange multipliers. One can find a discussions about the choice of such functional spaces (Radi, 1992).

The contact problem with the unilateral condition is equivalent to solve a quasi-variational inequality (Duvaut and Lions, 1972). At each iteration of the fixed point method, an intermediate problem must be solved.

After dualization of the constraint $u_n \leq 0$ and regularization of the euclidian norm, the intermediate problem is transformed in a search of a saddle point of a Lagrangian.

Using a finite element approximation. the saddle point have to satisfy the nonlinear system (12) (Radi, 1992):

$$\begin{cases} [A + B(u)]u + C\lambda &= f \\ (\mu - \lambda, C^t u) &\leq 0 \quad \forall \mu \in M_h \end{cases} \tag{12}$$

where M_h is the discret space relative to the Lagrange multipliers.

The same substitution-condensation technique as below is proposed. The obtained condensed inequality is express as:

$$(\mu - \lambda, C^t[A + B(u^{(n)})]^{-1}f - C^t[A + B(u^{(n)})]^{-1}C\lambda) \leq 0 \tag{13}$$

where $u^{(n)}$ is given.

So λ is a minimum of the following quadratic functional $\mathcal{J}$ on M_h:

$$\mathcal{J}(\mu) = 1/2(\mathcal{K}\mu, \mu) - (\mathcal{F}, \mu) \tag{14}$$

where

$$\mathcal{K} = C^t[A + B(u^{(n)})]^{-1}C \tag{15}$$

and

$$\mathcal{F} = C^t[A + B(u^{(n)})]^{-1}f \tag{16}$$

The Rosen algorithm (one can see the Cea's book (1971)) is choosen in order to have the value of λ, because it is only need the action of the condensed matrix on a given vector.

NUMERICAL RESULTS

Two examples are presented. The first one concerns a bilateral contact problem where the condensation-substitution technique is used in conjunction with a CG solver. The second one concerns a bilateral frictional contact problem between two deformable solids where the decomposition method is used in conjunction with a CG solver.

Example 1:

Let a geometrical domain with height $h = 5$ units and width $l = 10$ units. There is frictional contact between the bottom surface of the solid and the foundation. The Young's modulus and Poisson's ratio are respectively $E = 2.1 \times 10^7$ psi and $\nu = 0.29$. Assuming a constant friction coefficient $\nu_f = 0.3$, a uniform tension $t = 10^4$ psi is applied in the negative direction of the y axis without any body forces. Using the symmetry of the problem, only the right half portion is meshed, figure 2.

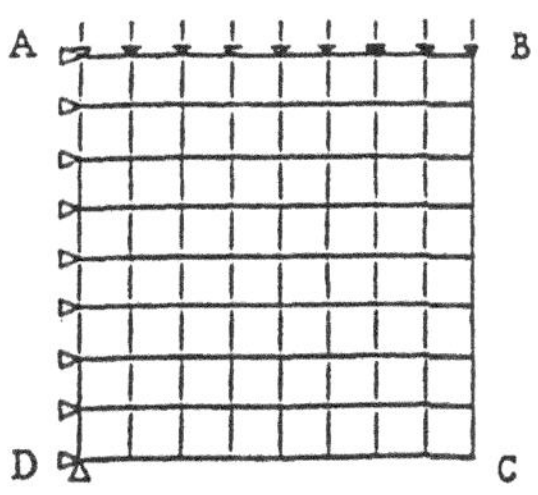

Figure 2. The right half portion

Two friction laws are tested assuming respectively:
(i) $g_h = \nu_f \lambda_h$ (local law (Licht and al., 1991))
(ii) $g_h = \nu_f S_h(\lambda_h)$, by adopting the analytical form of S given by Oden and Pires (1981).

The results obtained with the proposed method denoted by FPM are compared with the results given by Kikuchi and Oden (1988) founded by using the algorithm proposed by Campos and al. (1982) (this algorithm is denoted below as COK).

Local law:

Figure 3 and 4 show the contours of the normal and tangential stresses, respectively, on the contact surface Γ_c.

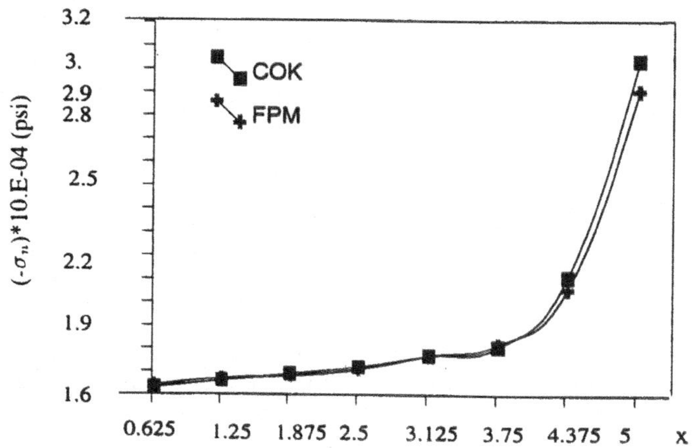

Figure 3. The contact pressure with local law

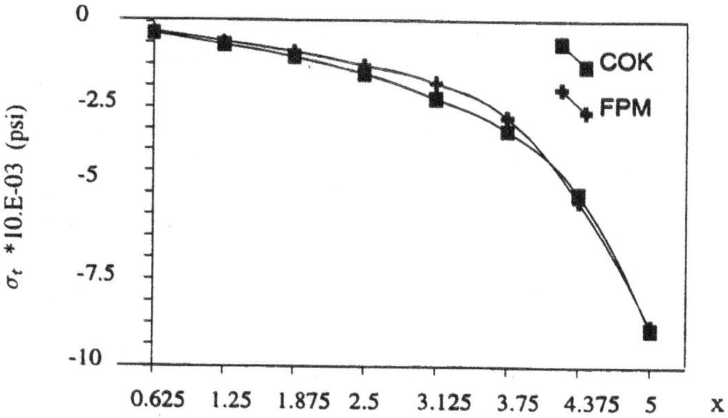

Figure 4. The friction stress σ_t with local law

Nonlocal law:

Figure 5 shows the contours of the normal stresses on the contact surface Γ_c. Concerning the contours of the tangential stresses, similar results are found except for the last node whose value is very small($< -9.10^3$ psi).

With the two friction laws, the behavior is the same: all the nodes are sticking except the last one which is sliding.

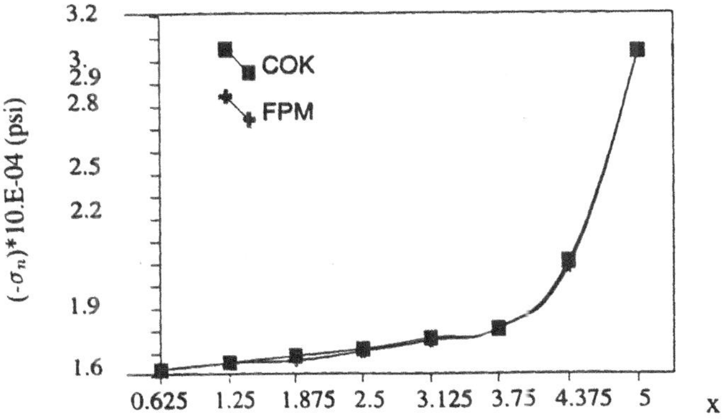

Figure 5. The contact pressure with nonlocal law

Example 2:

In this example, the contact with a friction between two elastic solids S_1 and S_2 is examined. The two bodies have respectively Young's modulus equal 2.10^5 kN/cm^2 and 5.10^4 kN/cm^2 and Poisson's ratio equal 0.3 and 0.25. Figure 6 shows the geometries as well as the deformed shapes.

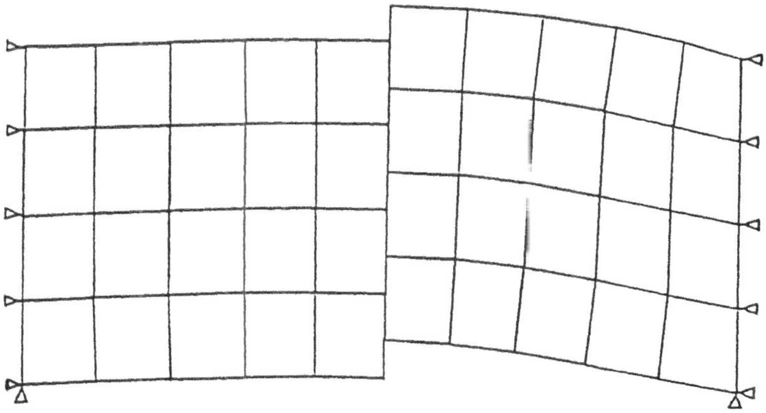

Figure 6. Geometry and the equilibrium state

CONCLUSIONS

The solution of the contact problem between two deformable solids or between a rigid and a deformable solids is studied.

The formulation adopted for the contact problem between two deformable solids presents a number of advantages, namely: (i) dissociating the problem into two coupled subproblems where the normal stresses are the coupled variables, thus allowing the use of parallel architecture computers, (ii) reducing the problem to one of rigid-elastic contact, thus allowing the contact problem on each solid to be treated separately.

The proposed iterative methods lead to the solution of a condensed system that is obtained using descent methods: conjugate gradient method or Rosen algorithm for which it is not necessary to compute the matrix of the condensed system.

The numerical results obtained prove the validity of the proposed approaches.

REFERENCES

L.T. Campos, J.T. Oden and N. Kikuchi, 1982, A numerical analysis of class of contact problems with friction in elastostatics, *Comp. Meth. Appl. Mech. and Eng.* 34:821.

J. Cea, 1971, "Optimisation: théorie et algorithmes" Dunod, Paris.

G. Duvaut, 1980, Equilibre d'un solide élastique avec contact unilatéral et frottement de Coulomb, *C.R. Acad. Sci. série A.*

G. Duvaut and J.L. Lions, 1972, "Les Inéquations en Mécanique et en Physique" Dunod, Paris.

N. Kikuchi and J.T. Oden, 1988, "Contact problems in elasticity. A study of variational inequalities and finite element methods" SIAM Studies in Applied Mathematics.

C. Licht, E. Pratt and M. Raous, 1991, Remarks on a numerical method for unilateral contact including friction, *Int. Series of Num. Math.* 101:129.

J.T. Oden and E. Pires, 1981, Contact problems in elastostatics with nonlocal friction laws, TICOM Report 81-12, Univ. Texas.

Y. Pinto and M. Raous, 1987, Direct mathematical programming method for the contact between two deformable solids, ICIAM-87,Paris, 29 Juin -3 Juillet.

B. Radi, 1992, The subdomain calculation methods. Application to the contact problem between two deformable solids, PhD Thesis, Franche-Comté University.

B. Radi, J.C. Gelin and A. Perriot, 1995, Solution of the contact problem between two deformable solids using a decomposition method, to be published in *Comp. Meth. in Appl. Mech. and Eng.*

B. Radi, J.C. Gelin and A. Perriot, 1994, Subdomain methods in structural mechanics, to be published in *Int. J. of Num. Meth. and Eng.*

J.C. Simo and T.A. Laursen, 1992. An augmented Lagrangian treatment of contact problems involving friction, *Comp. & Struc.* 42:97.

MODELLING OF FRICTIONAL CONTACT FROM MIXED VARIATIONAL PRINCIPLES

S. Cescotto

M.S.M. Department
Université de Liège
Quai Banning, 6
B-4000 Liège, Belgium

INTRODUCTION

In the numerical modelling of unilateral contact with friction in a finite element context, the contact inequalities are usually written in terms of nodal co-ordinates. For example, it is expressed that the finite element nodes belonging to the boundary of the discretized solid may not penetrate into a rigid second body, named here the foundation.

This approach can be qualified as 'compatible or displacement-type approach' since the discretized displacements of the solid are the basic variables.

In a recent paper, Cescotto and Charlier (1993) propose an alternative solution, in which the contact stresses and the displacement field on the solid boundary are discretized independently. It is based on mixed variational principles and allows to control the average overlapping between the solid boundary and the foundation.

The basic advantage of such mixed contact elements is that the contact condition is naturally smoother than with the compatible approach. In the latter method, a stiffness or an out-of-balance force is attributed to a node only when contact has taken place. On the contrary, in the mixed approach, a stiffness matrix and out-of-balance forces are computed for the mixed contact element, even when it is only partially in contact. In other words, a node which is not yet in contact but only close to contact is 'informed' by its neighbours that contact is going to occur soon.

The above mentioned paper uses the penalty method for solving the unitaleral contact and slip conditions and the Coulomb model for the friction strength. However, the choice of the penalty coefficients remains a controversial question. Generally, it is required to keep the numerical values of the bodies penetration small relative to the finite element. Therefore, the penalty coefficient may depend on the nature of the bodies in contact and on their discretization.

For these reasons, alternative approaches have been explored, including the Lagrange multipliers and the augmented Lagrangian method.

The goal of this paper is to present an extension of the mixed formulation to the case of the augmented Lagrangian method.

Contact Mechanics, Edited by M. Raous *et al.*
Plenum Press, New York, 1995

MIXED VARIATIONAL PRINCIPLE FOR THE AUGMENTED LAGRANGIAN METHOD

Hypotheses

For simplicity, the development is presented in a very restrictive framework: the deformable solid is linear elastic, its deformations are infinitesimal and the foundation is rigid, so that the only source of non linearity is unilateral contact. Furthermore, sticking contact is assumed.

It is clear that above restrictions are very easily removed by using standard procedures of large displacements and large strains formulation as explained in details in Cescotto and Charlier (1993).

Local and global co-ordinates

We consider a solid of volume V and boundary A. Let A_U, A_T and A_C be the parts of A on which displacements, surface tractions and unilateral contact conditions are imposed, respectively. At each point S of A_C, local co-ordinates are defined (Figure 1), with $\mathbf{e}_1$ normal and $\mathbf{e}_2$, $\mathbf{e}_3$ tangent to A_C, which is assumed to be smooth. At point S, the displacement of the material particle of the solid, with respect to some reference configuration, is

$$\boldsymbol{u}_S = u_{si}\boldsymbol{e}_i \tag{1}$$

and the stress tensor is denoted by $\boldsymbol{\sigma}_S$.

If p is the contact pressure and τ_1, τ_2 the contact shear stresses at point S and σ_{s11}, σ_{s21}, σ_{S31} are components of $\boldsymbol{\sigma}_S$ in the local frame (e_1, e_2, e_3), the surface equilibrium conditions give :

$$\boldsymbol{\sigma}_S + \boldsymbol{\sigma}_C = 0 \tag{2}$$

with:

$$\boldsymbol{\sigma}_s = \begin{bmatrix} \sigma_{s11} \\ \sigma_{s21} \\ \sigma_{s31} \end{bmatrix} , \qquad \boldsymbol{\sigma}_c = \begin{bmatrix} p \\ \tau_2 \\ \tau_3 \end{bmatrix} \tag{3}$$

If point S of the solid is in contact with point F of the foundation, let

$$\boldsymbol{u}_F = u_{Fi}\boldsymbol{e}_i \tag{4}$$

be the displacement of the latter.

Then, the "gap" or "overlapping" between solid and foundation is

$$\boldsymbol{\varepsilon}_C = \boldsymbol{u}_S - \boldsymbol{u}_F \tag{5}$$

with:

$$\boldsymbol{u}_S = \begin{bmatrix} u_{S1} \\ u_{S2} \\ u_{S3} \end{bmatrix}, \quad \boldsymbol{u}_F = \begin{bmatrix} u_{F1} \\ u_{F2} \\ u_{F3} \end{bmatrix} \tag{6}$$

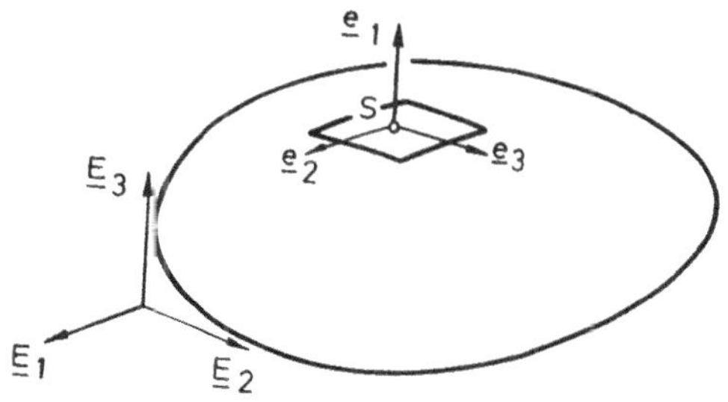

Figure 1

For points belonging to A_U, A_T or V, we will write, in a global Cartesian frame $\mathbf{E}_1$, $\mathbf{E}_2$, $\mathbf{E}_3$:

$$\boldsymbol{\sigma} = \begin{bmatrix} \sigma_{11} \\ \sigma_{22} \\ \sigma_{33} \\ \sigma_{23} \\ \sigma_{13} \\ \sigma_{12} \end{bmatrix}, \quad \boldsymbol{\varepsilon} = \begin{bmatrix} \epsilon_{11} \\ \epsilon_{22} \\ \epsilon_{33} \\ 2\epsilon_{23} \\ 2\epsilon_{13} \\ 2\epsilon_{12} \end{bmatrix}, \quad \partial \boldsymbol{u} = \begin{bmatrix} \partial u_1/\partial x_1 \\ \partial u_2/\partial x_2 \\ \partial u_3/\partial x_3 \\ \partial u_2/\partial x_3 + \partial u_3/\partial x_2 \\ \partial u_1/\partial x_3 + \partial u_3/\partial x_1 \\ \partial u_1/\partial x_2 + \partial u_2/\partial x_1 \end{bmatrix}$$

$$\boldsymbol{u} = \begin{bmatrix} u_1 \\ u_2 \\ u_3 \end{bmatrix}, \quad \boldsymbol{T} = \begin{bmatrix} T_1 \\ T_2 \\ T_3 \end{bmatrix}, \quad \boldsymbol{F} = \begin{bmatrix} F_1 \\ F_2 \\ F_3 \end{bmatrix}$$

$$\boldsymbol{\sigma}_n = \begin{bmatrix} n_1\sigma_{11} + n_2\sigma_{21} + n_3\sigma_{31} \\ n_1\sigma_{12} + n_2\sigma_{22} + n_3\sigma_{32} \\ n_1\sigma_{13} + n_2\sigma_{23} + n_3\sigma_{33} \end{bmatrix}, \quad \partial\boldsymbol{\sigma} = \begin{bmatrix} \partial\sigma_{11}/\partial x_1 + \partial\sigma_{21}/\partial x_2 + \partial\sigma_{31}/\partial x_3 \\ \partial\sigma_{12}/\partial x_1 + \partial\sigma_{22}/\partial x_2 + \partial\sigma_{32}/\partial x_3 \\ \partial\sigma_{13}/\partial x_1 + \partial\sigma_{23}/\partial x_2 + \partial\sigma_{33}/\partial x_3 \end{bmatrix}$$

where $\mathbf{n} = n_i\mathbf{E}_i$ is the normal to the solid boundary on A_T or A_U.

Contact constitutive law

Let

$$\boldsymbol{\lambda} = \begin{bmatrix} \lambda_1 \\ \lambda_2 \\ \lambda_3 \end{bmatrix}$$

be a Lagrangian multiplier which, ideally, should be equal to the contact stresses $\boldsymbol{\sigma}_C$ when equilibrium and contact compatibility is reached.
As the problem is non linear, it is solved incrementally and, at each time step, Newton-Raphson iterations are performed.
At time step N, the Lagrange multiplier for iteration (k + 1) is computed by :

$$\boldsymbol{\lambda}^{(k+1)} = \langle \boldsymbol{\lambda}^{(k)} - \mathbf{K}_C\, \boldsymbol{\varepsilon}_C^{(k)} \rangle \tag{7}$$

where the symbol < > means that the right hand side is computed only in case of contact, that is when there is some overlapping between the solid and the foundation : $(u_{S1} - u_{F1}) > 0$, and

$$\mathbf{K}_C = \begin{bmatrix} K_p & 0 & 0 \\ 0 & K_\tau & 0 \\ 0 & 0 & K_\tau \end{bmatrix} \tag{8}$$

is a matrix of penalty coefficients.
Equation (7) means that, for iteration (k + 1), the Lagrange multiplier is not an additional variable but is considered fixed.

On the other hand, let us compute "penalty stresses" $\mathbf{p}_C$ by

$$\boldsymbol{p}_C = \mathbf{K}_C \boldsymbol{\varepsilon}_C \tag{9}$$

Then, the contact stresses at iteration (k + 1) of step N are computed by the following relation :

$$\boldsymbol{\sigma}_C^{(k+1)} = \boldsymbol{\lambda}^{(k+1)} + \mathbf{K}_C\, \boldsymbol{\varepsilon}_C^{(k+1)} \tag{10}$$

Using (7), we get :

$$\boldsymbol{\sigma}_C^{(k+1)} = \boldsymbol{\lambda}^{(k+1)} + \mathbf{K}_C\, (\boldsymbol{\varepsilon}_C^{(k+1)} - \boldsymbol{\varepsilon}_C^{(k)}) \tag{11}$$

The physical meaning of this equation is that the Lagrange multipliers tend to the contact stresses when, during iterations, compatibility is achieved on the contact surface

$$\boldsymbol{\varepsilon}_C = 0 \tag{12}$$

The mixed functional Π_C^*

The following functional is defined :

$$\Pi_C^* = \int_{A_C} \{W_C(\varepsilon_C) + \sigma_C^T [u_S - u_F - \varepsilon_C] + \lambda^T \varepsilon_C\} \, dA_C$$
$$+ \int_V \{W(\varepsilon) + \sigma^T [\partial u - \varepsilon] - F^T u\} \, dV \tag{13}$$
$$- \int_{A_T} T^T u \, dA_T - \int_{A_U} \sigma_n^T [u - \tilde{u}] \, dA_U$$

In this functional, $\tilde{u}$ are the displacement imposed on A_U;

$$W_C(\varepsilon_C) = \frac{1}{2} \varepsilon_C^T K_C \varepsilon_C \tag{14}$$

is the contact penalty energy density ;

$$W(\varepsilon) = \frac{1}{2} \varepsilon^T C \varepsilon \tag{15}$$

is the strain energy density in the solid; C is Hooke's elastic tensor written in matrix form, so that

$$\sigma = C\varepsilon \tag{16}$$

In the functional Π_C^* , the independent fields are $\sigma, \varepsilon, u, \sigma_C, \varepsilon_C, u_C$.

It must be emphasized that all these independent variables are evaluated at iteration (k+1) of step N. Note also that the Lagrange multipliers given by (7) are not independent variables and that $\mathbf{u}_S$ is the restriction of $\mathbf{u}$ on the contact surface so that it is not an independent variable neither.

Variation of Π_C^*

We express that the above functional is stationary: $\delta\Pi_C^* = 0$.

Taking the variation of Π_C^* with respect to the independent fields is straightforward. The results are summarized in Table 1. It is seen that all the field equations are recovered in the solid as well as on its boundary.

During the derivation of the first equation, it must be noticed that

$$\delta\lambda^{(k+1)} = 0 \tag{17}$$

because the right hand side of (7) contains only quantities evaluated at iteration (k), that is known quantities.

Table 1. Variation of Π_C^*

Variable	Region	Equation	Interpretation
$\boldsymbol{\varepsilon}_C$	A_C	$\boldsymbol{\sigma}_C = \mathbf{K}_C\boldsymbol{\varepsilon}_C + \boldsymbol{\lambda}$	Constitutive law at contact
$\boldsymbol{\sigma}_C$	A_C	$\boldsymbol{\varepsilon}_C = \boldsymbol{u}_S - \boldsymbol{u}_F$	Compatibility condition at contact
$\mathbf{u}_S$	A_C	$\boldsymbol{\sigma}_C + \boldsymbol{\sigma}_S = 0$	Surface equilibrium at contact
$\boldsymbol{\varepsilon}$	V	$\boldsymbol{\sigma} = \boldsymbol{C\varepsilon}$	Constitutive law in the solid
$\boldsymbol{\sigma}$	V	$\boldsymbol{\varepsilon} = \partial\boldsymbol{u}$	Compatibility in the solid
$\mathbf{u}$	V	$\partial\boldsymbol{\sigma} + \boldsymbol{F} = 0$	Equilibrium in the solid
$\boldsymbol{\sigma}$	A_U	$\boldsymbol{u} = \bar{\boldsymbol{u}}$	Compatibility with the imposed displacements on A_U
$\mathbf{u}$	A_U	$0 = 0$	-
$\mathbf{u}$	A_T	$\boldsymbol{\sigma}_n = \boldsymbol{T}$	Surface equilibrium on A_T

Simplified functionals

By assuming that some of the equations of table 1 are satisfied a priori, it is possible to obtain simplified functionals deduced from (13).
For example, if we assume that the following equations :

$$\boldsymbol{\varepsilon} = \partial\boldsymbol{u} \quad \text{in } V$$

$$\boldsymbol{\sigma} = \boldsymbol{C\varepsilon} \quad \text{in } V$$

$$\boldsymbol{u} = \bar{\boldsymbol{u}} \quad \text{on } A_U$$

are satisfied a priori, we get a functional of the independent fields $\mathbf{u}$, $\boldsymbol{\sigma}_C$, $\boldsymbol{\varepsilon}_C$.

$$\begin{aligned} \Pi_{C1}^* = & \int_{A_C} \{W_C(\boldsymbol{\varepsilon}_C) + \boldsymbol{\sigma}_C^T [\boldsymbol{u}_S - \boldsymbol{u}_F - \boldsymbol{\varepsilon}_C] + \boldsymbol{\lambda}^T \boldsymbol{\varepsilon}_C\} dA_c \\ & + \int_V [W(\boldsymbol{u}) - \boldsymbol{F}^T \boldsymbol{u}] dV - \int_{A_T} \boldsymbol{T}^T \boldsymbol{u} dA_T \end{aligned} \tag{18}$$

where W(u) is the strain energy density in the solid, expressed as a function of the independent field $\mathbf{u}$.

If, in addition to (18) we assume that the constitutive relation (10) is satisfied a priori, that is (dropping the subscript k+1)

$$\boldsymbol{\varepsilon}_C = \mathbf{K}_C^{-1} (\boldsymbol{\sigma}_C - \boldsymbol{\lambda}) \tag{19}$$

the contact penalty complementary energy can be written

$$W_C^* (\boldsymbol{\sigma}_C) = (\boldsymbol{\sigma}_C^T - \boldsymbol{\lambda}^T) \boldsymbol{\varepsilon}_C - W_C (\boldsymbol{\varepsilon}_C) \tag{20}$$

and a new functional of the independent fields **u** and $\boldsymbol{\sigma}_C$ is obtained

$$\Pi_{C2}^* = \int_{A_C} \{\boldsymbol{\sigma}_C^T [\boldsymbol{u}_S - \boldsymbol{u}_F] - W_C^*(\boldsymbol{\sigma}_C)\} dA_C + \int_V [W(\boldsymbol{u}) - \boldsymbol{F}^T\boldsymbol{u}] dV - \int_{A_T} \boldsymbol{T}^T\boldsymbol{u} dA_T \tag{21}$$

Finally, if in addition to (18) and (19), contact compatibility is assumed, that is

$$\boldsymbol{u}_S - \boldsymbol{u}_F = \boldsymbol{\varepsilon}_C = \mathbf{K}_C^{-1} (\boldsymbol{\sigma}_C - \boldsymbol{\lambda}) \tag{22}$$

the only remaining independent variable is **u** and we recover the classical functional

$$\Pi_{C3}^* = \int_{A_C} W_C(\boldsymbol{u}_S) dA_C + \int_V [W(\boldsymbol{u}) - \boldsymbol{F}^T\boldsymbol{u}] dV - \int_{A_T} \boldsymbol{T}^T\boldsymbol{u} dA_T + \int_{A_C} \boldsymbol{\lambda}^T (\boldsymbol{u}_S - \boldsymbol{u}_F) dA_C \tag{23}$$

If the last integral is dropped, what remains is nothing else than the usual functional of a "compatible or displacement type" approach of the contact problem.

CONTACT FINITE ELEMENT FORMULATION

Basic idea

In the functionals developed above, the integral on V is used to develop solid elements, the integral on A_T is used to develop loading elements and the integral on A_C is used to develop contact elements (figure 2).

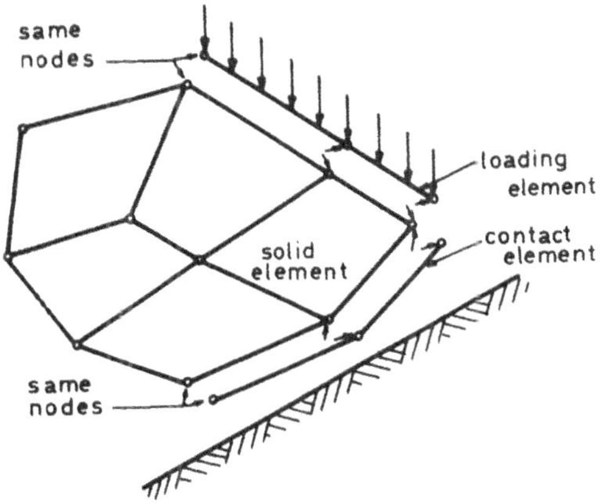

Figure 2

Solid and loading elements are classical and depend on the starting functional. For example, the volume integral of (13) is the basis of a four node mixed finite element presented by Jetteur and Cescotto (1991) or Belytschko et al (1984). On the contrary, if the starting point is the volume integral of (21) or (22), the solid element is the usual compatible one. Examples of loading elements can be found in Charlier (1987).
Hereafter, we present the development of a mixed contact element based on (13).

A three field contact element

The following discretization is assumed

$$\boldsymbol{u}_S = \boldsymbol{N}\,\boldsymbol{U} \tag{24}$$

$$\boldsymbol{\sigma}_C = \boldsymbol{P}\,\boldsymbol{q} \tag{25}$$

$$\boldsymbol{\varepsilon}_C = \boldsymbol{P}\,\boldsymbol{r} \tag{26}$$

where **U** are the nodal displacements of the contact element, **q** and **r** the discretization parameters of the contact stresses $\boldsymbol{\sigma}_C$ and compatibility mismatch $\boldsymbol{\varepsilon}_C$.

The nodal interpolation function **N** of the contact element are the restriction of those of the underlying solid element on its boundary because its nodes coïncide with those of the contact element.

The same interpolation functions **P** are chosen for $\boldsymbol{\sigma}_C$ and $\boldsymbol{\varepsilon}_C$.

Introducing (24), (25), (26) in the integral on A_c of (13) and taking the variations with respect to **U,** **q** and **r** give respectively
- the average surface equilibrium at contact:

$$\boldsymbol{M}^T\,\boldsymbol{q} = 0 \tag{27}$$

- the average compatibility at contact:

$$\boldsymbol{M}\,\boldsymbol{U} - \boldsymbol{V} - \boldsymbol{Q}\,\boldsymbol{r} = 0 \tag{28}$$

- the average constitutive equation at contact:

$$\boldsymbol{R}\,\boldsymbol{r} - \boldsymbol{Q}\,\boldsymbol{q} + \boldsymbol{L} = 0 \tag{29}$$

with:

$$\boldsymbol{M} = \int_{A_C} \boldsymbol{P}^T\boldsymbol{N}\,dA_C \tag{30}$$

$$\boldsymbol{R} = \int_{A_C} \boldsymbol{P}^T\,\mathbf{K}_C\,\boldsymbol{P}\,dA_C \tag{31}$$

$$Q = \int_{A_C} P^T P \, dA_C \tag{32}$$

$$V = \int_{A_C} P^T \, u_F \, dA_C \tag{33}$$

$$L = \int_{A_C} P^T \, \lambda \, dA_C \tag{34}$$

From (28)

$$r = Q^{-1} \, (M \; U \; - \; V) \tag{35}$$

and from (29)

$$q = Q^{-1} \, R \, r + Q^{-1} \, L \tag{36}$$

Let

$$S = Q \, R^{-1} \, Q \tag{37}$$

$$W = V - Q \, R^{-1} \, L \tag{38}$$

Introducing (36) into (35) gives

$$q = S^{-1} \, (M \; U \; - \; W) \tag{39}$$

Finally, introducing (39) into (27) gives

$$k_C \, U = f_C \tag{40}$$

where

$$k_C = M^T \, S^{-1} \, M \tag{41}$$

is the contact element stiffness matrix and

$$f_c = M^T \, S^{-1} \, W \tag{42}$$

is the corresponding nodal force vector which takes account of the displacement of the foundation $\mathbf{u}_F$ via V (eq. 33) and of the value of the Lagrange multipliers $\boldsymbol{\lambda}$ via **L** (eq. 34).

Example of application

To illustrate the method. we apply the preceeding results to a contact element with linear displacement field $\mathbf{u}$, constant contact stress $\boldsymbol{\sigma}_C$ and constant compatibility mismatch $\boldsymbol{\varepsilon}_C$, in the two dimensional case (figure 3).

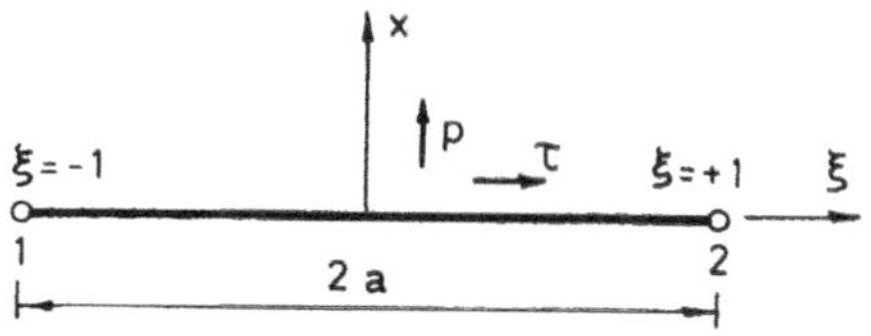

Figure 3

Then, the assumed displacement field (24) becomes

$$\begin{bmatrix} u_x \\ u_y \end{bmatrix} = \begin{bmatrix} N_1 & 0 & N_2 & 0 \\ 0 & N_1 & 0 & N_2 \end{bmatrix} \begin{bmatrix} U_x^1 \\ U_y^1 \\ U_x^2 \\ U_y^2 \end{bmatrix} \tag{43}$$

with

$$N_1 = \frac{1}{2}(1 - \xi); \quad N_2 = \frac{1}{2}(1 + \xi) \tag{44}$$

the assumed contact stress field (25) writes

$$\begin{Bmatrix} p \\ \tau \end{Bmatrix} = \begin{bmatrix} 1 & 0 \\ 0 & 1 \end{bmatrix} \begin{Bmatrix} p \\ \tau \end{Bmatrix} \tag{45}$$

and the assumed compatibility mismatch writes

$$\begin{Bmatrix} \varepsilon_p \\ \varepsilon_\tau \end{Bmatrix} = \begin{bmatrix} 1 & 0 \\ 0 & 1 \end{bmatrix} \begin{Bmatrix} \varepsilon_p \\ \varepsilon_\tau \end{Bmatrix} \tag{46}$$

with p, τ the contact pressure ans shear stress; ε_p, ε_τ the corresponding mismatch.

We assume for simplicity that the motion of the foundation is a simple translation with components u_{Fx}, u_{Fy}.

In this case, a straightforward calculation gives

$$\boldsymbol{k}_C = \frac{a}{2}\begin{bmatrix} K_p & 0 & K_p & 0 \\ 0 & K_\tau & 0 & K_\tau \\ K_p & 0 & K_p & 0 \\ 0 & K_\tau & 0 & K_\tau \end{bmatrix}, \quad \boldsymbol{f}_C = a \begin{bmatrix} K_p u_{Fx} - \lambda_p \\ K_\tau u_{Fy} - \lambda_\tau \\ K_p u_{Fx} - \lambda_p \\ K_\tau u_{Fy} - \lambda_\tau \end{bmatrix} \tag{47}$$

Applying (35), we get

$$r = \begin{bmatrix} \varepsilon_p \\ \varepsilon_\tau \end{bmatrix} = \begin{bmatrix} \frac{1}{2}\ (u_x^1 + u_x^2) - u_{Fx} \\ \frac{1}{2}\ (u_y^1 + u_y^2) - u_{Fy} \end{bmatrix} \tag{48}$$

This means that the discretization parameters of the compatibility mismatch can be interpreted as the difference of displacement between the foundation and the center of the element. The same result would be obtained with a compatible contact element derived from (23) in which the integrals are calculated numerically with only one integration point at the element center.
The application (39) gives

$$q = \begin{bmatrix} p \\ \tau \end{bmatrix} = \begin{bmatrix} K_p\ [\frac{1}{2}\ (U_x^1 + U_x^2) - u_{Fx}] + \lambda_p \\ K_\tau\ [\frac{1}{2}\ (U_y^1 + U_y^2) - u_{Fy}] + \lambda_\tau \end{bmatrix} = \begin{bmatrix} K_p \varepsilon_p + \lambda_p \\ K_\tau \varepsilon_\tau + \lambda_\tau \end{bmatrix} \tag{49}$$

This shows that the difference between the contact stresses p, τ and the Lagrange multipliers λ_p, λ_τ is compensated by the penalty stresses (9) evaluated at the center of the element. In other words, the compatibility mismatch at iteration (k+1) is due to the fact that the Lagrange multipliers computed by (7) from the results of iteration k are not exactly equal to the contact stresses which ensure equilibrium at iteration (k+1). Clearly, as the Newton-Raphson iterative process tends to convergence, the modification of the Lagrange multipliers between two successive iterations becomes negligible and, consequently, the compatibility mismatch tend to zero, even for small values of the penalty coefficients.

CONCLUSION

This paper has presented the general formulation of mixed contact elements in the frame of the augmented Lagrangian approach. Functionals with different degrees of complexity are proposed and the general derivation of mixed contact elements is developed. As an illustration, it is applied to the simple case of a three field contact element with linear displacement field, constant contact stresses and constant compatibility mismatch.. It is remarked that this element is equivalent to a compatible contact element in which a one point reduced integration scheme is used.

REFERENCES

T. Belytschko, J. Ong, W.K. Liu, J.M. Kennedy, (1984), Hourglass control in linear and nonlinear problems,Comp. Meth. Appl. Mech. Eng., vol. 43, 251-276.

S. Cescotto and R. Charlier, (1993), Frictional contact elements based on mixed variational principles, Int. Jl. Num. Meth. Eng., vol. 36, 1681-1701.

R. Charlier, (1987), Approche unifiée de quelques problèmes non linéaires de mécanique des milieux continus par éléments finis, Ph.D. Thesis, Dept. M.S.M. Université de Liège.

Ph. Jetteur and S. Cescotto, (1991), A mixed finite element for the analysis of large inelastic strains, Int. Jl. Num. Meth. Eng., vol. 31, 229-239.

DECOMPOSITION METHODS AND MIXED FINITE ELEMENT APPROXIMATIONS OF ADHERENCE AND CONTACT PROBLEMS IN FINITE VISCOELASTICITY

Gonzalo Alduncin

Instituto de Geofísica, UNAM
Coyoacán, 04510-México D.F., Mexico

INTRODUCTION

In this article, we are concerned with the application of decomposition methods and iterative numerical algorithms for mixed finite element approximations of constrained problems in finite elasticity. As model problems, we consider adherence and contact problems in viscoelasticity, in accordance with the constitutive model studied by Le Tallec et al. (1993), and the mechanical interaction models treated by Le Tallec and Lotfi (1988) and Curnier et al. (1992). Other constitutive models, like elastoplasticity (Eve et al., 1990; Eve and Reddy, 1992) could also be treated by the same methods.

Operator decomposition methods based on resolvent mappings and fixed point problem characterizations have been proposed as generalizations of Gabay's algorithms (Gabay, 1982) for discrete mixed variational inequalities. For maximal monotone mixed models, with subdifferential dual component, such methods produce in a systematical manner iterative algorithms of the Uzawa's and penalty-duality type, generalizing processes in particular derived from augmented Lagrangian formulations. They have been recently applied to contact problems in elastoviscoplasticity (Alduncin, 1992), with small displacements and strains.

As is well known (Ciarlet, 1988), one of the difficulties of finite elasticity problems is that their deformation constraints and constitutive potentials are not convex. Hence, special strategies have to be conceived for their mixed variational formulations and numerical treatment (Glowinski and Le Tallec, 1989). Accordingly, without a maximal monotone structure, the resolvent methodology (Alduncin, 1992) can not be applied directly and we should be content with the strategy of combining penalty-duality formulations and resolvent-proximal point approximations.

On the other hand, in order to extend the decomposition methodology to the spatial geometry of the problems, we are also interested in domain decomposition formulations. To that purpose, we characterize interface continuity constraints of nonoverlapping domain decompositions in terms of synchronizing primal and dual monotone subdifferential equations. In this manner, we are able to costruct augmented primal mixed variational formulations of decomposed problems, the basis for mixed finite element approximations and iterative numerical algorithms.

Contact Mechanics, Edited by M. Raous *et al.*
Plenum Press, New York, 1995

PHYSICAL MODELS

Let us consider a physical body whose material points are identified with spatial positions of a reference configuration, Ω, embedded in the three-dimensional Euclidean space $\mathcal{E}$. Since we shall be interested only in quasistatic processes, we assume that the reference configuration Ω coincides with an initial stress-free configuration of the body, which undergoes finite deformations described by smooth invertible mappings $\chi(\cdot, t) : \Omega \longrightarrow \mathcal{E}$, at each time $t \in (0, T)$, $T > 0$. The material gradient of χ will be denoted by $\mathbf{F}$ and assumed to satisfy the local impenetrability constraint

$$J(x,t) \equiv \det \mathbf{F}(x,t) > 0, \quad \forall\, (x,t) \in \Omega \times (0,T). \tag{1}$$

Defining the trajectory of the body by $\mathcal{T} = \{(y,t) : y \in \Omega_t = \chi(\Omega, t),\ t \in (0,T)\}$, let $\mathbf{b} : \mathcal{T} \longrightarrow \mathcal{V}$ be a prescribed body force vector field and $\mathbf{T} : \mathcal{T} \longrightarrow \mathcal{L}(\mathcal{V}, \mathcal{V})$ the Cauchy stress tensor field. Let $\rho_o : \Omega \longrightarrow (0, +\infty)$ denote the reference mass density of the body. Then, introducing the reference fields of body force, $\mathbf{b}_o = J\mathbf{b}(\chi, \cdot)$, and Piola-Kirchhoff stress, $\mathbf{S} = \mathbf{T}(\chi, \cdot)\mathrm{Cof}\, \mathbf{F}$, where $\mathrm{Cof}\, \mathbf{F} = J\mathbf{F}^{-T}$, the fundamental principles of balance of momentum for quasistatic processes are given locally by the field equations

$$\left.\begin{aligned} -\mathrm{div}\mathbf{S} &= \mathbf{b}_o, \\ \mathbf{S}\mathbf{F}^T &= \mathbf{F}\mathbf{S}^T, \end{aligned}\right\} \quad \text{in } \Omega \times (0,T). \tag{2}$$

We are then interested in determining quasistatic processes $\{\chi, \mathbf{S}\}$, under constraint (1), governed by equations (2), characterized by viscoelastic constitutive equations, and subjected to boundary mechanical contact conditions and constraints. As boundary conditions, we introduce the classical ones of prescribed place and material surface traction on disjoint parts of the boundary, $\partial\Omega$,

$$\mathbf{u} = \hat{\mathbf{u}}, \quad \text{on } \partial\Omega_D \times (0,T), \tag{3}$$

$$\mathbf{S}\mathbf{n}_o = \hat{\mathbf{s}}, \quad \text{on } \partial\Omega_N \times (0,T), \tag{4}$$

where $\mathbf{u} \equiv \chi - x$ corresponds to the displacement vector field and $\mathbf{n}_o$ denotes the material outward unit normal vector.

A Viscoelastic Constitutive Model

We consider the finite viscoelastic constitutive model studied by Le Tallec (1993). Therein, as internal variables, especific free energy potential and dissipation function, one has, respectively:

$$\left.\begin{aligned} &\{\mathbf{C}, \mathbf{C}_v\}, \\ &\mathcal{F}(\mathbf{C}, \mathbf{C}_e) = \mathcal{F}_o(\mathbf{C}) + \mathcal{F}_e(\mathbf{C}_e), \\ &\mathcal{D}(\dot{\mathbf{C}}, \dot{\mathbf{C}}_v) = \mathcal{D}(\dot{\mathbf{C}}_v). \end{aligned}\right\} \tag{5}$$

Here $\mathbf{C} = \mathbf{F}^T\mathbf{F}$ is the Cauchy-Green strain tensor and, with respect to the usual microscopic multiplicative splitting of deformation $\chi = \chi_e \circ \chi_v$, $\mathbf{C}_v$ corresponds to the viscoelastic strain component. Also, assuming viscous rotations $\mathbf{R}_v \equiv \mathbf{0}$, the elastic strain component $\mathbf{C}_e = \sqrt{\mathbf{C}_v^{-1}}\, \mathbf{C}\, \sqrt{\mathbf{C}_v^{-1}}$. $\mathcal{F}_o$ is the elastic stored energy function, which can be identified with hyperelastic potentials, and $\mathcal{F}_e$ is the viscous stored energy function.

Thus, for incompressible materials, upon the introduction of the viscoelastic variable

$$\mathbf{A} \equiv \mathbf{C}_v^{-1}, \tag{6}$$

and the particular thermodynamically admissible dissipation function

$$\mathcal{D}(\dot{\mathbf{C}}_v) = -\nu \dot{\mathbf{A}}, \tag{7}$$

a final form of the viscoelastic constitutive model for large deformations turns out to be (Le Tallec, 1993):

$$\left.\begin{aligned} &\mathbf{S} = \mathbf{F}\{2\partial\mathcal{F}/\partial\mathbf{C}(\mathbf{C},\mathbf{A}) - p\mathbf{C}^{-1}\}, \\ &-\partial\mathcal{F}/\partial\mathbf{A}(\mathbf{C},\mathbf{A}) + \nu(\dot{\mathbf{A}})^{-1} - q\mathbf{A}^{-1} = 0, \\ &\det\mathbf{F} = 1 = \det\mathbf{A}, \end{aligned}\right\} \text{ in } \Omega \times (0,T). \tag{8}$$

Here p and q correspond to the elastic and viscoelastic pressure fields. For compressible materials, where function p and constraint $\det\mathbf{F} = 1$ are eliminated, viscoelastic incompressibility must be retained.

Unilateral Contact without Friction

Next, let us assume that the trajectory of the viscoelastic body is constrained by the presence of a rigid target body. Let $\partial\Omega_C$ denote the potential contact surface of the reference configuration of the body, disjoint to the surfaces $\partial\Omega_D$ and $\partial\Omega_N$, and Γ_C the potential target surface with outward unit normal vector $\mathbf{m}$. Also let $\mathbf{n}$ denote the outward unit normal vector to the actual potential contact surface $\chi(\partial\Omega_C, t)$ and τ the corresponding tangential unit vector. Thus, following Curnier et al. (1992), we utilize the projection operator $Proj_{\Gamma_C}$ onto the target surface to model the unilateral contact constraint without friction, defining the signed distance functions, for $(x,t) \in \partial\Omega_C \times (0,T)$,

$$\left.\begin{aligned} &d_{\mathbf{m}}(\chi(x,t),\Gamma_C) = \{Proj_{\Gamma_C}(\chi(x,t)) - \chi(x,t)\} \cdot \{-\mathbf{m}(Proj_{\Gamma_C}(\chi(x,t))\}, \\ &\overline{d}_{\mathbf{m}}(x,\Gamma_C) = \{Proj_{\Gamma_C}(\chi(x,t)) - x\} \cdot \{-\mathbf{m}(Proj_{\Gamma_C}(\chi(x,t))\}. \end{aligned}\right\} \tag{9}$$

In subdifferential form, I_{R^+} standing for the indicator function of the convex R^+, the model is then expressed by

$$\left.\begin{aligned} &\mathbf{S}\mathbf{n}_o + \{\mathbf{S}\mathbf{n}_o \cdot \mathbf{m}(Proj_{\Gamma_C} \circ \chi)\}\mathbf{m}(Proj_{\Gamma_C} \circ \chi) = \mathbf{0}, \\ &-\mathbf{S}\mathbf{n}_o \cdot \mathbf{m}(Proj_{\Gamma_C} \circ \chi) \in \partial I_{R^+}(d_{\mathbf{m}}(\chi,\Gamma_C)), \end{aligned}\right\} \text{ on } \partial\Omega_C \times (0,T). \tag{10}$$

In terms of the normal version of the boundary displacement, $u_{\mathbf{m}} \equiv \mathbf{u} \cdot \mathbf{m}(Proj_{\Gamma_C} \circ \chi)$, the above subdifferential equation can be rewritten in the following mixed form of primal type

$$\left.\begin{aligned} &d = u_{\mathbf{m}} + \overline{d}_{\mathbf{m}}(\cdot,\Gamma_C), \\ &\lambda_{\mathbf{m}} = -\mathbf{S}\mathbf{n}_o \cdot \mathbf{m}(Proj_{\Gamma_C} \circ \chi), \\ &-\lambda_{\mathbf{m}} \in \partial I_{R^+}(d), \end{aligned}\right\} \text{ on } \partial\Omega_C \times (0,T). \tag{11}$$

An Adherence Model

As another kind of mechanical boundary interaction, let us introduce the adherence model of Frémond (1987) and treated numerically in finite elasticity by Le Tallec and Lotfi (1988). Thus, we assume in addition that the body is glued to a rigid body through the complementary part of its boundary $\partial\Omega_A = \partial\Omega \backslash \{\partial\Omega_D \cup \partial\Omega_N \cup \partial\Omega_C\}$, in accordance with an adhesion intensity function, β, such that

$$\left.\begin{array}{l} 0 \leq \beta \leq 1, \quad \text{on } \partial\Omega_A \times (0,T), \\ \beta(\cdot,0) = \beta_o, \quad \text{on } \partial\Omega_A, \end{array}\right\} \tag{12}$$

and under the impenetrability constraint

$$u_{\mathbf{n}_o} \equiv \mathbf{u} \cdot \mathbf{n}_o \leq 0, \quad \text{on } \partial\Omega_A \times (0,T). \tag{13}$$

For this model an adhesion potential energy is associated, relative to the boundary displacement, by

$$J_a(\mathbf{u},\beta) = -\int_{\partial\Omega_A} w\ \beta\ d\partial\Omega + \frac{1}{2\epsilon}\int_{\partial\Omega_A} \|\mathbf{u}\|^2\ \beta^2\ d\partial\Omega, \tag{14}$$

where w is the Dupré superficial energy and ϵ the Frémond regularizing factor that works as an adhesion force limit. Also a potential of dissipation is associated via the convex functional

$$J_d(\dot{\beta}) = \frac{1}{2}\int_{\partial\Omega_A}\int_{\partial\Omega_A} \tilde{d}(x-z)\ \dot{\beta}(x,\cdot)\ \dot{\beta}(z,\cdot)\ dx\ dz. \tag{15}$$

DOMAIN DECOMPOSITION FORMULATIONS

One of the central aspects of this paper is to show how nonoverlapping domain decomposition formulations can be systematically produced in constrained finite elasticity. The essential idea, suitable for numerical applications, is to model interface continuity constraints in terms of a dual subdifferential equation. This technique can be regarded as an extension of dual decomposition methods (Le Tallec, 1994) to general nonlinear constrained problems.

Let the reference configuration of the problem be decomposed by means of disjoint and connected subdomains,

$$\bar{\Omega} = \bigcup_{l=1}^{L} \bar{\Omega}_l, \tag{16}$$

and let the internal boundaries or interfaces be denoted by

$$\Gamma_l = \partial\Omega_l \backslash \partial\Omega, \quad \Gamma_{lm} = \Gamma_l \cap \Gamma_m, \quad 1 \leq l \neq m \leq L. \tag{17}$$

In the context of finite elasticity, the interface transition conditions of continuity are imposed to the displacement and Cauchy surface traction fields, $(\{\mathbf{u}_l\}, \{\mathbf{t}_l\})$. From the relation between the Cauchy and Piola-Kirchhoff stress vectors, $\mathbf{t}(\chi,\cdot) = j^{-1}\mathbf{s}$, where $j \equiv \|\text{Cof } \mathbf{F}\mathbf{n}_o\|$, these are expressed as follows:

$$\{\mathbf{u}_l\} \in \mathbf{C}_D = \{\{\mathbf{u}_l\} \in \prod_{l=1}^{L} \mathbf{B}(\Gamma_l) : \ \mathbf{u}_l = \mathbf{u}_m \ \text{ on } \Gamma_{lm},\ 1 \leq l \neq m \leq L\}, \tag{18}$$

$$\{j_l^{-1}\mathbf{s}_l\} \in \mathbf{C}_N^* = \{\{\mathbf{r}_l\} \in \prod_{l=1}^{L} \mathbf{B}'(\Gamma_l) : \ \mathbf{r}_l = -\mathbf{r}_m \ \text{ on } \Gamma_{lm}, \ 1 \le l \ne m \le L\}. \tag{19}$$

Here $\mathbf{C}_D$ and $\mathbf{C}_N^*$ are then the admissibility subspaces of interface displacements and suface tractions, relative to displacement local spaces $\mathbf{V}(\Omega_l)$ and corresponding Dirichlet and Neumann interface spaces $\mathbf{B}(\Gamma_l)$ and its dual $\mathbf{B}'(\Gamma_l)$, for $l = 1, 2, ..., L$.

Therefore, taking into account that the conjugate of the indicator functional $(I_{\mathbf{C}_D})^* = I_{\mathbf{C}_N^*}$, a natural way to incorporate transition constraints (18) and (19) to general constrained problems is via the equivalent subdifferential models

$$\{-j_l^{-1}\mathbf{s}_l\} \in \partial I_{\mathbf{C}_D}(\{\mathbf{u}_l\}) \iff \{\mathbf{u}_l\} \in \partial I_{\mathbf{C}_N^*}(\{-j_l^{-1}\mathbf{s}_l\}), \ \text{ on } \Gamma = \prod_{l=1}^{L} \Gamma_l. \tag{20}$$

An appropriate form of such a model for mixed formulations is its primal mixed version

$$\left.\begin{array}{l} \left.\begin{array}{l} \mathbf{w}_l = \mathbf{u}_l, \\ \mathbf{g}_l = -j_l^{-1}\mathbf{s}_l, \end{array}\right\} \ \text{on } \Gamma_l, \ 1 \le l \le L, \\ \{\mathbf{g}_l\} \in \partial I_{\mathbf{C}_D}(\{\mathbf{w}_l\}), \ \text{ on } \Gamma \end{array}\right\} \tag{21}$$

DECOMPOSED VARIATIONAL FORMULATIONS

Proceeding as usual, the variational formulation of the proposed adherence and contact problem for incompressible viscoelastic bodies, decomposed spatially in accordance with dual model (20_2), becomes,

Find $((\{\mathbf{u}_l\}, \{\beta_l\}), \{p_l\}) \in \prod_{l=1}^{L} \mathbf{K}(\Omega_l, \partial\Omega_{l,A}) \times P(\Omega_l)$ and $(\{\mathbf{A}_l\}, \{q_l\}) \in \prod_{l=1}^{L} \mathbf{W}(\Omega_l) \times Q(\Omega_l)$: for $l = 1, ..., L$,

$$(\mathbf{V}_{l,E}) \left\{ \begin{array}{l} \int_{\Omega_l} \mathbf{F}_l\{2\partial\mathcal{F}/\partial\mathbf{C}(\mathbf{C}_l, \mathbf{A}_l) - p_l\mathbf{C}_l^{-1}\} \cdot \{\nabla\mathbf{v} - \nabla\mathbf{u}_l\} \, d\Omega \\ \quad + J_{a,l}(\mathbf{v}, \mu) - J_{a,l}(\mathbf{u}_l, \beta_l) + J_{d,l}(\dot{\mu}) - J_{d,l}(\dot{\beta}_l) \\ \quad \ge \int_{\Omega_l} \mathbf{b}_o \cdot \{\mathbf{v} - \mathbf{u}_l\} \, d\Omega + \int_{\partial\Omega_{l,N}} \hat{\mathbf{s}} \cdot \{\mathbf{v} - \mathbf{u}_l\} \, d\partial\Omega - \int_{\Gamma_l} j_l \mathbf{g}_l \cdot \{\mathbf{v} - \mathbf{u}_l\} \, d\Gamma, \\ \quad \forall \ (\mathbf{v}, \mu) \in \mathbf{K}(\Omega_l, \partial\Omega_{l,A}), \\ \int_{\Omega_l} \{\det\mathbf{F}_l - 1\} \ w \ d\Omega = 0, \ \ \forall \ w \in L^2(\Omega_l), \\ \beta_l(0) = \beta_{l,0}; \end{array}\right.$$

$$(\mathbf{V}_{l,VE}) \left\{ \begin{array}{l} \int_{\Omega_l} \{-\partial\mathcal{F}/\partial\mathbf{A}(\mathbf{C}_l, \mathbf{A}_l) + \nu(\dot{\mathbf{A}}_l)^{-1} - q\mathbf{A}_l^{-1}\} \cdot \mathbf{B} = 0, \ \ \forall \ \mathbf{B} \in \mathbf{W}(\Omega_l), \\ \int_{\Omega_l} \{\det\mathbf{A}_l - 1\} \ s \ d\Omega = 0, \ \ \forall s \in L^2(\Omega_l), \\ \mathbf{A}_l(0) = \mathbf{A}_{l,0}; \end{array}\right.$$

with dual synchronizing condition

$$(\mathbf{DS}) \left\{ 0 \ge \sum_{l=1}^{L} \int_{\Gamma_l} \mathbf{u}_l \cdot \{\mathbf{c}_l - \mathbf{g}_l\} \, d\Gamma, \ \ \forall \{\mathbf{c}_l\} \in \mathbf{C}_N^*. \right.$$

Here, the local admissible displacements and adhesion intensities set is defined by

$$\begin{aligned}\mathbf{K}(\Omega_l, \partial\Omega_{l,A}) = \{(\mathbf{v}, \mu) \in \mathbf{V}(\Omega_l) \times Y(\partial\Omega_{l,A}) :\ \mathbf{v} = \hat{\mathbf{u}}_l \text{ on } \partial\Omega_{l,D}, \\ v_{\mathbf{m},l} + \overline{d}_{\mathbf{m}}(\cdot, \Gamma_{C,l}) \geq 0 \text{ on } \partial\Omega_{l,C},\ v_{\mathbf{n}_o,l} \leq 0 \text{ on } \partial\Omega_{l,A},\ 0 \leq \mu \leq 1 \text{ on } \partial\Omega_{l,A}\},\end{aligned} \tag{22}$$

and $\mathbf{V}(\Omega_l)$, $Y(\partial\Omega_{l,A})$, $P(\Omega_l)$, $\mathbf{W}(\Omega_l)$ and $Q(\Omega_l)$ are appropriate local Sobolev-Banach spaces.

For numerical purposes, we next give an equivalent mixed variational formulation, dualizing, and penalizing exactly the adherence and unilateral contact constraints, as well as the interface transition constraints. Toward this end, on the basis of primal mixed models (11) and (21), we incorporate augmented versions with fixed parameter

$$r > 0. \tag{23}$$

The heuristic of such a process comes from the methodologies of augmented Lagrangians (Le Tallec and Lotfi, 1988; Glowinski and Le Tallec, 1989; Alduncin, 1992), but we here proceed in a direct manner without evoking convex techniques.

Find $(\{\mathbf{u}_l\}, \{p_l\}) \in \prod_{l=1}^{L} \mathbf{K}_{\hat{\mathbf{u}}_l}(\Omega_l) \times P(\Omega_l)$, $(\{\mathbf{e}_l\}, \{\beta_l\}) \in \prod_{l=1}^{L} \mathbf{M}(\partial\Omega_{l,A})$, $(\{d_l\}, \{\mathbf{w}_l\}) \in \prod_{l=1}^{L} C(\partial\Omega_{l,C}) \times \mathbf{L}^2(\Gamma_l)$, $(\{\mathbf{h}_l\}, \{\lambda_{\mathbf{m},l}\}, \{\mathbf{g}_l\}) \in \prod_{l=1}^{L} \mathbf{L}^2(\partial\Omega_{l,A}) \times L^2(\partial\Omega_{l,C}) \times \mathbf{L}^2(\Gamma_l)$ and $(\{\mathbf{A}_l\}, \{q_l\}) \in \prod_{l=1}^{L} \mathbf{W}(\Omega_l) \times Q(\Omega_l)$: for $l = 1, ..., L$,

$$(\mathbf{V}^r_{l,E}) \begin{cases} \int_{\Omega_l} \mathbf{F}_l\{2\partial\mathcal{F}/\partial\mathbf{C}(\mathbf{C}_l, \mathbf{A}_l) - p_l\mathbf{C}_l^{-1}\} \cdot \nabla\mathbf{v}\, d\Omega \\ \quad = \int_{\Omega_l} \mathbf{b}_o \cdot \mathbf{v}\, d\Omega + \int_{\partial\Omega_{l,N}} \hat{\mathbf{s}} \cdot \mathbf{v}\, d\partial\Omega - \int_{\partial\Omega_{l,A}} \{\mathbf{h}_l + r(\mathbf{u}_l - \mathbf{e}_l)\} \cdot \mathbf{v}\, d\partial\Omega \\ \quad - \int_{\partial\Omega_{l,C}} \{\lambda_{\mathbf{m},l} + r(u_{\mathbf{m},l} + \overline{d}_{\mathbf{m},l} - d_l)\}\ (\mathbf{v} \cdot \mathbf{m})\, d\partial\Omega \\ \quad - \int_{\Gamma_l} j_l\{\mathbf{g}_l + r(\mathbf{u}_l - \mathbf{w}_l\} \cdot \mathbf{v}\, d\Gamma, \quad \forall\, \mathbf{v} \in \mathbf{K}(\Omega_l), \\ \int_{\Omega_l} \{\det\mathbf{F}_l - 1\}\ \delta\, d\Omega = 0, \quad \forall\, \delta \in L^2(\Omega_l), \\ \beta_l(0) = \beta_{l,0}; \end{cases}$$

$$(\mathbf{V}^r_{l,A}) \begin{cases} J_{a,l}(\mathbf{e}_l, \beta_l) + J_{d,l}(\dot{\beta}_l) - \int_{\partial\Omega_{l,A}} \mathbf{h}_l \cdot \mathbf{e}_l\, d\partial\Omega + r/2 \int_{\partial\Omega_{l,A}} \|\mathbf{u}_l - \mathbf{e}_l\|^2\, d\partial\Omega \\ \quad \geq J_{a,l}(\mathbf{b}, \mu) + J_{d,l}(\dot{\mu}) - \int_{\partial\Omega_{l,A}} \mathbf{h}_l \cdot \mathbf{b}_l\, d\partial\Omega + r/2 \int_{\partial\Omega_{l,A}} \|\mathbf{u}_l - \mathbf{b}_l\|^2\, d\partial\Omega \\ \quad \forall\, (\mathbf{b}, \mu) \in \mathbf{M}(\partial\Omega_{l,A}); \end{cases}$$

$$(\mathbf{V}^r_{l,P}) \left\{\ 0 \geq - \int_{\partial\Omega_{l,C}} \{\lambda_{\mathbf{m},l} + r(u_{\mathbf{m},l} + d_{\mathbf{m},l} - d_l)\}\ \{c - d_l\}\, d\partial\Omega, \quad \forall\, c \in C(\partial\Omega_{l,C}), \right.$$

$$(\mathbf{V}^r_{l,D}) \begin{cases} \int_{\partial\Omega_{l,A}} \{\mathbf{u}_l - \mathbf{e}_l\} \cdot \mathbf{y}\, d\partial\Omega = 0, \quad \forall\, \mathbf{y} \in \mathbf{L}^2(\partial\Omega_{l,A}), \\ \int_{\partial\Omega_{l,C}} \{u_{m,l} + \overline{d}_{m,l} - d_l\}\ \eta\, d\partial\Omega = 0, \quad \forall\, \eta \in L^2(\partial\Omega_{l,C}), \\ \int_{\Gamma_l} \{\mathbf{u}_l - \mathbf{w}_l\}\ \mathbf{z}\, d\Gamma = 0, \quad \forall\, \mathbf{z} \in \mathbf{L}^2(\Gamma_l); \end{cases}$$

$$(\mathbf{V}_{l,VE})(\{\mathbf{A}_l\}, \{q_l\});$$

with augmented primal synchronizing condition

$$(\mathbf{PS}_r)\left\{0 \geq \sum_{l=1}^{L} \int_{\Gamma_l} \{\mathbf{g}_l + r(\mathbf{u}_l - \mathbf{w}_l)\} \cdot \{\mathbf{c}_l - \mathbf{w}_l\}\, d\Gamma, \quad \forall\{\mathbf{c}_l\} \in \mathbf{C}_D,\right.$$

and local admissibility sets

$$\left.\begin{array}{l} \mathbf{K}(\Omega_l) = \{\mathbf{v} \in \mathbf{V}(\Omega_l) : \ \mathbf{v} = \hat{\mathbf{u}}_l \text{ on } \partial\Omega_{l,D}\}, \\ \mathbf{M}(\partial\Omega_{l,A}) = \{(\mathbf{b}, \mu) \in \mathbf{L}^2(\partial\Omega_{l,A}) \times Y(\partial\Omega_{l,A}) : b_{\mathbf{n}_o,l} \leq 0, 0 \leq \mu \leq 1 \text{ on} \partial\Omega_{l,A}\}, \\ C(\partial\Omega_{l,C}) = \{c \in L^2(\partial\Omega_{l,C}) : \ c \geq 0 \text{ on } \partial\Omega_{l,C}\}. \end{array}\right\} \quad (24)$$

MIXED FINITE ELEMENT APPROXIMATIONS

On the basis of the above augmented mixed variational formulation of the problem, finite element approximations can be introduced systematically taking advantage of its decomposed structure. Here, we shall just mention that, for example, an important strategy for subproblems $(\mathbf{V}_{l,A}^r)$ and $(\mathbf{V}_{l,VE}^r)$, of adherence and viscoelasticity, is to introduce piecewise constant approximations (Le Tallec and Lotfi, 1988; Le Tallec et al., 1993) that permite a further spatial decomposition at the subdomain level. Also, for the elasticity subproblem $(\mathbf{V}_{l,E}^r)$, we can even introduce an additional operator decomposition of its hyperlastic constitutive component (Glowinski and Le Tallec, 1989), dualizing, and penalizing as done before.

On the other hand, following the spirit of resolvent methods (Alduncin, 1992), dual subproblems $(\mathbf{V}_{l,D}^r)$ can be interpreted as fixed point problems in terms of resolvent operators, which in turn correspond to proximation and, in particular, projection mappings. In this case the situation is trivial since the resolvent operators are simply the identity operators (the resolvents of the subdifferentials of the indicator functions of the whole vector spaces). So, denoting by $\mathbf{u}_l^c$ and $\mathbf{h}_l^c$ the finite element coordinate vectors of the local displacement and adherence traction reaction fields, respectively, and by $\mathbf{C}_{\mathbf{u},\mathbf{h}}^l$ the corresponding coupling matrix, and similarly for the other pairs of primal and dual local approximate fields and coupling matrices, semidiscrete dual subproblems $(\mathbf{V}_{l,D}^r)$ have the following interpretation:

$$(\mathbf{V}_{l,D}^r)_h \left\{ \begin{array}{l} \mathbf{h}_l^c = \mathbf{h}_l^c + r(\mathbf{C}_{\mathbf{u},\mathbf{h}}^l \mathbf{u}_l^c - \mathbf{C}_{\mathbf{e},\mathbf{h}}^l \mathbf{e}_l^c), \\ \lambda_{\mathbf{m},l}^c = \lambda_{\mathbf{m},l}^c + r(\mathbf{C}_{u\mathbf{m},\lambda\mathbf{m}}^l u_{\mathbf{m},l}^c + \mathbf{C}_{\overline{d}\mathbf{m},\lambda\mathbf{m}}^l \overline{d}_{\mathbf{m},l}^c - \mathbf{C}_{d,\lambda\mathbf{m}}^l d_l^c), \\ \mathbf{g}_l^c = \mathbf{g}_l^c + r(\mathbf{C}_{\mathbf{u},\mathbf{g}}^l \mathbf{u}_l^c - \mathbf{C}_{\mathbf{w},\mathbf{g}}^l \mathbf{w}_l^c). \end{array} \right.$$

ITERATIVE ALGORITHMS

Then, the remaning steps in the numerical resolution of the problem are the numerical time integration and introduction of iterative algorithms. For time discretization implicit schemes are the proper ones, as the implicit scheme of Euler and the general midpoint scheme for $\theta \in (0, 1]$.

From the very decomposed structure of the resulting fully discrete finite element models, and taking into account the semidiscrete interpretation $(\mathbf{V}_{l,D}^r)_h$ of the dual subproblems, iterative algorithms are easy to be proposed. For instance, for a fully discrete version of the augmented mixed model, with initial conditions at time $t_n = n\Delta t_n$, at time t_{n+1} we have the algorithm:

Known $(\{\mathbf{A}_l\}, \{q_l\})^m$, $(\{\mathbf{u}_l\}, \{p_l\})^m$ and $(\{\mathbf{h}_l\}, \{\lambda_{\mathbf{m},l}\}, \{\mathbf{g}_l\})^m$, $m \geq 0$, calculate $(\{\mathbf{A}_l\}, \{q_l\})^{m+1}$, $(\{\mathbf{e}_l\}, \{\beta_l\})^{m+1}$, $\{d_l\}^{m+1}$, $(\{\mathbf{u}_l\}, \{p_l\})^{m+1}$, $(\{\mathbf{h}_l\}, \{\lambda_{\mathbf{m},l}\}, \{\mathbf{g}_l\})^{m+1}$ and $\{\mathbf{w}_l\}^{m+1}$:

$$(\mathbf{V}_{l,VE})_h^{n+1}(\{\mathbf{A}_l\}, \{q_l\})^{m+1};$$

$$(\mathbf{V}_{l,A}^r)_h^{n+1}(\{\mathbf{e}_l\}, \{\beta_l\})^{m+1};$$

$$(\mathbf{V}_{l,P}^r)_h^{n+1}(\{d_l\})^{m+1};$$

$$(\mathbf{V}_{l,E}^r)_h^{n+1}(\{\mathbf{u}_l\}, \{p_l\})^{m+1};$$

$$(\mathbf{V}_{l,D}^r)_h^{n+1}(\{\mathbf{h}_l\}, \{\lambda_{\mathbf{m},l}\}, \{\mathbf{g}_l\})^{m+1};$$

with synchronizing condition

$$(\mathbf{PS}_r)_h^{n+1}(\{\mathbf{w}_l\})^{m+1}.$$

Similarly, other algorithms can be proposed following the structure of the penalty-duality algorithms in numerical convex analysis (Alduncin, 1992).

REFERENCES

Alduncin, G., 1992, Augmented Lagrangian methods for the quasistatic viscoelastic two-body contact problem with friction, *in*: "Contact Mechanics," A. Curnier, ed., PPUR, Lausanne: 337-359.

Ciarlet, P.G., 1988, "Mathematical Elasticity," North-Holland, Amsterdam.

Curnier, A., He, Q.-C., and Telega, J.J., 1992, Formulation of unilateral contact between two elastic bodies undergoing finite deformations, *C.R. Acad. Sci. Paris* 314-II: 1-6.

Eve, R.A., Gültop, T., and Reddy, B.D., 1990, An internal variable finite-strain theory of plasticity within the framework of convex analysis, *Quart. Appl. Math.* 48: 625-643.

Eve, R.A., and Reddy, B.D., 1992, Algorithms for the solution of problems of finite-strain elasto-plasticity, *in*: "Computational Plasticity,"D.R.J. Owen, E. Oñate and E. Hinton, eds., Pineridge Press / CIMNE, Barcelona: 103-115.

Frémond, M., 1987, Adhérence de solides, *J. Méc. Théor. Appl.* 6: 383-407.

Gabay, D., 1982, Application de la méthode des multiplicateurs aux inéquations variationnelles, *in*: "Méthodes de Lagrangien Augmenté," M. Fortin and R. Glowinski, eds., Dunod-Bordas, Paris: 279-307.

Glowinski, R., and Le Tallec, P., 1989, "Augmented Lagrangian and Operator-Splitting Methods in Nonlinear Mechanics," SIAM, Philadelphia.

Le Tallec, P., 1994, Domain decomposition methods in computational mechanics, *Comput. Mech. Adv.* 1: 121-220.

Le Tallec, P., and Lotfi, A., 1988, Decomposition methods for adherence problems in finite elasticity, *Comp. Meth. Appl. Mech. Engrg.* 68: 67-82.

Le Tallec, P., Rahier, C., and Kaiss, A., 1993, Three-dimensional incompressible viscoelasticity in large strains: Formulation and numerical approximation, *Comp. Meth. Appl. Mech. Engrg.* 109: 233-258.

AN EFFECTIVE PLATE ELEMENT FOR CONTACT PROBLEMS

Ferdinando Auricchio[1] and Elio Sacco[2]

[1]University of Rome "Tor Vergata" (Italy)
[2]University of Cassino (Italy)

INTRODUCTION

In the present work the attention is devoted to unilateral frictionless contact problems within the class of flat structures. We show that the use of inappropriate plate elements, such as the displacement-based Lagrangian isoparametric elements relative to a first-order shear deformation theory (Q4 and Q9 (Averill 1990), may induce numerical pathologies (Auricchio 1994b). As a consequence, effective and reliable elements must be used, such as the new quadrilateral plate element Q4-LIM (Auricchio 1994).

The augmented Lagrangian method for unilateral contact problems is developed for the two displacement-based Q4 and Q9 elements and for the new mixed Q4-LIM element. Two kind of unilateral problems are approached: a plate constrained through an unilateral edge support and a plate seating in its undeformed configuration at a given distance g from a rigid support. Numerical results are carried out and are compared with analytical solutions.

FORMULATION OF THE PROBLEM

The plate is modeled using the Reissner-Mindlin plate theory, which includes both bending deformation and the primary effects of transverse shear deformation. Moreover, a field a a boundary constraint are introduced , corresponding to the unilateral contact conditions. For the field constraint the transversal displacement w must satisfy the inequality:

$$w - g \leq 0 \quad \text{in} \quad \mathcal{A} \tag{1}$$

where $\mathcal{A}$ is the mid-plane of the plate. Furthermore, the mid-plane boundary $\partial\mathcal{A}$ is split in two parts $\partial_1\mathcal{A}$ and $\partial_2\mathcal{A}$, such that $\partial\mathcal{A} = \partial_1\mathcal{A} \cup \partial_2\mathcal{A}$ and $\partial_1\mathcal{A} \cap \partial_2\mathcal{A} = \emptyset$. On $\partial_1\mathcal{A}$ classical mixed boundary conditions are imposed, i.e. displacements or tractions are specified; on $\partial_2\mathcal{A}$ unilateral contact is possible, expressed in the form:

$$w \leq 0 \quad \text{on} \quad \partial_2\mathcal{A} \tag{2}$$

The model can be cast in a variational scheme, introducing the *augmented* functional (Auricchio 1994b, Alart 1991, Klarbring 1992):

$$\begin{aligned} \Lambda_{aug}(w, \boldsymbol{\theta}, \mathbf{S}, r, p) \quad = \quad & \Lambda(w, \boldsymbol{\theta}, \mathbf{S}, r, p) \\ & + \frac{k_\alpha}{2} \int_A (w-g)^2 dA - \frac{1}{2k_\alpha} \int_A \left\{ [p - k_\alpha (w-g)]^+ \right\}^2 dA \\ & + \frac{k_\beta}{2} \int_{\partial_2 A} w^2 ds - \frac{1}{2k_\beta} \int_{\partial_2 A} \left[(r - k_\beta w)^+ \right]^2 ds \end{aligned} \tag{3}$$

Contact Mechanics, Edited by M. Raous *et al.*
Plenum Press, New York, 1995

where Λ is the usual Lagrangian functional, $\boldsymbol{\theta}$ is the midplane rotation, $\mathbf{S}$ is the shear resultant, p and r are respectively the field and the edge pressures due the unilateral contact, introduced as the Lagrangian multiplier corresponding to the constraints 1 and 2. Note that k_α and k_β are penalty parameters and the superscript + indicates the positive part. From the augmented Lagrangian formulation it is possible to develop an efficient iterative solution algorithm as follow:

i) assume the pressures, p_{n-1} and r_{n-1}, as known and solve the nonlinear minimax problem 3 in terms of w, $\boldsymbol{\theta}$ and $\mathbf{S}$ (for example using the Newton-Raphson method);

ii) update the pressures: $p_n = [p_{n-1} - k_\alpha (w_n - g)]^-$, $r_n = (r_{n-1} - k_\beta w_n)^-$

iii) go to the step i) until convergence is reached.

The augmented Lagrangian formulation condenses in itself the advantages of both the Lagrangian and the penalty approach, without preserving the disadvantages of both methods. In fact, at each iteration, since the pressures are assumed to be known, we must solve a problem with a minor number of unknowns than in the Lagrangian approach. At the same time, k_α and k_β are not required to reach high values as in the classical penalty approach; hence, the problem to solve is well conditioned.

APPLICATIONS

The Q4-LIM element. This is a four-node element, based on a mixed formulation in terms of transverse displacement, rotations and shear resultants (Auricchio 1994). The transverse displacement interpolation is taken bi-linear in the nodal parameters, enriched with linked quadratic functions of the nodal rotations. The interpolation for the rotational field is bi-linear in the nodal parameters, with added internal degrees of freedom associated with bubble functions. The shear interpolation is linear within each element through internal parameters. Both the internal rotations and the shear parameters are statically condensed at the element level. Accordingly, as for Q4 and Q9, also the Q4-LIM stiffness matrix can be written only in terms of displacements and rotations.

Comparison between plate elements. We consider a square plate of side L, thickness h, subjected to a pointwise force F applied at the center, with $\partial_2\mathcal{A} = \partial\mathcal{A}$. The following non-dimensional parameters are adopted: $h/L = 0.01$, $FL/D = 0.4$, $\nu = 0.3$, $k_\beta h^3/D = 1E-03$, where D is the plate bending stiffness and ν is the Poisson ratio.

The computations are performed using the elements Q4, Q9 (with both selective and full integration formulas) and Q4-LIM. Due to the symmetry, only a quarter of the plate is considered. A regular 16×16 mesh is used for the analyses with the four node elements, while a 8×8 mesh is used with the nine-node element. In Figure 1 the vertical displacements along the boundary ($x = L/2$ and $y \in [0, L/2]$) is plotted after the first iteration of the first augmentation. If selective integration is used with the Q4 or Q9 elements, the displacement along the supported edge have the shape of an unreal wave. This result is a direct consequence of the zero-energy modes, present in the displacement-based elements when reduced integration formulas are used on the shear terms of the stiffness matrices, and it makes impossible to converge to the solution. Such behavior does not depend on the value of the penalty parameter. On the other hand, if full integration is used, the Q4 and Q9 elements do not produce the wave along the boundary, but they suffer the *locking* effect. On the contrary, the Q4-LIM element does not present any wave along the boundary, it is locking-free, reliable and computationally efficient (Auricchio 1994). Hence, in the remaining examples we consider only the Q4-LIM element.

Beam problem. We consider a beam sliding at $x = 0$ (i.e. $w'(0) = w'''(0) = 0$), simply supported at $x = L$ (i.e. $w(L) = w''(L) = 0$), subjected to a constant distributed load q. The beam lies at a gap distance g from a rigid support. We choose $g/h = 3$, $qL^3/D = 1.50$, $k_\alpha h^4/D = 1E-04$. The finite element computations are carried out adopting a 16×1 mesh.

Table 1. Edge contact plate problem.

It.#	w_1/h	w_5/h	w_6/h
1	5.4563E-01	3.0821E-02	-1.5228E-01
4	5.1659E-01	3.3329E-04	-1.5787E-01
7	5.1616E-01	-4.7225E-05	-1.5405E-01
10	5.1614E-01	-5.6596E-06	-1.5374E-01
13	5.1613E-01	-3.3759E-06	-1.5357E-01
16	5.1613E-01	-1.7898E-05	-1.5378E-01

Table 2. Field contact plate problem.

g/h	It.#	w_1/h	w_2/h	w_3/h	w_4/h
3	1	3.0959E+00	3.1354E+00	3.1368E+00	2.2831E+00
	4	3.0032E+00	2.9984E+00	2.9912E+00	2.2140E+00
	7	2.9999E+00	3.0000E+00	2.9873E+00	2.2144E+00
	10	3.0000E+00	3.0000E+00	2.9866E+00	2.2145E+00
	13	3.0000E+00	3.0000E+00	2.9865E+00	2.2146E+00
	16	3.0000E+00	3.0000E+00	2.9866E+00	2.2146E+00
2	1	2.0916E+00	2.1163E+00	2.1489E+00	1.6476E+00
	4	2.0023E+00	1.9992E+00	2.0060E+00	1.5785E+00
	7	2.0000E+00	2.0001E+00	2.0004E+00	1.5778E+00
	10	2.0000E+00	2.0000E+00	1.9998E+00	1.5786E+00
	13	2.0000E+00	2.0000E+00	1.9998E+00	1.5794E+00
	16	2.0000E+00	2.0000E+00	1.9999E+00	1.5799E+00
1	1	1.0930E+00	1.1032E+00	1.1371E+00	9.4510E-01
	4	1.0007E+00	1.0001E+00	1.0027E+00	8.7239E-01
	7	1.0000E+00	1.0000E+00	9.9985E-01	8.7137E-01
	10	1.0000E+00	1.0000E+00	9.9992E-01	8.7091E-01
	13	1.0000E+00	1.0000E+00	9.9997E-01	8.7124E-01
	16	1.0000E+00	1.0000E+00	9.9998E-01	8.7104E-01

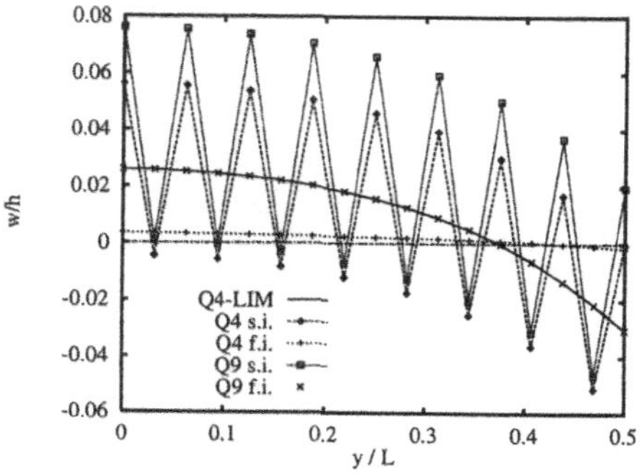

Fig.1. Simply supported plate: displacement along the boundary

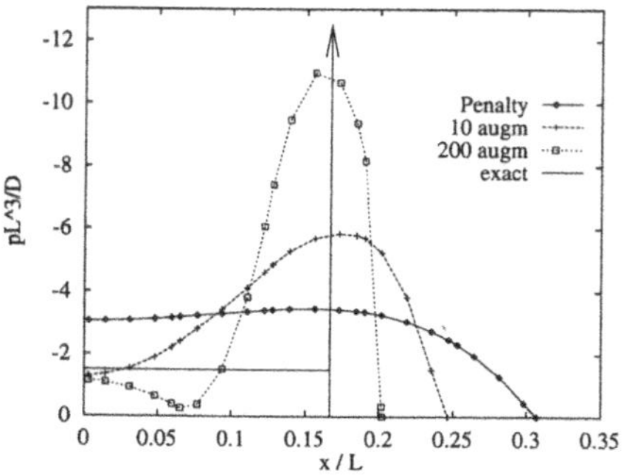

Fig.2. Sliding-simply supported beam: contact pressure

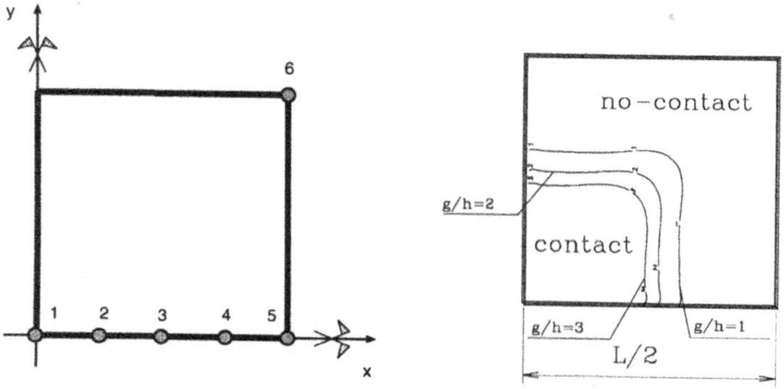

Fig.3. Simply supported plate: nodes of interest and countour plot of the contact no-contact zone.

The contact pressure along the beam axis is plotted in Figure 2 after 1, 10 and 200 augmentations. We report also the exact solution, consisting in the constant function $p = q$ in the contact zone and in a pointwise force. Note the progressive blowing-up of the pressure and the convergence to the the exact solution.

Edge contact plate problems. We consider the plate introduced in the first example with unilateral boundary constraint. The displacement of the central point (point 1 in Figure 3) and of other two points on the boundary (points 5 and 6 in Figure 3) are reported in Table 1 for $k_\beta h^3/D = 1E-03$. The solutions are obtained increasing progressively k_β ($k_\beta = 10 * k_\beta$ every three augmentations). Excellent enforcement of the constraints are obtained in just few iterations.

Field contact plate problems. We now consider the same plate, simply supported along the boundary ($w = 0$ on $\partial_1\mathcal{A} = \partial\mathcal{A}$), subject to a constant distributed load ($qL^3/D = 100$) and seating at a fixed distance g from a rigid support. The following three cases are considered: $g/h \in \{3, 2, 1\}$ The displacement of four nodes on one of the symmetry axis (point 1,2,3,4 in Figure 3) are reported in Table 2 for $k_\alpha h^4/D = 1E-03$. Even in this case, we increase progressively the penalty parameter ($k_\alpha = 10 * k_\alpha$ every three augmentations). As in the previous analysis, excellent enforcement of the constraints is obtained. In Figure 3 we plot the line defining the contact zone boundary for the three cases after 30 augmentations.

REFERENCES

R.C. Averill and J.N. Reddy, (1990) *Behavior of plate elements based on the first-order shear deformation theory*, Engineering Computations **7-1**, 57-74

F. Auricchio and E. Sacco, (1994) *An effective algorithm for plate contact problems*, to be published

F. Auricchio and R.L. Taylor, (1994) *A thick plate finite element with exact thin limit.* In press on Computer Methods in Applied Mechanics and Engineering.

P. Alart and A. Curnier, (1991) *A mixed formulation for frictional contact problems prone to Newton like solution methods*, Computer Methods in Applied Mechanics and Engineering **92**, 353-375.

A. Klarbring, (1992) *Mathematical programming and augmented Lagrangian methods for frictional contact problems*, in Proceedings Contact Mechanics International Symposium, ed. A. Curnier, Presses Polytechniques et Universitaires Romandes.

NUMERICAL APPROACH OF CONTACT USING AN AUGMENTED LAGRANGIAN METHOD

J-Ph. Bille , S. Cescotto, A.M. Habraken and R. Charlier

M.S.M. Department
University of Liège
Quai Banning, 6
B-4000 Liège, Belgium

INTRODUCTION

For most engineering applications, contact interactions between deformable bodies are usually modelled by means of given bilateral boundary conditions. Such models are too coarse when the contact region behaviour is subject to interest. In these cases where the boundary is evolving during the loading,, unilateral contact is needed. Finite elements modelling unilateral contact with friction in two-dimensional, axisymmetric and three-dimensional cases have been implemented in the finite element code LAGAMINE developed by the M.S.M. department of University of Liège (Charlier and Habraken 1990, Cescotto and Charlier 1993). They take into account large displacements and rotations between a deformed body and a so-called tool or between two deformed bodies. Until now, we have used a penalty method and a Coulomb dry friction law to describe the interface behaviour. But it is well known that penalty method suffers from ill-conditioning that worsens as penalty values are increased. Constraints are satisfied only in the limit of infinite penalty values. Thus, for many problems it may be desirable or even necessary to consider the augmented Lagrangian technique as an alternative approach capable of circumventing these difficulties.

The first approach of this method applied to contact seems to be realized by Wriggers, Simo and Taylor (1985). These authors used Uzawa's algorithm with up-dating at each iteration. Recently, Simo and Laursen (1992) have extended this technique to take into account friction and have studied the choice of the number of up-dating by step to insure a precise solution. In another way, Alart and Curnier (1991) are working on the same problem. They have pointed attention to the difficulties of convergence and they have proposed an algorithm named GNM (Generalized Newton Method). This new method uses a complete system of degrees of freedom (including Lagrangian multipliers). We will note that Alart and Curnier work with Node-Node and Node-Facet contact elements without friction. Simo and Laursen use also Node-Node contact element and give results with friction in large displacements.

What we propose here is an augmented Lagrangian method based on the Uzawa's algorithm.

UNILATERAL CONTACT PROBLEM

The unilateral contact problem between a deformable solid Ω with boundary Γ and a rigid foundation is expressed in the following variational form

$$G(\underline{u},\underline{\delta u}) - \int_{\Gamma_c} \underline{t}.\underline{\delta u} = 0 \tag{1}$$

for any virtual displacement $\underline{\delta u}$ such that $\underline{\delta u}.\underline{n} \leq 0$ on the contact surface Γ_c.
In (1), $G(\underline{u},\underline{\delta u})$ is given by

$$G(\underline{u},\underline{\delta u}) = \int_\Omega \underline{\sigma}^T.\underline{grad}(\underline{\delta u}) d\Omega - \int_\Omega \underline{F}^T.\underline{\delta u}\, d\Omega - \int_{\Gamma_\sigma} \underline{T}^T.\underline{\delta u}\, d\Gamma_\sigma \tag{2}$$

$\underline{t}$ with the contact stresses on Γ_c, $\underline{\sigma}$ the Gauchy stresses in Ω, $\underline{T}$ the applied surface tractions on Γ_σ and $\underline{n}$ the outward normal to Γ. It is the last term of (1) which is difficult to handle because Γ_c is unknown a priori and $\underline{\delta u}$ is subject to an inequality constraint on Γ_c.

PENALTY METHOD

In the penalty method, a penetration g_N between body and foundation is accepted but controled via a penalty coefficient ε_N. This coefficient is depicted at figure 1. The greater ε_N, the smaller the penetration. With this method, (1) is replaced by

$$G(\underline{u},\underline{\delta u}) - \int_{\Gamma_C} \varepsilon_N g_N\, \underline{\delta u}.\underline{n}\, d\Gamma_C = 0 \tag{3}$$

which must hold for all $\underline{\delta u}$.

In this way the inequality constraint is relaxed. This method is very simple to implement in a finite element code because it is not necessary to change the global structure of the code. A too small penalty coefficient induces imprecisions, but ill-conditionning of the tangent matrix occurs if it is too high.

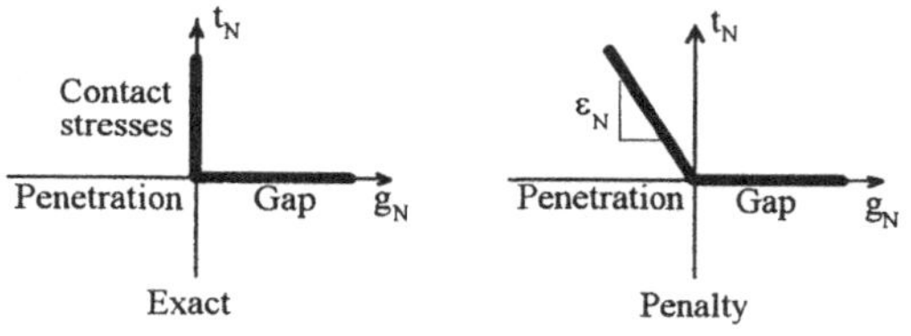

Figure 1

LAGRANGE MULTIPLIER

An additional variable λ_N called Lagrange multiplier is introduced.
Now, (1) is replaced by

$$\begin{cases} G(\underline{u},\underline{\delta u}) - \int_{\Gamma_c} \lambda_N\, \underline{\delta u}.\underline{n}\, d\Gamma_C = 0 \\ \int_{\Gamma_c} \delta\lambda_N\, \underline{u}.\underline{n}\, d\Gamma_C = 0 \end{cases} \tag{4}$$

which must hold for all $\underline{\delta u}$ and all $\delta\lambda_N$.

In this case, $\underline{\delta u}$ is not submited to condition $\underline{\delta u}.\underline{n} \leq 0$ because the contact condition is strictly verified if there is contact. If no contact exists the integral on Γ_c disappears. However, the use of this technique leads to additional variables and the structure of the resulting iteration matrix depends on the contact area size. Another problem is the semi positive definite character of this matrix which requires an appropriate system solver.

AUGMENTED LAGRANGIAN METHOD

This technique tries to circumvent the inconvenients of the preceeding methods. The purpose is to penalize the Lagrange multiplier problem. Now (1) is replaced by

$$G(\underline{u}, \underline{\delta u}) - \int_{\Gamma_c} \lambda_N^{(k+1)} \underline{\delta u}.\underline{n} d\Gamma_c = 0 \tag{5}$$

which must hold for all $\underline{\delta u}$ and with

$$\lambda_N^{(k+1)} = \left\langle \lambda_N^{(k)} + \varepsilon_N . g_N \right\rangle \tag{6}$$

which represents a Lagrangian multiplier and is an estimate of the correct contact pressure. To approach the real pressure, we use an iterative process described by eq(6). Here, λ_N is not an additional variable and is considered like fixed to solve the system at each iteration. The advantage of the current treatment is that satifaction of non penetration constraint can be improved even if ε_N is much smaller than in the penalty method through repeated application of the augmentation procedure. One augmentation corresponds to the application of (6).

EXAMPLE

In order to evaluate the augmented Lagrangian method we consider a very simple numerical test. It is a compression test of a single elastic solid element ($E = 205\ 10^9$ N/m^2 , $\upsilon = 0,3$). We upset the solid element with a rigid plate. The contact is modelled by a contact element with a friction coefficient of 0,2. We work with imposed displacement of the plate. The compression is realized in five equal steps to reach 25% deformation.
This example is interesting because it shows how we converge during the iterations.
Graph 1 show the evolution of the contact pressure with the penalty method at one integration point for different values of the penalty coefficient. The exact value is near 6.6E10N/m². We see immediately that we must choose a penalty coefficient 3.46E14 to evaluate correctely the pressure. Higher value induces numerical problem.

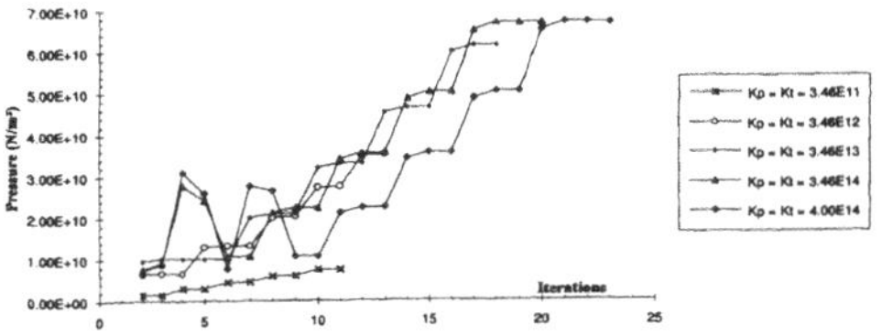

Graph 1

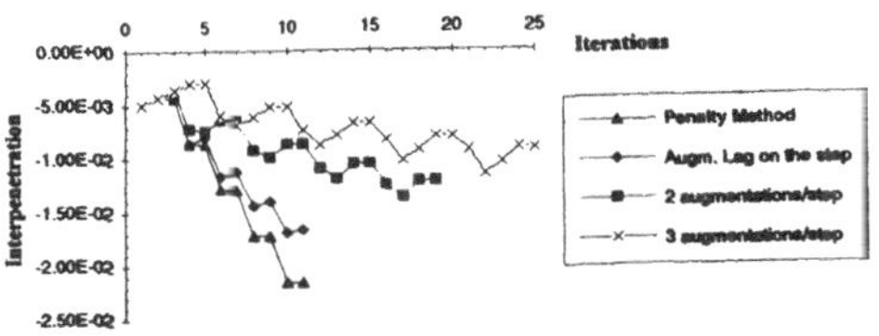

Graph 2

Note that the pressure is increasing from step to step. Therefore the penetration (which is existing which each method) is also increasing from step to step.
In order to compare the penalty and the augmented Lagrangian methods, graph 2 clearly presents the penetration evolution with the same penalty coefficient 3.46E11 for the two methods. We note a higher number of iterations for Uzawa's algorithm.

To see the convergence of the iterative technique we will present next the graphs obtained with a large number of augmentations.The use of a large number of augmentations with a small penalty coefficient is a good utilisation of the augmented Lagrangian technique. However the large number of augmentations leads to a large number of iterations and a higher CPU cost. Graph 3 shows the case where we use 12 augmentations per step and each of them allows to reduce the penetration between solid and tool.

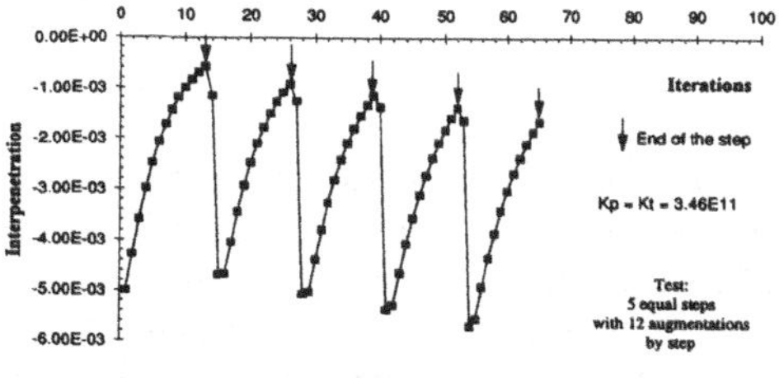

Graph 3

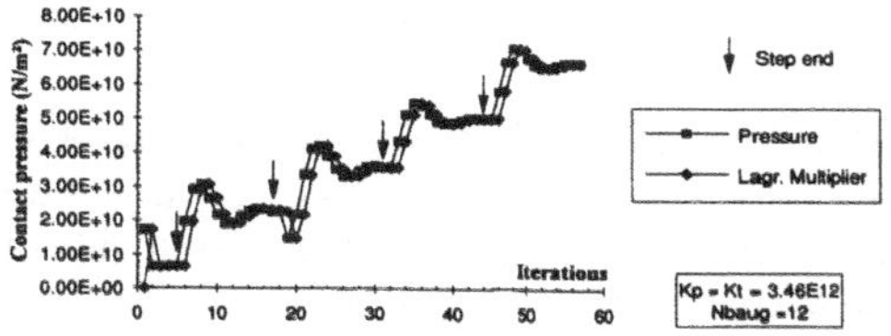

Graph 4

For the same simulation,we stabilize the pressure to a the correct value. We started with a small penalty coefficient which explain the large number of iterations to reach to the correct pressure. However if we increase the penalty coefficient we decrease the number of augmentations.

CONCLUSIONS

We have described the aumented Lagrangian method using Uzawa's algorithm. This method is expensive in CPU but improves contact solution. A good use of the method is to work with small penalty coefficient and a large number of augmentations. In this way you can approach the solution slowly and smoother than the penalty method which becomes more and more violent if we increase the penalty coefficient.

As a rule we have no problem of convergence with the augmented method except with high value of the penalty coefficient nomatter the number of augmentations..

REFERENCES

P. Alart and A. Curnier (1991) A mixed formulation for frictional contact problems prone to Newton like solution methods, Computer Methods in Applied Mechanics and Engineering 92, 353-375.

R. Charlier and A.M. Habraken (1990) Numerical modelisation of contact with friction phenomena by the finite element method, Computers and Geotechnics 9, 59-72.

S. Cescotto and R. Charlier (1993) Frictional contact finite elements based on mixed variational principles, Int. Journal for Numerical Methods in Engineering, vol. 36, 1681-1701.

J Simo, P. Wriggers and R. Taylor (1985) A perturbed Lagrangian formulation for the finite element solution of contact problems, Computer Methods in Applied Mechanics and Engineering 50, 163-180.

J. Simo and T. Laursen (1992) An augmented Lagrangian treatment of contact problems involving friction, Computers & Structures, Vol.42, n°1, 97-116.

AN AUGMENTED LAGRANGIAN FORMULATION FOR THE ANALYSIS OF NO-TENSION STRUCTURES WITH UNILATERAL SUPPORTS

Massimo Cuomo and Giulio Ventura

Istituto di Scienza delle Costruzioni
Facoltà di Ingegneria - University of Catania - Italy

INTRODUCTION

Object of the paper is a class of structural problems that will be abstractly defined by the following three sets of equations:

- the compatibility equation $C(u) = \varepsilon$, where the operator $C : \mathcal{U} \rightarrow \mathcal{D}$ is assumed linear (and so is its dual C' that relates the internal stress σ to the external actions $f \in \mathcal{U}'$);
- the constitutive equation $f(\varepsilon) = \sigma$, assumed to be a monotone lower semi-continuos map from $\mathcal{D}$ to $\mathcal{D}'$. Therefore the map is obtained from a generalized potential, that turns out to be lower semi-continuos and convex. This means that the material behaviour is reversible;
- the external force field, supposed to derive from a functional that is assumed to be lower semi-continuos and convex. The functional defines also the (non-linear) static and kinematic boundary conditions (eventually unilateral).

Main goals of the study are to formulate the appropriated mixed functionals ruling the problem, and to present a strategy of solution based on a generalized complementary energy formulation. This appears to present substantial advantages on the traditional displacement-based formulation, both as reduction of the size of the problem, and as rate of convergence. The numerical FE problem is solved using an algorithm that employs the augmented lagrangian technique, already presented in a particular case (Cuomo M. and Ventura G., 1992)

THE EQUATIONS OF THE MODEL

The material constitutive relations

A masonry-like material not supporting any tension is considered. Therefore the stress field must belong to the cone of negative semidefinite tensors, K_σ. For negative stress fields, the deformation is obtained from a convex potential $\hat{\phi}^c(\sigma)$. The complete stress potential is then

$$\phi_c(\sigma) = \hat{\phi}^c(\sigma) + \int_B \mathrm{ind} K_\sigma dB \qquad K_\sigma = \{\sigma : \sigma n \cdot n \leq 0 \quad \forall n\} \tag{1}$$

Contact Mechanics, Edited by M. Raous *et al.*
Plenum Press, New York, 1995

The force potential and the contact conditions

The boundary of the region B occupied by the structure is partitioned in three parts: ∂B_q, where external tractions are prescribed, ∂B_u, where conditions on the displacements, $u = \bar{u}$, are imposed, and ∂B_w, where unilateral contact can occur. Calling n the inward normal to the external obstacle, $g \geq 0$ the initial gap and p_n the constraint reaction, the local contact conditions are defined by the relations

$$\begin{array}{lll} -p_n \in \partial j_n(w_n) & j_n(w_n) = \text{ind}\, W_n & W_n = \{w_n : w_n - g \leq 0\} \\ w_n \in \partial j_n^c(-p_n) & j_n^c(-p_n) = -p_n \cdot g + \text{ind} P_n & P_n = \{-p_n : -p_n \geq 0\} \end{array} \tag{2}$$

Partitioning the displacement field $u \in \mathcal{V}$ in the subspaces of the displacements v defined in B or on ∂B_q; and in the subspace of the displacements w defined on ∂B_w, the concave force potential and its dual are given by

$$\psi(u) = \langle v, f \rangle - \int_{\partial B_w} j_n(Nw) ds \qquad \psi^c(\hat{f}) = - \int_{\partial B_w} j_n^c(-Np) ds \tag{3}$$

THE VARIATIONAL FORMULATION

The Hellinger-Reissner functional for generalized elasticity problems, with the present notations, is

$$\Pi_R(\sigma, u) = \phi^c(\sigma) + \psi(u) - \langle \sigma, Cu \rangle \tag{4}$$

convex on σ and concave on u. Extremizing the mixed functional for eliminating the displacements component it is obtained a generalized form of the complementary energy functional

$$\begin{array}{c} \Pi_c(\sigma) = \phi^c(\sigma) - \psi^c(\hat{f}) \\ \text{subject to } C'\sigma = f + \hat{p} \end{array} \tag{5}$$

that is convex on σ and p. It has to be noted that the stress variables and the unknown contact reactions are related through the equilibrium equations $C'\sigma = f + \hat{p}$. Substituting the expressions (1) and (3) of the potentials, the explicit representation of the functional (5) is obtained

$$\Pi_c = \frac{1}{2} \int_B E^{-1} \sigma \cdot \sigma dB - \int_{\partial B_u} \sigma n \cdot \hat{u} ds - \int_{\partial B_w} g N p ds + \int_{\partial B_w} \text{ind} P_n ds + \int_B \text{ind} K_\sigma(\sigma) dB \tag{6}$$

where the linear elastic stress potential has been introduced.

THE DISCRETIZED FORMULATION

Suitable interpolation functions are introduced for stresses and displacements on a FE mesh, say $\sigma = G_\sigma s \quad u = G_u \boldsymbol{u}$. The normal contact displacement w_n in discretized form then becomes

$$w_n = G_u(s) \boldsymbol{w} \cdot \boldsymbol{n}(s) = A(s) \boldsymbol{w} \tag{7}$$

where $A^T = G_w^T(s)\boldsymbol{n}(s) : R^m \to \mathcal{H}_1$ is an m-vector containing the interpolation functions for the normal component of the contact displacement. The discretized contact reactions are then defined by the subdifferential of the contact potential

$$\boldsymbol{p} = -\partial \int_{\partial B_\mu} j_n(Aw)ds = -\int_{\partial B_\mu} A^T \partial j_n(Aw)ds = \int_{\partial B_\mu} A^T p_n ds \tag{8}$$

The last expression puts into evidence that only nodal reactions can be directly evaluated, while their distribution over the boundary is undetermined.

Equilibrium equations discretized

The weak form of the equilibrium equations, obtained from the variation of (5) with respect to the displacements u, in discretized form, is

$$\begin{aligned}
&\int_B (CG_u)^T G_\sigma dB\ \boldsymbol{s}\cdot \boldsymbol{v} = \int_B G_u^T b dB \cdot \boldsymbol{v} + \int_{\partial B_t} G_u^T q ds \cdot \boldsymbol{v} \\
&\int_B (CG_u)^T G_\sigma dB\ \boldsymbol{s}\cdot \boldsymbol{w} = \int_{\partial B_w} G_u^T q ds \cdot \boldsymbol{w} + \int_{\partial B_w} A^T p_n ds \cdot \boldsymbol{w}
\end{aligned} \tag{9}$$

or

$$\begin{bmatrix} \boldsymbol{C} \\ \boldsymbol{Q} \end{bmatrix} [\boldsymbol{s}] = \begin{bmatrix} \boldsymbol{f} \\ \boldsymbol{q} \end{bmatrix} + \begin{bmatrix} \boldsymbol{0} \\ \boldsymbol{p} \end{bmatrix} \tag{10}$$

where the equilibrium matrix $\boldsymbol{C}$ has $\nu - m$ rows and τ columns, being ν the number of free nodal displacements, m the number of uilateral nodes and τ the dimension of the space of the stress components.

The stress variables are partitioned in the set of the self-equilibrated stresses s_0, of dimension $\tau - \nu + m$ and the set of the stresses in equilibrium with the external forces, s_1, of dimension $\nu + m$. The equilibrium equations are then written as

$$\begin{bmatrix} \boldsymbol{C}_0 & \boldsymbol{C}_1 \\ \boldsymbol{Q}_0 & \boldsymbol{Q}_1 \end{bmatrix} \begin{bmatrix} \boldsymbol{s}_0 \\ \boldsymbol{s}_1 \end{bmatrix} = \begin{bmatrix} \boldsymbol{f} \\ \boldsymbol{q} \end{bmatrix} + \begin{bmatrix} \boldsymbol{0} \\ \boldsymbol{p} \end{bmatrix} \tag{11}$$

The partition of the stresses must be performed in such a way that the submatrix $\boldsymbol{C}_1$ be invertible. Then the equilibrated stresses s_1 and the contact reactions can be obtained explicitly solving equations (11) with respect to the self equilibrated stresses s_0

$$\boldsymbol{s} = \begin{bmatrix} \boldsymbol{s}_0 \\ \boldsymbol{s}_1 \end{bmatrix} = \boldsymbol{R}\boldsymbol{s}_0 + \boldsymbol{t} \qquad \boldsymbol{R} = \begin{bmatrix} \boldsymbol{I} \\ -\boldsymbol{C}_1^{-1}\boldsymbol{C}_0 \end{bmatrix} \qquad \boldsymbol{t} = \begin{bmatrix} \boldsymbol{0} \\ \boldsymbol{C}_1^{-1}\boldsymbol{f} \end{bmatrix} \tag{12}$$

$$\boldsymbol{p} = \boldsymbol{T}\boldsymbol{s}_0 + \boldsymbol{y} \qquad \boldsymbol{T} = \boldsymbol{Q}_0 - \boldsymbol{Q}_1\boldsymbol{C}_1^{-1}\boldsymbol{C}_0 \quad \boldsymbol{y} = \boldsymbol{Q}_1\boldsymbol{C}_1^{-1}\boldsymbol{f} - \boldsymbol{q} \tag{13}$$

Similar expressions can be obtained for the nodal reactions $\boldsymbol{r}$ of the v bilateral constraints

$$\boldsymbol{H}\boldsymbol{s} = \boldsymbol{r} \tag{14}$$

The discretized complementary energy functional

A discretized expressions of the indicator functions is obtained introducing the lagrangian variables following the augmented lagrangian technique (Bertsekas)

$$\int_B \mathrm{ind}K_\sigma(\sigma)dB = \sum_{e=1}^{nel}\sum_{i=1}^{ng}\left[\lambda_i^e \sigma_{n_i}^{+e}(s) + \frac{1}{2}\alpha(\sigma_{n_i}^{+e}(s))^2\right] t_e W_i \tag{15a}$$

$$\int_{\partial B_w} \text{ind} P_n(p) ds = \sum_{i=1}^{m} \left[-p_{n_i}^{+}(s)\mu_i + \frac{1}{2}\beta (p_{n_i}^{+}(s))^2 \right] \tag{15b}$$

where

$$\sigma_{n_i}^{+e}(s) = \max(\sigma_{n_i}^{e}(s), -\frac{\lambda_i}{\alpha}) \qquad p_{n_i}^{+}(s) = \max(p_{n_i}(s), \frac{\mu_i}{\beta}) \tag{16}$$

being $\sigma_{n_i}^{e}$ the maximum principal stress at the Gauss point i of the element e and λ_i the corresponding Lagrange multiplier having the physical meaning of cracking deformation, α and β the penalty parameters for the two sets of constraints, t_e the thickness of the e-th element and W_i the weight for the Gauss point i.

It has to be noted that in (15) the lagrangian variables λ are defined at the Gauss points, since the stress variables have been directly interpolated. The contact displacement variables μ are instead directly interpolated. This is not the only possibility, but it derives from the choice of not introducing special contact elements.

Using expressions (12-14) for enforcing equilibrium, the discretized form of the functional Π_c in augmented lagrangian form is therefore

$$\begin{aligned} &\frac{1}{2}\boldsymbol{R}^T\boldsymbol{F}\boldsymbol{R}s_0 \cdot s_0 + \boldsymbol{R}^T\boldsymbol{F}\boldsymbol{t} \cdot s_0 - \boldsymbol{R}^T\boldsymbol{H}^T\bar{\boldsymbol{u}} \cdot s_0 + \sum_{i=1}^{m} \left[-p_{n_i}^{+}(s_0)(\mu_i + g_i) + \frac{1}{2}\beta(p_{n_i}^{+}(s_0))^2 \right] + \\ &\sum_{e=1}^{nel}\sum_{i=1}^{ng} \left[\lambda_i^e \sigma_{n_i}^{+e}(s_0) + \frac{1}{2}\alpha(\sigma_{n_i}^{+e}(s_0))^2 \right] t_e W_i \qquad \boldsymbol{F} = \int_B G_\sigma^T E^{-1} G_\sigma dB \end{aligned} \tag{17}$$

THE NUMERICAL SOLUTION METHOD

The solution of the structural problem, given by the optimum of functional (17), is obtained through the iterative scheme

$$\begin{aligned} s_0^{(k)} &= \arg \left[\inf_{\sigma_0} \Pi_{al}(s_0, \lambda^{(k-1)}, \mu^{(k-1)}) \right] \\ \lambda^{(k)} &= \lambda^{(k-1)} + \alpha^{(k-1)} \sigma_n^{+}(s_O^{(k)}) \\ \mu^{(k)} &= \mu^{(k-1)} - \beta^{(k-1)} p_n^{+}(s_O^{(k)}) \end{aligned} \tag{18}$$

in which the penalty parameters can be kept constant or incremented during the iterations. The saddle point problem is thus converted into a sequence of minimisation problems on direct variables s_0 while the Lagrange multipliers are locally updated through an iterative first order formula.

REFERENCES

Bertsekas D.P., 1982, "Constrained Optimization and Lagrange Multiplier Methods", Academic Press

Cuomo M. and Ventura G., 1994, An effective computational implementation of the notension model for masonry structures, *in* "Computer Methods in Structurtal Masonry", G.N. Pande and J. Middleton eds., B.J. Int., Swansea

Kikuchi N. and Oden J.T., 1988, "Contact Problems in Elasticity", SIAM, Philadelphia

A COMPETITIVE GAME ALGORITHM WITH FIVE PLAYERS FOR UNILATERAL CONTACT PROBLEMS INCLUDING THE ROTATIONAL AND THE THERMAL DEGREES OF FREEDOM

C.D. Bisbos

Institute of Steel Structures
Aristotle University, Thessaloniki 54006, Greece

INTRODUCTION

The unilateral frictional contact problem is a topic of vital research interest, yielded important results (see e.g. the classical books of Duvaut and Lions(1972) and Panagiotopoulos(1985) for the convex case as also the book of Panagiotopoulos(1993) for the nonconvex case). The mathematical frameworks of the Theory of Quasivariational Inequalities (Baiocchi and Capelo,1984) and of the Game Theory (Aubin and Ekeland,1984) have been used by Telega(1988), Curnier et. al(1992) and by Bisbos(1992,1993) respectively to obtain appropriate settings for the unilateral contact-Coulomb friction problem.

The present study extends previous work of the author in order to include in a Quasivariational Inequality (QVI)-Game formulation the additional interfacial effects: i) the thermal contact according to a simple phenomenological model, although more elaborate models have been proposed e.g. by Zavarize et al.(1992) and by Johansson(1992), ii) the rotational constraint condition that the tangent planes must coincide in case of contact, a condition used in rigid body vehicle dynamics (Eich,1993) and finally iii) the drilling Coulomb friction condition, known as soft finger condition in Robotics (already used by Panagiotopoulos and Al-Fahed(1994) in a Linear Complementarity model). These boundary contact conditions along with the classical normal unilateral contact and Coulomb friction conditions correspond to the five players in the Game formulation presented.

The framework used is the displacement FEM, transformed through matrix manipulations to a force method-like technique. This way the method preserves the constraint character of the various quantities involved.

Aspects of parallelization are finally discussed.

Contact Mechanics, Edited by M. Raous *et al.*
Plenum Press, New York, 1995

THE GOVERNING RELATIONS

Mechanical and thermal gaps

Let us consider a representative pair of nodes A and B with initial position vectors $\boldsymbol{\xi}^A, \boldsymbol{\xi}^B$ possibly in contact. We shall call it the j-th interfacial contact element with element displacement vector $\mathbf{u}^{(j)} = (\mathbf{u}^A, \mathbf{u}^B)$ and element absolute temperature vector $\boldsymbol{\theta}^{(j)} = (\boldsymbol{\theta}^A, \boldsymbol{\theta}^B)$. Each node has six mechanical DOFs (the three tranlations and the three rotations), denoted in the global system as $\mathbf{u} = (v_1, v_2, v_3, \psi_1, \psi_2, \psi_3)$ and in the local one as $\mathbf{u}_{loc}$ respectively. Let $\mathbf{R}^{(j)}$ be the element rotation matrix, transforming the local coordinate system (N, T) to the global one, inducing respectively the transformation $\mathbf{T}^{(j)}$ matrix ($\mathbf{u} = \mathbf{T}^{(j)}\, \mathbf{u}_{loc}$, $\mathbf{u}_{loc} = \mathbf{T}^{(j)T}\, \mathbf{u}$). The following matrices shall be needed ($\mathbf{I}_k$ is the $k \times k$ identity matrix):

$$\mathbf{H}_\Theta^{(j)T} = [\,-\mathbf{I}_1 : +\mathbf{I}_1], \qquad \mathbf{H}_W^{(j)T} = \frac{1}{2}[\,+\mathbf{I}_1 : +\mathbf{I}_1], \qquad \mathbf{H}_M^{(j)T} = \mathbf{T}^{(j)T}[\,-\mathbf{I}_6 : +\mathbf{I}_6] \quad (1)$$

If an isoparametric patch is used in place of node B, then $\boldsymbol{\theta}^B$, $\mathbf{u}^B$ list the quantities of the resp. patch nodes and (1) is replaced by:

$$\mathbf{H}_\Theta^{(j)T} = [\,-\mathbf{I}_1 : +\mathbf{N}_\Theta^B], \quad \mathbf{H}_W^{(j)T} = \frac{1}{2}[\,+\mathbf{I}_1 : +\mathbf{N}_\Theta^B], \quad \mathbf{H}_M^{(j)T} = \mathbf{T}^{(j)T}[\,-\mathbf{I}_6 : +\mathbf{N}_M^B] \quad (2)$$

where $\mathbf{N}_\Theta^B$, $\mathbf{N}_M^B$ are the values of the appropriate patch shape functions at the projection of node A to the patch surface. Then the local mechanical and thermal gaps $\mathbf{w}_M^{(j)}$, $\mathbf{w}_\Theta^{(j)}$ are given by:

$$\mathbf{w}_M^{(j)} = \mathbf{e}_M^{(j)} + \mathbf{h}_M^{(j)} \quad (3)$$

$$\mathbf{e}_M^{(j)} = \mathbf{u}_{loc}^B - \mathbf{u}_{loc}^A = \mathbf{T}^{(j)T}(\mathbf{u}^B - \mathbf{u}^A) = \mathbf{H}_M^{(j)T}\mathbf{u}^{(j)} \quad (4)$$

$$\mathbf{h}_M^{(j)} = (h_N^{(j)}, 0, 0, 0, 0, 0)^T, \quad h_N^{(j)} = \mathbf{R}_N^{(j)T}(\boldsymbol{\xi}^B - \boldsymbol{\xi}^A) \quad (5)$$

$$\mathbf{w}_\Theta^{(j)} = \mathbf{H}_\Theta^{(j)T}\, \boldsymbol{\theta}^{(j)} \quad (6)$$

It is convienent to partition the mechanical gap into the normal translational gap $w_N^{(j)}$, the tangential translational gap $\mathbf{w}_T^{(j)}$, the drilling rotational gap $w_D^{(j)}$, and the tangential rotational gap $\mathbf{w}_R^{(j)}$, i.e.:

$$\mathbf{w}_M^{(j)} = (w_N^{(j)}, \mathbf{w}_T^{(j)}, w_D^{(j)}, \mathbf{w}_R^{(j)}) \quad (7)$$

Contact stresses

Let $(\mathbf{p}^A, \mathbf{p}^B)$ be the external contact forces acting on the two nodes of the generic contact pair. The force $\mathbf{p}^B$ in the local coordinate system is considered as the element contact stress, $\mathbf{s}_M^{(j)} = (s_N^{(j)}, \mathbf{s}_T^{(j)}, s_D^{(j)}, \mathbf{s}_R^{(j)})$. Then the external nodal load vector $\mathbf{p}_c^{(j)}$ in global coordinates is conjugate to (4):

$$\mathbf{p}_C^{(j)} = \mathbf{H}_M^{(j)}\, \mathbf{s}_M^{(j)} \quad (8)$$

Equations (3) and (8) establish the connection with the rest of the structure.

Signorini - Fichera conditions

The classical unilateral contact conditions in the normal direction are:

$$w_N^{(j)} \geq 0\,, \quad s_N^{(j)} \geq 0\,, \quad w_N^{(j)} s_N^{(j)} = 0 \tag{9}$$

In our formulation we shall need the trivial (constant) point-to-set valued map:

$$s_N^{(j)} \mapsto K_N^{(j)} \subset \mathbf{R} \quad : \quad K_N^{(j)} = \{\, x \in \mathbf{R} \mid x \geq 0 \,\} \tag{10}$$

Coulomb tangential force friction conditions

If $c_T^{(j)}$ is the local translational friction coefficient, the friction force bound is defined as $b_T^{(j)} = (\, c_T^{(j)} s_N^{(j)} \,)^2$ and the Coulomb force cone is given by:

$$\varphi_T^{(j)} = 1/2\,(b_T^{(j)} - \mathbf{s}_T^{(j)T}\, \mathbf{s}_T^{(j)}\,) = 1/2\,(b_T^{(j)} - \mathbf{Q}_T^{(j)T}\, \mathbf{s}_T^{(j)}\,) \qquad with: \quad \mathbf{Q}_T^{(j)} = \mathbf{s}_T^{(j)} \tag{11}$$

Actually the normal contact force $s_N^{(j)}$ defines the permissible region for the tangential contact forces through the following point-to-set valued map:

$$s_N^{(j)} \mapsto K_T^{(j)} \subset \mathbf{R}^2 \quad : \quad K_T^{(j)} = \{\, \mathbf{s}_T^{(j)} \in \mathbf{R}^2 \mid \varphi_T^{(j)} \geq 0 \,\} \tag{12}$$

If $s_N^{(j)}$ is positive, the classical Coulomb friction conditions are:

$$\varphi_T^{(j)} \geq 0\,, \quad \mathbf{w}_T^{(j)} = -\mathbf{Q}_T^{(j)} \lambda^{(j)}\,, \quad \lambda^{(j)} \geq 0\,, \quad \lambda^{(j)} \varphi_T^{(j)} = 0 \tag{13}$$

Coulomb drilling moment friction conditions

In a completely analogous way the rotational Coulomb friction conditions hold. The friction torque bound $b_D^{(j)} = (\, c_D^{(j)} s_N^{(j)} \,)^2$ defines the Coulomb torque cone:

$$\varphi_D^{(j)} = 1/2\,(b_D^{(j)} - s_D^{(j)T}\, s_D^{(j)}\,) = 1/2\,(b_D^{(j)} - \mathbf{Q}_D^{(j)T}\, s_D^{(j)}\,) \qquad with: \quad \mathbf{Q}_D^{(j)} = s_D^{(j)} \tag{14}$$

and similarly to (12) and (13) the following relations hold:

$$s_N^{(j)} \mapsto K_D^{(j)} \subset \mathbf{R} \quad : \quad K_D^{(j)} = \{\, s_D^{(j)} \in \mathbf{R} \mid \varphi_D^{(j)} \geq 0 \,\} \tag{15}$$

$$\varphi_D^{(j)} \geq 0\,, \quad w_D^{(j)} = -\mathbf{Q}_D^{(j)} \mu^{(j)}\,, \quad \mu^{(j)} \geq 0\,, \quad \mu^{(j)} \varphi_D^{(j)} = 0 \tag{16}$$

Rotational constraint condition

The tangential moments $\mathbf{s}_R^{(j)}$ are the constraint forces, which are energetically conjugate to the rotational tangential gaps $\mathbf{w}_R^{(j)}$. They cannot yield mechanical work and the following complementarity condition holds:

$$\mathbf{w}_R^{(j)} \mathbf{s}_R^{(j)} = 0 \tag{17}$$

The constraint moments can develop, only if the normal contact force is positive. In other words $s_N^{(j)}$ defines a permissible region for $\mathbf{s}_R^{(j)}$ through the point-to-set valued map:

$$s_N^{(j)} \mapsto K_R^{(j)} \subset \mathbf{R}^2 \quad : \quad K_R^{(j)} = \begin{cases} \mathbf{s}_R^{(j)} = \mathbf{0} & \text{if} \quad s_N^{(j)} = 0 \\ \mathbf{s}_R^{(j)} \in \mathbf{R}^2 & \text{if} \quad s_N^{(j)} \neq 0 \end{cases} \tag{18}$$

Thermal contact

Let $s_W^{(j)}$ be the dissipated mechanical work, due to frictional slipping. It is given by:

$$s_W^{(j)} = -\mathbf{s}_T^{(j)T}\mathbf{w}_T^{(j)} - s_D^{(j)T} s_D^{(j)} = \lambda^{(j)} b_T^{(j)} + \mu^{(j)} b_D^{(j)} \tag{19}$$

Since $s_W^{(j)}$ is the thermal analogon to the external element loads in Structural Mechanics, we assume that it is equidistributed to the nodes A and B. Let us denote further by $s_\Theta^{(j)}$ the heat flux transferred from node A to node B. Then the respective source and heat transfer terms to be added to the heat balance equations are:

$$\mathbf{q}_W^{(j)} = \mathbf{H}_W^{(j)} s_W^{(j)}, \qquad \mathbf{q}_\Theta^{(j)} = \mathbf{H}_\Theta^{(j)} s_\Theta^{(j)} \tag{20}$$

Physically $s_\Theta^{(j)}$ has a constraint character, being complementary $w_\Theta^{(j)}$:

$$w_\Theta^{(j)} s_\Theta^{(j)} = 0 \tag{21}$$

i.e. either the thermal gap is closed (equal node temperatures) and heat transfer can develop, or the thermal gap is open and the heat transfer is zero.

As in case of the rotational constraint condition, the normal contact force defines a permissible region for the constraint heat flux $s_\Theta^{(j)}$ through the point-to-set valued map:

$$s_N^{(j)} \mapsto K_\Theta^{(j)} \subset \mathbf{R} \quad : \quad K_\Theta^{(j)} = \begin{cases} s_\Theta^{(j)} = 0 & \text{if} \quad s_N^{(j)} = 0 \\ s_\Theta^{(j)} \in \mathbf{R} & \text{if} \quad s_N^{(j)} \neq 0 \end{cases} \tag{22}$$

The structural relations

Let m be the number of the interfacial elements. In the sequel the structural quantities shall be denoted through the same symbols used for the individual interfacial elements with the (j)-superscript dropped, e.g. $\mathbf{s}_N$ collects all $s_N^{(j)}$. Let further $\mathbf{s}_M = (\mathbf{s}_N, \mathbf{s}_T, \mathbf{s}_D, \mathbf{s}_R)$ be the vector of the interfacial generalized contact stresses. Then the contact conditions for the whole structure take the form:

$$\mathbf{w}_M = \mathbf{H}_M^T \mathbf{u} + \mathbf{h}_M \tag{23}$$

$$\mathbf{p}_C = \mathbf{H}_M \mathbf{s}_M \tag{24}$$

$$\varphi_T = 1/2\,[\,\mathbf{b}_T\,(\mathbf{s}_N) - \mathbf{Q}_T^T(\mathbf{s}_T)\,\mathbf{s}_T\,] \tag{25}$$

$$\varphi_D = 1/2\,[\,\mathbf{b}_D\,(\mathbf{s}_N) - \mathbf{Q}_D^T(\mathbf{s}_D)\,\mathbf{s}_D\,] \tag{26}$$

$$\mathbf{s}_N^T\mathbf{w}_N = 0, \qquad \mathbf{s}_N \geq \mathbf{0}, \qquad \mathbf{w}_N \geq \mathbf{0} \tag{27}$$

$$\boldsymbol{\lambda}^T\boldsymbol{\varphi}_T = 0, \qquad \boldsymbol{\varphi}_T \geq \mathbf{0}, \qquad \boldsymbol{\lambda} \geq \mathbf{0}, \qquad \mathbf{w}_T = -\mathbf{Q}_T(\mathbf{s}_T)\,\boldsymbol{\lambda} \tag{28}$$

$$\boldsymbol{\mu}^T\boldsymbol{\varphi}_D = 0, \qquad \boldsymbol{\varphi}_D \geq \mathbf{0}, \qquad \boldsymbol{\mu} \geq \mathbf{0}, \qquad \mathbf{w}_D = -\mathbf{Q}_D(\mathbf{s}_D)\,\boldsymbol{\mu} \tag{29}$$

$$\mathbf{s}_R^T\mathbf{w}_R = 0 \tag{30}$$

$$\mathbf{w}_\Theta = \mathbf{H}_\Theta^T\,\boldsymbol{\theta} \tag{31}$$

$$\mathbf{q}_\Theta = \mathbf{H}_\Theta\,\mathbf{s}_\Theta \tag{32}$$

$$\mathbf{q}_W = \mathbf{H}_W\,\mathbf{s}_W\,(\mathbf{s}_N, \boldsymbol{\lambda}, \boldsymbol{\mu}) \tag{33}$$

$$\mathbf{s}_\Theta^T\mathbf{w}_\Theta = 0 \tag{34}$$

This relation set is to be completed with the momentum balance equations and the heat balance ones:

$$\mathbf{K}_M\mathbf{u} + \mathbf{K}_\Theta\boldsymbol{\theta} + \mathbf{p}_{\Theta-ref} = \mathbf{p}_{ext} + \mathbf{p}_C \tag{35}$$

$$\mathbf{C}_\Theta\boldsymbol{\theta} = \mathbf{q}_{ext} + \mathbf{q}_\Theta + \mathbf{q}_W \tag{36}$$

THE MECHANICAL PROBLEM

The element level point-to-set valued mappings (10), (12), (15) and (18) become now the respective Cartesian products:

$$\mathbf{s}_N \mapsto K_N \subset \mathbf{R}^m \;:\; K_N(\mathbf{s}_N) = K_N^{(1)} \times \cdots \times K_N^{(j)} \times \cdots \times K_N^{(m)} \tag{37}$$

$$\mathbf{s}_N \mapsto K_T \subset \mathbf{R}^{2m} \;:\; K_T(\mathbf{s}_N) = K_T^{(1)} \times \cdots \times K_T^{(j)} \times \cdots \times K_T^{(m)} \tag{38}$$

$$\mathbf{s}_N \mapsto K_D \subset \mathbf{R}^m \;:\; K_D(\mathbf{s}_N) = K_D^{(1)} \times \cdots \times K_D^{(j)} \times \cdots \times K_D^{(m)} \tag{39}$$

$$\mathbf{s}_N \mapsto K_R \subset \mathbf{R}^{2m} \;:\; K_R(\mathbf{s}_N) = K_R^{(1)} \times \cdots \times K_R^{(j)} \times \cdots \times K_R^{(m)} \tag{40}$$

inducing the combined point-to-set valued map:

$$\mathbf{s}_M \mapsto K_M \subset \mathbf{R}^{6m} \;:\; K_M(\mathbf{s}_M) = K_N \times K_T \times K_F \times K_R \tag{41}$$

Now assuming the temperatures fixed (isothermal conditions) and setting:

$$\mathbf{F}_M = \mathbf{H}_M^T\mathbf{K}_M^{-1}\mathbf{H}_M \tag{42}$$

$$\mathbf{p}_{eff} = \mathbf{p}_{ext} - \mathbf{K}_\Theta\boldsymbol{\theta} - \mathbf{p}_{\Theta-ref} \tag{43}$$

$$\mathbf{u}_0 = \mathbf{K}_M^{-1}\mathbf{p}_{eff} \tag{44}$$

$$\mathbf{w}_{M0} = \mathbf{H}_M^T\mathbf{u}_0 + \mathbf{h}_M \tag{45}$$

yields the mechanical gaps as linear functions of the interfacial contact stresses:

$$\mathbf{w}_M = \mathbf{F}_M\mathbf{s}_M + \mathbf{w}_{M0} \tag{46}$$

being the gradient of the strain energy functional:

$$J_M(\mathbf{s}_M) = 1/2\, \mathbf{s}_M^T \mathbf{F}_M \mathbf{s}_M + \mathbf{s}_M^T \mathbf{w}_{M0} \tag{47}$$

It is easy to see that the following QVI solves the mechanical subproblem:

$$\left.\begin{array}{ll} \textit{Find} \;\; \mathbf{s}_M \in \mathbf{R}^{6m} \;\; \textit{such that} : & \\ \text{i)} \;\; \mathbf{s}_M \in K_M(\mathbf{s}_M) & \\ \text{ii)} \;\; (\mathbf{x}_M - \mathbf{s}_M)^T \, \nabla J_M(\mathbf{s}_M) \geq \mathbf{0} \quad \forall\, \mathbf{x}_M \in K_M(\mathbf{s}_M) & \end{array}\right\} \tag{48}$$

THE THERMAL PROBLEM

For the thermal subproblem we assume that all mechanical quantities are given. Consequently the dissipation $\mathbf{s}_W$ and the sets $K_\Theta^{(j)}$ are constant. Now the structural analogon of (22) is the point-to-set valued map:

$$\mathbf{s}_M \mapsto K_\Theta \subset \mathbf{R}^m \, : \, K_\Theta(\mathbf{s}_M) = K_\Theta^{(1)} \times \cdots \times K_\Theta^{(j)} \times \cdots \times K_\Theta^{(m)} \tag{49}$$

Setting:

$$\mathbf{F}_\Theta = \mathbf{H}_\Theta^T \mathbf{C}_\Theta^{-1} \mathbf{H}_\Theta \tag{50}$$

$$\mathbf{q}_{eff} = \mathbf{q}_{ext} + \mathbf{q}_W \tag{51}$$

$$\boldsymbol{\theta}_0 = \mathbf{C}_\Theta^{-1} \mathbf{q}_{eff} \tag{52}$$

$$\mathbf{w}_{\Theta 0} = \mathbf{H}_\Theta^T \boldsymbol{\theta}_0 \tag{53}$$

yields the thermal gaps as linear functions of the interfacial contact stresses:

$$\mathbf{w}_\Theta = \mathbf{F}_\Theta \mathbf{s}_\Theta + \mathbf{w}_{\Theta 0} \tag{54}$$

being the gradient of the heat functional:

$$J_\Theta(\mathbf{s}_\Theta) = 1/2\, \mathbf{s}_\Theta^T \mathbf{F}_\Theta \mathbf{s}_\Theta + \mathbf{s}_\Theta^T \mathbf{w}_{\Theta 0} \tag{55}$$

We can directly verify that the following Variational Equality solves the thermal problem:

$$\left.\begin{array}{ll} \textit{Find} \;\; \mathbf{s}_\Theta \in \mathbf{R}^{m} \;\; \textit{such that} : & \\ \text{i)} \;\; \mathbf{s}_\Theta \in K_\Theta(\mathbf{s}_M) & \\ \text{ii)} \;\; (\mathbf{x}_\Theta - \mathbf{s}_\Theta)^T \, \nabla J_\Theta(\mathbf{s}_\Theta) = 0 \quad \forall\, \mathbf{x}_\Theta \in K_\Theta(\mathbf{s}_M) & \end{array}\right\} \tag{56}$$

SOLUTION AND THE GAME POINT OF VIEW

The Game view of the solution techniques

The mechanical and the thermal problems constitute a fixed point problem, which can be solved in a serial or parallel way. Since the first problem constitutes a QVI and the second one a linear system of equations, a parallelization would lead to unbalanced tasks for the computer processors and a serial solution is preferable.

The mechanical subproblem can be decomposed in four problems, each possessing its own structure. These problems correspond to the four players, i.e the N-player (normal unilateral contact), the T-player (Coulomb force friction), the D-player (Coulomb torque friction), and the R-player (Rotational constraint condition). Their behaviour is determined by the linear functions:

$$\left.\begin{array}{lllllllll}
\mathbf{d}_N & = & \mathbf{w}_{N0} & & & + & \mathbf{F}_{NT}\mathbf{s}_T & + & \mathbf{F}_{ND}\mathbf{s}_D \; + \; \mathbf{F}_{NR}\mathbf{s}_R \\
\mathbf{d}_T & = & \mathbf{w}_{T0} & + & \mathbf{F}_{TN}\mathbf{s}_N & & & + & \mathbf{F}_{TD}\mathbf{s}_D \; + \; \mathbf{F}_{TR}\mathbf{s}_R \\
\mathbf{d}_D & = & \mathbf{w}_{D0} & + & \mathbf{F}_{DN}\mathbf{s}_N & + & \mathbf{F}_{DT}\mathbf{s}_T & & \qquad\qquad\;\; + \; \mathbf{F}_{DR}\mathbf{s}_R \\
\mathbf{d}_R & = & \mathbf{w}_{R0} & + & \mathbf{F}_{RN}\mathbf{s}_N & + & \mathbf{F}_{RT}\mathbf{s}_T & + & \mathbf{F}_{RD}\mathbf{s}_D
\end{array}\right\} \quad (57)$$

and the resp. strategy selection rules:

$$\mathbf{s}_N = Argmin \; [\; J_N(\mathbf{x}_N) = 1/2\, \mathbf{x}_N^T \mathbf{F}_{RR}\, \mathbf{x}_N + \mathbf{x}_N^T \mathbf{d}_N \;\; | \;\; \mathbf{x}_N \in K_N \;] \quad (58)$$

$$\mathbf{s}_T = Argmin \; [\; J_T(\mathbf{x}_T) = 1/2\, \mathbf{x}_T^T \mathbf{F}_{TT}\, \mathbf{x}_T + \mathbf{x}_T^T \mathbf{d}_T \;\; | \;\; \mathbf{x}_T \in K_T(\mathbf{s}_N) \;] \quad (59)$$

$$\mathbf{s}_D = Argmin \; [\; J_D(\mathbf{x}_D) = 1/2\, \mathbf{x}_D^T \mathbf{F}_{DD}\, \mathbf{x}_D + \mathbf{x}_D^T \mathbf{d}_D \;\; | \;\; \mathbf{x}_D \in K_D(\mathbf{s}_N) \;] \quad (60)$$

$$\mathbf{s}_R = Argmin \; [\; J_R(\mathbf{x}_R) = 1/2\, \mathbf{x}_R^T \mathbf{F}_{RR}\, \mathbf{x}_R + \mathbf{x}_R^T \mathbf{d}_R \;\; | \;\; \mathbf{x}_R \in K_R(\mathbf{s}_N) \;] \quad (61)$$

The easiest task corresponds to the R-Player (linear system) and the heaviest onek corresponds generally (i.e. in 3D-case) to the T-Player (in the 2-D case the D-Player has to accomplish an equivalent work). Although task balance can be obtained in some degree (Bertsekas and Tsitsiklis,1989), this decomposition is best suited for serial application.

The Variational Section (VS) technique described by Bisbos(1992,1993) can be applied in any case. It is best suited for parallelization In the form of a decomposition-coordination algorithm, described in the sequel.

Parallelization

Let us assume that the interfacial elements have been grouped in some number of groups and let the subscript K denote the representative group. Due to the Cartesian product nature of the set $K_M\,(\mathbf{s}_M)$ we can consider each group as a player, which acts in parallel with the other groups aiming to minimize its own cost functional:

$$J_K\,(\mathbf{x}_K) = 1/2\;\; \mathbf{x}_K^T\, \mathbf{F}_{KK}\, \mathbf{x}_K \;+\; \mathbf{x}_K^T\, \mathbf{d}_K \quad (62)$$

where $\mathbf{F}_{KK}$ denotes the resp. block-diagonal submatrix of the flexibility matrix $\mathbf{F}_M$. The respective feasible set K_K, to be used in the VS of the group arises naturally from the Cartesian product of K_M.

It remains to compute the appropriate gaps $\mathbf{d}_K$. This computation is incorporated in the decomposition phase. It is noteworthy that the group off-diagonal submatrices of $\mathbf{F}_M$ are not used, resulting in considerable memory savings.

We assume that the initial mechanical gaps $\mathbf{w}_{M0}$ and the decomposed structural matrix $\mathbf{K}$ are available during the whole iteration.

Coordination phase of the l-th iteration In this phase we compute the correction of structural displacements, due to the actual values of the contact forces (denoted as $\mathbf{s}_M^{Old}$). Depending on the hardware and software available, the coordination phase can be executed in parallel. The relative formulas are:

$$\mathbf{p}_C = \mathbf{H}_M \, \mathbf{s}_M^{Old} \tag{63}$$

$$\Delta \mathbf{u}_C = \mathbf{K}_M^{-1} \, \mathbf{p}_C \tag{64}$$

Decomposition phase of the $(l+1)$-th iteration Now the respective gap corrections are computed and the the strategies of the individual groups are selected in parallel. This phase can be executed as concurrent programming code.

$$\mathbf{w}_{KC} = \mathbf{H}_K^T \, \Delta \mathbf{u}_C \tag{65}$$

$$\mathbf{w}_K = \mathbf{w}_{K0} + \mathbf{w}_{KC} \tag{66}$$

$$\mathbf{d}_K = \mathbf{w}_K - \mathbf{F}_{KK} \, \mathbf{s}_K^{Old} \tag{67}$$

$$\mathbf{s}_K^{New} = Argmin \, [\ 1/2 \ J_K(\mathbf{x}_K) \ \mid \ \mathbf{x}_K \in K_K(\mathbf{s}_K^{Old})\ ] \tag{68}$$

Obviously (65) computes the group gap correction due to all contact stresses, while (66) actualizes the group total gap estimation. (67) eliminates the influence of the group contact forces on the respective group total gap: $\mathbf{d}_K$ represents the group total gap due to sources other than its contact forces. Finally (68) is the strategy selection rule of the group players.

REFERENCES

Aubin, J.-P. and I.Ekeland, I., 1984, "Applied Nonlinear Analysis," Wiley, New York.

Baiocchi, G. and Capelo, A., 1984, "Variational and Quasivariational Inequalities," Wiley, New York.

Bertsekas, D. and Tsitsiklis, C., 1989, "Parallel and Distributed Computation. Numerical Methods," Prentice Hall, Englewood Cliffs.

Bisbos, C.D., 1992, A Nash - game formulation for frictional unilateral contact problems, in: "Contact Mechanics," A.Curnier, ed., PPUR, Lausanne.

Bisbos, C.D., 1993, A Competitive game formulation for large scale frictional contact problems, in: "Contact Mechanics. Computational Techniques," M.H.Aliabadi and C.A.Brebbia, eds., Computational Mechanics Publications, Southampton.

Curnier, A., He, Q.-C., and Telega, J.J., 1992, Formulation of unilateral contact between two elastic bodies undergoing finite deformations, *C.R.Acad.Sci.Paris*, 314, Series II:1.

G.Duvaut, G. and Lions, J.L., 1972, "Les Inéquations en Mécanique et en Physique," Dunod, Paris.

Eich, E., 1993, Convergence results for a coordinate projection method applied to mechanical systems with algebraic constraints, *SIAM J. on Numer. Anal.* 30:1467.

Johansson, L., 1992, Sliding contact between two elastic half-planes with frictional heat generation and wear, in: "Contact Mechanics," A.Curnier, ed., PPUR, Lausanne.

Panagiotopoulos, P.D., 1985, "Inequality Problems in Mechanics and Applications. Convex and Nonconvex Energy Functions," Birkhäuser, Boston-Basel.

Panagiotopoulos, P.D., 1993, "Hemivariational Inequalities. Applications in Mechanics and Engineering," Springer-Verlag, Berlin-Heidelberg.

Panagiotopoulos, P.D., Al-Fahed, A.M., 1994, Robot Hand Grasping and Related Problems: Optimal Control and Identification *Int.J. of Robotics Research* 13:127.

Telega, J.T., 1988, Topics on unilateral contact problems of elasticity and inelasticity, in: " Nonsmooth Mechanics and Applications. CISM Courses and Lectures No 302 ," J.J.Moreau, P.D.Panagiotopoulos, ed., Springer-Verlag, Berlin-Heidelberg.

Tzaferopoulos, M.A., Mistakidis, E.S., Bisbos, C.D., and Panagiotopoulos, P.D., 1994, On two algorithms fon nonconvex nonsmooth optimization problems in structural mechanics, in: "Large Scale Optimization. State of the Art," W.W.Hager *et al.*, ed., Kluwer Academic Publ., New York.

Zavarise, G., Wriggers, P., Stein, E., and Shrefler, B.A., 1992, Real contact mechanisms and finite element formulation - a coupled thermomechanical approach, *Int.J. Numer. Methods Eng.* 36:767.

A NEW NUMERICAL APPROACH TO OBSTACLE PROBLEMS

Alain Cimetière and Thierry Texier

Laboratoire de Mécanique Théorique
40, avenue du Recteur Pineau - 86022 POITIERS FRANCE

INTRODUCTION

Unilateral problems and variational inequalities arise in various areas of Mechanics, for instance obstacle and contact problems, plasticity... Sometimes industrial contact problems require an accurate calculation of the contact area, as in problems of heat transfer between joined metallic structures.

Usually frictionless elastic contact problems are solved using mathematical programming and the position of the contact area is obtained as a subproduct of the solving process.

Contact problems are nonlinear problems and in many practical applications, the finite element method leads to a large number of unknowns whose numerical treatment may prove expensive.

The aim of the paper is to introduce a new numerical approach to the solutions of obstacle problems. For simplicity, the main ideas of the new method are presented in the case of a thin elastic membrane fixed along its boundary, loaded in the transversal direction and constrained to lie above a given rigid plane support. The contact is frictionless.

The method is based on the fact that the solution of the variational inequality governing the problem is regular (C^1 regularity) when the loading is sufficiently regular (see Kinderlehrer and Stampacchia, 1980). Thanks to that C^1 regularity, the unilateral problem can be turned into a nonlinear equation, whose unknows are the deflection of the membrane and the position of the contact area boundary. Another application of a regularity result can be found in Cimetière and Léger (1993), where a differentiability result is obtained for a class of obstacle problems.

THE MECHANICAL MODEL PROBLEM

Before loading, the membrane occupies a bounded domain ω, located in the x - y plane. The membrane deflection u is limited by the rigid plan z = -1. The transversal regular loading λ is supposed great enough for the contact to exist. For simplicity, the contact zone ω_c is supposed simply connected, but multiconnected contact zones could be taken into account. The membrane is fixed on the exterior boundary Γ_e. The contact zone boundary is denoted Γ_i.

Let u be the solution of the unilateral problem and w the solution of the bilateral problem (bending problem without support). Thanks to a regularity result for the solutions of variational inequalities, the solution u possesses the C^1 regularity and hence the difference v = u - w is the solution of the following problem :

$$\begin{aligned} \Delta v &= 0 \quad \text{on } \Omega = \omega \setminus \omega_c \\ v &= 0 \quad \text{on } \Gamma_e \\ v &= -1 \quad \text{on } \Gamma_i \\ v' &= 0 - w'\big|_{\Gamma_i} \qquad (v' = \text{normal derivative}) \end{aligned}$$

The unknown is changed only to eliminate the loading λ, but the method also stands when u is kept as the unknown.

THE UNILATERAL PROBLEM REDUCED TO A NONLINEAR EQUATION

The usual boundary element method (C.A. Brebbia, 1984) is based on the following formula which gives a relation between the unknown v and its normal derivative v' on $\Gamma_i \cup \Gamma_e = \partial\Omega$:

$$\tfrac{1}{2}\, v(x) = -\int_{\partial\Omega} v E'(x)\, ds + \int_{\partial\Omega} E(x)\, v'\, ds \qquad x \in \partial\Omega$$

In the previous formula, E(x)(y) denotes the fundamental solution for the laplacian operator. In usual boundary element method, that relation enables to calculate v' on the boundary when v is given. Here the unilateral character of the problem introduces a new difficulty : the boundary position of Γ_i is an unknown of the problem. However, as on Γ_i two conditions must be satisfied, the determination of the boundary Γ_i will be possible.

To discretize the previous equation, we first choose a point 0 in the "middle" of the presumed contact area and then we choose m points $P_{m+1}, \dots, P_{m+j}, \dots, P_{2m}$ on Γ_e. The intersection of the segment OP_{m+j} with Γ_i defines the point P_j. As in the simplest boundary element method, the boundary Γ_e is replaced by the segments $P_{m+j} P_{m+j+1}$, with $1 \le j \le m-1$. In the same way, Γ_i is replaced by the segments $P_j P_{j+1}$, with $1 \le j \le m-1$. The respective positions of the m points $P_1, \dots, P_j, \dots, P_m$ on the segments OP_{m+j} $(1 \le j \le m-1)$ are characterized by m values $\rho_1, \dots, \rho_j, \dots, \rho_m$, which are m unknows of the discretized problem. On each segment interpolating the boundary $\partial\Omega = \Gamma_i \cup \Gamma_e$, we approximate v (resp. v') by constant values $v_1, \dots, v_{2m}$ (resp. $v'_1, \dots, v'_{2m}$). When we write the integral equation for the middles $a_1, \dots, a_{2m}$ of the 2m interpolating segments, we obtain the discretized equation :

$$\sum_{n=1}^{2m}(\tfrac{1}{2}\,\delta_{kn} + B_{kn})\,v_n - \sum_{n=1}^{2m} A_{kn}\,v'_n = 0 \qquad k=1,\ldots,2m$$

The coefficients A_{kn} , B_{kn} are line integrals which depend on the position of Γ_i and hence on the values $\rho_1,\ldots,\rho_m$. On Γ_i, v_n and v'_n are known fonctions of $\rho=(\rho_1,\ldots,\rho_m)$ and on Γ_e. v_n is given. Finally the 2m equations involve 2m unknows exactly : m values v'_n and m values ρ'_n. So the resolution of the unilateral problem is turned into the resolution of a nonlinear system :

$$F_i\,(\rho_1,\ldots,\rho_m, v'_{m+1},\ldots,v'_{2m}) = 0 \qquad i = 1,\ldots, 2m$$

A TEST PROBLEM

We consider a circular membrane fixed on its boundary and subjected to a uniform transversal loading λ. Its deflection is limited by the rigid plane $z = -1$. The contact exists for $\lambda > 4/R^2$, where R is the membrane radius. The radius r of Γ_i is linked to R and λ by the nonlinear equation :

$$\frac{1}{\lambda} = \frac{1}{4}\,(R^2 - r^2) - \frac{1}{2}\,r^2\,\mathrm{Ln}\,(\frac{R}{r})$$

The nonlinear system is solved by the Newton method. The integrals A_{kn} , B_{kn} are evaluated by the standard nine points Gauss formula. Numerical results are given for different contact area radii r and for different initial shapes of contact area. All results have been obtained with R = 6.

NUMERICAL RESULTS WITH AXISYMMETRIC INITIALIZATIONS

First case : $\lambda = 0{,}1723$

The analytical calculation gives r = 2 and v' = 0,4595. For the initialization $(\rho, v') = (1, 0)$, or in other words $\rho_1 = \ldots = \rho_m = 1$ and $v'_{m+1} = \ldots = v'_{2m} = 0$ we obtain, after four iterations (I = 4) and for different discretizations :

m	I	ρ	Δr/r	v'
24	4	2.0118	6.10^{-3}	0,45942
72	4	2.0013	6.10^{-4}	0,45954
120	4	2.0004	2.10^{-4}	0,45950

We now give some results for an initialization far from the solution. We choose $(\rho, v) = (5, 0)$

m	I	ρ	Δr/r	v'
24	5	2.0118	6.10^{-3}	0,45942
72	5	2.0013	6.10^{-4}	0,45954
120	5	2.0004	2.10^{-4}	0,45950

Second case : $\lambda = 0{,}1111$ $r = 0{,}01$ $v' = 0{,}33335$

Now we consider a very small contact area. For the initial vector $(\rho , v') = (1{,}0)$, we obtain :

m	I	ρ	$\Delta r/r$	v'
36	6	$1{,}003.10^{-2}$	3.10^{-3}	0,333345

NUMERICAL RESULTS WITH NONAXISYMMETRIC INITIALIZATIONS

In order to test the capacity of the method, the nonlinear system has also been solved for initializations quite different from the axisymmetric ones. The convergence has always been obtained in less than ten iterations and with the same accuracy.

CONCLUDING REMARKS

The numerical method introduced in this paper is very accurate, especially to find the position of the contact boundary area. The size of the nonlinear system to be solved is very small. The method can be extended to multiconnected contact areas. In many situations, very good results are obtained after two iterations only.

REFERENCES

Brebbia, C.A., 1984, "Topics in Boundary Element Research" Springer Verlag.

Cimetière, A., Léger, A., 1993, un résultat de différentiabilité dans un problème d'obstacle pour des poutres en flexion. *C.R. Acad. Sci. Paris, t 316, Série I, p. 749-754.*

Kinderlehrer, D. Stampacchia, G., 1980, "An Introduction to variational inequalities and their applications" Academic Press - New York.

THE ACTIVE SET ALGORITHM FOR SOLVING FRICTIONLESS UNILATERAL CONTACT PROBLEMS

Georges Dumont*

EDF-DER
Mécanique et Modèles Numériques
1 Avenue du Général De Gaulle, 92141 Clamart Cédex

INTRODUCTION

In this paper is presented a method taking into account the frictionless unilateral contact phenomenon within a FEM program. This method is based on the active set algorithm for solving the constrained minimization problem associated with the unilateral contact model. Three main families of contact models are usually used : the first one is based on the penalty method regularizing the contact conditions. The second one (see for example : Bathe and Chaudary (1985), Klarbring (1986a) and Kalker (1990)), in which category falls our algorithm, is concerned with the treatment of the contact geometrical conditions by duality and requires the development of algorithms to detect the frontiers where contact may occur during the calculation steps. In this case, the variational equality turns into a variational inequality and thus involves constrained minimization methods to be solved. The third one, as proposed by Heegaard and Curnier (1993) and by Cescotto and Charlier (1993), often called mixed approach, is based either on the use of performing algorithms dedicated to the resolution of the problem, as the augmented lagrangian method, or on mixed variational formulations. Our approach lies on a node to node linearized contact modelization. The associated variational formulation leads equivalently to a constrained minimization problem, for further reference in Boot (1968), which is presented, with an unique solution characterized by the Kuhn-Tucker relations. The principle of the so called active set algorithm, which differs from the projected gradient algorithm, proposed by Rosen (1960), in the mean of updating the set of constraints, is then presented and discussed on a theoretical point of view, that is without any computer implementation considerations. This iterative algorithm ensures, at each step, the minimization in a convex set of the deformation energy associated to the structure and gives a condition to modify the set of active constraints in order to perform the next step. Eventually, the basis of an original demonstration of convergence in a finite number of steps is given. As a conclusion, an industrial study using the implementation of the active set algorithm in our FEM industrial software, Code *Aster*, is mentioned and the future investigations aiming at improving the behavior of this algorithm and extending it to the friction contact problem are presented.

*Now at École Normale Supérieure de Cachan, Avenue R. Schumman, Antenne de Bretagne, Campus de Ker Lann, 35170 Bruz

Contact Mechanics, Edited by M. Raous *et al.*
Plenum Press, New York, 1995

FORMULATION OF THE CONTACT PROBLEM

Let us consider the classical problem of solids mechanics (see figure 1), with the usual notations :

$$\begin{cases} -\sigma_{ij,j} & = f \quad \text{in } \Omega \\ \sigma_{ij} \cdot n_j & = t_i \quad \text{sur } \Gamma_f \\ u_i & = u_0 \quad \text{sur } \Gamma_d \end{cases} \tag{1}$$

with $\sigma_{ij} = \frac{1}{2} A_{ijkl}(u_{k,l} + u_{l,k})$ in Ω for the elastic behaviour.

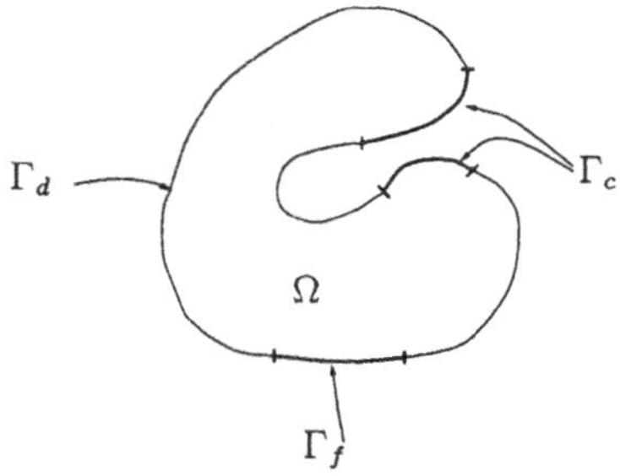

Figure 1 . Position of the contact problem

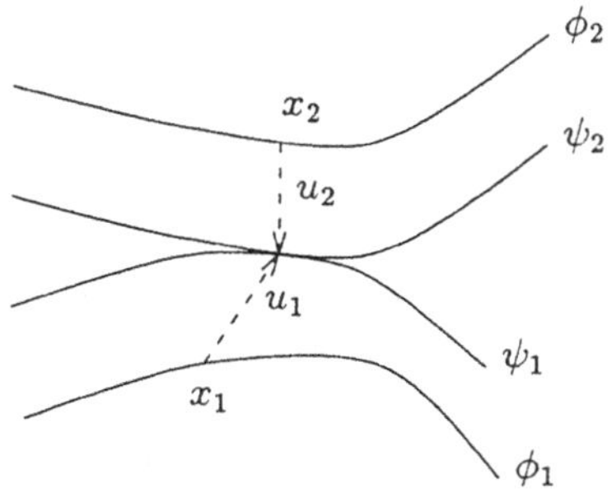

Figure 2 . Frontiers description

The contact condition for a point x of the Γ_c frontier is written as $G(x) \leq 0$ associated with the reaction $R(x) \geq 0$ and with the complementarity condition $R(x)\,G(x) = 0$. The $G(x)$ function cannot be explicitly written and can be linearized at the first order, in accordance with the description of figure 2 where x is an initial position of a particle and u the displacement of this particle :

$$u_1 N_1 + u_2 N_2 \leq -(x_1 N_1 + x_2 N_2) \tag{2}$$

The contact problems then leads to a variational inequality which has a unique solution, as demonstrated by Duvaut and Lions (1972) or by Ekeland and Temam (1974). By the application of the Stampacchia theorem, this solution is equivalently the solution of a minimization problem :

$$\bar{u} \in K \quad | \quad F(\bar{u}) = \min_{v \in K} F(v) \tag{3}$$

This minimization approach has been notably used by Klarbring (1986 and 1992) or by Kalker (1990).

Remark : *Considering the hardening plasticity, the incremental formulation presents the same minimum property, which is to be found in the paper by Kikuchi and Oden (1988) : then, the presented algorithm should be used.*

PRESENTATION OF THE ALGORITHM

Let us consider the resolution of the problem (see equation 3) :

$$(\mathcal{P}) \left\{ \; \bar{u} \in K \quad | \quad F(\bar{u}) = \min_{v \in K} F(v) \quad \text{with } K = \{v \in I\!R^n | Av \leq d\} \right.$$

The active set algorithm is dedicated to the resolution of this minimization problem under inequality constraints $\mathcal{P}$ by an iterative method, replacing it by successive minimizations under equality constraints $\mathcal{P}_I$. Let v a feasible, i.e. v lies in the convex K, point of $\mathcal{P}$. The idea, as proposed by Best (1983), stands in the minimization of F on an affine variety, intersection of the active constraints in v : $K_I = \{(v+\delta) \in I\!R^n | A_i(v+\delta) = d_i, i \in I\}$. Let us note $\Delta_I = \{\delta \in I\!R^n | A_i\delta = 0, i \in I\}$ the associated vectorial space, we have to solve the following sequence of problems :

$$(\mathcal{P}_I) \begin{cases} \min\limits_{\delta \in \Delta_I} F(v+\delta) \\ A_i(v+\delta) = d_i, i \in I \end{cases}$$

Description of the algorithm

Let us describe the algorithm in details. Let v^0 an initial feasible point, I_0 a regular set (I_0 is a set of indexes referring to linear independent constraints) and let start the algorithm in **1** with the iterations counter k equal to zero.

1 :

- we dispose of $v^k \in \mathcal{P}$, of I_k regular and of Δ_{I_k},
- let us calculate $\bar{\delta}^k$ and $\bar{\mu}^k$ associated to the problem $(\mathcal{P}_{I_k})$.
- If $v^k + \bar{\delta}^k \in \mathcal{P}$ go to **2**
- else go to **3**

2 : (iter− : a constraint will be deleted from the set)

- If $\bar{\mu}_i^k \geq 0$, $v^{k+1} = v^k + \bar{\delta}^k$ is the solution of $(\mathcal{P})$. STOP
- Else, find $s \in I$ with $\bar{\mu}_s^k < 0$. Let :

  $v^{k+1} = v^k + \bar{\delta}^k$
  $I_{k+1} = I_k \setminus s$

  Go to **1** with iteration counter set to $k+1$.

3 : (iter+ : a constraint will be added in the set)

- Find v^{k+1} so that $[v^k, v^{k+1}] = [v^k, v^k + \bar{\delta}^k] \cap K$
- Find $r \in L \setminus I$ solution of

  $A_r v^{k+1} = d_r$
  $A_r \bar{\delta}^k > 0$

- Set $I_{k+1} = I_k \cup r$ and go to **1** with iteration counter set to $k+1$.

The finite convergence of this algorithm is proven, in a paper by Dumont (1994), by using the following arguments :

- The sequence of the so constructed points lies in the convex K :
  - either the minimization leads to a point inside K
  - or the constructed point is brought back to the convex frontier in the minimization direction
- by construction, the sequence of $F(v)$ is a minimizing sequence : indeed, for each iteration k, we have $F(v^k) \geq F(v^{k+1})$
- if, during one iteration k, a constraint is deleted from the active set, then $v^{k+2} \neq v^{k+1}$, this ensures a strict decrease of the function F : $F(v^{k+1}) > F(v^{k+2})$
- the set of possible problems of minimization under constraints is finite.

CONCLUSION AND FUTURE WORK

For an industrial study, a thermomecanical simulation of a vessel head penetrator (see figure 3) has been realized. It consists in the fretting and welding of an elastoplastic penetrator. It validates the displacement and stress computations with respect to measures and also the applicability of the algorithm for industrial cases for which it is now in current use inside the firm.

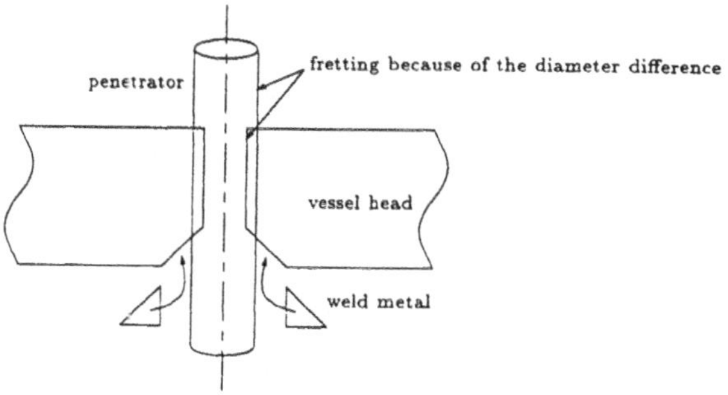

Figure 3 . Plan of a vessel head penetrator

In the future, we will develop the applicability of the algorithm to the frictional case. Because this case can be transformed into a minimization problem, we foresee to use two imbricated steps of minimization by means of the presented algorithm. Concerning the large displacements, it is to be noticed that the linearized equation of contact holds inside a nonlinear step and so a method with geometry actualization is planed.

REFERENCES

K. J. Bathe and A. Chaudary. A solution method for planar and axisymetric contact problems. *International Journal for Numerical Methods in Engineering*, 21 : 65–88, 1985.

A. Klarbring. A mathematical programming approach to three-dimensional contact problems with friction. *Computer Methods in Applied Mechanics and Engineering*, 58(2) : 175–200, 1986.

J. J. Kalker. *Three-dimensional Elastic Bodies in Rolling Contact.* Kluwer Academic, 1990.

J. H. Heegaard and A. Curnier. An augmented lagrangian method for discrete large-slip contact problems. *International Journal for Numerical Methods in Engineering*, 36(4) : 569–593, 1993.

S. Cescotto and R. Charlier. Frictional contact finite elements based on mixed variational principles. *International Journal for Numerical Methods in Engineering*, 36 : 1681–1701, 1993.

J. C. G. Boot. *Programmation quadratique : Algorithmes, anomalies, applications.* Dunod, 1968.

J. B. Rosen. The gradient projection method for nonlinear programming : Part 1- linear constraints. *Journal of Applied Mathematics*, 8 : 181–217, 1960.

G. Duvaut and J. L Lions. *Les inéquations en mécanique et en physique.* Dunod, 1972.

I. Ekeland and R. Temam. *Analyse convexe et problèmes variationnels.* Dunod, 1974.

A. Klarbring. Mathematical programming and augmented lagrangian methods for frictional contact problems. In A. Curnier, editor, *Proceedings of Contact Mechanics International Symposium*, pages 409–422. Presses Polytechniques et Universitaires Romandes, October 1992.

N. Kikuchi and J. T. Oden. *Contact Problems in Elasticity : A Study of Variational Inequalities and Finite Element Methods.* SIAM, 1988.

M. J. Best. Equivalence of some quadratic programming algorithms. *Mathematical Programming*, 30 : 71–87, 1984.

G. Dumont. Algorithme de contraintes actives et contact unilatéral sans frottement. *To appear in Journal européen des éléments finis*, 1994.

MODELLING OF SLIDING WAVE PHENOMENON, ON THE CONTACT BOUNDARY BETWEEN TWO BODIES, BY THE BOUNDARY INTEGRAL ELEMENT METHOD : NUMERICAL VISUALIZATION OF ISOCHROMS

Danielle Fortuné and Bruno Deshoullières

Université de Poitiers
Laboratoire de Mécanique Théorique EA 1217
40, Avenue du Recteur Pineau
Poitiers Cedex 86022

INTRODUCTION

The aim of the present numerical study is to model a phenomenon of propagating stress waves on the contact boundary between two bodies. The photoelasticimetry studies achieved by PROGRI R , VILLECHAISE B (1984,1984), MOUWAKEH M (1989), ZEGHLOUL T (1992) show this phenomenon. In particular, it is displayed that the global sliding is due to the crossing of a localised perturbation of the isochroms field over all the contact zone. The tools of this numerical modelling are : an incremental formulation of the evolution problem, the boundary integral element method to solve the problem, the Coulomb friction law or non linear law, non local law with a variable coefficient of friction.

CHARACTERISTICS OF THE EVOLUTION PROBLEM STUDIED

The experimental studies have been carried out on a plane rectangular polyurethan test plate Ω standing on an araldit foundation (fig 1). The boundary Γ of Ω is constitued of the side $\Gamma_c = (AB)$ which will be a possible contact ligne, Γ_F free of stresses sides and the upper part Γ_u on which a mechanism commands a vertical displacement u_2 constant while one commands an horizontal displacement from 0 to u_ℓ during a period of time T.

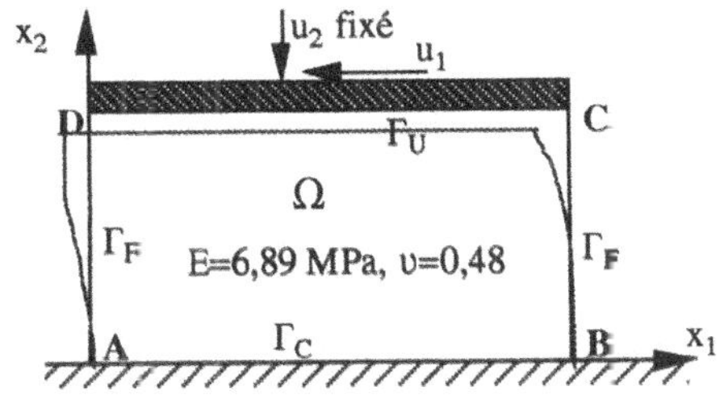

Fig. 1. The evolution problem studied.

Contact Mechanics, Edited by M. Raous *et al.*
Plenum Press, New York, 1995

On each point of the boundary Γ, we denote the cartesian components of
the displacement : $u = (u_i)$
the outward unit vector : $n = (n_i)$
the stress vector : $\sigma n = (\sigma_i) = (\sigma_{ij} n_j)$
and the tangential and normal components of u and σn with indices T and N

FORMULATION OF INCREMENTAL PROBLEM

The quasi-static problem of evolution is approached by an incremental form by means of a temporal semi-discretization. For a given loading history, on a period of time (0,T) splitted into K + 1 intervals (t_p, t_{p+1}) p ranging from 0 to K, the time derivative is approximated by finit differences. We denote $u(t_p) = u_p$ sum of incremental solutions

$$u_{p+1} = u_p + \Delta u_p \quad \text{et} \quad u_p = \sum_{p=0}^{p-1} \Delta u_p + u_0$$

u_{p+1} is calculated from u_p through solving the incremental problem

- div $\sigma(\Delta u_p) = 0$
- $\sigma_{ij}(\Delta u_p) = a_{ijkl}\, \varepsilon_{kl}$
- $\sigma_i(\Delta u_p) = 0$ on Γ_F
- $u_o = (0,u_2)$ on Γ_u, at $t = 0$
- unilatéral contact onΓ_c

$$\sigma_N(u_{p+1}) \leq 0 \quad u_{p+1N} \leq 0 \quad \sigma_N(u_{p+1})(u_{p+1N}) = 0$$

- friction laws on Γ_c

$$|\sigma_T(u_{p+1})| \leq f|\sigma_N(u_{p+1})|$$

$$\text{if } |\sigma_T(u_{p+1})| < f|\sigma_N(u_{p+1})| \quad \text{then} \quad \Delta u_{pT} = 0 \quad \text{(adherence)}$$

$$|\sigma_T(u_{p+1})| = f|\sigma_N(u_{p+1})| \quad \text{then there exist } \lambda \geq 0 \text{ such as}$$

$$\Delta u_{pT} = -\lambda \sigma_T(u_{p+1}) \quad \text{(sliding)}$$

we can use non local and non linear friction laws instead of local friction laws as ODEN JT and PIRES E (1983) and variations of friction coefficient as RAOUS M and TOPIN S (1989).

BOUNDARY INTEGRAL EQUATION METHOD. DISCRETIZATION OF THE BOUNDARY

The foundamental solutions E_{ij} of the Lamé operator for plane elasticity, introduced in the equation of three potentials (BREBBIA and WALKER , 1979) involve an expression of the displacement at a point x of Ω

$$u_i(x) = \int_\Gamma E_{ij}(x,y)\, \sigma_j(y) dS_y - \int_\Gamma T_{ij}(x,y)\, u_j(y) dS_y \qquad (1)$$

T_{ij} are the stresses associated to the displacements E_{ij}

A sharp study of the singularity on the boundary, it comes an boundary integral equation linking the displacements and the stresses on Γ

$$\frac{1}{2}\delta_{ij} u_j(a) = \int_\Gamma E_{ij}(a,y)\, \sigma_j(y)\, dS_y - \int_\Gamma T_{ij}(a,y)\, u_j(y) dS_y \qquad (2)$$

Γ is divided into straight segments Γ_m, the wedges being located at the extremity of segments. The discretization of the previous boundary integral equation under the assumption that the functions u_j and σ_j are to be replaced by their mean values on Γ_m involves the solving of a linear system

$$SX_{inc} = S_0 X_{don}$$

The dimension of the matrices S and S_o is reasonable because it is linked at the discretization of the boundary (one hundred segments). The vectors X_{inc} and X_{don} are constituted with the values of the components of displacements and stresses on each segment of discretization. The difficulty to resolve this problem is due to the fact that one doesn't discern the knowns and unknowns on the zone supposed to be in contact.

NUMERICAL ALGORITHM FOR UNILATERAL CONTACT AND FRICTION

A computer program BIEM-CONTACT is developed at laboratory in the case of plane elasticity. The conditions linked to unilateral contact and to Coulomb friction laws have been treated by an iterative scheme which lead the numerical solution to fullfill the different inequalities. Each filtring associated to a sliding state, adherence, non contact, involves a reestablishment of matrices S and S_o. Different modules have been implanted in this calcul code in order to smooth the values of tangential stress, less stable than the normal one. The non-local laws bring finit values at the wedge of the domain. The non linear laws bring to light micro-displacements which have same lenght as the relative sliding. The decreasing of the friction coefficient while sliding improves the quality of the numerical modelling.

We obtain a non contact boundary part located at the right of the plate, local slidings arise on the left side and propagate to the right part. Finally, it is present on all the zone (AB). The global sliding is obtained for u_1 = -1,8 mm. For a better analysis of this modelling of phenomenon studied experimentaly by photoelasticimetry, we are going to build the isochroms network in the domain Ω.

NUMERICAL VISUALIZATION OF ISOCHROMS NETWORK

The numerical modelling of the contact problem give the mean values of displacements (u_j) and stresses (σ_j) on each element of discretization of the boundary. The functions piecewise constant have been smoothed by an interpolation method by cubic splines. We inject these approached values in the integral equation (1). It gives the value of displacements at each point of the domain Ω. The integrals don't show anymore singularities. They are approximated by the Simpson-method by means of a finit element discretization of the domain independently of the finit element discretization of the boundary. The system is solved once for each incrementation. The building of the level curves for the subtraction of the principal stresses permeats us to display numericaly the evolution of isochroms network. We obtain the following isochroms network for f = 0,6 and u_1 = -1,7 mm.

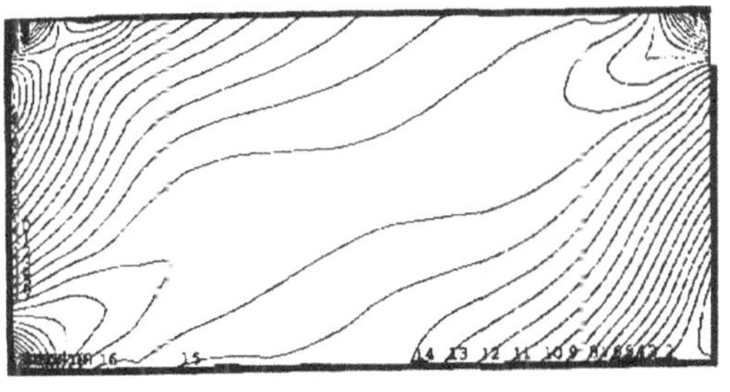

Fig. 2. Isochroms network for u_1= -1.7mm and constant friction coefficient.

As recommended by RAOUS M and TOPIN S (1989) the use of friction law with friction coefficient decreasing according to te speed of sliding assures a quality numerical modelling of the phenomenon observated. The trigging of local sliding, their propagation along the contact and the global sliding appeare sooner which match well with the experimental results (fig. 3).

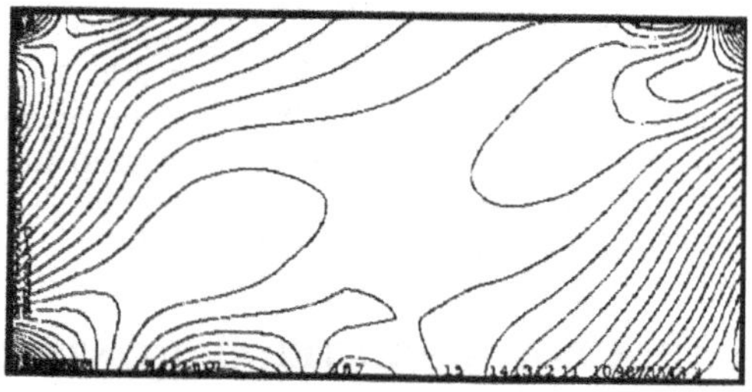

Fig. 3. Isochroms network for u_1 = -1,7 mm and variable friction coefficient.

If the numerical study effectuated by BIEM in our laboratory, properly speaking don't bring new results on the phenomenon of sliding waves already modeled by the finit element method and the method of interfacial fissuration, it brings an flexibility to use means for solving the problems relatively complex of contact friction. The linear systems which are to be traited have small dimensions and the finely discretization of the boundary is made only in the zone where contact will be happen. The time of computation around the minute is very reasonable. The computations inside the domain are effected without iteration, like for a problem with boundary conditions given. The study displayed show the quality of the code of calculus, and her performances with adaptable modules for different modelling of contact friction.

REFERENCES

Brebbia C.A. and Walker S., 1979, "The boundary element method techniques in Engineering". Butterwords, London.

Deshoullières B. and Fortuné D., 1988, "A boundary integral equation approach to friction contact problems of classical elastostatics", Proccedings of the first European Boundary Element Meeting, Brussels.

Deshoullières B., 1990, "Contact avec frottement sec entre solides. Résolution pour la loi de COULOMB et des lois non classiques par la méthode des équations intégrales de frontière" Thèse Doct, Université de Poitiers, pp. 116.

Deshoullières B. and Fortuné D., 1991, "Application de la méthode des équations intégrales de frontière aux problèmes de contact avec frottement . Lois non classiques et problème quasi-statique, 10e Congrès Français de Mécanique, n° 2, pp. 129-132.

Mouwakeh M., 1989, "Etude quantitative des phénomènes de glissements dans un contact sec à deux corps par comparaison avec la propagation d'une fissure interfaciale", Thèse Doct INSA Lyon, p. 206.

Oden J.T. and Pires E., 1983, "Non local and non linear friction laws and variational principes for contact problems in elasticity". J. Appl. Mech., n° 50, pp. 67-76.

Progri R. and Villechaise B., 1984, "Physique des surfaces et des interfaces - Analyse des glissements dans un contact sec", C.R.A.S. Paris, T 299, Serie II, n° 12, pp 763-768.

Progri R., 1984, "Contribution à l'étude théorique et expérimentale des conditions aux limites d'un contact sec avec frottement", Thèse Doct Etat, Université P. et M. Curie, Paris VI, p 325.

Raous M. and Topin S., 1989, "Modélisation théorique et numérique d'ondes de glissement dans un contact adhésif". Rapport de contrat DRET n° 87/029, Marseille.

Zeghloul T., 1992, "Etude des phénomènes d'adhérences et de glissements dans un contact entre solides : Approche expérimentale et modélisation". Thèse Doct, Université de Poitiers, p. 188.

NUMERICAL METHODS OF STATICS AND DYNAMICS OF CONTACT SYSTEMS WITH ARBITRARY NUMBER OF 3D ELASTIC BODIES

Edward Gol'nik[1], Ivan Radchenko[1]

[1]State Technical University
Voronezh

We analyse systems with any finite number of linear-elastic bodies which interact in contact and have, in a certain sense, arbitrary massive, thin-wall and combined configurations, mutual displacements, variants of loading, initial (design and changeable with time) clearances, interference fits and their combinations in large and small contact zones on different surfaces. It is accepted a priori that there is no dissipation of energy, and the appropriate stability of balance or movement takes place.

In the present research we were guided by the purposes created by the development of the machine-building industry and the related branches where analysis and forecasting of the states of part systems from the point of view of the contact mechanics is a necessary stage in the technological design.

Discrete description of presumable contact zones is based on the principle of the method of forces (Conry and Seirey,1971; Haug a. o., 1977) that we consider to be the most efficient for minimization of the calculations at the stage of iterative search for contact zones. The bodies under modelling are approximated according the finite element method (FEM) in displacements. The structure of the algorithms for problems of statics and dynamics of contact systems is invariant to the methods of digitisation.

In the system under modelling we distinguish between the two bodies μ and λ interacting with each other. In the contact joint zone with initial clearance $\Delta^{\mu\lambda}$ in each pair i of the conjugate nodes we introduce the basic unknowns – the nodal contact forces X_i ($i = 1,\dots,N^{\mu\lambda}$) directed along the same normals to the contact surfaces as the nodal displacements q_i^{μ}, q_i^{λ} to be found which satisfy the condition of consistency of contact

$$q_i^{\mu} + q_i^{\lambda} \leq \Delta_i^{\mu\lambda}. \tag{1}$$

In the general case, contact forces X_f and X_h from the "third" body side affect each body μ and λ.

Developing the condition (1) for the static and dynamic states of the system leads to the solving equations that differ in principle but, at the same time, have the structural similarity mentioned below.

Conditions (1) of consistency of static contact are constructed considering the fact that each of the bodies: may have from zero to six kinematic degrees of freedom; it is approximated by an ensemble of elements at separate coding in the movable coordinate system connected with the present body ; in every node i it has the displacement consisting of the two components – due to the deformation of the body and due to its kinematic mobility. Static coefficients of influence ξ_{ij} as well as weight coefficients ξ_{iP} are calculated from the corresponding single and external specified forces with imposing conventional constraints on the kinematically free body.

In the above-mentioned pioneer publications the dependence of coefficients of influence on conventional constraints of the bodies was not given attention to. We (Гольник, Радченко, 1987) proved the invariance of solutions using such coefficients that allow us to consider the equations of the offered mixed method of contact forces and transfer displacements as correct:

$$\begin{gathered}\sum_{j=1}^{N^{\mu\lambda}}(\xi_{ij}^{\mu}+\xi_{ij}^{\lambda})X_j+\sum_{f=1}^{N^{\mu}}\xi_{if}^{\mu}X_f+\sum_{h=1}^{N^{\lambda}}\xi_{ih}^{\lambda}X_h+ \\ +\sum_{k=1}^{S^{\mu}}A_{ik}^{\mu}q_k^{o\mu}+\sum_{r=1}^{S^{\lambda}}A_{ir}^{\lambda}q_r^{o\lambda}=\Delta_i^{\mu\lambda}-\xi_{iP}^{\mu}-\xi_{iP}^{\lambda}; \qquad i=1,\ldots,N^{\mu\lambda}.\end{gathered} \tag{2}$$

where $q_k^{o\mu}$ and $q_r^{o\lambda}$ are transfer (kinematic) displacements of bodies μ and λ which together with the contact forces are the basic unknowns of the method; A_{ik}^{μ}, A_{ir}^{λ} are the components of the kinematic matrices (Haug a. o.,1977; Гольник, Радченко, 1987).

Unlike the procedure of quadratic programming (Conry and Seirey,1971; Haug a. o., 1977) for search and possible optimisation of contact zone which, at the same time, complicates the solutions of 3D problems for multi-body systems by the volume of calculations, in the algorithm of the mixed method the system of N equations (2) written for all contact zones is added by S equations of balance of all kinematically movable bodies. So, for body μ:

$$\sum_{l=1}^{N_r^{\mu}}A_{lm}X_l^{\mu}=-P_m; \qquad m=1,\ldots,s^{\mu}, \tag{3}$$

where N_r^{μ} is the number of all nodes of body μ in which the contact forces affecting the body from the side of all adjacent bodies are applied; P_m is the sum of views (m = 1,2,3) and moments (m = 4,5,6) of the specified forces with respect of the chosen coordinate axes.

The united system of N equations (2) and S equations (3) is the basis for the algorithm of the mixed method. We realise the iteration procedure of solving the problems with successive exclusion of nodes with $X_j > 0$ and checking the conditions of mutual non-interference of the bodies outside the analysed contact zones.

To apply this method in industrial design of heavy machines, we developed a purposeful software package MACS intended for Modeling and Analysis of Contact Systems operating in MS-DOS and OS UNIX on PC-386 and more powerful computers.

The architecture of the package is open: the module principle of program design realised for different combinations of hard and flexible module interface allows one to include new modules and solve new types of problems. The library contains wide-spread types of solid and shell elements. Multi-terminal mode of simultaneous operation is provided. The alternative approach allows the user to choose: the way of data preparation including that with the help of preprocessor, the text editor and the procedure of problem solution – single run or with step-by-step breaks. The hierarchical menu contains the appropriate functions of the package. The service tools guarantee efficient diagnostics, step-by-step protocoling and recording of the results. The package is enhanced by a special graphic postprocessor.

The dynamic states of contact systems are described by the equations of kinetostatics:

$$[M]\{\ddot{q}\} + [K]\{q\} = \{P\} + \{X\}, \tag{4}$$

in which global matrices of inertia $[M]$ and rigidity $[K]$ are multiplied by vectors of nodal accelerations $\{\ddot{q}\}$ and displacements $\{q\}$, and the load is expanded in the specified external $\{P\}$ and the unknown contact $\{X\}$ forces.

Equation (4) is solved according to Newmark's implicit scheme (Бате, Вильсон, 1982) with presentation of the dependencies between the variables relating to moments of time t and $t + \Delta t$. The contact forces are written according to the general scheme

$$\{X_{t+\Delta t}\} = \{X_t\} + \{\Delta X\}, \tag{5}$$

where $\{\Delta X\}$ is the vector of additional contact forces.

The search for vectors $\{\Delta X\}$ at each step of integration is executed by the method of contact forces. Unlike the problems of statics, the dynamic analysis of contact systems does not require that kinematic displacements of free bodies be taken into special account because the inertia of the latter allows one to determine dynamic coefficients of influence without imposing conventional constraints.

The offered algorithm is obtained by generalising some well-known results (Шапошников a. o., 1985) for the class of contact systems with any finite number of bodies.

At each step of integration the following operations are executed.

For moment of time $t + \Delta t$ without considering at this step additional forces $\{\Delta X\}$, the so-called "efficient" load $\{\hat{R}^s\}$ is calculated. It consists of the sum $\{P_{t+\Delta t}\}$, $\{X_t\}$ and the members of expansion by Newmark and depends on $\{q_t\}$, $\{\dot{q}_t\}$ and $\{\ddot{q}_t\}$.

"Starting" – without considering $\{\Delta X\}$ – nodal displacements are obtained from the equation

$$[\hat{K}]\{q^s_{t+\Delta t}\} = \{\hat{R}^s\}, \tag{6}$$

where $[\hat{K}]$ is the efficient matrix of rigidity of the body presented similar to $\{\hat{R}^s\}$.

"Normal" displacements $\{q\}$ of the nodes conjugate in pairs are calculated, and for all pairs i of bodies μ and λ the approximations $\delta_i = q_i^\mu + q_i^\lambda$ counted from the initial clearances are found, then the "starting" residual clearances $\Delta^s_i = \Delta_i - \delta_i$ are obtained.

Vector $\{\Delta^s_{t+\Delta t}\}$ does not satisfy the condition of mutual non-interference of the bodies because additional forces $\{\Delta X\}$ were not taken into account at calculation.

To determine forces $\{\Delta X\}$, we write the condition of contact for nodes of pair i

$$\Delta q_i^\mu + \Delta q_i^\lambda = \Delta^s_i. \tag{7}$$

Additional displacements Δq_i^μ and Δq_i^λ from additional contact forces $\{\Delta X\}$ – taking into account the linear character of deformation in the limits of step of integration – are expressed by the dependencies showing that not only additional contact forces of interaction ΔX_j but also additional forces ΔX_f and ΔX_h from the "third" body side contribute to the change of the clearance between the bodies μ and λ.

Conditions of consistency of dynamic contact (7) (necessary but not enough) are presented by the equations of the following type

$$\sum_{j=1}^{N^{\mu\lambda}} (\hat{\xi}^\mu_{ij} + \hat{\xi}^\lambda_{ij})\Delta X_j + \sum_{f=1}^{N^\mu} \hat{\xi}^\mu_{if}\Delta X_f + \sum_{h=1}^{N^\lambda} \hat{\xi}^\lambda_{ih}\Delta X_h = \Delta^s_i, \tag{8}$$

written for all pairs i of the conjugate nodes of contact zones for bodies μ and λ. Dynamic coefficients of influence ξ^μ_{ij}, ξ^λ_{ij}, ξ^μ_{if}, ξ^λ_{ih} of bodies μ and λ in the nodes of contact zones along the normals are obtained by partial converting the efficient matrices of rigidity $[\hat{K}]$ (similar to obtaining the static coefficients of influence by partial converting matrices of rigidity $[K]$).

The structure of equation (8) for a dynamic problem is considered to be a substructure of static equation (2), if in the latter the complete contact forces are replaced by the additional ones at the analysed step of integration and, besides, if the static coefficients are replaced by the dynamic ones.

Additional forces are obtained in the iterative way from the systems of equations (8). At each step we analyse the nodes in the zones of boundary superposition of the adjacent bodies. The obtained correcting vectors $\{\Delta X\}$ allow us to calculate the contact forces $t + \Delta t$ according to (5). The peculiarities of the iterations lie in the fact that the nodes in which $X_i > 0$ are not excluded (like in statics) but $\Delta X_i = -X_{it}$ is assigned in these nodes. The conditions of mutual non-interference of the bodies outside the analysed dynamic contact zones are checked according to the same method that in statics but for each step of integration. Then, with the help of vector $\{\Delta X\}$ the efficient load is calculated for each body at the end of the step $\{\hat{R}^f\} = \{R^s\} + \{\Delta X\}$. After that the "finishing" displacements $\{q^f_{t+\Delta t}\}$ are obtained from the system of equations of type (6) where indices "s" are replaced by "f".

Nodal velocities and accelerations are calculated from the values of the displacements with the help of the equations in Newmark's scheme.

The correct conditions of consistency of dynamic contact are ensured not only by equations (8) but also by equalities of velocities and accelerations of the conjugate nodes along the normals to the contact surfaces. In order to do this, we correct the equated velocities (using the wave theory) and accelerations (by dynamic averaging). To preserve kinetostatic balance of the conjugate nodes we also correct their contact forces.

Dynamic modelling is realised using the software created to develop the software package MACS.

Test, illustrative and industrial examples of solving problems of statics and dynamics of contact systems show the level of capabilities and efficiency of the considered methods. For instance, on the basis of the software package MACS we have analysed a 3D structure of a heavy crankshaft press (9 components, 18 contact joints). Fig.1 presents a 2D system consisting of two prismatic bodies; the foot of the lower body is fixed, and the upper body is loaded on the face ends by the constant forces after removing of which the process of dynamic contact starts.

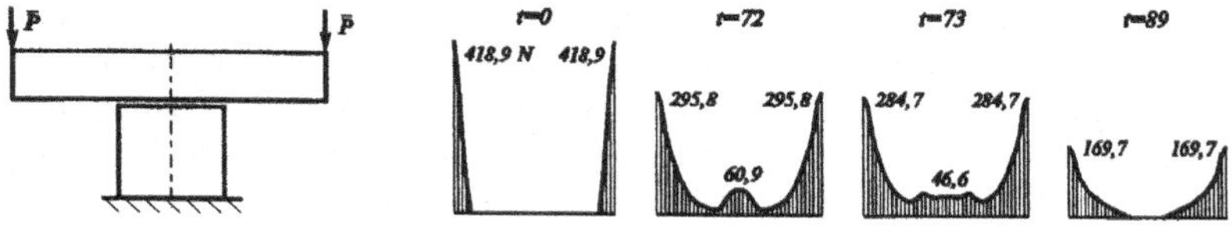

Fig.1. Curves of contact forces for characteristic moments of time (μ s) for dynamic interaction of bodies.

REFERENCES

Conry, T.F., Seirey, A.A., 1974, Mathematical programming method for design of elastic bodies in contact, *Appl. Mech.* 38:387.

Haug, E. J., Chand, R., and Pan, K.C., 1977, Multibody elastic contact analysis by quadratic programming, *Opt. Th. &Appl.* 21:189.

Bate, K., Wilson, E., 1982, Numerical analysis and final element method, Stroiizdat, Moscow

Gol'nik, E.R., Radchenko, I.G., 1987, Discrete modeling of elastic bodies under contact interactions at arbitrary static loads in gaps and ties, Izv. Vuz. Mash. 12:11

Shaposhnikov, N.N., Babayev, V.B., and Senyushchenkov, M.A., 1985, Solution of contact dynamic problems by the final element method using an implicit scheme in the system of strength calculations "SPRINT", in: *Strength Calculations, Issue 26*, editor: N.D. Tarabasova, Mashinostroyenie, Moscow

PLASTIC DEFORMATION OF SURFACE ASPERITIES ASSOCIATED WITH BULK DEFORMATION OF METAL WORKPIECE IN CONTACT WITH RIGID TOOL

Hiroshi Ike*

*Materials Fabrication Laboratory
The Institute of Physical and Chemical Research (RIKEN)
2-1 Hirosawa, Wako, Saitama 351-01, Japan

INTRODUCTION

A metalworking process or plastic working of metals is a basis of modern industrialized lives. The processes are usually composed of tool and workpiece. This means that the technology makes use of contact phenomena between tool and workpiece as a means of adapting the macroscopic shape of a workpiece to the tool at some contact area. Meanwhile, the surface asperities of the workpiece are subjected to plastic straining much severer than the bulk material.

Among various metalworking processes, sheetmetal forming ones, (for instance, deep drawing, punch stretching, and ironing) are especially interested in contact interface between tool and workpiece, because an effective control of overall deformation by surfaces is possible due to bulk material locating adjacent to surfaces.

From tribological point of view, it is well-known that the plastic deformation of surface asperities of workpiece in contact with tool plays an important role in lubrication of deformation processing. For that reason, specially roughened surfaces (usually called "dull-finished") are provided to autobody steel sheets by rolling using roughened temper rolls. Shot-blasted or electro-discharged rolls create irregular 3-D patterns of surface asperities, whereas laser-textured rolls create ring- or crescent-like cavities with projections aside.

In addition, the deformed surface itself is often evaluated directly or indirectly after post-processing from a view point of surface finish.

CHARACTERISTICS OF ASPERITY DEFORMATION OF METALWORKING PROCESSES

The asperity deformation of metalworking processes is characterized by associated elastic and plastic deformation of the workpiece bulk because of gross plastic deformation to be achieved. One typical contact mode in metal working is flattening of surface asperities by a relatively flat tool. It is because difference in hardness between tool and workpiece (usually tool is harder than workpiece more than three to five times) and because of difference in surface roughness.

Lubricating Conditions

In usual metalworking processes, some lubricant is provided to the interface between tool and workpiece to avoid excess friction and surface damage. However, the lubricant must be generally removed in the post processes. Moreover, excess lubricant in some cases,

Contact Mechanics, Edited by M. Raous *et al.*
Plenum Press, New York, 1995

degrades the surface of workpiece by segregating tool from workpiece, resulting in roughened workpiece surface caused by free plastic deformation of the surface layer. Therefore, for technological, ecological and economical reasons, minimum amount of lubricant is supplied in most of the cases. On the other hand, if the sliding distance of the workpiece surface against tool surface is small, usually no lubricants are provided because resultant hydrocarbon films on the workpiece surface, more or less, plays the role of lubricant.

For the above reasons, filling of the surface cavities by liquid lubricant is not always attained. In that case the tool - workpiece interface belongs to boundary lubrication. In the following, attention is focused on boundary lubrication, where vacant cavities freely decrease their capacity until some stages.

On the contrary, if the surface cavities (or micropools) of a workpiece is filled with liquid lubricant and pressurized by the shrinkage of the cavities, mixed lubrication occurs. In that case, some lubrication from the deforming micropool is expected. The surface cavities can substantially survive until the end of metalforming due to the resistance of contained liquid lubricant against hydrostatic pressure. This category is quite different from the boundary lubrication and it should be discussed elsewhere.

At the beginning of tool-workpiece contact, in all cases, asperities with irregular shape are partially conformed to tool surfaces before establishing pressurized micropools. Therefore, wedges of model asperities can be replaced by truncated ones for ease of computation if necessary.

RELATED STUDIES - A HISTORICAL REVIEW

Hill (1950) gave a basic slipline field to the flattening of the wedge and predicted the contact pressure as a function of semi-angle of the wedge. Shindo (1962) constructed another field with rigid fan region. Johnson (1968) obtained a velocity field and a changing profile consistent with an experiment of drawn copper. However, all these modellings lack the bulk material supporting the wedge. For that reason, the models are not directly connected with surface asperity deformation in metalworking processes, where the asperity deformation is strongly affected by the underlying bulk material.

Wanheim (1973), Wanheim *et al.* (1974), and Bay and Wanheim (1975) were interested in deformation of surface asperities of metalworking workpiece. A relation between contact ratio and mean contact pressure was obtained under the constant friction factor. Though they analyzed model asperities with underlying bulk, they did not allow for the bulk plasticity, presumably because of the limitation of slipline field theory.

Wilson and Sheu (1988) dealt with asperity flattening on a workpiece surface undergoing bulk plastic deformation by upper-bound models. They treated a simplified mixed lubrication category by introducing the idea of effective hardness characterizing the pressure difference in fluid and boundary contact regions. This treatment was succeeded by boundary contact analysis of Sutcliffe (1988) by using slipline-field and a related model experiment.

POTENTIAL OF FEM MODELLING IN ASPERITY DEFORMATION

Comparing with the classical theory, upper bound analysis, and slipline field method, the merits of finite element (FEM) simulation of asperity deformation in metalworking processes consist in the possibility of introducing various factors, for example:

1. bulk plastic deformation,
2. strain hardening of workpiece,
3. actual asperity geometry,
4. actual contact geometry,
5. elastic deformation,
6. various frictional relation,
7 non-steady deformation, and
8. heterogeneity of material properties and geometry.

As a results of introducing the above factors, realistic solutions and more information, *e.g.*., stress and strain distributions, are obtainable by FEM simulations.

According to the above merits, since 1990's on, FEM simulations occupy the major seats in numerical analyses of asperity deformation.

Practical Limitations in FEM Modelling

FEM modelling, where both microscopic surface asperities and macroscopic bulk deformation should be analyzed in the same model, number of elements easily increases up to thousands or even to ten-thousands. Owing to the limited memories or computation time, some kind of simplification or hypothesis must be introduced to conduct the calculation in most of the cases.

For the numerical simulation of asperity deformation by FEM, there are three major strategies applicable: namely, 2-D, quasi 3-D, and 3-D.

Naturally pure 3-D analysis without any special hypothesis on deformation mode and periodicity seems preferable. However, it is practically impossible since it requires numerous elements and resulting long computation time. To reduce the number of elements required, some additional assumptions (symmetry and / or mode of deformation *e.g.*, plane-strain) are introduced in actual 3-D modelling.

Quasi 3-D modelling is introduced to reduce the number of elements dramatically. The Quasi 3-D treatment is based on the assumption that each periodic asperity constitutes one unit cell which can cover the whole body of the workpiece in total under the following assumptions throughout the deformation process.

1. All the unit cells cover the body without any separation or overlap. (geometrical continuity)
2. All the interfaces of unit cells fulfill the continuity of normal stresses. (stress continuity)
3. All the interfaces of unit cells fulfill the continuity of velocity across the interface. (velocity continuity)
4. All the unit cells are strained in the same manner. (periodicity of deformation)

Under the above assumptions of *quasi 3-D* model, the deformation of *a single unit cell* is analyzed 3-dimensionally.

Plane-strain assumption or axi-symmetric assumption in some cases yields applicability of 2-D modelling, for example, if the parallel grooves can be assumed. The 2-D analysis sometimes also requires reduced number of asperities actually analyzed, or even limiting to only one asperity to be actually analyzed with additional assumption of periodicity.

THE AUTHOR'S FEM MODELLING OF FLATTENING ASPERITIES

In sheet metal forming, plane-strain deformation, shrink-flanging deformation, and stretch-flanging deformation are the three major modes of deformation. Among them plane-strain is the simplest and easiest for numerical simulation. Therefore, a simple 2-D model of parallel wedges with plane-strain assumption is adopted to analyze the asperity deformation with deformable bulk. In the first model, three asperities on both sides of a sheet workpiece are assumed (hereafter designated by Three-Asperity Model, see Fig. 1(a)). This model was introduced mainly to compare the results between calculation and experiment.

In the next stage after finding a good agreement between calculation and experiment in the Three-Asperity Model, the authors try to exclude the edge effect of the model, because most asperities lie far from sheet edge in a microscopic sense. For that purpose Five Asperity Model, which has five asperities on both sides of the sheet was introduced (Fig. 1(b)). Finally, a model of infinite number of asperities with an assumption of periodic deformation (Infinite-Asperity Model) was adopted (Fig.1 (c))

To analyze the boundary contact, no filling of surface cavities by lubricant was assumed and the friction was neglected on the tool - workpiece interface.

Three-Asperity Model (Makinouchi *et al.*, 1988)

Theory and Computer Code An incremental elastic-plastic finite element computer code is developed and used for the calculation. The code is based on the rate-type updated Lagrangian formulation (McMeeking and Rice, 1975) which can deal properly with the finite deformation. The J2 flow constitutive relationship is used, assuming isotropic elastic-plastic workpiece material with von Mises' yield function and the associative flow rule. The variational principle and the constitutive equation are described elsewhere. (Makinouchi *et al.*, 1988). The R-minimum method (Yamada *et al.* (1968)) is used to determine the

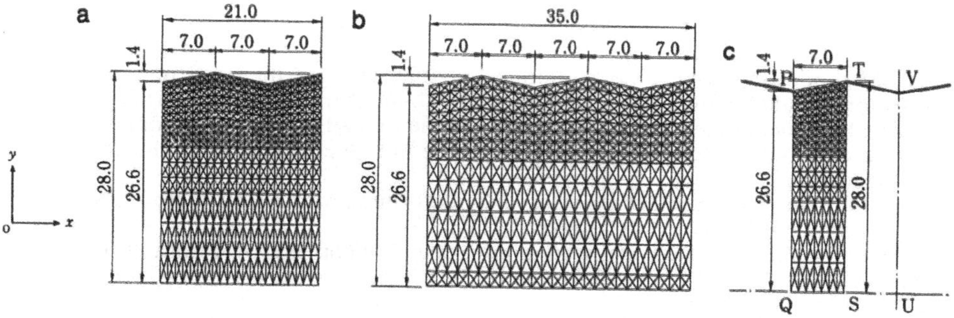

Fig. 1 Cross-sectional geometry of models. (a) Three-Asperity Model. (b) Five-Asperity Model. (c) Infinite-Asperity Model.

incremental step size in computation. In this method, the size of each step is limited by yielding of an element or change in the boundary conditions as explained below.

Computational Conditions The geometry of a Three-Asperity Model specimen used for the computation is shown in Fig. 2. The height-to-width ratio of each asperity is 1:10 (asperity slope: 1/5), so that the wedge angle of an asperity is about 157°. Though the size is expressed in terms of mm to compare with the experiment, it can be regarded in micron to image the actual scales. The model specimen is compressed between two parallel dies with flat and perfectly lubricated surfaces. The dies are assumed to be rigid throughout the calculation. The computations are carried out under the plane-strain condition. The deformation of only one quarter of the workpiece is analyzed owing to the symmetry of the process with respect to the two orthogonal center lines. The horizontal symmetry line of the workpiece (the base line of Fig. 2) is assumed to be stationary throughout the flattening.

Material Properties The model material used in the calculation is a commercially pure soft aluminium whose work-hardening property is expressed by eqn.(1).

$$\bar{\sigma}=118.9\ (0.000\ 69+\bar{\varepsilon})^{0.163} \tag{1}$$

Young's modulus is assumed to be 68.6 GPa and Poisson's ratio is 0.3.

Boundary Conditions Since no friction between die and workpiece is assumed, the boundary conditions for all the nodes in contact with the die surface, which lie along the surfaces G'D'H' and I'F' in Fig. 2(c), are given by eqn.(2).

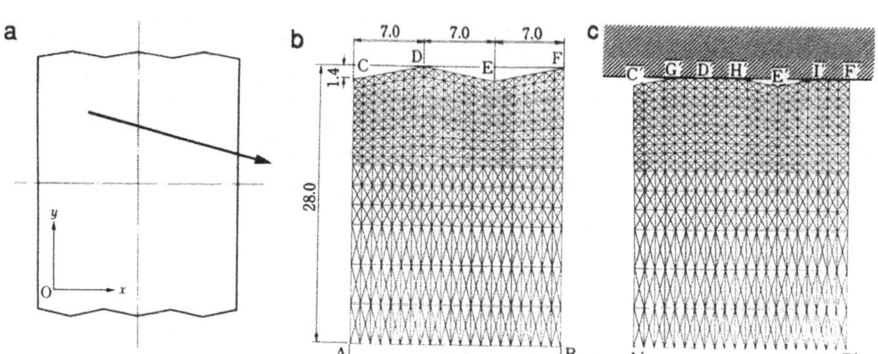

Fig. 2 Cross-sectional geometry of the Three-Asperity Model used in calculation and experiment. (mm). (a) Total cross-section. (b) A quarter of the undeformed specimen. (c) A quarter of the deformed specimen.

$$\dot{f}_x=0 \text{ and } v_y=\bar{v}_0 \ (<0, \text{ constant}) \quad (2)$$

Here, $\dot{f}$ denotes the rate of nodal force, v denotes the nodal velocity, and $\bar{v}_0$ is the downward displacement increment of the die. If the y -coordinate of any node on the free surfaces C'G' and H'E'I' becomes equal to that of the surfaces G'H' and I'F', the boundary condition of the node is changed to eqn.(2) to calculate the next incremental step, assuming that the node is in contact with the die surface thereafter.

Development of Plastic Regions Fig. 3 shows several stages of the calculated flattening of a quarter of a deformed workpiece under load. The stages are primarily specified by "height reduction δ " defined by the normal displacement of the upper die (and hence of the asperity tops) after the initial contact. The height reduction δ is not equal to the actual height reduction of the asperities, because of the displacement of the valley bottom. The darkened parts in the figure show the plastically deformed regions and the white parts are the elastic regions. To have the general expression of flattening stages, "contact ratio R", defined by sum of G'H' and I'F' divided by C'F' in Fig. 2, is also used.

The two independent plastic regions under the tops of the asperities show similar shape and size up to the height reduction δ of 0.10 mm (contact ratio: 10%). At δ=0.20 mm (contact ratio: 15%), the plastic regions under the two asperities are connected with each other, and then the plastic region spreads towards the centre of the specimen. By δ of 0.3 mm (contact ratio: 20%) to 0.5 mm (contact ratio: 31%), an isolated elastic region is formed just under a valley. It should be noted that it still exists at the height reduction of 2.0 mm, where ε_{yy} at the central bulk amounts 4.5 %.

Comparison with Experiment An associated experiment was conducted (Makinouchi *et al.*, 1988). The development of plastic regions was experimentally obtained by a moire method. A qualitative agreement between calculation and experiment was quite good, namely moire results show quite similar pattern of plastic regions at somewhat larger height reductions. This delay can be attributed to the sensitivity of the experiments.

The profiles of the specimens were also compared. The agreement between numerical and experimental was quite good even quantitatively. The lateral bulging of the bulk material is more active at the top region of side surface than at the central.

To sum up elastic-plastic deformation of flattening of three asperities is well simulated by the 2-dimensional Three-Asperity Model.

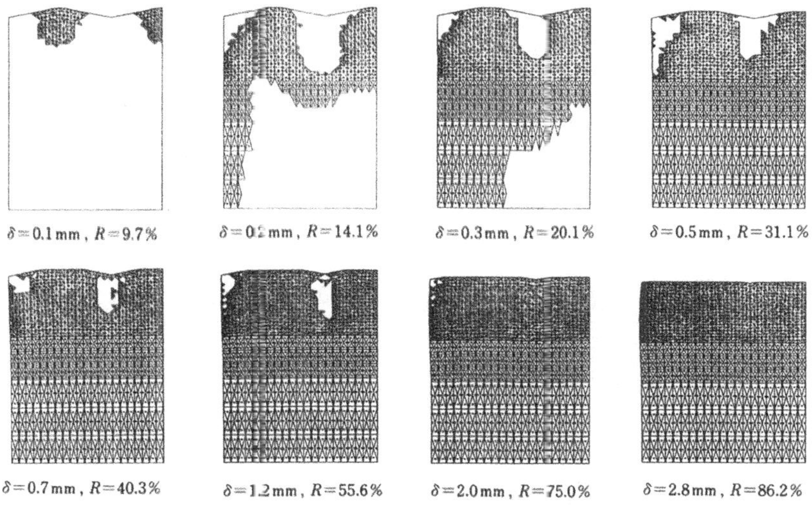

Fig. 3 Development of the plastically deformed regions in a quarter of the Three-Asperity Model workpiece. Commercially pure soft aluminium. Gray area: plastic. White area: elastic. δ denotes the downward displacement of the upper die after initial contact. R denotes contact ratio.

Further Modelling (Ike and Makinouchi, 1990a)

Five-Asperity Model In the next trial, the authors analyzed so-called Five-Asperity Model (Fig. 1(b)) to know the deformation behaviour of the asperities locating far from side edge. It was found that the central asperities both in the Three- and Five- Asperity Models behave in the similar manner, suggesting that the deformation behaviour of general asperities far from edge can be approximated by these central asperities.

Infinite-Asperity Model Based on these findings, a model of infinite number of asperities (hereafter to be denoted by the Infinite-Asperity Model) with additional assumption of periodic deformation was introduced

In the Infinite Asperity Model (Fig. 1(c)), planes PQ and ST are assumed to remain normal to a center-line QS throughout the flattening process. This assumption seems reasonable since surface asperities are generally located far away from the side surface, and therefore, all the cells are subjected to almost the same (periodic) deformation. At the same time, no friction between die and workpiece is again assumed.

The repeated boundary condition used for the Infinite-Asperity Model assumed a uniform but unknown displacement increment in the x-direction for all the nodes on PQ in Fig. 1(c). In order to solve this problem, the slave node method (Yamada and Yokouchi, 1981) is employed as follows. Assuming that no friction exists at the die-workpiece interface, and that the x-component of resultant force acting on plane PQ is zero, the following relations are obtained:

$$\Sigma \dot{f}_x = 0 \quad \text{(to be summed for all the nodes on PQ) and} \tag{3}$$

$$v_x = v_x^P. \tag{4}$$

for all the nodes on PQ where v^P denotes the velocity of node P. The stiffness equation should be solved under the constraint conditions (3) and (4). Because of the symmetry and periodicity, the deformation of the left half of one asperity with the underlying bulk (PQST in Fig. 1(c)) is actually analyzed.

Comparison of the Three Models The distribution of equivalent plastic strain in percentage is shown in Fig. 4. The center asperity in the Three- and Five- Asperity models show a distribution pattern of equivalent plastic strain similar to that of the Infinite-Asperity model. The most remarkable feature of the results under the assumptions (unconstrained, plane-strain, and no friction), is the existence of area of minimum strain under the valley.

Effect of Bulk Plasticity on Asperity Flattening (Ike and Makinouchi, 1990a)

Boundary Conditions Further trials to make clear the effect of bulk plasticity (or constraint of bulk deformation) were made by using the Infinite Asperity Model. Here, among the four conditions simulated, typical two conditions of LF and LC cases are described. Laterally Free case (LF) is a boundary condition at which no lateral force acts on the interface of cells (PQ in Fig. 1(c)), as employed in the preceding calculation. Laterally

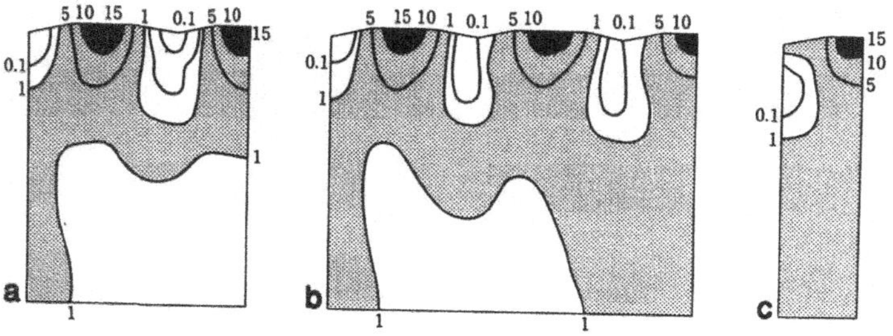

Fig. 4 Distribution of equivalent plastic strain in the three models. δ=0.8μm. Mild steel. (a) Three-Asperity Model. (b) Five-Asperity Model. (c) Infinite-Asperity Model. Figures show equivalent plastic strain in %.

Constrained case (LC) is a condition at which displacement increment in x-direction u_x along PQ in Fig. 1(c) is assumed to be zero throughout the process. LC represents asperity flattening without bulk plasticity or a forming condition similar to closed die forging.

Material Properties The model material used in the following calculation is, in this case, a mild steel. The relation between equivalent stress in MPa and equivalent plastic strain is described as follows:

$$\bar{\sigma}=612.9\ (0.0078+\bar{\varepsilon}^{P})^{0.235} \tag{5}$$

The initial yield stress is 196 MPa. Young's modulus is assumed to be 205.9 GPa, and Poisson's ratio 0.3. The tool material is assumed to be rigid.

Metal Flow Fig. 5 shows the velocity distribution at the three stages of die displacement δ. A comparison at a given stage of δ shows that LC model yields the most marked bulging over non-contacting valleys. The flattening of asperity tops necessarily cause bulging of non-contacting valleys on account of the volume constancy in plastic deformation. LF model, on the contrary, shows nearly uniform deformation of vertical compression and lateral expansion over the whole body.

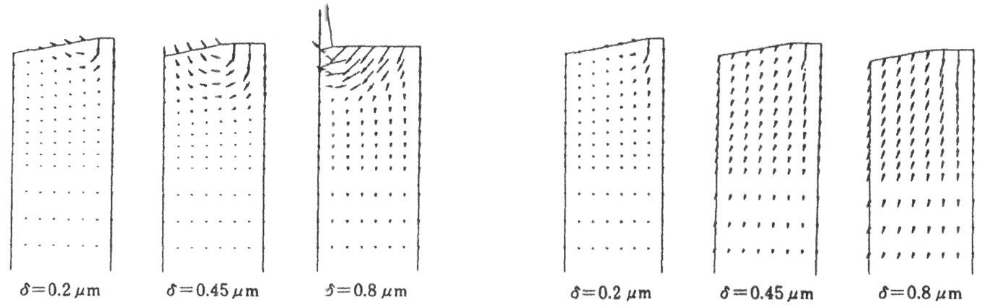

(a) Laterally Constrained (LC) condition. (b) Laterally Free (LF) condition.

Fig. 5 Velocity distribution at three stages of the downward displacement of the upper die δ.

Equivalent Strain Fig. 6 shows the distribution of equivalent plastic strain $\bar{\varepsilon}^P$ at the same stages as in Fig. 5. The distributions of $\bar{\varepsilon}^P$ are not so different at the height reduction of 0.2 μm, because plastic deformation is localized inside the asperity. At δ=0.8 μm, where the difference in deformation patterns becomes quite remarkable as shown in Fig. 5, $\bar{\varepsilon}^P$ is localized over the surface layer under the condition LC, whereas it is distributed in a relatively uniform manner down to the underlying bulk under the condition LF. Regions of very small $\bar{\varepsilon}^P$ still exist at δ=0.8 μm over the surface layer of non-contacting valleys under LF. The generation of these low strained regions can be attributed to the bulk plasticity which suppresses bulge deformation under non-contacting valleys.

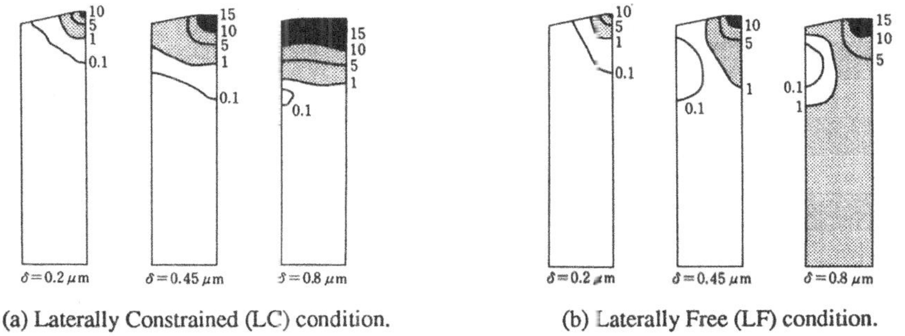

(a) Laterally Constrained (LC) condition. (b) Laterally Free (LF) condition.

Fig. 6 Distribution of equivalent plastic strain at the same δ as in Fig. 5.

Mean Contact Pressure Fig. 7 shows the relation between mean contact pressure p_m (normalized by the initial yield stress) and the downward displacement δ of the die. To reach the same δ, LC model exerts a quite high contact pressure because of lateral constraint of the workpiece.

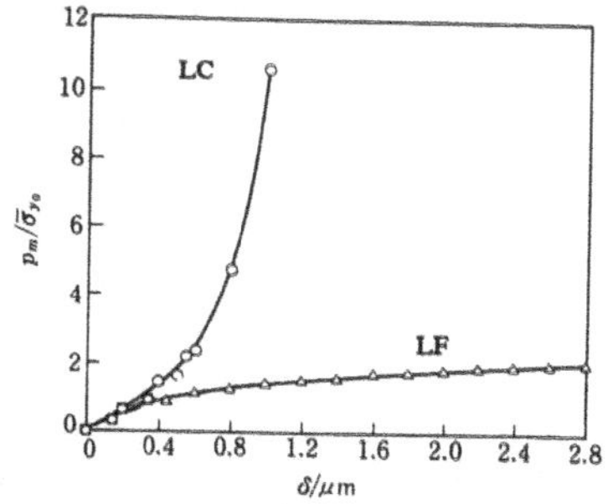

Fig. 7 Evolution of mean contact pressure p_m as a function of δ.

Contact Ratio Fig. 8 (a) shows the evolution of the contact ratio R as a function of δ. An interpolation method is employed in calculating the contact ratio R. The contact ratio in LC is the higher than that of LF at the same δ. This means that the deformation localized in the contact region in LC yields the higher contact ratio with small amount of interfacial slip between die and workpiece. The LC condition is suitable for coining and other metalworking processes whose objectives are to form a precise replica of the tool surface.

Fig. 8(b), on the contrary, shows the evolution of R as a function of mean contact pressure p_m. In this case, LF shows the higher contact ratio under a given p_m in contrast to Fig. 8(a). This means bulk plasticity is connected with promotion of flattening.

Exactly speaking, however, the idea should be expanded because *lateral tension* promotes flattening (Ike and Makinouchi, 1990a). Therefore, a more appropriate idea will be that the lateral tension promotes asperity flattening, whereas lateral compression controls the asperity flattening. Laterally Free condition locates in between.

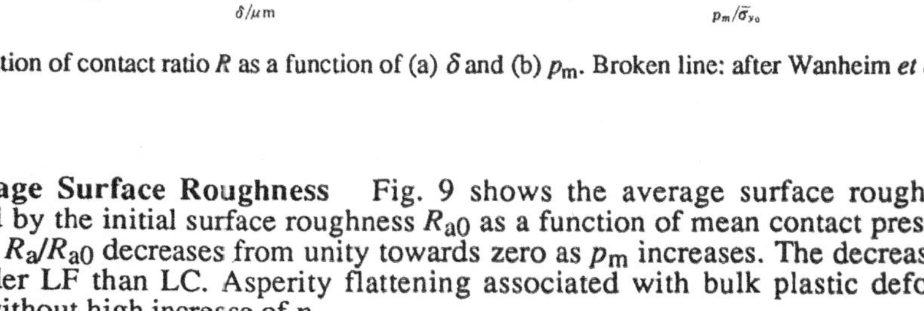

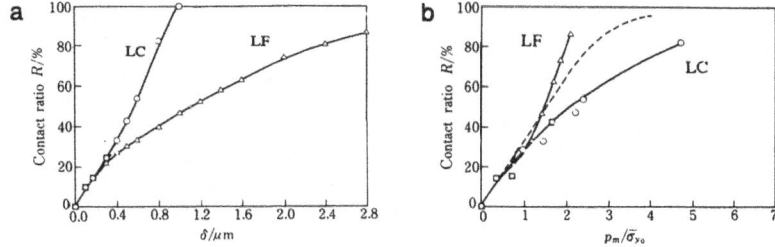

Fig. 8 Evolution of contact ratio R as a function of (a) δ and (b) p_m. Broken line: after Wanheim *et al.* (1974).

Average Surface Roughness Fig. 9 shows the average surface roughness R_a normalized by the initial surface roughness R_{a0} as a function of mean contact pressure p_m. In general, R_a/R_{a0} decreases from unity towards zero as p_m increases. The decrease rate is higher under LF than LC. Asperity flattening associated with bulk plastic deformation proceeds without high increase of p_m.

Real Contact Pressure Fig. 10 shows that the real contact pressure p_r normalized by the initial yield stress $\bar{\sigma}_{y0}$ ranges from 4 to 5 at a contact ratio R of 15% independently of the lateral boundary condition. Under LF, p_r decreases monotonously with increasing

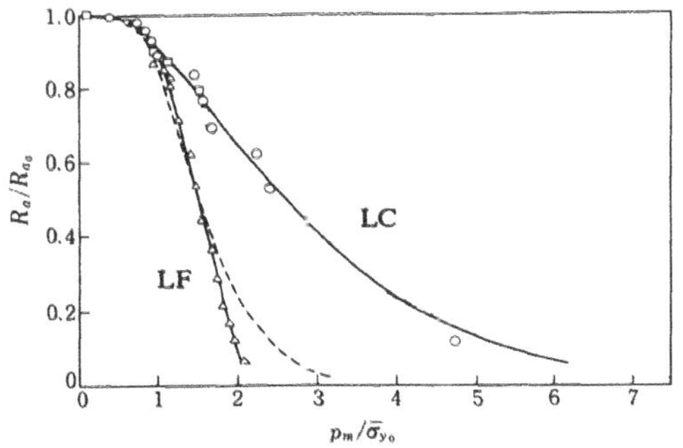

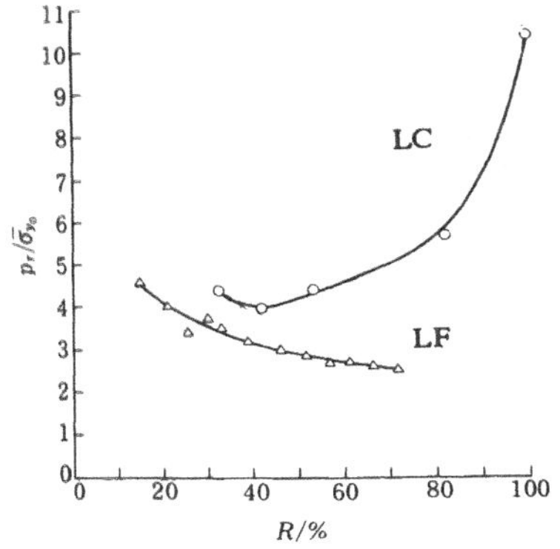

Fig. 9 (left) Change in average surface roughness Ra as a function of mean contact pressure p_m. Broken line: obtained by Bay *et al.* (1975) by slip-line field theory.
Fig. 10 (right) Real contact pressure p_r as a function of downward displacement of the die δ.

R, whereas under LC real contact pressure p_r increases with R at high contact ratio because of increased constraint of metal flow.

Factors

Effect of Workhardening of Workpiece An imaginary material having the same initial yield stress, Young's modulus and Poisson's ratio as the mild steel in the previous Infinite Asperity Model, but with substantially no workhardenability was used in an additional simulation. The relation between equivalent stress and equivalent plastic strain is expressed by eqn.(6) in contrast to eqn.(5):

$$\bar{\sigma}=220.6\ (0.02+\bar{\varepsilon}^{p})^{0.03}\,. \qquad (6)$$

The Infinite Asperity Model as in Fig. 1(c) is used and the lateral boundary conditions of LF (Laterally Free) and LC (Laterally Constrained) are again adopted.

Fig. 11 shows the relation between contact ratio R and mean contact pressure p_m. It shows that p_m is higher for workhardenable material at the same R for both LC and LF conditions.

Fig. 12 shows the distribution of equivalent plastic strain at the height reduction of δ=0.6 µm. In Fig. 12, the workhardenable material shows more uniform plastic deformation than the other at the same height reduction, resulting in (1) wide plastic region, (2) lower maximum equivalent strain in contact region, and (3) lower contact ratio.

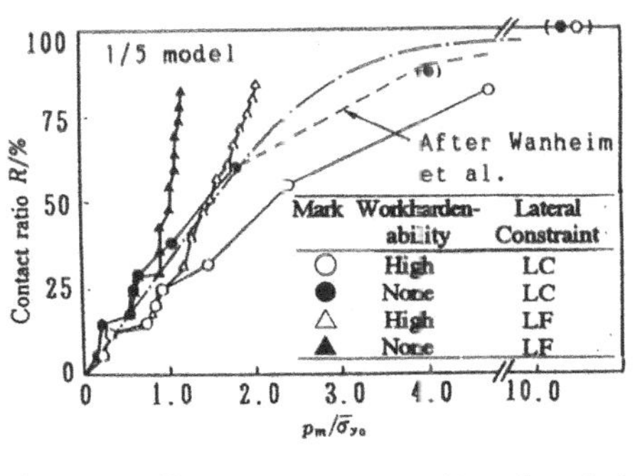

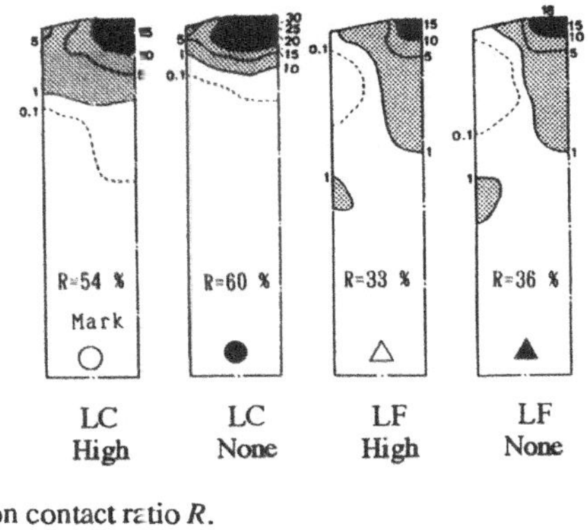

Fig. 11 (left) Effect of workhardenability of workpiece on contact ratio R.
Fig. 12 (right) Effect of workhardenability of workpiece on equivalent plastic strain (%). δ=0.6 µm.

Effect of Asperity Shape (Ike *et al.*, 1988) Previous numerical simulations were made for the asperity with 1 to 10 height-to-width ratio (= asperity slope of 1/5, hence hereafter called 1/5 model). To see the effect of asperity slope, models of different asperity slope (2/5 , and 5/5) as shown in Fig. 13 is introduced. The Infinite-Asperity Model and boundary condition of LC and LF are adopted. The workhardenable mild steel is used.

Fig. 14 shows the distribution of equivalent plastic strain at the same specific displacement of the die, δ / H_0=0.43. In the 5/5 model, plastic deformation is more localized on asperity regions than in the 2/5 model. Basically, the features of LC model and LF model are still preserved in both 2/5 and 5/5 models. In detail, bulging in the non-contact region in LC condition becomes more active in 5/5 model, resulting in increased slope of the valleys. It means that narrow but still deep valleys may remain after considerable flattening in the 5/5 model. On the other hand, in both 1/5 and 2/5 models, the bottom of the valleys comes up by the bulging action in LC boundary condition.

Fig. 15 shows contact ratio R as a function of mean contact pressure p_m. As the asperity slope increases, the effect of workhardenability plays a more important role, because the deformation within asperity is the major deformation pattern rather than the bulk plastic deformation. In 2/5 model, where asperity slope is intermediate, the workhardenability of workpiece plays a major role under low mean contact pressure range, whereas lateral boundary condition plays a major role at the high mean contact pressure range. The linear relation between R and p_m in rather wide pressure range is successfully utilized to develop a pressure sensor to be applied between tool and workpiece (Ike, 1990c).

Fig. 16 shows mean contact pressure p_m required to obtain the contact ratio of 70% as a function of asperity slope. Under Laterally Constrained boundary condition, p_m decreases with asperity slope. This is because high asperity slope expands the stage of actually unconstrained asperity deformation. On the contrary, under Laterally Free boundary condition, p_m increases with asperity slope *with workhardenable workpiece*. This is attributed to the predominant deformation in asperity region, which results in increased contribution of workhardenability of the workpiece.

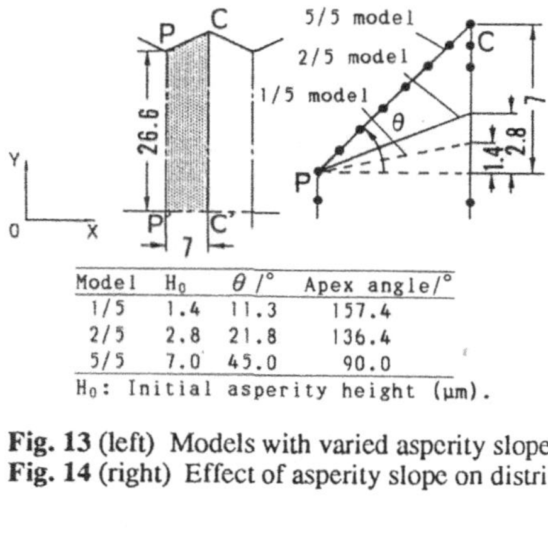

Model	H_0	θ /°	Apex angle/°
1/5	1.4	11.3	157.4
2/5	2.8	21.8	136.4
5/5	7.0	45.0	90.0

H_0: Initial asperity height (μm).

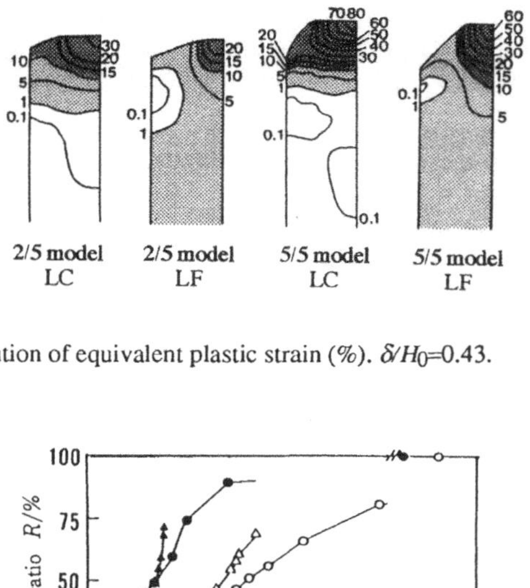

Fig. 13 (left) Models with varied asperity slope.
Fig. 14 (right) Effect of asperity slope on distribution of equivalent plastic strain (%). δ/H_0=0.43.

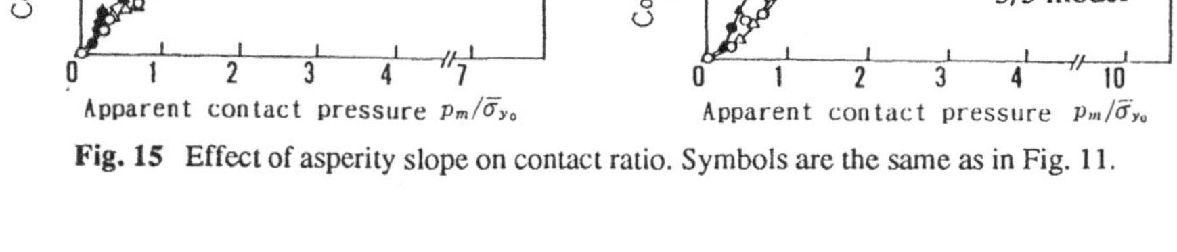

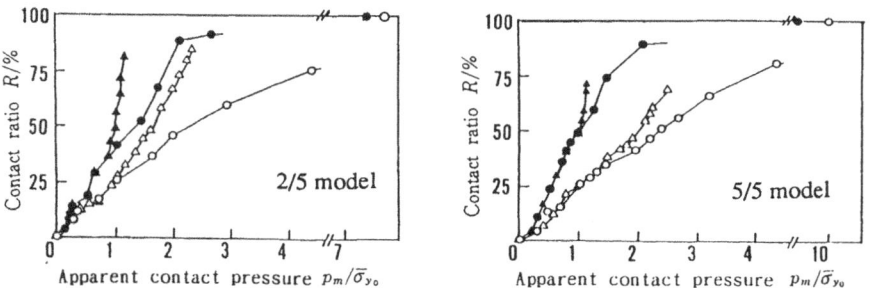

Fig. 15 Effect of asperity slope on contact ratio. Symbols are the same as in Fig. 11.

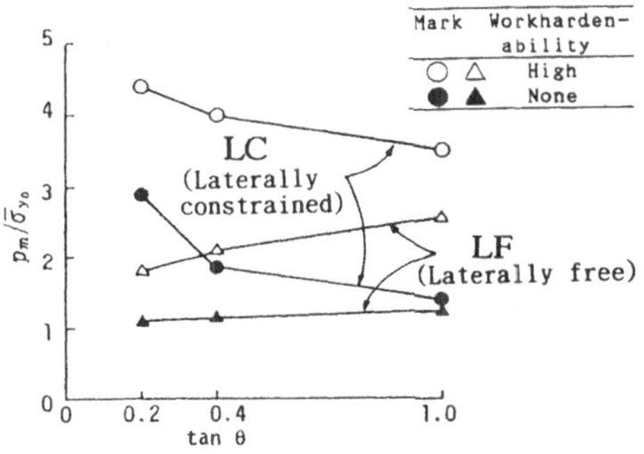

Fig. 16 Effect of asperity slope tanτ on the mean contact pressure p_m at contact ratio of 70 %.

Effect of Sliding and Friction in Asperity Deformation (Ike and Makinouchi, 1990b) In the above numerical simulations, no friction was introduced. Theoretically, it is not realistic because in the real contact, friction is more or less involved. So, an additional simulation was made , where a friction over boundary contact area is introduced by the relation expressed by eqn.(7).

$$\tau_a = R t_f = Rmk \ . \tag{7}$$

Here k=18.2 MPa, and m=0.77. Material properties were assumed to be the same as in eqn.(1). Three asperities on the top and bottom surfaces were assumed. The results show that the basic results obtained for a perfectly lubricated workpiece holds good at least for cases where slight friction is introduced over the boundary contact region.

SUMMARY OF FEM ANALYSES OF ASPERITY FLATTENING

From the above results, the constraint or bulk plasticity of a workpiece plays a decisive role in asperity flattening of metalworking. The results can be summarized as follows.

1. Lateral tension (or bulk plasticity) promotes the asperity flattening by helping the bulk metal flow, whereas lateral compression (or lateral constraint) controls asperity flattening by forcing top surface bulging. At the same contact ratio R, mean contact pressure required is *ca.* 50 to 100% higher in Laterally Constrained (LC) condition than in Laterally Free (LF) condition. The deformation mode varies from uniform compression (in LF) to surface bulging (in LC). In LF, low strained region (even elastic one) survives in surface layer just under valleys *e.g.*, until R reaches around 80%.
2. The workhardenability of workpiece plays an important role of making deformation uniform. The effect is dominant in the initial stage and/or in asperities with high slope.
3. Asperity shape affects the flattening process in the following manner: Increased slope of asperities results in enhanced deformation within asperity region including bulging. The effect of bulk plasticity is more predominant for asperities with a small asperity slope.

RECENT AND FUTURE DEVELOPMENTS

Owing to the rapid progress in computers and computer soft ware, (quasi) 3-D finite element analyses of asperity deformation with plastically deformable bulk became possible.

On the boundary contact where valleys are assumed practically vacant, Hira *et al.* (1989) made a quasi 3-D finite element analysis of sheet deformation under biaxial stresses in plane and subsequent asperity flattening with the assumption of periodic deformation in plane. Korzekwa *et al.* (1992) analyzed asperity flattening using a similar quasi 3-D modelling under various biaxial stress states in plane. Quite recently, Bunten and Kopp (1994) presented a 3-D finite element simulation on temper rolling with laser-textured rolls taking actual surface microgeometry into account.

Raous and Sage (1992) modeled the asperity deformation with tangential shear against identical shape. In the future, the asperity flattening modelling should be expanded to contact and sliding problems with roughened tools, which will simulate surface damage.

Asperity flattening finds another field of application in pressure welding (Takahashi *et al.*, 1993), where a viscoplastic workpiece is analyzed by a 2-D infinite-asperity model.

On the other hand, another concept of mixed lubrication, where bulky liquid lubricant entrapped into surface cavities plays an important role, should be numerically simulated, though a pioneer work was already presented by Azushima *et al.* (1989).

ACKNOWLEDGMENT

This article is based on the cooperative works with Dr. A. Makinouchi. The author is also grateful to Mr. Mikio Kimura for cooperation in computation. The author wishes to express his appreciation to Springer Verlag, Berlin, Elsevier Sequoia S.A., Lausanne, Japan Thermal Spraying Soc. and Japan Soc. for Technol. Plasticity for permission of figure reproduction.

REFERENCES

Azushima, A., Tsubouchi, M., Kudo, H., Furuta, N. and Minemura, K., 1989, Experimental confirmation of the Micro-Plasto-Hydrodynamic lubrication mechanism at the interface between workpiece and forming die, *J. Japan Soc. Technol. Plasticity*, 30:1631.

Bay, N., Wanheim, T. and Petersen, A.S., 1975, *R*a and the average effective strain of surface asperities deformed in metal-working processes, *Wear*, 34: 77.

Bunten, R. and Kopp, R., 1994, Simulation of roughness changings during metal forming processes by FEM, Paper presented at The 3rd World Congress on Computational Mechanics, Chiba, Japan.

Hill, R., 1950, Compression of a wedge by a flat die, *in* Mathematical Theory of Plasticity, Hill, R., Oxford Univ. Press.

Hira, T., Tamari, T., Isobe, K. and Yarita, I., 1989, Deformation of metal surface in sheet forming by finite element method, Paper presented at the North American D.D.R.G. / Japan D.D.R.G. Joint Meeting.

Ike , H., Makinouchi, A. and Kimura, M. , 1988, Plastic deformation behaviour of metal surface layer with varied surface geometery under compression, Proc. Surface Engng. Intern. Conf., Japan Thermal Spraying Society, 387.

Ike, H. and Makinouchi, A., 1990a, Effect of lateral tension and compression on plane strain flattening processes of surface asperities lying over a plastically deformable bulk, *Wear*, 140 : 17.

Ike, H. and Makinouchi, A., 1990b, Effect of plastic deformation of specimen bulk on the flattening process of surface asperities by an unlubricated flat tool, *Proc. Japan Intern. Tribology Conf.*, Nagoya, 557.

Ike, H., 1990c, Structure and properties of pressure sensing foil and its application to measurements of distribution of contact pressures over tool surfaces, *Advanced Technology of Plasticity 1990*, 3 : 1257.

Ike, H. and Makinouchi, A., 1991, Finite element analyses of factors influencing flattening process of surface asperities of workpiece under boundary contact, *J. Jpn. Soc. Technol. Plasticity*, 32: 848. (in Japanese).

Johnson, K.L., 1968, Deformation of a plastic wedge by a rigid flat die under the action of a tangential force, *J. Mech. Phys. Solids*, 16: 395.

Korzekwa, D.A., Dawson, P.R. and Wilson, W.R.D., 1992, Surface asperity deformation during sheet forming, *Int. J. Mech. Sci.*, 34: 521.

McMeeking, R.M. and Rice, J.R., 1975, Finite-element formulations for problems of large elastic-plastic deformation, *Int. J. Solids Struct.*, 11: 601.

Makinouchi, A., Ike, H., Murakawa, M., Koga, N. and Ciupik, L.F., 1987, Finite element analysis of flattening of surface asperities by rigid dies in metal working processes, in "*Advanced Technology of Plasticity 1987*", K. Lange, ed., Springer Verlag, 1: 59.

Makinouchi, A., Ike, H., Murakawa, M. and Koga, N., 1988, A finite element analysis of flattening of surface asperities by perfectly lubricated rigid dies in metal working processes, *Wear*, 128:109.

Raous, M. and Sage, M., 1992, Numerical simulation of the behaviour of surface asperities for metal forming, *in*: "Numerical Methods in Industrial Forming Processes," Chenot, Wood and Zienkiewicz, ed., Balkema, Rotterdam, 175.

Shindo, A., 1962, General considerations on the compression of a wedge by a rigid flat die, *Bull. J. S. M. E.*, 5-17: 21, & 30.

Sutcliffe, M.P.F., 1988, Surface asperity deformation in metal forming processes, , *Int. J. Mech. Sci.*, 30: 847.

Takahashi, Y., Koguchi, T., and Nishiguchi, N., 1993, Effect of bulk deformation on viscoplastic adhering process - A numerical study of solid state pressure welding, *J. Engng. Mater. Technol.*, 115: 171.

Wanheim, T., 1973, Friction at high normal pressures, *Wear*, 25:225.

Wanheim, T., Bay, N. and Petersen, A.S., 1974, A theoretically determined model for friction in metal working processes, *Wear*, 28: 251.

Wilson, W.R.D. and Sheu, S., 1988, Real area of contact and boundary friction in metal forming, *Int. J. Mech. Sci.*, 30: 475

Yamada, Y., Yoshimura, N. and Sakurai, T., 1968, *Int. J. Mech. Sci.*, 10: 343.

Yamada, Y. and Yokouchi, Y., 1981, Introduction to the elastic plastic finite element code EPIC-IV, Baihukan, Tokyo. (in Japanese)

COMPARISONS OF FRICTION MODELS FOR FINITE ELEMENT MODELLING OF CLOSED-DIE FORGING

L. Baillet and J.C. Boyer

Laboratoire de Mécanique des Solides
Institut National des Sciences Appliquées de Lyon
France

INTRODUCTION

Theoretical advances in finite element modelling of contact and friction phenomena are very important but from the software user point of view the available solutions still remain restricted to some improvements of classical friction laws for the particular case of bulk metal forming.

In order to study the application range of the Coulomb-Orowan model, the Tresca model, the power law model and the plastic wave theory model, numerical comparisons are proposed for the ring test, a cylinder upsetting, a backward extrusion and a combined forward and backward extrusion.

FRICTION MODELLING FOR SLOW BULK FORMING PROCESSES

As stated by W.R.D WILSON (1979), four different lubrication regimes can be associated with a given set of contact normal stress and relative velocity :

- a thick film regime when the mean lubricant film thickness is greater than about ten times the R.M.S róughness of the surface of the tool and of the workpiece
- a thin film regime when lubricant thickness is between three and ten times the R.M.S roughness
- a mixed lubrication regime when the lubricant film thickness is less than three times the R.M.S roughness of the surfaces with a significant fraction of the load carried by contact between asperities
- a boundary lubrication regime when the load is only carried by the asperities of the surfaces.

A thick film regime and a thin film regime need relative velocity greater than 500.mm/s and very high lubricant viscosity which seem to be unusual conditions for slow bulk forming processes. As experimental works show that the tool asperities play an important part in

Contact Mechanics, Edited by M. Raous *et al.*
Plenum Press, New York, 1995

the tool-workpiece interface behaviour with or without any kind of lubricant, the friction modelling has to depict the boundary lubrication regime and the mixed lubrication regime.

The Coulomb model assumes the shear stress τ applied on the tool-workpiece interface proportional to the contact normal stress σ_n and acting in the opposite direction of the tool-workpiece relative velocity V_t :

$$\tau = -\mu|\sigma_n|\frac{V_t}{\|V_t\|} \tag{1}$$

The "macroscopic" Coulomb coefficient μ has to include all the local instantaneous parameters as the roughness, the temperature, the local velocity, the lubricant effects and the mechanical property changes of the workpiece material as well as tool wear.

In order to take into account the physical limit of the magnitude of the frictional stress, Coulomb-Orowan model considers the maximum shear stress τ_i carried by the tool-workpiece interface as the upper bound of the Coulomb law :

$$\tau = -\mathrm{Min}\left(\mu|\sigma_n|, \tau_i\right)\frac{V_t}{\|V_t\|} \tag{2}$$

This friction law is frequently associated with elasto-plastic formulation of general purposes finite element softwares. Tresca or constant shear model is more suited to rigid or visco-plastic formulation of dedicated software for bulk metal forming. The frictional stress is proportional to the shear yield stress k of the workpiece material :

$$\tau = -mk\frac{V_t}{\|V_t\|} \tag{3}$$

The m coefficient has to include all the local instantaneous parameters. Some kind of improvements of this simple law is the power or viscoplastic friction model formulated with the tool-workpiece tangential relative velocity V_t and the material consistency k :

$$\tau = -mkV_t\|V_t\|^{q-1} \tag{4}$$

with m the friction coefficient
and q the sensitiveness to the sliding velocity

These simplified views of the interface activity do not include variations of the real contact area when the tool asperities strain the workpiece surface. S. STANCU-NIEDERKORN, U. ENGEL, and M. GEIGER (1994) developed an in-situ ultrasound set-up for measuring the real contact area and the friction coefficient during a ring test and a closed-die upsetting. These important results seem to be in good qualitative agreement with the predictions of the plastic wave theory proposed by AVITZUR (1986) and J.M. CHALLEN, J.M. LEAN, P.L.B. OXLEY (1984).

The works of these authors predict the steady state motion of plastic ridges created in the workpiece by perfectly rigid asperities of the tool. The surface geometry of the tool is simplified by a wedge shape of angle α deduced from the random distribution of the actual profile with statistical formulation. The wedge angle α is the roughness parameter of the tool. A local constant shear coefficient m_0 is assumed for the contact between the tool asperities and the plastic ridges formed at the workpiece surface.

For a perfectly rigid plastic workpiece material, the plastic wave theory leads to a friction stress τ which rises linearly as the contact normal stress increases from zero up to values close to the level of the workpiece material yield stress. With further increase of the contact normal stress, the friction stress rises non-linearly and slowly as an asymptotic function.

Fig.1 and Fig.2 show the normalized friction stress plotted as a function of the normalized contact normal stress for different roughness parameters α and two local constant shear coefficients. Normalized stress stands for the ratio of the stress to the yield stress.

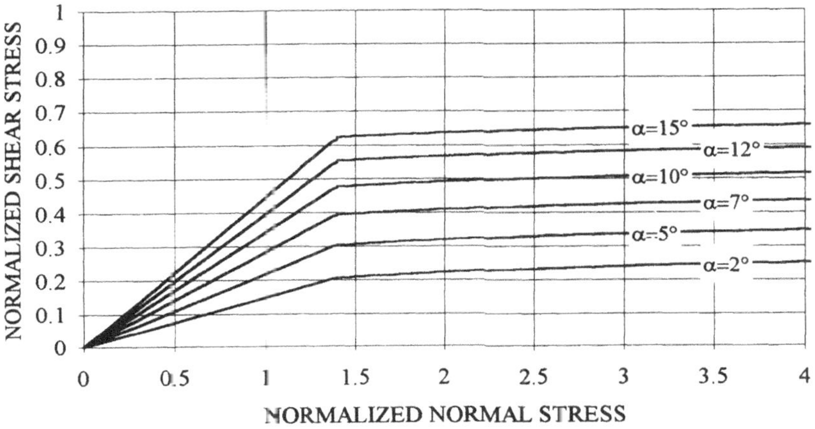

Fig.1 . Plastic wave theory model for m_0=0.2

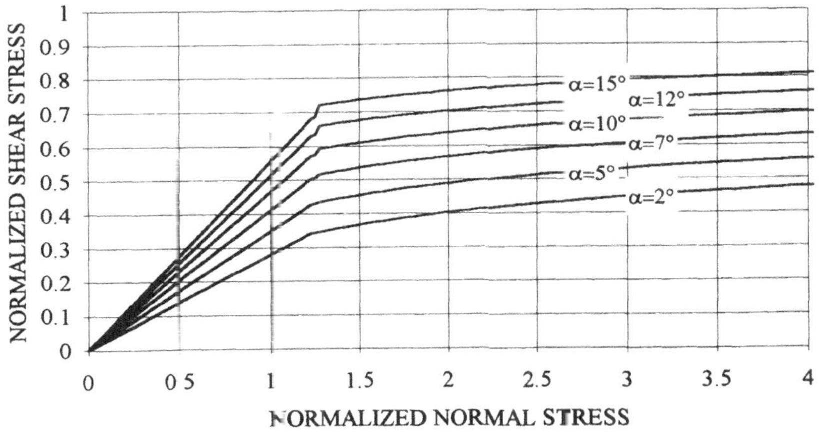

Fig.2 . Plastic wave theory model for m_0=0.5

These predicted trends are in good agreement with experimental studies in the low tangential relative velocity range for low and high contact normal stress.

The plastic wave theory model can be extended to mixed lubrication regime with an hydrostatic contribution of a fluid trapped between the plastic ridge of the workpiece and the tool asperities as proposed by L. BAILLET and J.C. BOYER (1994) with several transient stages depending upon the thickness of the lubricant.

ABOUT COMPARISON OF FRICTION MODELS

The constitutive formulation of friction models in finite element programs may have significant influence on numerical results as they are dependant upon a contact algorithm.

For example, the upsetting modelling, with ABAQUS implicit code and with POLLUX, a code developed by B. MICHEL (1993), of an elastic-plastic cylindrical workpiece without hardening and with a Coulomb-Orowan friction model gives the same interface stress distributions for friction coefficients lower than 0.4 but quite different friction stresses for greater values. The normal stress and the friction stress plotted versus the radius of the contact area on Fig.3 are representative of sticking prediction for ABAQUS and of relative sliding for POLLUX.

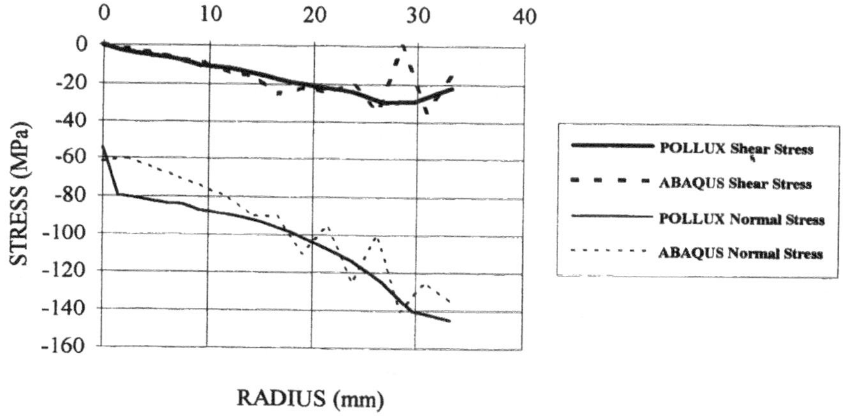

Fig.3 . The normal stress and the friction stress plotted versus the radius of the contact area

For such cases the adjustments of the sticking parameters have a great influence on the numerical response of the interface.

So the four different friction models presented above have been implemented in POLLUX, a finite element program with an elastic-visco-plastic formulation and an initial stiffness scheme. The contact is controlled explicitly with rigid tools and introduces as normal boundary condition by the penalty method. The nodal equivalent friction forces are expressed with the unknown nodal displacements and introduced as friction stiffness added to the global stiffness of the workpiece.

EXPERIMENTAL METHODS OF FRICTION PARAMETER IDENTIFICATION

Each friction mode needs the determination of some constitutive parameter. Three tests are most quoted in the literature for estimating friction in bulk forming processes : the ring test, the spike test and the double cup extrusion test. The limitation of these tests is that the measure of friction is a global value which does not relate the instantaneous conditions which exist at all contact points of the workpiece. E.DOEGE and CH.BEDERNA (1994) proposed an indirect analysis of boundary stress with an elastic measuring tool. The Solid Mechanic Laboratory of the Applied Sciences National Institute (Lyon, FRANCE) is also developing a dynamometric tool for the same purpose, but at present, local contact experimental data are not yet available and the conventional identification still remains in use to evaluate the friction behaviour.

The friction parameters of the Tresca law, the power law and the plastic wave theory model can be globally determined with the ring test carried for example on a 6-3-2 hollow cylindrical test piece under axial compression.

The experimental relative changes in inner contact diameter compared to numerical curves obtained by finite element modelling with different values of the friction law parameter allow model adjustments for similar conditions. Numerical results of the ring test simulation with the Tresca model for a workpiece elastic-plastic without hardening material are presented on Fig.4 for constant shear coefficient varying from m=0 (perfect sliding) to m=0.8 (quasi-sticking friction).

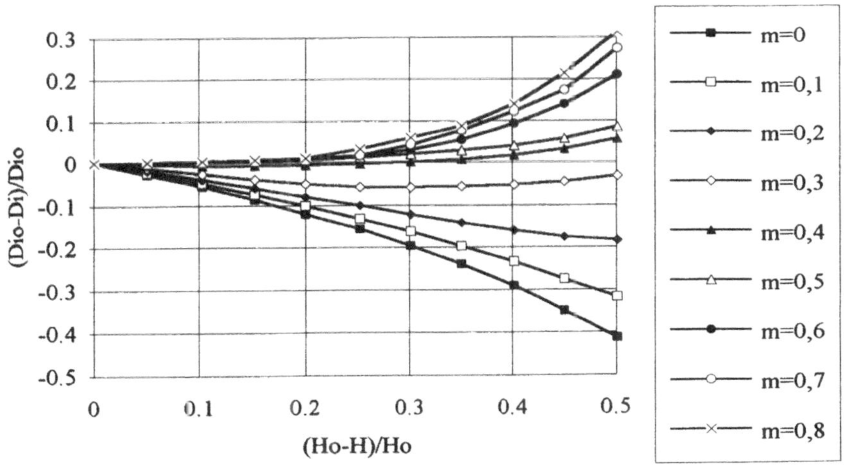

Fig.4 . Ring test simulation with the Tresca model

The same contact algorithm used with the plastic wave theory model give the results, see Fig.5 for different local constant shear coefficient m_0=0. to m_0=0.5 and a roughness angle α=5° which is a tool geometry constant not dependent on the lubrication conditions. The ring test allows the evaluation of the local shear coefficient of the plastic wave theory model

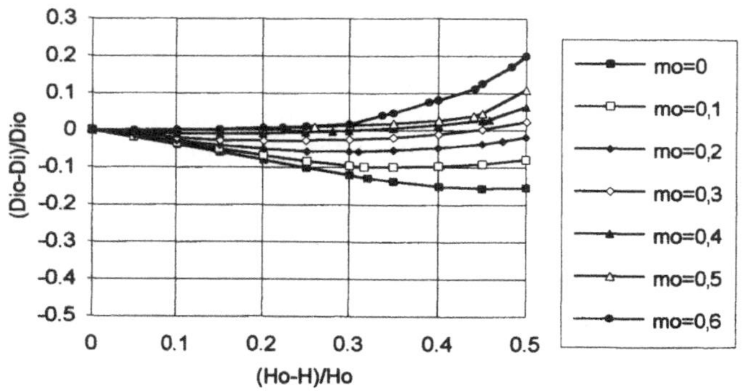

Fig.5 . Ring test simulation with the plastic wave theory model with α=5°

The comparison of the curves sets of Fig.4 and Fig.5 shows discrepancies for low values of the constant shear coefficients. A zero local constant shear coefficient and a roughness angle α=5° predict an inner diameter change similar to a Tresca shear coefficient close to 0.25. The roughness parameter has significant influence on the solution curves corresponding to different values of the roughness angle α for a low local constant shear coefficient m_0=0.2 as presented on Fig.6. The actual value of the roughness angle α has consequences on the friction stress at low and high contact normal stress levels, see Fig.1.

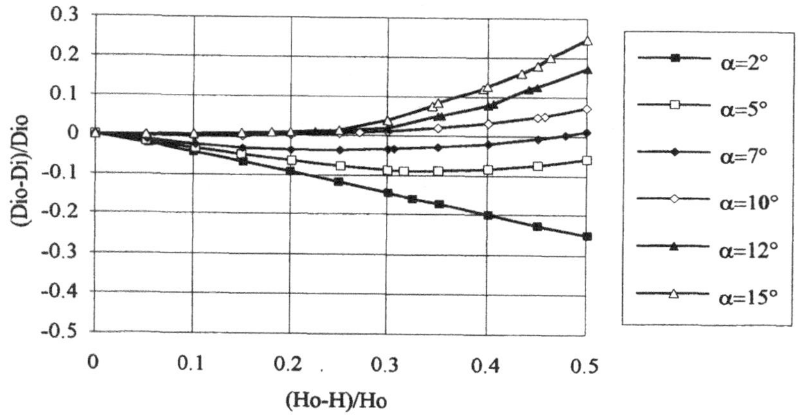

Fig.6 . Ring test simulation with the plastic wave theory model with m_0=0.2

An other drawback of the ring test is its inability for low contact normal stress levels, as the flat punch induces pressures on the workpiece surface greater than the yield strength of the material without any low pressure transient state. Such high local stress distribution is not present on each point of the workpiece surface at every moment of the forming process.

FINITE ELEMENT METHOD SIMULATION OF BACKWARD EXTRUSION

In practical metal working operations, the frictional conditions at the workpiece-tool interface affect the material flow and the applied loads. A numerical analysis of the frictional conditions in backward extrusion is presented for the Tresca model, the Coulomb-Orowan model and the plastic wave theory model. The rigid punch geometry and the rigid outer cylinder geometry are defined together with the initial shape of the workpiece on Fig.7.

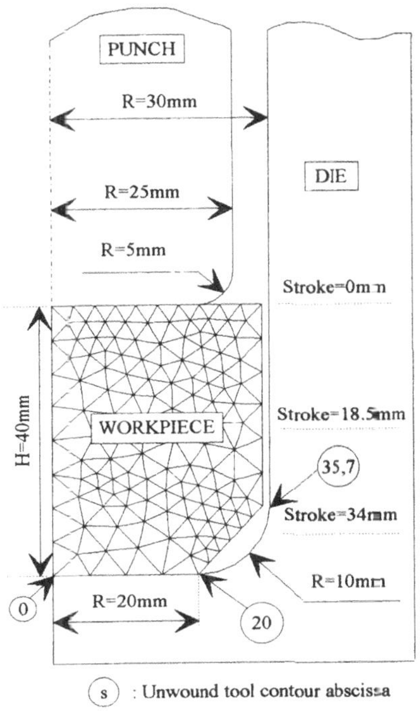

Fig.7 . Geometry of the workpiece, die, and punch in the backward extrusion simulation

The workpiece material is considered as elastic-plastic with no hardening. The friction models are calibrated with the same constant friction stress at the high contact stress level, see Fig. 8, the normalized friction stress-normal stress curves.

The f.e.m. analysis shows that the loads, see Fig. 9, needed for the backward extrusion are lower for the plastic wave theory model than for the two other conventional friction laws as the frictional energy losses on the wall of the cylinder are strongly related to the normal contact stress distribution.

The analysis of the contact stress distribution along the inner cylinder contour points out some coupling between the low normal stress on the vertical region of the cylinder, the resulting low friction stress in the same region and the normal stress under the punch as it can be observed on Fig. 10 and Fig. 11 where respectively the normal stress and shear stress on the workpiece interface are plotted versus the cylinder contour abscissa for a punch stroke of 18.5 mm.

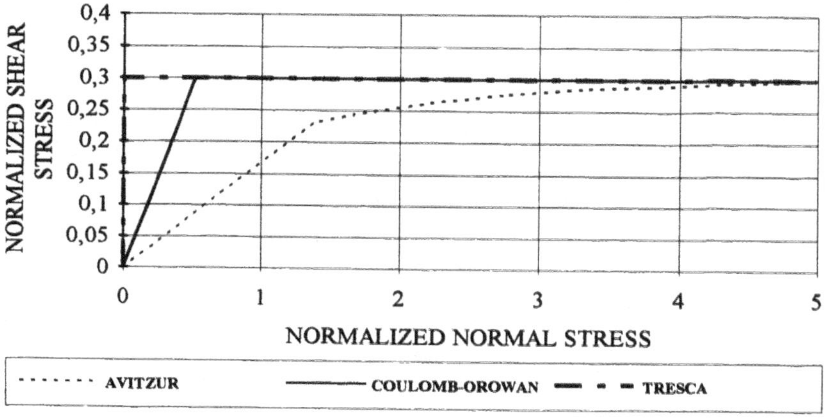

Fig.8 . Backward extrusion,normalized friction stress-normal stress curves for the three friction models

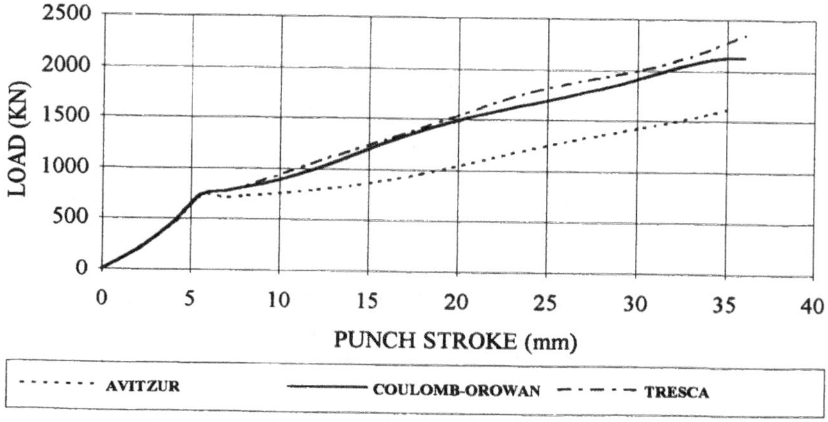

Fig.9 . Backward extrusion, load versus punch stroke curves for the three conditions of friction

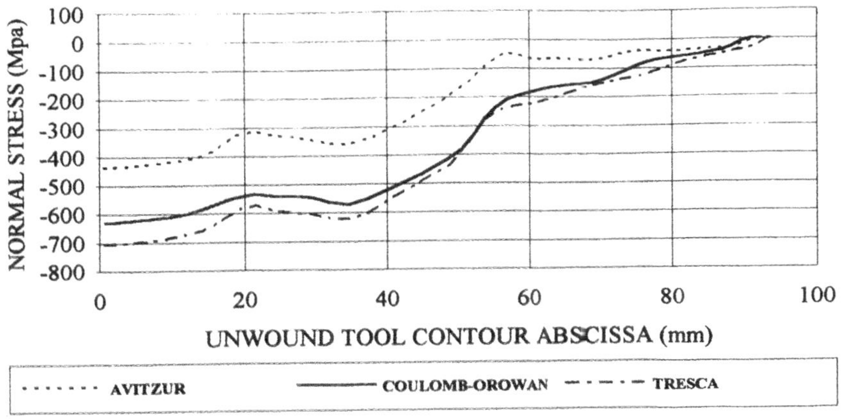

Fig.10 . Normal stress versus contour abscissa for the three conditions of friction and for a punch stroke of 18.5mm

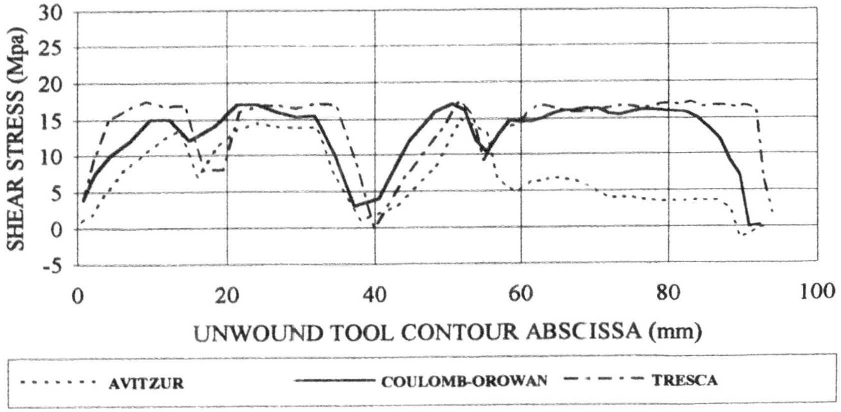

Fig.11 . Shear stress versus countour abscissa for the three conditions of friction and for a punch stroke of 18.5mm

The contour abscissa from 0 to 20.mm is the horizontal region of the inner cylinder which is followed by the first vertical region below the punch till the contour abscissa 60.mm where the high level of normal stress induced a friction stress equal to the maximum shear stress allowed on the interface by the three laws.

The region of the vertical wall of the cylinder above the level of the punch-workpiece interface between the contour abscissa 60.mm and 90.mm supports a low level of normal stress but the Tresca law and the Coulomb-Orowan predict no change for the friction stress when the plastic wave theory model takes into account the normal stress state changes as it is suited to low and high normal stress ranges.

As in the backward extrusion, friction operates mainly on the flat surfaces of the tools, wrong results can be obtained if the friction model is not normal stress dependent for the tool region where the normal stresses are low. To assess more realistic frictional conditions for extrusion, A. FORCELLESE, F. GABRIELLI and A. BARCELLONA (1994) studied the double cup extrusion test but industrial process can be more complicated with heterogeneous friction conditions on the tools.

A real combined backward-forward extrusion is presented below with three different friction distributions. The constitutive law of the material is temperature dependent, strain dependent ,and strain rate dependent and given by the Norton-Hoff law :

$$\sigma_0 = 1471.*\sqrt{3}*0.0572\varepsilon^{0.022}*(\sqrt{3}\dot{\varepsilon}^{0.113}) \tag{5}$$

The friction law is of the power type with :

$$\tau = 0.15*\frac{\sigma_0}{\sqrt{3}}*\|V_t\|^{0.113} \tag{6}$$

The cylindrical workpiece is set in a die which is conical at its low end and cylindrical at its high end, the punch is cylindrical, see Fig.12 for the tool geometry after a 171.mm stroke. The intermediate shape of the workpiece is presented for three different friction conditions. As the tools are copiously lubricated a first simulation considers no friction on the tools (model 1).

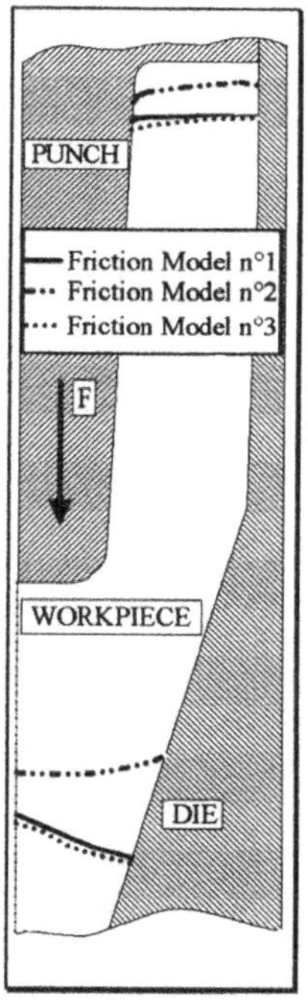

Fig.12 . Geometry of the workpiece, die, and punch in the backward extrusion simulation after a 171.mm punch stroke

The second considers the friction power law defined by Eq[4] everywhere on the punch and on the die.

The third case presents the case of no friction under the flat region of the punch, no friction on the conical region of the die where the lubricant is gathered and the power law friction elsewhere. Such an example shows the influence of friction "heterogeneity" on the plastic flow of the workpiece. The maximum equivalent plastic strain is respectively for model 1, model 2 and model 3 : 2.59, 2.76 and 2.28.

CONCLUSIONS

Simulation using friction models is becoming an integral part of the design of bulk forming processes. The knowledge of the effects of different friction conditions is required with experimental studies in order to obtain reliable data for calibration of friction constitutive model over all the normal stress range for different lubrication regimes. The classical laws seem to be restricted to high normal stress levels and dry friction behaviour. The plastic wave theory can be validate as well as the previous models and some extensions let mixed lubrication conditions to be considered with corresponding calibration.

ACKNOWLEDGEMENTS

The authors wish to tank the French Industrial Group for Research Advances in forging for the financial support of this work.

REFERENCES

B. AVITZUR and Y. NAKAMURA. Analytical determination of friction resistance as a function of normal load and geometry of surface irregularities. WEAR. 107. 1986. pp.367-383

L. BAILLET and J.C. BOYER. A friction model for closed-die forging FEM simulation. Baden-Baden. Congress Center Germany.28-31 Septembre 1994

J.M. CHALLEN. J.M. LEAN and P.L.B. OXLEY. Plastic deformation of a metal surface in sliding contact with a hard wedge: its reaction to friction and wear. Proc. R. Soc. Lond A 394. 1984. pp161-181

E. DOEGE and CH. BEDERNA. Indirect analysis of boundary stresses. Proceedings of th 5th International Conference on Metal Forming. Metal Forming 94. The University of Birmingham. UK. September 13-15. 1994

A. FORCELLESE. F. GABRIELLI and A. BARCELLONA. Evaluation of friction in cold metal forming. Proceedings of th 5th International Conference on Metal Forming. Metal Forming 94. The University of Birmingham. UK. September 13-15. 1994

B. MICHEL. Modélisation thermo-elasto-visco-plastique de procédés de formage des métaux. Thesis. INSA LYON (1993)

S. STANCU-NIEDERKORN. U.ENGEL and M. GEIGER. Ultrasonic investigation of friction mechanism in metal forming. Proceedings of th 5th International Conference on Metal Forming. Metal Forming 94. The University of Birmingham. UK. September 13-15. 1994

W.R.D. WILSON. Friction and lubrification in bulk Metal-Forming Processes. Journal of Applied Metalworking. Vol 1.No. 1. 1979. pp7-19

GEOMORPHOLOGICAL AND FRACTAL APPROACHES IN CONTACT MECHANICS, CASE OF PLASTIC DEFORMATION

Hassan Zahouani, Thomas Mathia, and Jacques Rousseau

Laboratoire de Tribologie et Dynamique des Systèmes - U.R.A. C.N.R.S 855
Département de Technologie des Surfaces, Ecole Centrale de Lyon
B.P. 163
69131 ECULLY - FRANCE

INTRODUCTION

When two surfaces come in contact, the presence of surface roughness morphology causes an imperfect contact at their interface. Due to the multi scale nature of rough surface, the imperfect contact is composed of a large number of contact spots with different sizes.

To understand the mechanics of static and dynamic contact, which are responsible of phenomena of contact conductance, friction, wear and lubrication, it's very important to have a knowledge of the size distribution of contact spots which are correlated to the rheological aspect of the asperities and to the morphological type of roughness.

Two numerical approaches are developed to study, through experiments, the mechanism of asperities deformation. The first one is based upon the geomorphological analysis which consist to classify the morphology of surface roughness as a family of summits, crests, passes, valleys, ridges and flats; the second upon fractal approach using the concept developed by Mandelbrot concerning the size distribution of islands on earth's surface.

The Mandelbrot s work showed that the cumulative size distribution of islands follows the power law $N \sim a^{-D/2}$ where N is the total number of islands with area larger than a and D is the fractal dimension of its cost line ($0 < D < 1$). The adaptation of this approach to the distribution of contact spots shows that this concept is suitable for modelling the spot contact evolution.

Contact Mechanics, Edited by M. Raous *et al.*
Plenum Press, New York, 1995

EXPERIMENTAL STUDY

The experiment of contact is realised between the flat surface and an ideally smooth and rigid counter surface. The specimens are held in a specially designed measuring device which is placed inside our laboratory precision hydraulic press, the investigations being carried out for wide range nominal load. The specimen used in the experiment is an aluminium alloy (Cu 4%, Si & Mg < 1%) which exhibits low strain hardening under deformation (Vickers hardeness: 147 daN/mm², Young's modulus: 7824 daN/mm²). The surface roughness is a random and homogenous real surface elaborated according to sand blasted process. The smooth hard plane is in steel (Hv = 885 daN/mm²).

The morphological characterisation of the surface topography before and after asperities deformation was carried out using a tridimensionnel profilometer. The figures (1a,1b) illustrate the morphology of random surface before and after contact with smooth hard plane.

Morphological approach

The conceptual method developed to characterise the surface morphology, is based on the geographical approach developed initially by Peuker and Douglass in 1975, in order to classify the geomorphological features of the earth relief as family of summits, crests, passes, valleys, ridges, slopes etc...

This approach has been adjusted and applied to real engineering surfaces (Zahouani, Mathia and Guinet,1992; Zahouani, Mathia and Rousseau, 1994).The geomorphological analysis supposes that every point on a surface can be identified and classified by analysis of its neighbouring. Let us consider a point Ω, whose elevation with reference to a mean plane, is Z_Ω. This elevation can be subtracted from those of its neighbours in either a clockwise or counter clockwise sequence around Ω, (see Table 1).

The result is an array of positive or negative numbers (differences $Z_{ij}-Z_\Omega = \Delta_+$ or Δ_-) allowing the characteristic features to be recognised. The following quantities are usually introduced : n = number of Ω neighbouring points, Δ_i = height difference between Ω and one of its neighbours (i=1, n), Δ_+= sum of all positive differences, Δ_- = sum of all negative differences, N_Ω = number of sign changes, L_Ω = number of points between two sign changes. The different topological points are identified as follows:

summit	$\Delta_+ = 0$	$\Delta_- > t_s$	$N_\Omega = 0$	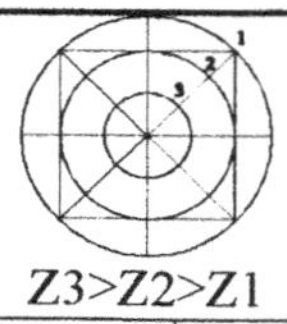 Z3>Z2>Z1
crest	$\Delta_- - \Delta_+ > t_c$	$L_\Omega \neq n/2$	$N_\Omega = 2$	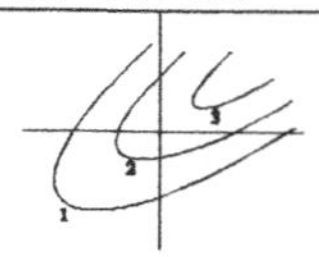 Z3>Z2>Z1

Table 1. Algorithm approach of relief identification

pit	$\Delta_+ > t_p$	$\Delta_- = 0$	$N_\Omega = 0$	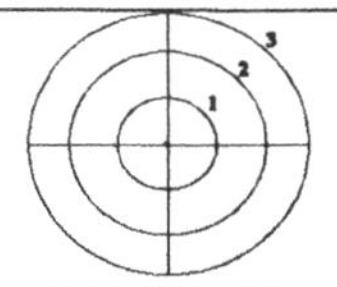 Z3>Z2>Z1
pass	$\Delta_+ + \Delta_- > t_{ps}$		$N_\Omega = 4$	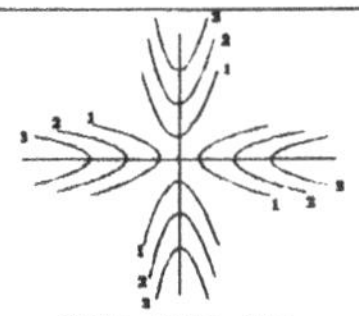 Z3>Z2>Z1
slope	$\lvert \Delta_+ - \Delta_- \rvert < t_s$	$L_\Omega = n/2$	$N_\Omega = 2$	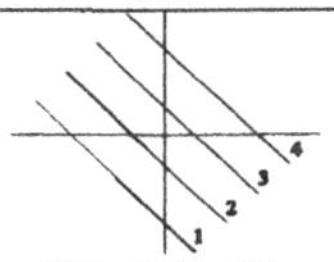 Z3>Z2>Z1

Where t_s, t_c, t_p, t_{ps}, are the thresholds. Each family of particular topology is statistically analysed by determining the three-dimensional Cartesian coordinates (x,y,z) of each detected elements. This concept is clearly illustrated by the cartography of the height position (x,y) of summits before contact fig 2a, and the cartography of the contact spots after deformation against the smooth hard plane fig 2b.

Fractal approach

The use of fractal theory in this work has two motivations. The first one concerns the simulation of surface roughness at any scale as a random process, using the fracal algorithm which make approximation of fractional brownian motion (Lopez, Hansali, Zahouani, Lebosse and Mathia, 1994) fig (3).

The second goal is to analyse the behaviour law of the contact spots distribution under fractality notion. It can be seen, in the example of the cartography of contact areas, that a large number of contact spots of different sizes coexist at the surface plane of contact and are spread randomly over this surface.

The fractal approach is applied in this work by considering the distribution of contact spots for engineering surface to be the same as that of islands on the earth's surface. Using the cartography of contact spots determined by the geomorphological algorithm and the statistical analysis of spots at the plane surface, the experimental results allow to extract the effective power law of the area of spots contact fig (4), with the fractal dimension D = 1.64+/-0.02.

From the cumulative distribution: $N(A>a) \sim a^{-D/2} = (a_m / a)^{D/2}$ where a_m is the area of the largest spot (Mandelbrot, 1979; Manjundar and Bhushan, 1989), the distribution for n(a) can be obtained by differentiation to get:

$$n(a) = \frac{D}{2} \frac{a_m^{D/2}}{a^{D/2+1}}$$

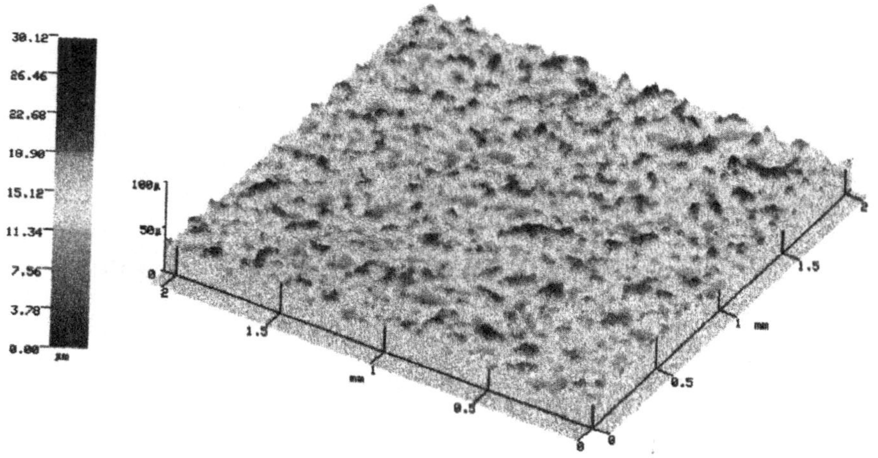

Fig. 1a. Morphology of random surface before contact with smooth plane

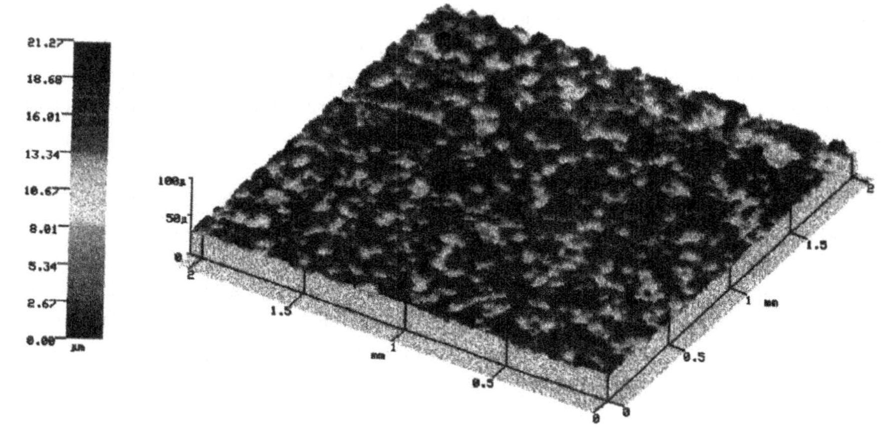

Fig. 1b. Morphology of asperities deformation

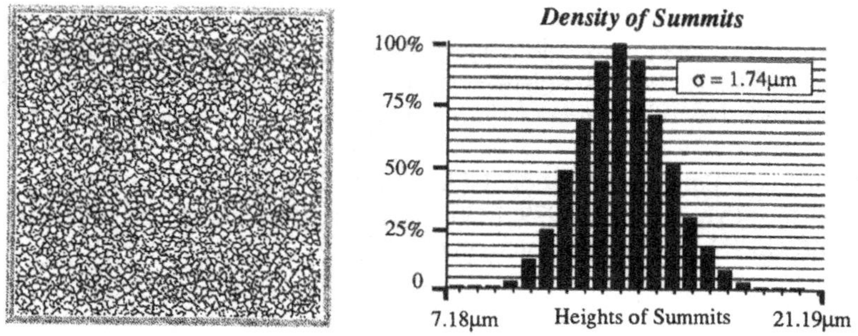

Fig. 2a. Summits distribution of random morphology before contact

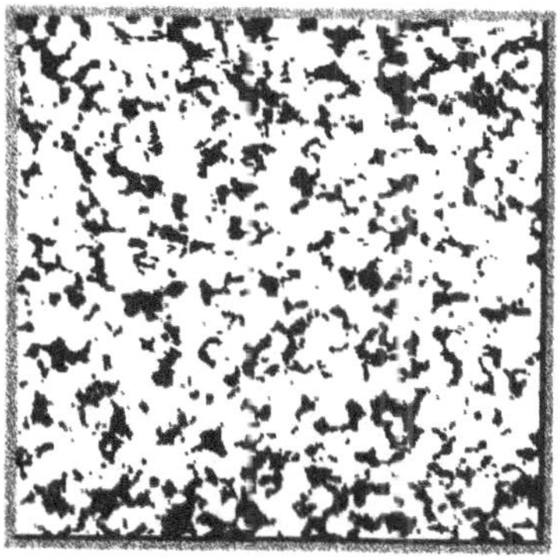

STATISTICS OF SPOTS AREA

Minimum area	128.00 µm²
Maximum area	36416.00 µm²
Mean area	2069.07 µm²
σ	3754.38

Fig. 2b. Cartography of contact spots after deformation

Fig. 3. 3D fractal simulation of random surface roughness

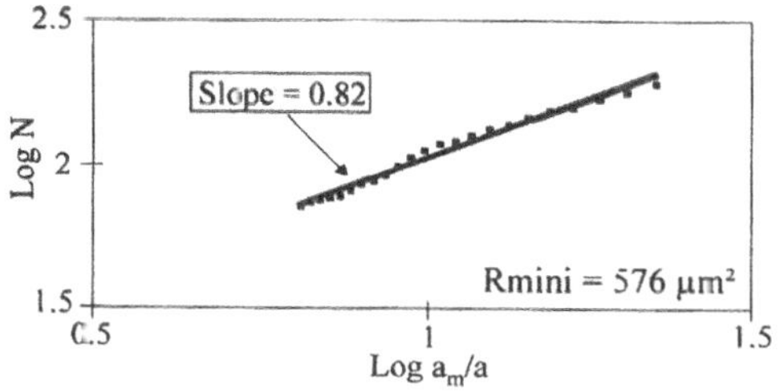

Fig. 4. Power low of the area of spots contact

Therefore the real area of contact A_r is given as:

$$A_r = \int_0^{a_m} n(a)\, a\, da = \frac{D}{2 - D}\, a_m$$

CONCLUSION

One of most fundamental question in contact mechanic of rough solids is the knowledge of the size distribution of contact spots, which determine the distribution of pression and the thermal contact resistance. The knowledge of the asperities interaction in the interface contact zone is a complex process calling for simultaneously rheological and topographical approaches.

According to the morphological characterisation point of view, the two approaches developed in this work show clearly that the investigations must be conducted in two ways.

The first one is experimental, using the geomorphological approach as tool to understand the asperities deformation mechanism.

The second approach which uses the fractal theory shows the existence of the power law of distribution of contact spots in relation to the fractal dimension D. Our results show that the use of these two approaches are very promising and have to be confirmed with another type of surface morphology and with other material.

REFERENCES

Lopez J., Hansali G., Zahouani H., Lebosse J.C. and Mathia T.G., "3D fractal based characterisation for engineered surface topography", accepted for J. Mach Tools, 1994.

Majundar A., Bhushan B., "Role of fractal geometry in roughness characterisation and contact mechanics of surfaces", ASME Journal of Tribology, Vol. 112, pp. 205-216, 1989.

Mandelbrot B.B, Stochastic models for the earth's relief, the shape and the fractal dimension of coastlines, and Number-area rule for Islands", Proceeding of the National Academy of Science, U.S.A, Vol .72, pp 3825-3838, 1979.

Peuker K.,Douglas D.,"Detection of surface specific points by local parallel processing of discrete terrain elevation data" Comput Graph.and Image process.Vol 4, 1975, pp375-387.

Zahouani H.,Mathia T.G.,Guinet A., "Evolution of 3D morphology surface motifs in plastic contact", Tribologica- Finish Journal of Tribology, Vol 11/1992, N° 4,pp83-88.

Zahouani H.,.Mathia T.G and Rousseau J. "Morphology of Engineering surfaces in deformation mechanism" Proc. 6th Nordic Symposium on Tribology, Nortrib'94 12-15 june 1994, vol 2, pp 369-379.

NUMERICAL ANALYSIS OF MICROSCOPICALLY ELASTIC CONTACT PROBLEMS

G. Zavarise and B.A. Schrefler

Istituto di Scienza e Tecnica delle Costruzioni
Università di Padova,
Via Marzolo 9, 35131 Padova - Italy

INTRODUCTION

Numerical treatment of contact problems is more and more growing in complexity, due to the efforts spent to represent real contact mechanisms. Numerical formulations considering adaptation to the Finite Element framework of contact constitutive laws based on microscopical characterisation of contact zones have been proposed recently. Microscopical contact laws actually available consider the mechanisms of force and heat transmission between contacting asperities. The global behaviour of the contact is determined using statistical relationships and mechanical hypotheses.

Such formulations imply some basic assumptions that differentiate the models. Starting from a two- or three-dimensional experimental determination of contact surface real shape different statistical parameters can be extracted. Some models (Cooper et al., 1969) are based on statistical parameters of the surface profile, e.g. the mean plane and the standard deviation, other formulations (Greenwood and Williamson, 1966) consider statistical parameters of the summits. Additional hypotheses are involved concerning the shape of the ideal statistically equivalent asperities. Cylindrical asperities with the top perfectly plane (Cooper et al., 1969), spherical (Greenwood and Williamson, 1966) or conical shapes and other ones (Shai and Santo, 1982) have been considered both for the representation of the mechanical and thermal response.

Finally different mechanical behaviour of the contacting asperities can be reasonably assumed. Two distinct hypotheses are usually made for the deformation of the asperities under applied loads. These can be assumed either as plastic or elastic. It is not easy to fix the range of mechanical pressures in which the hypotheses should be applied. Experience has shown that plastic models can be applied during first cycles of contact between clean surfaces, and elastic ones can be more properly applied for low pressure contact or for repeated contact between clean surfaces when the surface profiles have been stabilised.

We remark that real contact area tends to zero if no pressure is applied, and it increases more and more with increasing pressure. The contact area is always so small that it is reasonable to assume that the yielding limit is always reached on it. On the contrary it is

Contact Mechanics, Edited by M. Raous *et al.*
Plenum Press, New York, 1995

also reasonable to assume that under repeated contact a steady state is reached where no plastic deformations are involved, and the deformation of contacting asperities is purely elastic. Quite surprisingly both these models are able to produce good and closely similar results. It means that the models are still rough to describe the effective contact mechanisms. We can consider them a sort of meso-level models giving us a mediate information of the real microscopical behaviour.

The most relevant difference between the models concerns the relationship between contact force and real contact area. Within the plastic hypothesis such dependence is obviously linear, while within elastic hypothesis Hertz's relationships state a nonlinear law with exponent 2/3. However Archard (1957) has shown that also the elastic model can produce linear dependence as limit case.

REVIEW OF THE GREENWOOD-WILLIAMSON MODEL

The elastic model proposed by Greenwood-Williamson (1966) consists of a statistical extension of basic Hertz relationships for contact between two elastic spheres

$$a_H = (\beta w)^{1/2} \qquad A_H = \pi \beta w \qquad P_H = \frac{4}{3} E^* \beta^{1/2} w^{3/2} \tag{1}$$

where a_H is the contact radius, β is the original radius of the spheres, w is the compliance, A_H is the contact area and P_H the contact force. The modified elastic modulus E^* takes into account the possibility of contact between different materials.

$$\frac{1}{E^*} = \frac{1-\nu_1^2}{E_1} + \frac{1-\nu_2^2}{E_2} \tag{2}$$

We remark that this starting point contains two basic assumptions on the asperities shape and on the mechanical behaviour. Eqs. (1) are applied to the contact between two asperities that can be assumed as representative from the statistical point of view of the whole contacting asperities.

Most existing models apply statistical relationships in similar way. Disregarding the modification of surface profile, hence considering that the materials interpenetrate, the probability that the contact takes place for a certain compliance, d, is given by

$$\text{prob. of contact} = prob(z_s > d) = \int_d^\infty \phi(z_s) dz_s \tag{3}$$

where $\phi(z_s)$ is a chosen statistical function and z_s is the summit height. It has to be remarked that this model considers the summits, and not the profile height.

Another step which is common to different contact models is the determination of the contacting asperities, n_c, with dependence on a certain compliance, when the total number of asperities, N is known

$$n_c = N \int_d^\infty \phi(z_s) dz_s \tag{4}$$

The basic equations of Hertz particularise the model in determining the real contact area, A_r, and the normal contact force, P

$$A_r = N\pi\beta\int_d^{\infty}(z_s - d)\phi(z_s)dz_s \tag{5}$$

$$P = N\frac{4}{3}E^*\beta^{1/2}\int_d^{\infty}(z_s - d)^{3/2}\phi(z_s)dz_s \tag{6}$$

A typical statistical function considered for height and summits distributions is the normal distribution density function

$$\frac{1}{\sigma_s\sqrt{2\pi}}e^{-\frac{z_s^2}{2\sigma_s^2}} \tag{7}$$

where σ_s represents the standard deviation of the summits. Sometimes exponential distributions are also used. We remark that other models consider standard deviation and height of the surface profile. All these values can be extracted from experimental analysis of the surfaces.

A tranformation with a normalisation using standard deviation is usually carried out to simplify hand calculations. Due to the fact that we will plug the model in a numerical scheme this normalisation is not mandatory in our case, but it is still convenient for numerical reasons. We define hence a new normalised variable

$$x_s = \frac{z_s}{\sigma_s} \tag{8}$$

which implies

$$z_s = d \quad \Rightarrow \quad x_s = \frac{d}{\sigma_s}, \quad dz_s = \sigma_s dx_s \tag{9}$$

Expressing the integral parts of eqs. (4-6) particularised with the normal distribution in compact way and applying the normalisation we obtain

$$G_n(d_\sigma) = \sigma_s^n\int_{d_\sigma}^{\infty}(x_s - d_c)^n\frac{1}{\sqrt{2\pi}}e^{-\frac{x_s^2}{2}}dx_s = \int_d^{\infty}(z_s - d)^n\frac{1}{\sigma_s\sqrt{2\pi}}e^{-\frac{z_s^2}{2\sigma_s^2}}dz_s \tag{10}$$

where, with reference to eq. (8), $d_\sigma = d/\sigma_s$. Previous relationships (4-6) now becomes

$$n_c = NG_0(d_\sigma) \tag{11}$$

$$A_r = N\pi\beta G_1(d_\sigma) \tag{12}$$

$$P = N\frac{4}{3}E^*\beta^{1/2}G_{3/2}(d_\sigma) \tag{13}$$

We have to remark that this basic structure of the model has been kept also for modified models. Most of modification proposals regard different techniques to determine the statistical parameters (McCool, 1986; Handzel-Powierza et al., 1992; McWaid and Marschall, 1992).

NUMERICAL SET UP

Formulation for FEM Computations

One of the main problems for the application of the reviewed model is related to the fact that the closed form of the integrals involved does not exist, hence tables with numerical integration are available in literature.

To set up a contact constitutive law suitable for FE discretisation we need to determine a relationship between nodal forces and surfaces approach (see also Zavarise, 1991; Schrefler and Zavarise, 1993). The Greenwoon and Williamson model in eq. (13) presents a nonlinear relationship between such parameters. The equation should be translated at element level

$$F_N = \eta A \frac{4}{3} E^* \beta^{1/2} G_{3/2}(d_\sigma) \tag{14}$$

where F_N is the element contact force, η is the known density of summits per unit of contact area and A is the element contact area.

The constitutive law should be linked with a chosen contact geometry, which defines the surfaces approach. We remark that the contact geometrical relationships can be defined without any limitation due to the constitutive law. We couple the constitutive law with the so-called master-slave contact geometry (Wrigger and Simo, 1985). The chosen geometry permits to detect contact closure or opening, current element contact area (it may change if large deformations are involved) and current surface approach. The local nodal value of the approach is considered as representative of the approach of all the contact area. Hence we obtain constant pressure within the element and discontinuity between elements are involved for pressure and approach fields.

The contribution of contact force can be explicitly added to the global virtual work equation (Zavarise et al., 1992)

$$\delta W_{contact} = \bigcup_{\text{active contacts}} F_N \delta g_N \tag{15}$$

where simply

$$\delta g_N = -\delta d \tag{16}$$

The proposed approach should be used to solve problems where a high precision is required. For such kind of problems implicit strategy is commonly adopted, hence linearisation of the equation set (15) should be carried out. Consistent linearization of the relationship requires the determination of the following terms

$$\Delta \delta W = \frac{\partial F_N}{\partial A} \Delta A \delta g_N + \frac{\partial F_N}{\partial d} \Delta d \delta g_N + F_N \Delta \delta g_N \tag{17}$$

Considering eq. (14) the first term of eq. (17) becomes

$$\frac{\partial F_N}{\partial A} = \eta \frac{4}{3} E^* \beta^{1/2} G_{3/2}(d_\sigma) \tag{18}$$

The variation of contact area, which is a geometrical term, depends totally on the geometry chosen for the contact element. It is important only when large changes are involved.

The linearization of the second term is more complicated, due to the fact that the derivation involves an integral equation having the derivation variable within the equation and at the lower integration limit. The linearisation of such kind of equation is the following

$$\frac{\partial}{\partial d}\left(\int_{d}^{\infty} f(d,x)dx\right) = -f(d,d) + \int_{d}^{\infty}\frac{\partial f(d,x)}{\partial d}dx \tag{19}$$

From eq. (14) then we obtain

$$\frac{\partial F_N}{\partial d} = \eta A\frac{4}{3}E^*\beta^{1/2}\frac{\partial G_{3/2}(d_\sigma)}{\partial d_\sigma}\frac{\partial d_\sigma}{\partial d} =$$

$$= \left(\eta A\frac{4}{3}E^*\beta^{1/2}\right)\frac{\partial}{\partial d_\sigma}\left[\sigma_s^{3/2}\int_{d_\sigma}^{\infty}(x_s - d_\sigma)^{3/2}\frac{1}{\sqrt{2\pi}}e^{-\frac{x_s^2}{2}}dx_s\right]\frac{1}{\sigma_s} \tag{20}$$

using the rule stated in eq. (19) we have

$$\frac{\partial F_N}{\partial d} = \left(\eta A\frac{4}{3}E^*\beta^{1/2}\right)\sigma_s^{3/2}\left[-(d_\sigma - d_\sigma)^{3/2}\frac{1}{\sqrt{2\pi}}e^{-\frac{x_s^2}{2}} + \int_{d_\sigma}^{\infty}-\frac{3}{2}(x_s - d_\sigma)^{1/2}\frac{1}{\sqrt{2\pi}}e^{-\frac{x_s^2}{2}}dx_s\right]\frac{1}{\sigma_s} \tag{21}$$

The first term within square brackets is clearly zero, hence arranging the equation in a suitable form we have

$$\frac{\partial F_N}{\partial d} = -2\eta AE^*\beta^{1/2}\sigma_s^{1/2}\int_{d_\sigma}^{\infty}(x_s - d_\sigma)^{1/2}\frac{1}{\sqrt{2\pi}}e^{-\frac{x_s^2}{2}}dx_s = -2\eta AE^*\beta^{1/2}G_{1/2}(d_\sigma) \tag{22}$$

Finally the linearization of the third term of eq. (17) pertains to a purely geometrical term, and it is available in literature (Zavarise, 1991; Zavarise et al., 1992).

Integration of Constitutive Law

Eqs. (14) and (22) that are involved within Newton iterations imply numerical integration. In this specific case the integration limits can be reconducted to the range (-1, 1) operating a change of variables

$$t_s = e^{-x_s}, \quad x_s = -\ln(t_s), \quad dx_s = -\frac{1}{t_s}dt_s \tag{23}$$

In such a way both the upper and lower integration limits are finite:

$$x_s = \infty \;\Rightarrow\; t_s = 0, \quad x_s = d_\sigma \;\Rightarrow\; x_s = e^{-d_\sigma} \tag{24}$$

and the tranformed integral becomes

$$G_n(d_\sigma) = \sigma_s^n\int_0^{e^{-d_\sigma}}(-\ln(t_s) - d_\sigma)^n\frac{1}{\sqrt{2\pi}}e^{-\frac{(-\ln(t_s))^2}{2}}\frac{1}{t_s}dt_s \tag{25}$$

Gauss integration is hence applied using the standard transformation

$$\int_{\alpha}^{\beta} f(x)dx = \frac{\beta-\alpha}{2} \sum_{k=1}^{\text{N Gauss Points}} B_k f\left(\frac{\beta-\alpha}{2}x_k + \frac{\beta+\alpha}{2}\right) \tag{26}$$

which gives

$$G_n(d_\sigma) = \sigma_s^n \frac{e^{-d_\sigma}}{2} \sum_{k=1}^{\text{N Gauss Points}} B_k \left[-\ln\left(\frac{e^{-d_\sigma}}{2}x_k + \frac{e^{-d_\sigma}}{2}\right) - d_\sigma\right]^n \frac{e^{-\frac{1}{2}\left[-\ln\left(\frac{e^{-d_\sigma}}{2}x_k + \frac{e^{-d_\sigma}}{2}\right)\right]^2}}{\left(\frac{e^{-d_\sigma}}{2}x_k + \frac{e^{-d_\sigma}}{2}\right)} \tag{27}$$

where B_k are the weights and x_k the gauss points.

The numerical integration performed has shown a not satisfactory numerical precision, even if a high number of Gauss points has been used. Table 1 reports the number of significative digits for $G_n(d_\sigma)$ computed at $d = 3.0$ with $\sigma_s = 1.0$. It is evident that the numerical integration of $G_0(3.0)$ and $G_1(3.0)$ presents an almost satisfactory number of correct digits. On the opposite, $G_{1/2}(3.0)$ and $G_{3/2}(3.0)$, which are both involved in the proposed contact numerical model, do not gain a satisfactory number of exact digits, even if a high number of Gauss points is used. Switching of the integration limits between zero and infinite and using Laguerre expressions has also shown poor precision. Work is in progress to obtain a better integration scheme using the technique proposed in Morandi-Cecchi et al. (1991).

Table 1. Gauss integration of $G_n(3.0)$.

N. points	$G_0(3.0)$ (10^{-4})	$G_1(3.0)$ (10^{-3})	$G_{1/2}(3.0)$ (10^{-3})	$G_{3/2}(3.0)$ (10^{-3})
2	0.13	0.3	0.6	0.2
4	0.1349	0.382	0.64	0.26
5	0.13498	0.382	0.64	0.263
6	0.134989	0.3821	0.64	0.263
7	0.1349898	0.3821	0.64	0.2639
8	0.1349898	0.382154	0.64	0.2639
9	0.1349898	0.382154	0.64	0.26396
10	0.13498980	0.382154	0.64	0.26396
15	0.1349898031	0.38215431	0.64	0.263967
20	0.13498980316	0.38215431	0.641	0.263967
25	0.13498980316	0.38215431704	0.641	0.2639675
30	0.1349898031630	0.382154317047	0.6418	0.2639675
40	0.13498980316300	0.38215431704771	0.6418	0.26396755
50	0.13498980316300495	0.382154317047710	0.64185	0.26396755
100	0.13498980316300495	0.382154317047710	0.64185	0.26396755

Geometrical Discretisation

The complete FE treatment requires the definition of contact geometry and the link with the constitutive law. Since constitutive law presents the same dependencies on the geometrical variables as the plastic model the complete FEM discretisation is carried out using the same guidelines proposed in Zavarise (1991) and Zavarise et al. (1992). Moreover the simplified contact geometry proposed in Wriggers and Zavarise (1993) can also be used to solve simple problems and to check the constitutive law without geometrical nonlinearities effects.

EXAMPLES

Numerical experiments have been carried out to check the performance of the proposed procedure. The simplified contact geometry has been used to show the numerical behaviour of the constitutive law. Implementation of the fully nonlinear version is in progress.

The well known standard test of the elastic block resting on rigid foundation has been used both for the simplicity and for the possibility of comparing the results obtained using a plastic constitutive law (Zavarise, 1991, Zavarise et al., 1992, Zavarise et al, 1994). The following idealised mechanical characteristics have been used for the elastic block: elastic modulus $E = 1000 \quad N/m^2$, Poisson's ratio $\nu = 0.3$. The geometry of the sample is shown in Figure 1. Typical contact data of surface geometry have been used: summit density $\eta = 4.0E+9\ summit/m^2$, summit radius $\beta = 6.5E-5\ m$, standard deviation of summits height $\sigma_s = 2.5E-7\ m$. The equivalent elastic modulus has been chosen taking into account the magnitude of the elastic modulus adopted for continuum: $E^* = 4E+7\ N/m$. Selected input data permits a close comparison with the results obtained using plastic contact models. Contact pressure is generated by imposing a vertical displacement of $0.32\ m$ at the top of the block.

The solution is carried out using a full-Newton iterative scheme. Due to the absence of both friction and thermal coupling, and due to the simple contact geometry a symmetric tangent stiffness is obtained. The selected parameters allow to avoid ill-conditioning problems, however the modified augmentation technique proposed in Zavarise et al (1994) can be applied also in this case.

The obtained results show a behaviour similar to the case with plastic contact law. In Figure 2 the evolution of normal contact force of the element at the middle of the contact zone is reported. The rate of convergence is shown in Figure 3 where it is evident the achievement of quadratic rate of convergence near the solution. It has to be remarked that the load condition has been applied in a single step, hence some efforts are required during the firsts iterations. Nonlinear behaviour of the constitutive law for the selected values of the parameters are evidenced in Figure 4. Here variation of contact pressure and of the derivative versus mean plane distance is reported in logarithmic scale.

The algorithm has been tested varying the contact parameters and using other simple but representative geometries. The results confirm again the possibility to deal with contact problems of elastic type with the high accuracy allowed by using contact constitutive relationships.

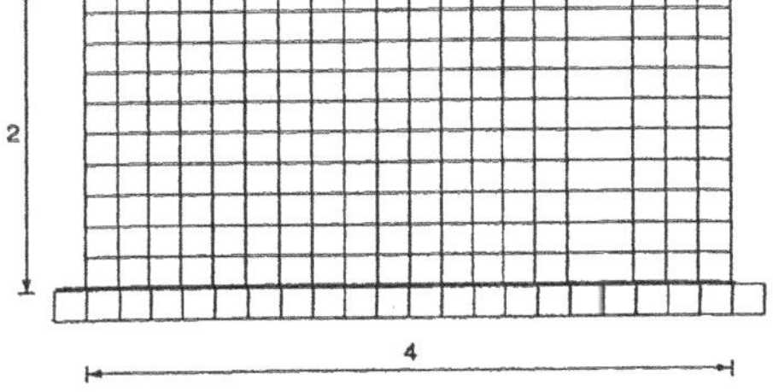

Figure 1. Discretisation of the block and the foundation.

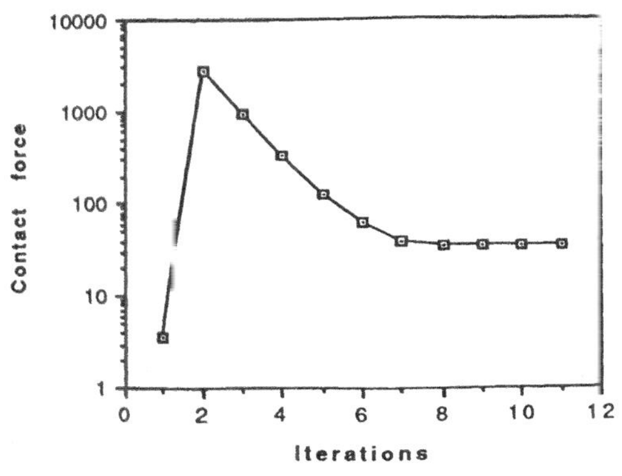

Figure 2. Normal contact force at the center of the contact zone.

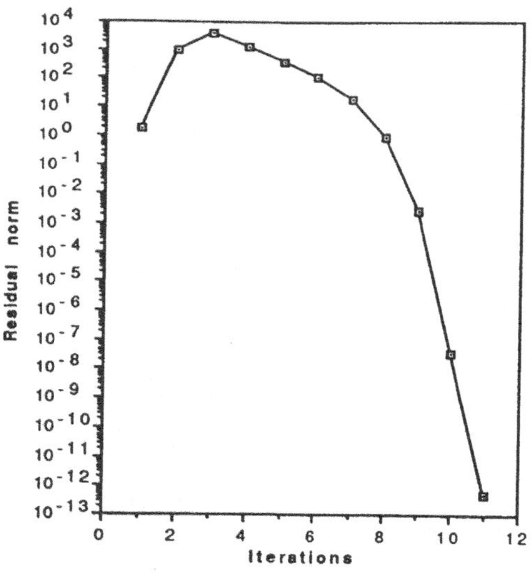

Figure 3. Evolution of the residual during iteration process.

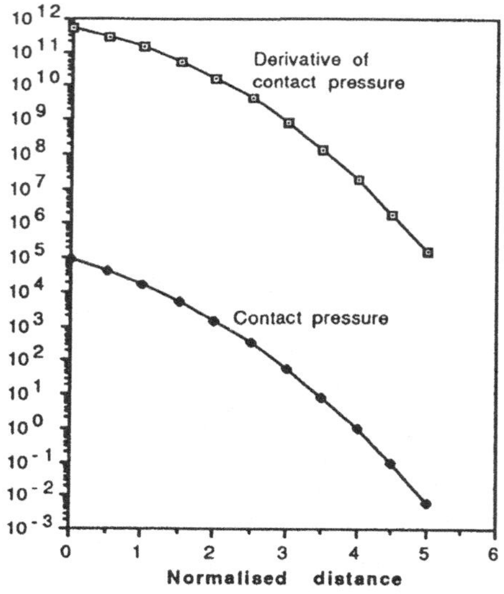

Figure 4. Nonlinear behaviour of the constitutive law.

REFERENCES

Archard, J. F., 1957, Elastic deformation and the laws of friction, Proc. R. Soc. London, 243-A:190.

Cooper, M.G., Mikic, B. B., Yovanovich, M.M., 1969, Thermal contact conductance, Int. J. of Heat and Mass Transfer, 12:279.

Greenwood, J.A., Williamson, J.B.P., 1966, The contact of nominally-flat surfaces, Proc. R. Soc. London, 295-A:300.

Handzel-Powierza, Z., Klimczak, T., Polijaniuk, A., 1992, On the experimental verification of the Greenwood-Williamson model for the contact of rough surfaces, Wear, 154:115.

McCool, J.I., 1986, Comparison of models for the contact of rough surfaces, Wear, 107:37.

McWaid, T.H., Marschall, E, 1992, Application of the modified Greenwood and Williamson contact model for the prediction of thermal contact resistance, Wear, 152:263.

Morandi Cecchi, M., Redivo Zaglia, M., 1991, A new recursive algorithm for a gaussian quadrature formula via orthogonal polynomials, Orthogonal Polynomials and their applications (C. Brezinski, L. Gori, A. Ronveaux eds.), J.C. Baltzer AG, Scientific Publishing Co, IMACS, 353.

Schrefler, B.A., Zavarise, G., 1993, Constitutive laws for normal stiffness and thermal resistance on contact element, Microcomputers in Civil Engineering, 8:299.

Shai, I., Santo, M., 1982, Heat transfer with contact resistance, Int. J. of Heat and Mass Transfer, 24-4:465.

Wriggers, P., Simo, J.C., 1985, A note on tangent stiffness for fully nonlinear contact problems, Comm. in App. Num. Methods, 1:199.

Wriggers, P., Zavarise, G., 1993, Thermomechanical contact - a rigorous but simple numerical approach, Computers & Structures, 46-1:47.

Zavarise, G., 1991, Problemi termomeccanici di contatto - aspetti fisici e computazionali, Ph.D. Thesis, Ist. di Scienza e Tecnica delle Costruzioni, Univ. of Padua, Italy.

Zavarise, G., Wriggers, P., Stein, E., Schrefler, B.A., 1992, A numerical model for thermomechanical contact based on microscopic interface laws, Mech. and Res. Communications, 19-3:173.

Zavarise, G., Schrefler, B.A., Wriggers, P., 1994, On augmented lagrangian algorithms for thermomechanical contact problems with friction, Int. J. for Num. Meth. in Engineering, to appear.

TESTING AND MODELLING OF DYNAMIC MICROSCALE CONTACT BEHAVIOUR

Josef Beneš, Eduard Veselý, Helena Šebková and Dušan Gabriel

Institute of Thermomechanics
Dolejškova 5, Prague 8
Czech Republic

INTRODUCTION

Presented investigation was initiated by the necessity to solve stress waves propagation through the impact loaded structural connections as connection of die block with ram of machine hammer (Beneš et al (1994)) or contact between acoustic emission transducer (AE) and structure (Beneš et al (1994)). Contact between these parts isn't perfect due to machining and sometimes also wearing occurs during working life. Numerical simulations of model experiments shown that the correspondence of calculations and measurements is strongly influenced by neglecting the mentioned imperfections. To explain this process clearly we can speak about two tasks :

1) The macroscale behaviour of contact surfaces. The contact area can be significantly changed during the loading process if the distance of parts of body surfaces in contact area is comparable to normal displacements of corresponding surface particles.

2) The microscale behaviour of contact interface. Usually there is not a perfect connection of individual particles in the contact of two surfaces of the same shape (in macroscale). This imperfection of contact is caused by surface roughness due to machining process. Variation of the normal and tangential stiffness of surface layer is caused by this effect. The situation is presented in Figure 1.

Contact with minimal normal prestress - Figure 1a, contact with higher pressure - Figure 1b, and Figure 1c - contact interface fulfilled by a contact medium (a special sound-transmitted grease is used for connection of AE - transducers). Roughness depending on machining is about (10^{-4} - 10^{-1}) mm, the distance of particular asperities is of about two orders higher, see Oden and Martins (1985), Martins et al (1990). There is another important effect connected with microscale behaviour of contact, namely the transmission of the tangential forces T between two surfaces in contact is limited by the friction forces depending on normal prestress N, material and quality of these surfaces and eventually on existence of contact interface media. If tangential forces exceed the friction forces the slipping arises.

Contact Mechanics, Edited by M. Raous *et al.*
Plenum Press, New York, 1995

The mentioned contact-impact phenomena can be simulated by finite difference method (FDM) or finite element method (FEM) by professional codes for example MSC/DYNA but the microscale behaviour can however be simulated phenomenologically for example by definition of normal and tangential stiffnesses and friction coefficients. These quantities can be determined experimentally.

Our work is devoted to determination of contact normal stiffness and measurements of dynamic friction coefficients. This task is complicated by the strong non-linear dependencies of investigated quantities on prestress in contact interface. A hybrid procedure based on calculation by FDM or FEM and on Hopkinson pressure bar type measurements for investigation of normal stiffness of contact interface area was developed.

INVESTIGATION OF NORMAL STIFFNESS

The scheme of the solution of normal contact stiffness parameters is shown in Figure 2. The contact of two bars of the same material is prestressed by static axial force N. The contact surfaces are standardly machined. Sufficiently short stress pulse, which was introduced into the first bar and propagates through it, is measured by the pair of strain gauges (SG1). The pulse partially propagates through the contact and it is partially reflected due to contact´s imperfection. The reflected part of the pulse is again recorded by the strain gauges (SG1) and the transmitted part by the pair of strain gauges (SG2). Ratio of single pulse components (transmitted and reflected) is dependent on quality of contact interface and on normal prestress. Quantitative evaluation of this dependency is enabled by frequency analysis of particular pulses and evaluation of their transfer characteristics. The result effect of wave transmission or reflection should be more or less similar to the experiment results when we simulate the same process by means of FDM or FEM through changing of the respective mesh parameters (used for substitution of interface) or parameters of FE model (number of elements used for interface substitution, stiffness, mass etc.) and find them suitable for both FE and FD models.

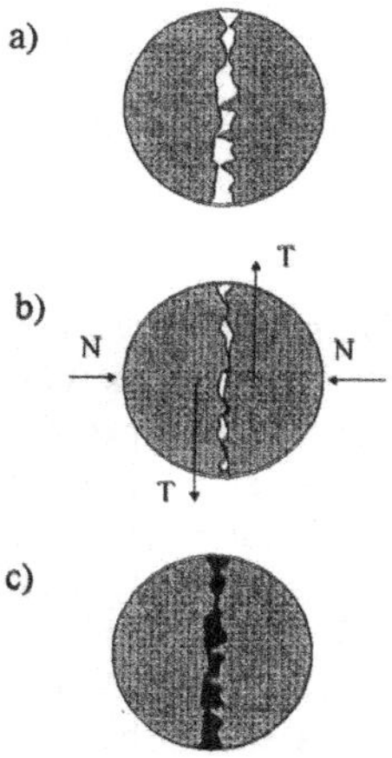

Figure 1. Schema of contact in microscale

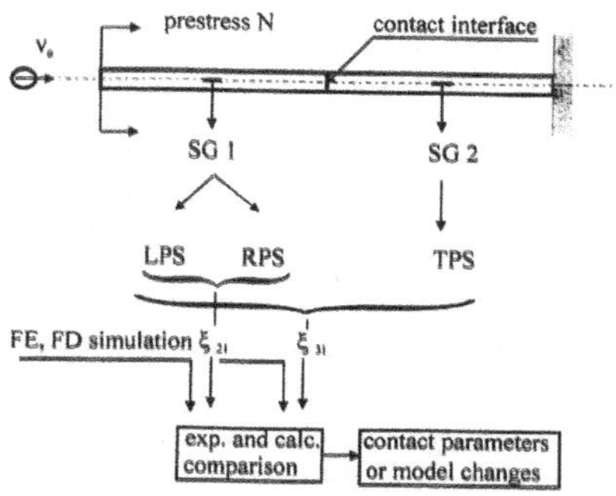

Figure 2. Identification of contact stiffness

The whole process is illustrated in the Figures 3 and 4. Measured and calculated pulses are shown in Figure 3 - loading pulse (LP) and reflecting pulse (RP) are in Figure 3a, transmittied pulse is in Figure 3b. Boundary conditions were formulated respecting the measured loading pulse. It is obvious that the results of FE and FD simulation are in graphic

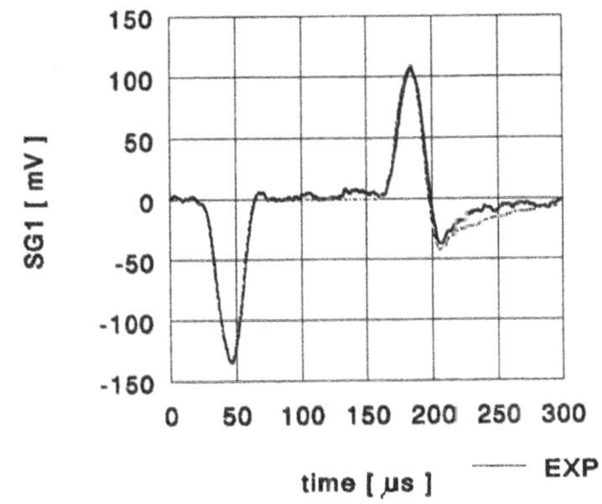

Figure 3a. Pulses in point SG1

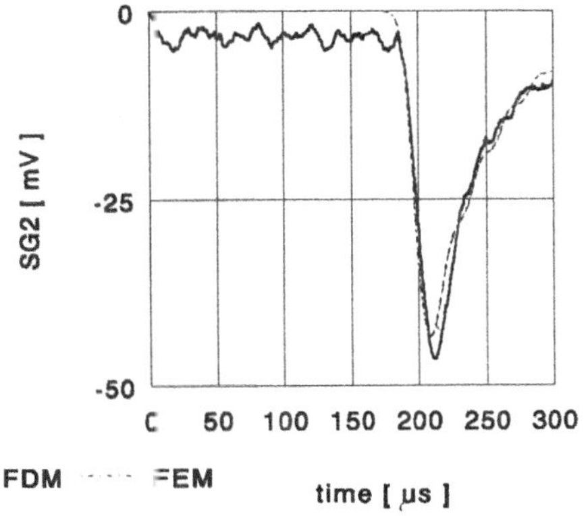

Figure 3b. Pulses in point SG2

interpretation only hardly distinguished and they approximate relevant measured pulses very well. There were used steel bars (E_1 =2,1.10^{-5} MPa, ρ = 7800 kgm^{-3}) with roughness of ends machining 3,2. Prestress pressure p ≐ 3 MPa. Only one element was used for subtitution of contact interface (thickness = 1 mm, E = 1,21.10^3 MPa, ρ = 11300 kgm^{-3}). The analysis of single pulse propagation was used to obtain quantitative evaluation (Beneš at al (1992)). The outcome transfer characteristics are presented in Figure 4. Parameters of FE or FD models found by mentioned process can be used for the solution of stress pulse propagation

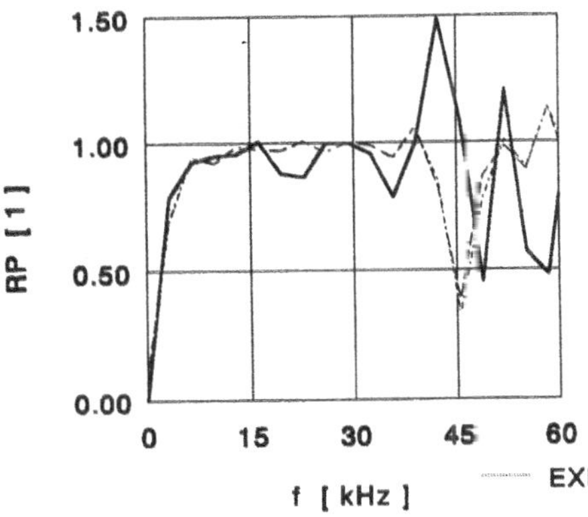

Figure 4a. RP - transfer characteristics

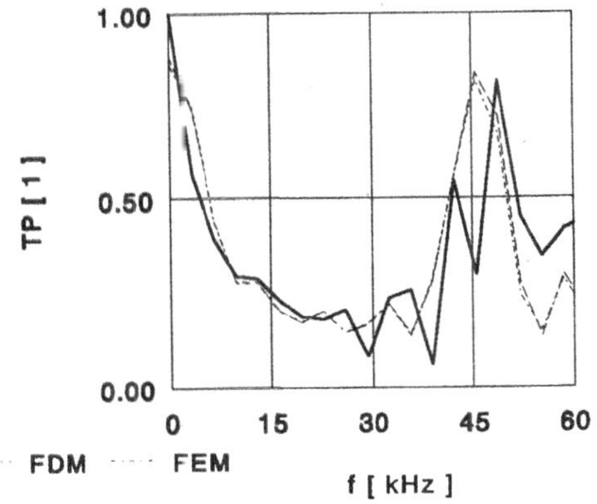

Figure 4b. TP - transfer characteristics

through the connection of elements of complicated shape. It is clear that these parameters must differ with changing of normal prestress because the energy distribution between RP and TP depends on this prestress.

INVESTIGATION OF FRICTION

Tangential forces occur during dynamic loading of general connection. Increasing of these forces to the level higher then the level of friction forces causes consequential slip. The nature of dynamic friction forces coming into existence in contact interface is extremely complex and is affected by a long list of factors for example by material of contact pair, state of contact interface, values and space rate of surface roughness, and so on. We supposed that the adhesion and friction coefficients exist and values of these quantities depend on contact normal prestress. Hopkinson's type of experimental device was designed for these purposes and it is presented in Figure 5.

Figure 5. Schema of friction measurements

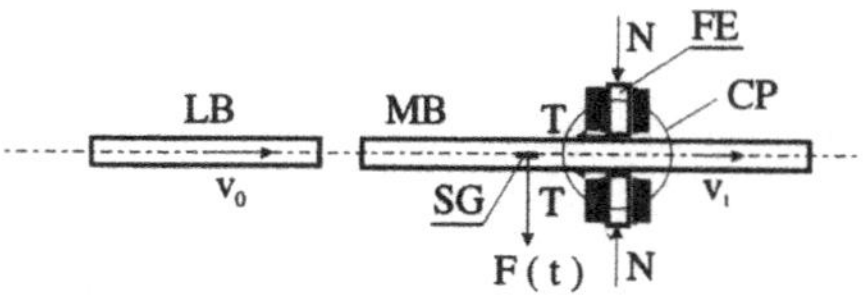

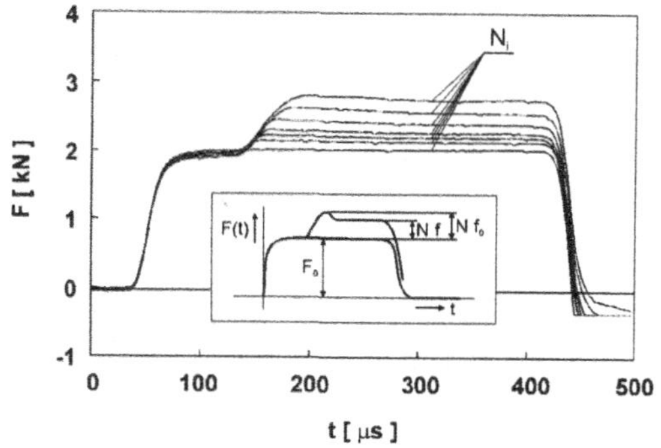

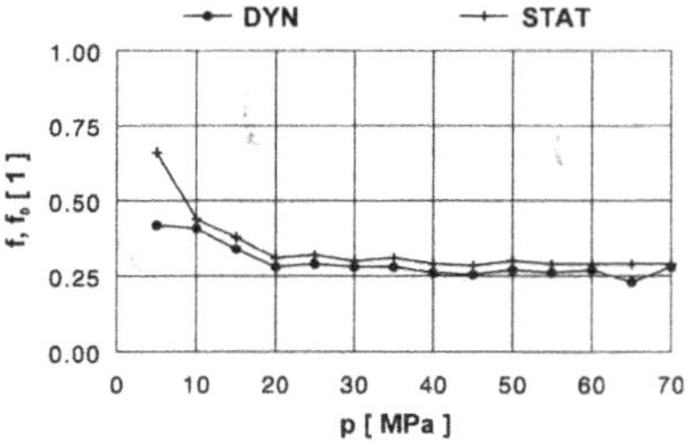

Measuring bar (MB) is loaded by the long striker (LB) in this case. "Rectangular" pulse of the constant stress level propagates through MB and a part of energy is reflected on the contact pair (CP). Reflected force pulse equals tangential forces in contact which cannot superpass the level of maximum adhesion forces. The following slip decreases the tangential reaction to the level of friction forces during the moving. The whole situation for various levels of normal prestress in CP is illustrated in Figure 6 (MB and FE are made of steel). Values of friction forces are dependent on the material of contact pair, quality of machining of contact surfaces and on the state of contact interface together with the value of normal prestress N. Evaluated friction coefficients f and adhesion coefficients f_0 valid for the investigated contact pair are shown in Figure 7.

ACKNOWLEDGEMENT

The research was suffered by Grant No 101/93/0277 of Grant Agency of the Czech Republic.

REFERENCES

Beneš, J., Šebková, H. and Veselý, E.,1992, "Determination of the dynamic characteristics of bars and chains," *Journal of Czech and Slovak Mechanical Engineering* 1 (3), 128-134

Beneš, J., Veselý, E. and Šebková, H., 1993 "Dynamical testing of material and contact parameters," Abstracts of lectures of the *EUROMECH 306 Colloquium - Mechanics of contact impact, Prague, September, 1993, 1*

Beneš, J., Veselý, E., Šebková, H. and Gabriel, D., 1994, "Dynamic identification of bar connection behaviour," *32th Conference of experimental stress analysis,* Liberec May-June 1994, 14-17

Martins, J.A.C., Oden, J.T. and Simões, F.M.F., 1990, "A study of static and kinetic friction," *Int. Jour. Engng Sci.* Vol. 28, No. 1. 29-92

Oden, J.T. and Martins, J.A.C., 1985, "Models and computational methods for dynamic friction phenomena," *Computer methods in applied mechanics and engineering* 52 , North-Holland, 527-634

CONTACT PROBLEMS FOR POWER-LAW CREEPING SOLIDS

Sergei A. Grishin

The Institute for Problems in Mechanics
101, prospect Vernadskogo
Moscow 117526, Russia

INTRODUCTION

All problems of non-linear creep are very difficult. We have developed an approximate approach to solve the contact problems of power-law creep in the case, when one of bodies in contact is a rigid punch with corners and the other one - thin deformable layer or half-space. The main merit of this approach consists in the simpliciy of calculations and in the clearity of assumptions. First of all the local problem of contact between a rigid corner and creeping half-space is investigated. It is similar to the old one of Cherepanov - Rice - Hutchinson for a crack. Hence we have the stress-strain state near the punch corner in function of one free parameter. Internal solution for a thin layer has been found by asymptotic analysis, for a half-space - by Arutyunyan's technics, both also as functions of another free parameter. The size of boundary layer may be regarded as a third free parameter. The integral equilibrum condition and the claim of continuity of main contact problem characteristics give a non-linear algebraic equation to determine free constants by calculations. Its solution may be given by explicit formulae in particular cases. Besides mentioned problems, a periodical problem, modeling the forming by poinsoning of a plate with ribes, has been solved. Moreover, we have completely studied as a separed situation the local solution near a creeping edge tip (arbitrary angle, various boundary conditions), and we have found the internal asymptotics for a thin layer of Nadai or of Imbert materials.

MODEL AND NOTATIONS

We investigate one of the most popular model of creeping solid (Rabotnov, 1966; Kachanov, 1960), so-called power-law creep, described with comportment law

$$\varepsilon_i = A\sigma_i^m \ ; \ (\sigma_i = K\varepsilon_i^\mu) \ ; \ \sum_k \varepsilon_{kk} = 0 \ ; \ s_{ij} = \frac{\sigma_i}{\varepsilon_i}\,\varepsilon_{ij} \tag{1}$$

Contact Mechanics, Edited by M. Raous *et al.*
Plenum Press, New York, 1995

where s_{ij} are stress deviator components with respect to some orthonormal basis, ε_{ij} are the components of strain rate tensor, σ_i is stress intensity, ε_i is strain rate intensity, A, m (or K, μ) are material constants, $A = K^{-m}$, $m = 1/\mu \geq 1$.

The equilibrum equations, Cauchy relationships and boundary conditions are similar to those of classical elasticity (Sneddon and Berry, 1961). If we regard ε_{ij} as strain tensor components, we obtain immediately the formulation of the problem of deformation theory of plasticity (Iljushin, 1948).

LOCAL PROBLEM

Let us consider a small neighbourhood of a rigid punch corner under assumption of plane strain. We see the rigid quarter of plane contacting with a half-plane of material (1). If contact interface is smooth, this problem is almost identical to the old one of Cherepanov(1967), Rice and Rosengren(1968) and Hutchinson(1968) for a crack tip. All the difference consists in the sign of stress components: we have compression for the punch there where we had tension for the crack.

Hence, the method is the same. We assume Airy stress function has a form

$$\Phi(r, \vartheta) = Q\ r^s\ \varphi(\vartheta) \tag{2}$$

where r and ϑ are polar coordinates with centre of the punch corner apex, Q and s are unknown real constants and φ is an unknown function of the variable ϑ. Stress and strain components are

$$\begin{aligned}
&\sigma_{ij}(r, \vartheta) = Q\ r^{s-2}\ \tilde{\sigma}_{ij}(\vartheta) \\
&\tilde{\sigma}_{rr} = s\varphi + \varphi''\ ;\ \tilde{\sigma}_{\vartheta\vartheta} = s(s-1)\varphi\ ;\ \tilde{\sigma}_{r\vartheta} = (1-s)\varphi' \\
&4\tilde{\sigma}_i^2 = (\varphi'' + s(2-s)\varphi)^2 + 4(1-s)^2\varphi' \\
&\varepsilon_{ij}(r, \vartheta) = AQ^m r^{m(s-2)}\tilde{\varepsilon}_{ij}(\vartheta) \\
&\tilde{\varepsilon}_{rr} = -\tilde{\varepsilon}_{\vartheta\vartheta} = \tilde{\sigma}_i^{m-1}(\varphi'' + s(2-s)\varphi)/2\ ;\ \tilde{\varepsilon}_{r\vartheta} = \tilde{\sigma}_i^{m-1}(1-s)\varphi'
\end{aligned} \tag{3}$$

Prime denotes the differentiation with respect to ϑ as well as the operator ∂_ϑ in the next formula.

The factorization (2) separates variables in the compatibility equation, which takes the form of ordinary differential equation for the function $\varphi(\vartheta)$ (where s occurs as a parameter):

$$\begin{aligned}
(\partial_\vartheta^2 - m(s-2)(m(s-2)+2))\left(\tilde{\sigma}_i^{m-1}\ \frac{1}{2}(\partial_\vartheta^2\ \varphi + s(2-s)\varphi)\right) + \\
2(s-1)(m(s-2)+1)\partial_\vartheta\tilde{\sigma}_i^{m-1}\partial_\vartheta\varphi = 0
\end{aligned} \tag{4}$$

Boundary conditions for load-free surface are $\varphi = \varphi' = 0$. We have $\varphi' = \varphi''' = 0$ on absolutely smooth punch base. Therefore we obtain an homogeneous boundary problem with eigennumber s. It is solved numerically like as in the papers of Hutchinson(1968) or of Aleksandrov and Grishin(1987). The minimal value of s, corresponding to the solution with finite energy, is given by the formula

$$s = (2m+1)/(m+1) \tag{5}$$

That figure is the main interesting characteristic in crack problem, but for our goals we need also values of φ'' and φ''' on load-free surface. They had never been published. Because of that we had to carry out all Hutchinson's calculations once more. The formula (5) and samples of stress-strain fields, published by Hutchinson(1968),

Table 1

m	$-\varphi''$	φ'''	m	$-\varphi''$	φ'''	m	$-\varphi''$	φ'''
1.1	.003	1.237	1.8	.141	1.337	5	.461	.534
1.15	.005	1.431	1.9	.164	1.275	6	.501	.432
1.2	.007	1.573	2	.185	1.223	7	.526	.387
1.25	.013	1.585	2.33	.247	1.059	8	.549	.334
1.3	.022	1.589	2.66	.296	.932	9	.567	.301
1.4	.042	1.575	3	.332	.842	10	.585	.259
1.5	.067	1.522	3.5	.376	.732	11	.598	.227
1.6	.091	1.468	4	.410	.653	12	.608	.221
1.7	.117	1.402	4.5	.438	.586	13	.618	.198

were confirmed. Values of φ'' and φ''' on load-free surface in dependence of material constant m are collected in Table 1.

Thus, normalized local solution may be determined by formulae (3) with s from (5) and $\varphi(\vartheta)$ found numerically. Constant factor in stress-strain fields holds unknown and plays role of first free parameter.

INTERIOR SOLUTION FOR A THIN LAYER (PLANE STRAIN)

Let us consider a rectangular rigid punch pressing a thin layer ($-\infty < x < \infty$, $0 \leq y \leq h$) of material (1) (Grishin and Manzhirov, 1986). "Thin" means that punch width is much more than layer thickness. If friction is neglected, boundary conditions are ($q(x)$ — contact pressure):

$$\begin{aligned} y = h: &\quad \sigma_{yy} = q(x)\ ;\ \sigma_{xy} = 0 \\ y = 0: &\quad u_y = 0\ ;\ \sigma_{xy} = 0 \end{aligned} \tag{6}$$

Far asymptotics (far from corner) may be found by putting the linear representation $\sigma_{xy} = a(x) + b(x)y$ in (1), (6) and equilibrum equations (σ being hydrostatic pressure):

$$\begin{aligned} \sigma_{yy} = q(x)\ ;\ \sigma_{xx} = f(y)\ ;\ \sigma_{xy} = \varepsilon_{xy} = 0\ ;\ \sigma_{xx} = \sigma + K\varepsilon_i^{\mu-1}\varepsilon_{xx} \\ \sigma = q(x) - K\varepsilon_i^{\mu-1}\varepsilon_{yy}\ ;\ q(x) = K\varepsilon_i^{\mu-1}(\varepsilon_{yy} - \varepsilon_{xx}) + f(y) \end{aligned} \tag{7}$$

The function $f(y) \equiv 0$ because stresses must be zero when $x \to \infty$. Strain intensity $\varepsilon_i = |\varepsilon_{yy}|$ because $\varepsilon_{xx} = -\varepsilon_{yy}$ and $\varepsilon_{xy} = 0$. We have from the last equation (7):

$$q(x) = 2K|\varepsilon_{yy}|^{\mu}\,\mathbf{sgn}(\varepsilon_{yy}) \tag{8}$$

Cauchy relationship, third boundary condition (6) and (8) give:

$$u_y = y(2K)^{-m}|q(x)|^m\,\mathbf{sgn}(q(x)) \tag{9}$$

Note, that far asymptotics may be obtained for any function $\sigma_i(\varepsilon_i)$ which may be inversed as $\varepsilon_i(\sigma_i)$. The method is the same. Formulae for u_y have the structure, similar to comportment law. For Imbert law (Bell, 1984) $\varepsilon_i = \exp(n\sigma_i) - 1$ (n –

const) we have $u_y = y\ (\exp(|nq/2|) - 1)\,\mathbf{sgn}(q)$ and for Nadai law (Iljushin, 1948) $\varepsilon_i = \varepsilon_0\,\mathbf{sh}(\sigma_i/\sigma_0)$ $(\varepsilon_0,\ \sigma_0\ -\ const)$ $u_y = y\ \varepsilon_0\ \mathbf{sh}\ (|q/(2\sigma_0)|)\,\mathbf{sgn}(q)$.

INTERIOR SOLUTION IN AXISYMMETRIC CASE

The same technique was used to find interior solution when a rigid cylinder with plane base is pressed in a thin layer of material (1) (Grishin and Manzhirov, 1986). In cylindrical coordinates $R\Theta Z$ we have

$$\sigma = q(R) - K\varepsilon_i^{\mu-1}\varepsilon_{ZZ}\ ;\ \sigma_R = q(R) + K\varepsilon_i^{\mu-1}(2\varepsilon_{RR} + \varepsilon_{\Theta\Theta}) \tag{10}$$

Equilibrum equations give a complicated non-linear ordinary differential equation (prime denotes the differentiation with respect to R):

$$q' + \left(K\varepsilon_i^{\mu-1}(2\varepsilon_{RR} + \varepsilon_{\Theta\Theta})\right)' + K\varepsilon_i^{\mu-1}(\varepsilon_{RR} - \varepsilon_{\Theta\Theta})/R = 0 \tag{11}$$

Utilizing formulae $\varepsilon'_{\Theta\Theta} = (\varepsilon_{RR} - \varepsilon_{\Theta\Theta})/R$; $\varepsilon_{RR} = \varepsilon_{\Theta\Theta} + O(h^2)$ $(h \to 0)$ we obtain from (11):

$$q' + (3\mu + 1)K3^{(\mu-1)/2}|\varepsilon_{\Theta\Theta}|^{\mu-1}\varepsilon'_{\Theta\Theta} = 0 \tag{12}$$

and because $\mu|\varepsilon_{\Theta\Theta}|^{\mu-1}\varepsilon'_{\Theta\Theta} = (\varepsilon_{\Theta\Theta}|\varepsilon_{\Theta\Theta}|^{\mu-1})'$ we have

$$q + (3\mu + 1)\mu^{-1}K3^{(\mu-1)/2}|\varepsilon_{\Theta\Theta}|^{\mu}\,\mathbf{sgn}(\varepsilon_{\Theta\Theta}) = 0 \tag{13}$$

Analog of (9) is:

$$u_Z = 2ZK^{-m}(3\mu + 1)^{-m}(\mu|q(R)|3^{(1-\mu)/2})^m\,\mathbf{sgn}(q(R)) \tag{14}$$

The structure of (9) and (14) is the same, but factors before contact pressure are different. These factors coincide one to another and equal to well known ones (Manzhirov, 1983) only in the case of linear creep: $m = \mu = 1$.

INTERIOR SOLUTION FOR A HALF-PLANE IN PLANE STRAIN

The most popular situation of contact between rectangular rigid punch and creeping half-plane has been studied by Arutyunyan(1959). He found an approximate solution in hypergeometric functions with the help of a not completely clear technics. That solution is published also in the paper of Aleksandrov and Sumbatyan(1983) or in the book of Johnson(1985). Notably, that Arutyunyan's technics gives results close (Aleksandrov and Brudnij, 1986) to exact solution (Amazigo, 1974) for the case of anti-plane shear always exept neighbourhoods of punch corners. This fact permited to regard it as far asymptotics and to develope (by Aleksandrov and Sumbatyan(1983) on this problem example) main principles of our method.

SINGLE PUNCH PRESSING THIN LAYER

We return to the problem when rectangular (in cross-section) punch presses thin layer of material (1) under plane-strain assumption (Grishin and Manzhirov, 1986). Far asymptotics is given by (9) and first formula (7) (contact pressure q being constant on x), near-corner solution — by (3). We normalize linear dimensions to obtain $x = \pm 1$ as punch corners coordinates and investigate symmetric situation when the force P, acting on the punch, coincides with axis $0y$. This yields that we may regard only right half-plane. The contact pressure near the corner is given by the next formula (in polar coordinates with centre in right corner and ray $\vartheta = 0$ being drawn on the left):

$$\sigma_{\vartheta\vartheta}(r,0) = -Qs(s-1)r^{s-2} \tag{15}$$

All stresses are normalized on K/t_0^{μ} here and below, t_0 being some time-dimension constant.

We introduce the unknown parameter ε as the distance from the corner to the point where far and near asymptotics will be glued. Integral condition of the system equilibrum may be written in the form

$$\int_0^{1-\varepsilon} q\,dx + Qs(s-1)\int_{\varepsilon}^{0} r^{s-2}\,dr = \frac{P}{2} \tag{16}$$

The claim of contact pressure continuity in the point $r = \varepsilon$, $\vartheta = 0$ gives

$$q = -Qs(s-1)\varepsilon^{s-2} \tag{17}$$

We obtain by integration of (3) utilizing Cauchy relationships near the corner on load-free surface (Aleksandrov and Sumbatyan, 1983):

$$u_{\vartheta}(r,\pi) = -\delta + g(m)Q^m r^{1/(m+1)} \quad ; \quad g(m) \equiv 2^{-m}(m+1)^2|\varphi''(\pi)|^{m-1}\varphi'''(\pi) \tag{18}$$

when δ means vertical rate of corner (and values of derivatives of $\varphi(\vartheta)$ on load-free surface $\vartheta = \pi$ are shown in Table 1). Left-hand side of (18) must be zero far from the corner, so that the claim of joint of solutions in the point $r = \varepsilon$, $\vartheta = \pi$ yields

$$\delta = g(m)\, Q^m\, \varepsilon^{1/(m+1)} \tag{19}$$

From the other side, δ may be found from (9) and (17):

$$\delta = 2^{-m} h Q^m (s(s-1))^m \varepsilon^{m(s-2)} \tag{20}$$

Thus, from (19) and (20) we have the next formula:

$$\varepsilon = h\,\frac{s^m(s-1)^m}{2^m\, g(m)} \tag{21}$$

Finally, we obtain the local solution normalizing factor, putting (21) in (16):

$$Q = -P\frac{\varepsilon^{1/(m+1)}(m+1)}{2s(m+\varepsilon)} \tag{22}$$

Now if P is given (and δ is unknown), we use successively (21), (22), (20) or (19) to obtain δ. The same formulae are also enough to find P, when δ is known. Free constants being fixed, all elements of approximate solution may be computed. Simple structure of formulae permit easy parametric analysis, published by Grishin and Manzhirov(1986). Exemplary computations and some remarks about method limitations were also made there.

We have even simpler formulae in the case of linear creep ($m = 1$, $g(1) = 3/2$):

$$\varepsilon = \frac{h}{4} \quad ; \quad Q = \frac{-4R\sqrt{h}}{3(4+h)} \quad ; \quad -P = k\delta^{\mu} \quad ; \quad k = \frac{4+h}{h} \tag{23}$$

Axisymmetric analog of this problem (with plane-strain near-corner solution) was also solved in the cadre of the same ideas and shown in author's thesis (1987).

The solution also is given by explicit formulae. Its structure is the same as one of (21), (19), (22), but its coefficients are another:

$$\varepsilon = h\,\frac{C}{g(m)} \;\;;\;\; \delta = CQ^m \varepsilon^{m(s-2)} \;\;;\;\; C \equiv 2\,\frac{3^{(m-1)/2}}{(m+3)^m}\,s^m(s-1)^m$$

$$Q = -P\,\frac{1}{2\pi}\,\varepsilon^{1/(m+1)}\left(\frac{1}{2}s(s-1)(1-\varepsilon)^2 + s\varepsilon - (s-1)\varepsilon^2\right)^{-1}$$

PERIODICAL PROBLEM

Let us consider a periodical system of punches pressing a thin layer of material (1) (Grishin, 1988). We assume the strain to be planar and neglect friction on interfaces. Geometry of deformable body and boundary conditions may be described in cartesian coordinates x, y by next formulae

$$\begin{aligned}
y = 0\;;\; 0 \le x \le 1+\Delta/2 \;:&\quad u_y = 0\;;\; \sigma_{xy} = 0\\
y = h\;;\; 0 \le x \le 1 :&\quad u_y = \mathbf{const} \equiv -\delta\;;\; \sigma_{xy} = 0\\
y = h\;;\; 1 < x < 1+\Delta/2 :&\quad \sigma_{yy} = \sigma_{xy} = 0\\
0 < y < h\;;\; x = 0\;;\; x = 1+\Delta/2 :&\quad u_x = 0\;;\; \sigma_{xy} = 0
\end{aligned} \tag{24}$$

They show that the problem is symmetrical with respect to the axis $0y$ and periodical along the axis $0x$ with period equal to $2+\Delta$. Punch corner is in the point $x = 1$; $y = h$, punch base is on the left hand and load-free surface — on the right hand. We assume that the layer thickness h and neighbouring punches distance Δ are jointed by the formula $h \sim \Delta \ll 1$ and that contact region is known. Hence we can predict the strain field to be singular only in corner apex. It seems to be fast varying running from this point and low varying far from the corner. Unlike the case of single punch, we cannot put $f(y) = 0$ in formulae (7), but we can average that function far from the corner and suppose $f(y) = \mathbf{const} \equiv p_0 < 0$. Thus p_0 means average value of σ_{xx} far from the corner. Like as for single punch $\varepsilon_i = |\varepsilon_{yy}|$. So, we may rewrite the last formula (7) as

$$q(x) - p_0 = 2|\varepsilon_{yy}|^{\mu-1}\varepsilon_{yy} \tag{25}$$

We solve (25) with respect to ε_{yy}, we use Cauchy relationships, boundary conditions (24) and we obtain that

$$\begin{gathered}
u_y(x,\,h) = h\,2^{-m}\,|q(x) - p_0|^m\;\mathbf{sgn}(q(x) - p_0)\\
(0 < x < 1-\varepsilon\;;\; 1+\varepsilon < x < 1+\Delta/2)
\end{gathered} \tag{26}$$

This yields $q = \mathbf{const} \equiv q_0 < 0$ under the punch far from the corner because (24) $u_y = \mathbf{const}$.

To use the local solution (3) we must claim

$$\varepsilon < h\;;\; \varepsilon < \Delta/2 \tag{27}$$

where ε is unknown radius of half-circle with centre at corner apex. Because $\Delta \sim h \ll 1$ we may assume the rate of load-free surface $u_y = \mathbf{const} \equiv u_0 > 0$ far from the corner: $y = h$, $1+\varepsilon < x < 1+\Delta/2$. Let us claim the continuity of contact pressure and of free surface to glue asymptotics in the points $y = h$, $x = 1 \pm \varepsilon$:

$$q_0 = -Qs(s-1)\varepsilon^{s-2} \; ; \; u_0 = -\delta + Q^m \varepsilon^{1/(m+1}g(m) \tag{28}$$

From the other side (26) yields $u_0 = h2^{-m}|p_0|^m$ because $q \equiv 0$ when $1+\varepsilon < x < 1+\Delta/2$, and $\delta = h2^{-m}|q_0 - p_0|^m$ because $u_y(x,\ h) = -\delta$ when $x < 1-\varepsilon$. Excluding p_0 from two last equations we obtain

$$q_0 = -2h^{-\mu}\left(\delta^\mu + u_0^\mu\right) \tag{29}$$

Claims of system equilibrium and of incompressibility give

$$\int_0^1 q(x)\ dx \equiv q_0(1-\varepsilon) - Qs\varepsilon^{s-1} = \frac{P}{2} \tag{30}$$

$$\begin{gathered}\int_0^{1+\Delta/2} u_y(x,\ h) \equiv -\delta(1+\varepsilon) + \int_0^\varepsilon u_\vartheta(r,\pi)\ dr + u_0(\frac{\Delta}{2}-\varepsilon) = \\ u_0(\frac{\Delta}{2}-\varepsilon) - \delta(1+\varepsilon) + \frac{m+1}{m+2}g(m)Q^m\varepsilon^{(m+2)/(m+1)} = 0\end{gathered} \tag{31}$$

Five equations (28) — (31) permit to find five constants: ε, Q, q_0, u_0 and δ or P. That system may be reduced to only one complicated non-linear algebraic equation for ε (Grishin, 1988). Explicit solution may be written in particular case of linear creep $m = 1$, $g(1) = 3/2$ (here we suppose P being given and δ unknown):

$$\begin{gathered}\varepsilon = \frac{h}{4} \; ; \; q_0 = \frac{2P}{4+h} \; ; \; Q = \frac{-4P\sqrt{h}}{3(4+h)} \\ \delta = \frac{Ph(h-6\Delta)}{3(4+h)(4+2\Delta)} \; ; \; u_0 = -\frac{Ph(h+12)}{3(4+h)(4+2\Delta)}\end{gathered} \tag{32}$$

A serie of computations has been carried out by Grishin(1988) for various m. We began with $m = 1$ and increased m step by step taking ε of previous step as initial iteration for the next one. Results were compared with those for single punch.

DISCUSSION

To sum up, we have
— approximate solutions by explicit formulae for single punch on creeping layer (both plane strain and axisymmetric),
— approximate solutions by simple calculation (non-linear scalar algebraic equation must be solved numerically) in the cases of planar periodic problem for thin layer and of single punch on half-plane.

All asymptotics are glued *only partially*. We can make nothing more to obtain its proximity in the deep of the body: the number of free parameters is not enough.

The collection of local solutions may be widely enlarged: Cherepanov – Rice – Hutchinson approach gives results for wedges of other than π apreture angles and for other than "smooth punch — free surface" boundary conditions. Eigennumber s cannot be given explicitly in this case, but a lot of numerical results has been shown by Aleksandrov and Grishin(1987).

The question of interior solution finding is more difficult. For instance, axisymmetric problem, when a punch presses a half-space, has been solved by Kuznetsov (1962) with Arutyunyan's technics , but later investigations (Martynenko and Svirskij (1984); Galanov(1984)) showed that it has some physically unclear properties. The question of legality of planar boundary layer use in axisymmetric problems also holds open. The comparison of results in plane and axisymmetric problems for thin layer shows signifiant differences for $m \geq 3$.

Thus, it would be very useful to compare our solutions (far from perfect, but easy to compute) with direct contact-creep experiments.

ACKNOWLEDGMENTS

Grateful acknowledgments to 2th Contact Mechanics International Symposium organizers and to International Science Foundation for the symposium visit financement.

REFERENCES

Aleksandrov, V.M., and Brudnij, S.R., 1986, On the method of generalized superposition in contact problem of anti-plane shear, *Mech.of Solids U.S.S.R.* 21: No 4.

Aleksandrov, V.M., and Grishin, S.A., 1987, State of stress and strain of a small neighbourhood of the apex of a wedge for a physical non-linearity and different boundary conditions, *PMM U.S.S.R.* 51: No 4.

Aleksandrov, V.M., and Sumbatyan, M.A., 1983, On some solution of contact problem of non-linear steady creep for a half-plane, *Mech.of Solids U.S.S.R.* 18: No 1.

Amazigo, J.C., 1974, Fully plastic crack in an infinite body under anti-plane shear, *Int.J.Solids Structures* 10:1003.

Arutyunyan, N.Kh., 1959, Plane contact problem of creep theory, *PMM U.S.S.R.* 23: No 5.

Bell, J.F., 1984, "Experimental Fundations of Mechanics of Solids, "Nauka, Moscow. (Russian translation from: Enciclopedia of Phisics, vol.VIa/1, Mechanics of Solids I, Springer–Verlag, Berlin–Heidelberg–New York (1973))

Cherepanov, G.P., 1967, On crack propagation in a continuous medium, *PMM U.S.S.R.* 31: No 3.

Galanov, B.A., 1984, Numerical solution to the problem about concentrated force acting onto the boundary of the half-space of power-law hardening material, *in:* "Strength of Materials and Building Theory, issue 44, Budivelnik, Kiev.

Grishin, S.A., 1988, Periodical contact problem of non-linear steady creep for a thin layer, *Mech.of Solids U.S.S.R.* 23: No 2.

Grishin, S.A., and Manzhirov, A.V., 1986, Contact problems for a thin layer under conditions of non-linear creep, *Mech.of Solids U.S.S.R.* 21: No 6.

Hutchinson, J.W., 1968, Singular behaviour at the end of a tensile crack in a hardening material, *J.Mech.Phys.Solids* 16:13.

Hutchinson, J.W., 1968, Plastic stress and strain fields at a crack tip, *J.Mech.Phys.Solids* 16:337.

Iljushin, A.A., 1948, "Plasticity, "Gostekhizdat, Moscow–Leningrad.

Johnson, K.L., 1985, "Contact Mechanics, "Cambrige University Press, Cambrige.

Kachanov, L.M., 1960, "Creep Theory, "Fizmatgiz, Moscow.

Kuznetsov, A.I., 1962, Pressing of rigid punches into half-space under conditions of power-law hardening and non-linear creep of material, *PMM U.S.S.R.* 26: No 3.

Manzhirov, A.V., 1983, Plane and axisymmetric problems about load acting onto thin viscoelastic layer, *Journ.Appl.Mech.and Tech.Phys. U.S.S.R.* No 5.

Martynenko, M.D., and Svirskij, E.A., 1984, Axisymmetric stress-strain fields in the problem about concentrated force acting onto non-linear half-space, *Differential Equations U.S.S.R.* 20:2007.

Rabotnov, Yu.N., 1966, "Creep of Structure Elements, "Nauka, Moscow.

Rice, J.R., and Rosengren, G.F., 1968, Plane strain deformation near a crack tip in a power-law hardening material, *J.Mech.Phys.Solids* 16:1.

Sneddon, J.N., and Berry, D.S., 1961, "The Classical Theory of Elasticity, "Fizmatgiz, Moscow. (Russian translation of Handbuch der Physic, band VI, Springer–Verlag, Berlin–Gottingen–Heidelberg (1958))

A GENERALISED STANDARD MODEL FOR CONTACT, FRICTION AND WEAR

Niclas Strömberg, Lars Johansson and Anders Klarbring

Department of Mechanical Engineering
Division of Mechanics, Linköping Institute of Technology
S-581 83 Linköping, Sweden

INTRODUCTION

In this paper a continuum thermodynamic model for interfacial phenomena including contact, friction and wear is proposed. The framework is that of small displacements, implying small slip. Consequently the model is mainly one for fretting, which is a wear phenomenon arising when contacting surfaces undergo oscillatory displacements with small amplitudes.

Starting from the method of virtual power and the laws of thermodynamics, stated in forms covering a wide class of small displacement contact problems, equilibrium equations and guiding principles for constitutive laws of interfacial phenomena are derived. The internal variable approach is used. An internal state variable is defined for each wear mechanism and is used in the formulation of a free energy and a pseudo-potential for the contact interface. The free energy together with the identified state laws and the pseudo-potential define the generalised standard model. The derivation, essentially, follows Klarbring (1990), and Johansson and Klarbring (1993).

Concerning references related to the general domain of problems being considered we mention the following: A general thermodynamic theory of elastic material interfaces was developed by Murdoch (1976). Later on Fremond (1988) formulated a thermodynamic model for contact with adhesion and in three extensive papers, Zmitrowicz (1987) derived a thermodynamic model of contact, friction and wear.

Within in the framework of the generalised standard model a particular interfacial model for contact, friction and wear is suggested in the present work. Signorini's classical constitutive assumption of unilateral contact is extended to take wear into account and a coupling between Coulomb friction and Archard's law of wear is proposed.

Concerning previous work related to our particular model we note that Curnier (1984) proposed a general theory of friction with a specific wear law and recently, Mróz and Stupkiewicz (1994) proposed wear models by assuming that the wear rate is proportional to the dissipation rate.

Contact Mechanics, Edited by M. Raous *et al.*
Plenum Press, New York, 1995

DERIVATION OF A GENERAL MODEL

Let the open, disjoint regions $\Omega^r\ (r = 1,2) \subset \Re^n\ (n = 2,3)$ with piecewise smooth boundaries $\partial\Omega^r$ be occupied by two continuous, deformable bodies (Figure 1). Both

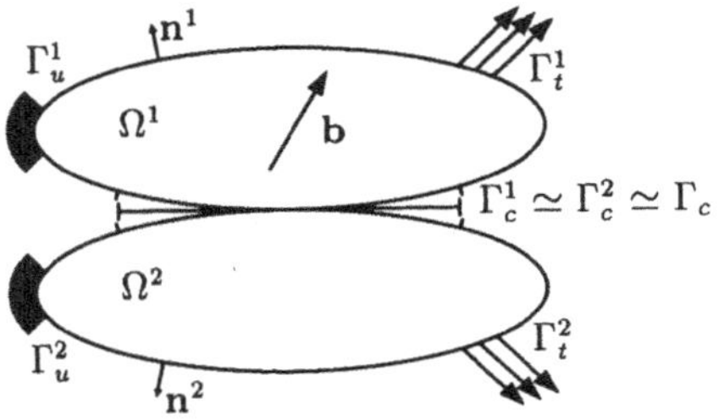

Figure 1. The two bodies considered, defined by the regions $\Omega^r\ (r = 1,2) \subset \Re^n\ (n = 2,3)$.

bodies are subjected to body forces $\mathbf{b}$, prescribed tractions $\mathbf{t}^r$ on $\Gamma_t^r \subset \partial\Omega^r$ and fixed displacements on $\Gamma_u^r \subset \partial\Omega^r$. The displacement field of the bodies is denoted by $\mathbf{u}$.

The material boundaries $\Gamma_c^r \subset \partial\Omega^r$ with outward unit normal directions $\mathbf{n}_c^r$ represent the potential contact surfaces. Only problems with small displacements are considered. Therefore, the potential contact surfaces and the corresponding normal directions have to be almost identical, i.e. $\Gamma_c^1 \simeq \Gamma_c^2$ and $\mathbf{n}_c^1 \simeq -\mathbf{n}_c^2$. This makes it possible to define a common contact surface $\Gamma_c \simeq \Gamma_c^1 \simeq \Gamma_c^2$ with outward unit normal direction $\mathbf{n}_c \simeq \mathbf{n}_c^1 \simeq -\mathbf{n}_c^2$, i.e. each particle on Γ_c^1 is coupled with a particle on Γ_c^2 in a one-to-one correspondence.

The Method of Virtual Power

The method of virtual power, in the sense of Germain (1973), is used to derive the equilibrium equations and to identify the internal forces to be Cauchy's stress tensor and the contact traction vector. For any part $\mathcal{D} \subset \Omega^1 \cup \Omega^2$ such that $\partial\mathcal{D} \cap \Gamma_c^1 \simeq \partial\mathcal{D} \cap \Gamma_c^2$ where $\simeq$ is in the sense indicated above, the virtual power of inertial forces balances the virtual power of all internal and external forces for any virtual velocity field $\hat{\mathbf{v}}$. Restricting ourselves to quasi-static problems, the principle of virtual power can be written as

$$\hat{P}_i + \hat{P}_x = 0 \quad \forall \hat{\mathbf{v}}, \tag{1}$$

where $\hat{\mathbf{v}}$ is a kinematically admissible virtual velocity field. The virtual power of internal and external forces are defined as

$$\hat{P}_i = -\int_{\mathcal{D}} \boldsymbol{\sigma} : \hat{\dot{\boldsymbol{\epsilon}}} dV - \int_{\partial\mathcal{D}\cap\Gamma_c} \mathbf{p} \cdot \hat{\dot{\mathbf{w}}} dA,$$

$$\hat{P}_x = \int_{\mathcal{D}} \mathbf{b} \cdot \hat{\mathbf{v}} dV + \int_{\partial\mathcal{D}-\Gamma_c} \mathbf{t} \cdot \hat{\mathbf{v}} dA, \tag{2}$$

where $\boldsymbol{\sigma}$ and $\mathbf{p}$ are internal forces, $\boldsymbol{\epsilon}$ is the infinitesimal strain tensor, $\mathbf{v} = \dot{\mathbf{u}}$ is the velocity field and $\mathbf{w} = \mathbf{u}^1 - \mathbf{u}^2$ is the relative displacement between coupled particles on Γ_c. A superimposed hat denotes a virtual quantity, a superimposed dot stands for right-hand time derivative and : and · are the inner products between tensors and vectors, respectively. Notice that all occurring time derivatives in the text are interpreted as right-hand derivatives. This is of importance since solutions to contact problems may only have such derivatives.

From (1), the equilibrium equations and Cauchy's theorem can be derived. The symmetry of the infinitesimal strain tensor $\boldsymbol{\epsilon} = \boldsymbol{\epsilon}^T$ implies that only the symmetric part of the internal force $\boldsymbol{\sigma}$ gives a contribution to the virtual power. Therefore, $\boldsymbol{\sigma}$ is considered to be symmetric. With suitable choices of $\hat{\mathbf{v}}$, the following equations are obtained from (1)

$$div\boldsymbol{\sigma} + \mathbf{b} = \mathbf{0} \quad in\ \mathcal{D}, \tag{3}$$

$$\boldsymbol{\sigma}\mathbf{n} = \mathbf{t} \quad on\ \partial\mathcal{D} - \Gamma_c, \tag{4}$$

$$\boldsymbol{\sigma}^1\mathbf{n}_c^1 = -\boldsymbol{\sigma}^2\mathbf{n}_c^2 = -\mathbf{p} \quad on\ \partial\mathcal{D} \cap \Gamma_c, \tag{5}$$

where $\boldsymbol{\sigma}^r$ is the limit of $\boldsymbol{\sigma}$ when approaching Γ_c from within $\mathcal{D} \cap \Omega^r$ and $\mathbf{n}$ is the outward unit normal on $\partial\mathcal{D} - \Gamma_c$. The symmetry of $\boldsymbol{\sigma}$, (3) and (4) imply that $\boldsymbol{\sigma}$ can be interpreted as Cauchy's stress tensor and (5) implies that $\mathbf{p}$ can be interpreted as the contact traction vector.

The Principles of Thermodynamics

By introducing Helmholtz free energies, the first and second law of thermodynamics give the Clausius-Duhem inequalities for the bodies Ω^r and the material interface Γ_c. The two basic principles of thermodynamics are postulated as

$$\dot{\mathcal{E}} = P_x + \mathcal{Q} \quad \forall\mathcal{D},$$

$$\dot{\mathcal{S}} \geq \int_{\mathcal{D}} \frac{r}{T} dV - \int_{\partial\mathcal{D}-\Gamma_c} \frac{\mathbf{q}\cdot\mathbf{n}}{T} dA \quad \forall\mathcal{D},$$

where $\mathcal{E}$ is the internal energy, P_x is the power of external forces obtained by evaluating (2) for the real velocity, $\mathcal{Q}$ is the heat supply per unit time, $\mathcal{S}$ is the entropy, r is the internal heat production, $\mathbf{q}$ is the heat flux vector and T is the absolute temperature in the bodies Ω^r. The internal energy $\mathcal{E}$, the entropy $\mathcal{S}$ and the heat supply per unit time $\mathcal{Q}$ are defined as

$$\mathcal{E} = \int_{\mathcal{D}} \rho e dV + \int_{\partial\mathcal{D}\cap\Gamma_c} E dA,$$

$$\mathcal{S} = \int_{\mathcal{D}} \rho s dV + \int_{\partial\mathcal{D}\cap\Gamma_c} S dA,$$

$$\mathcal{Q} = \int_{\mathcal{D}} r dV - \int_{\partial\mathcal{D}-\Gamma_c} \mathbf{q}\cdot\mathbf{n} dA,$$

where e is the specific internal energy, s is the specific entropy, E is the surface density of internal energy on Γ_c and S the surface density of entropy on Γ_c.

Noting that $\mathcal{D}$ is arbitrary, it is possible to express the first and second law on local form as

$$\left.\begin{aligned} \rho\dot{e} &= \boldsymbol{\sigma} : \dot{\boldsymbol{\epsilon}} + r - \mathrm{div}\mathbf{q} \\ \rho\dot{s} &\geq \tfrac{r}{T} - \mathrm{div}(\tfrac{\mathbf{q}}{T}) \end{aligned}\right\} \quad in\ \Omega^1 \cup \Omega^2, \tag{6}$$

$$\left.\begin{array}{l}\dot{E} = \mathbf{p}\cdot\dot{\mathbf{w}} + \mathbf{q}^1\cdot\mathbf{n}_c^1 + \mathbf{q}^2\cdot\mathbf{n}_c^2 \\ \dot{S} \geq \frac{\mathbf{q}^1\cdot\mathbf{n}_c^1}{T^1} + \frac{\mathbf{q}^2\cdot\mathbf{n}_c^2}{T^2}\end{array}\right\} \quad on\ \Gamma_c, \tag{7}$$

where $\mathbf{q}^r$ and T^r are the limits of $\mathbf{q}$ and T, respectively, when approaching Γ_c from within $\mathcal{D}\cap\Omega^r$. Next, we introduce Helmholtz free energies ψ and Ψ as

$$\psi = e - sT, \quad \Psi = E - S\mathcal{T}, \tag{8}$$

where $\mathcal{T}$ is the intrinsic temperature on Γ_c. Moreover, the contact traction vector $\mathbf{p}$ and the relative displacement $\mathbf{w}$ are decomposed into a normal component and a tangential vector as

$$p_N = \mathbf{p}\cdot\mathbf{n}_c, \quad \mathbf{p}_T = (\mathbf{I} - \mathbf{n}_c\otimes\mathbf{n}_c)\mathbf{p}, \quad w_N = \mathbf{w}\cdot\mathbf{n}_c, \quad \mathbf{w}_T = (\mathbf{I} - \mathbf{n}_c\otimes\mathbf{n}_c)\mathbf{w},$$

where the normal direction $\mathbf{n}_c$ was defined previously, $\mathbf{I}$ is the identity tensor and $\otimes$ is the tensor product.

By combining (6), (7) and (8), the Clausius-Duhem inequalities for the bodies Ω^r and the interface Γ_c are obtained as

$$\rho\dot{\psi} \leq \boldsymbol{\sigma} : \dot{\boldsymbol{\epsilon}} - \rho s\dot{T} - \mathbf{q}\cdot\frac{\nabla T}{T} \quad in\ \Omega^1\cup\Omega^2, \tag{9}$$

$$\dot{\Psi} \leq p_N\dot{w}_N + \mathbf{p}_T\cdot\dot{\mathbf{w}}_T - S\dot{\mathcal{T}} + \sum_{r=1}^{2}\frac{\mathbf{q}^r\cdot\mathbf{n}_c^r}{T^r}\theta^r \quad on\ \Gamma_c, \tag{10}$$

where $\theta^r = T^r - \mathcal{T}$ are the temperature differences between each body Ω^r and the interface Γ_c. From here, it is assumed that constitutive laws for the bodies Ω^r, satisfying (9), exist and the attention will instead be focused on the Clausius-Duhem inequality for the interface (10).

The Internal State Variables

Contact, friction and wear are influenced by several interfacial phenomena, depending on kinematics, material and geometry of the bodies and the environment. The interfacial phenomena of wear can normally be explained by four major wear mechanisms, namely, adhesive, abrasive and corrosive wear and surface fatigue, but several minor mechanism also exist, such as, the mechanism explained by the delamination theory of wear suggested by Suh (1973). One should notice that interfacial phenomena described by different wear mechanisms do not only affect the wear but also influence the friction and to some extent the contact conditions. We will base the modelling of wear at the interface Γ_c on the following assumption:

For each wear mechanism, it exists an internal state variable w_i^w. The total wear is measured in the normal direction $\mathbf{n}_c$ by the function $\delta(w_i^w)$ called the wear gap, which is the sum of all internal state variables w_i^w, i.e.

$$\delta(w_i^w) = \sum_i^{\zeta} w_i^w,$$

where ζ is the total number of wear mechanisms at the interface Γ_c.

The internal state variables w_i^w and the wear gap $\delta(w_i^w)$ are interpreted in Figure 2.

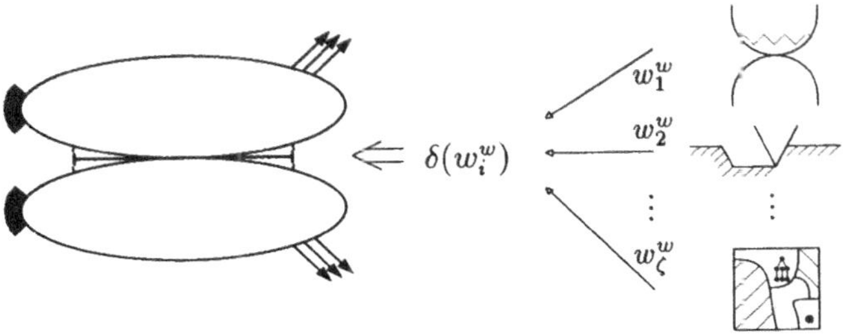

Figure 2. Interpretation of the internal state variables w_i^w and the wear gap $\delta(w_i^w)$.

The Free Energy and the Pseudo-Potential

We assume that the free energy Ψ introduced in (8) can be expressed as a continuous function of the observable state variables w_N, $\mathcal{T}$, θ^r and the internal state variables w_i^w, i.e. $\Psi = \Psi(w_N, w_1^w, \ldots, w_\zeta^w, \mathcal{T}, \theta^1, \theta^2)$. In addition, all variables can be considered as continuously right-handed differentiable functions of time t. The function $\Psi(w_N, w_1^w, \ldots, w_\zeta^w, \mathcal{T}, \theta^1, \theta^2)$ is convex but not necessarily differentiable with respect to w_N and w_i^w. On the other hand, it is concave and differentiable with respect to $\mathcal{T}$ and θ^r.

Next, the functions $\tilde{p}_N$ and $\mathcal{W}_i$ are defined as the components of any subgradient belonging to the subdifferential of Ψ, i.e.

$$(\tilde{p}_N, -\mathcal{W}_1, \ldots, -\mathcal{W}_\zeta) \in \partial\Psi(w_N, w_1^w, \ldots, w_\zeta^w, \mathcal{T}, \theta^1, \theta^2), \tag{11}$$

where $\partial\Psi$ denotes the subdifferential with respect to w_N and w_i^w, keeping $\mathcal{T}$ and θ^r fixed. Furthermore, the functions $\tilde{S}$ and Θ^r are defined as

$$-\tilde{S} = \frac{\partial\Psi}{\partial\mathcal{T}}, \quad -\Theta^r = \frac{\partial\Psi}{\partial\theta^r} \quad (r = 1, 2). \tag{12}$$

The time rate of change of Ψ, at a state determined by the state variables w_N, w_i^w, $\mathcal{T}$ and θ^r, is given as

$$\dot{\Psi} = \lim_{\Delta t \to 0^+} \frac{\Psi(t + \Delta t) - \Psi(t)}{\Delta t}, \tag{13}$$

where $\Psi(t) = \Psi(w_N(t), w_1^w(t), \ldots, w_\zeta^w(t), \mathcal{T}(t), \theta^1(t), \theta^2(t))$. A useful expression for $\dot{\Psi}$ is obtained by using the convexity. The definition of the subdifferential implies

$$\begin{aligned}&\Psi(w_N(t + \Delta t), w_1^w(t + \Delta t), \ldots, w_\zeta^w(t + \Delta t), \mathcal{T}(t), \theta^1(t), \theta^2(t))\\ &\geq \Psi(w_N(t), w_1^w(t), \ldots, w_\zeta^w(t), \mathcal{T}(t), \theta^1(t), \theta^2(t))\\ &+\tilde{p}_N(w_N(t + \Delta t) - w_N(t)) - \textstyle\sum_{i=1}^{\zeta} \mathcal{W}_i(w_i^w(t + \Delta t) - w_i^w(t)).\end{aligned}$$

By dividing this inequality with a positive time increment Δt and letting Δt approach zero, one can finally obtain the following inequality

$$\dot{\Psi} \geq \tilde{p}_N \dot{w}_N - \sum_{i=1}^{\zeta} \mathcal{W}_i \dot{w}_i^w - \tilde{S}\dot{\mathcal{T}} - \sum_{r=1}^{2} \Theta^r \dot{\theta}^r. \tag{14}$$

The Clausius-Duhem inequality in (10) and the inequality in (14) put together give that the dissipation per unit surface, defined as

$$D = (p_N - \tilde{p}_N)\dot{w}_N + \mathbf{p}_T \cdot \dot{\mathbf{w}}_T + \sum_{i=1}^{\zeta} \mathcal{W}_i \dot{w}_i^w - (S - \tilde{S})\dot{\mathcal{T}} + \sum_{r=1}^{2} \frac{\mathbf{q}^r \cdot \mathbf{n}_c^r}{T^r}\theta^r + \sum_{s=1}^{2} \Theta^s \dot{\theta}^s,$$

must be greater than or equal to zero for all possible evolutions of the system. The normal contact pressure p_N and the entropy per unit surface S are assumed not to depend on $\dot{w}_N$ and $\dot{\mathcal{T}}$, respectively. Moreover, the functions Θ^r do not depend on $\dot{\theta}^r$, see (12). The following are sufficient conditions to fulfill the above dissipation inequality

$$p_N = \tilde{p}_N, \quad S = \tilde{S}, \quad \Theta^r = 0 \quad (r = 1,2) \;\Rightarrow\; \Psi = \Psi(w_N, w_1^w, \ldots, w_\zeta^w, \mathcal{T}). \tag{15}$$

Together, (11), (12) and (15) define the state laws. The final expression for the dissipation per unit surface becomes

$$D = \mathbf{p}_T \cdot \dot{\mathbf{w}}_T + \sum_{i=1}^{\zeta} \mathcal{W}_i \dot{w}_i^w + \sum_{r=1}^{2} \frac{\mathbf{q}^r \cdot \mathbf{n}_c^r}{T^r}\theta^r \geq 0. \tag{16}$$

This inequality must be fulfilled for all evolutions of the relative tangential velocity $\dot{\mathbf{w}}_T$, the rate of each wear mechanism $\dot{w}_i^w$ and the temperature differences θ^r. The associated forces $\mathcal{W}_i$, defined by the state law in (11), are the wear driving forces for the respective wear mechanism.

In order to satisfy the dissipation inequality in (16), we assume that it exists a pseudo-potential $\Phi = \Phi(\mathbf{p}_T, \mathcal{W}_1, \ldots, \mathcal{W}_\zeta, \theta^1, \theta^2; p_N, w_1^w, \ldots, w_\zeta^w, \mathcal{T}, T^1, T^2)$, which is convex but not necessarily differentiable with respect to the tangential contact traction vector $\mathbf{p}_T$, the wear driving forces $\mathcal{W}_i$ and the temperature differences θ^r. The normal contact pressure p_N, the wear mechanisms w_i^w and the temperatures $\mathcal{T}$ and T^r serve as parameters. Let $\mathcal{P} = (p_N, w_1^w, \ldots, w_\zeta^w, \mathcal{T}, T^1, T^2)$. The relative tangential velocity $\dot{\mathbf{w}}_T$, the rate of each wear mechanism $\dot{w}_i^w$ and the heat terms $\frac{\mathbf{q}^r \cdot \mathbf{n}_c^r}{T^r}$ are assigned the values of the components of any subgradient belonging to the subdifferential $\partial\Phi(\mathbf{p}_T, \mathcal{W}_1, \ldots, \mathcal{W}_\zeta, \theta^1, \theta^2; \mathcal{P})$, i.e.

$$(\dot{\mathbf{w}}_T, \dot{w}_1^w, \ldots, \dot{w}_\zeta^w, \frac{\mathbf{q}^1 \cdot \mathbf{n}_c^1}{T^1}, \frac{\mathbf{q}^2 \cdot \mathbf{n}_c^2}{T^2}) \in \partial\Phi(\mathbf{p}_T, \mathcal{W}_1, \ldots, \mathcal{W}_\zeta, \theta^1, \theta^2; \mathcal{P}). \tag{17}$$

Moreover, the pseudo-potential is equal to zero with zero arguments and the subdifferential of the pseudo-potential contains the zero subgradient at the origin, i.e.

$$0 = \Phi(\mathbf{0}, 0, \ldots, 0, 0, 0; \mathcal{P}), \qquad (\mathbf{0}, 0, \ldots, 0, 0, 0) \in \partial\Phi(\mathbf{0}, 0, \ldots, 0, 0, 0; \mathcal{P}).$$

If the pseudo-potential $\Phi(\mathbf{p}_T, \mathcal{W}_1, \ldots, \mathcal{W}_\zeta, \theta^1, \theta^2; \mathcal{P})$ satisfies these conditions, then the dissipation inequality in (16) will always be fulfilled by the relations defined in (17). The free energy, the state laws and the pseudo-potential defined above, together constitute the generalised standard model for contact, friction and wear.

SIGNORINI'S CONTACT CONDITION, COULOMB FRICTION AND ARCHARD'S WEAR LAW

A classical constitutive assumption of unilateral contact is Signorini's unilateral contact condition. We will propose an extension of Signorini's unilateral contact condition, which besides the initial gap g also takes the wear mechanisms w_i^w $(i = 1, \ldots, \zeta)$ at the interface into consideration. The wear gap $\delta(w_i^w)$ is used to update the gap between the bodies due to the wear mechanisms. The extension is formulated with the free energy $\Psi(w_N, w_1^w, \ldots, w_\zeta^w, \mathcal{T})$ by taking

$$\Psi(w_N, w_1^w, \ldots, w_\zeta^w, \mathcal{T}) = I_{C_1}(w_N, w_1^w, \ldots, w_\zeta^w) - \frac{\mathcal{C}}{2\mathcal{T}_o}(\mathcal{T} - \mathcal{T}_o)^2, \tag{18}$$

$$C_1 = \{(w_N, w_1^w, \ldots, w_\zeta^w) \mid w_N - g - \delta(w_i^w) \le 0\},$$

where I_{C_1} is the indicator function of the set C_1, $\mathcal{C}$ is the heat capacity per unit area of the contact interface and $\mathcal{T}_o$ is a reference temperature. Notice the assumption of a decoupled form of the free energy into one mechanical part depending on w_N and w_i^w, and one thermal part depending only on $\mathcal{T}$. Furthermore, for this particular free energy all internal state variables can be replaced by one single wear mechanism $w^w = \delta(w_i^w)$ and thus the above formulation belongs to a class of free energies expressed as $\Psi = \Psi(w_N, w^w, \mathcal{T})$. The associated wear driving force with the wear mechanism w^w is denoted as $\mathcal{W}$.

The subdifferential $\partial I_C(\mathbf{x})$ of a convex indicator function is equal to the normal cone $N_C(\mathbf{x})$ of the set C. Thus, the state laws in (11) can be expressed by saying that $(p_N, -\mathcal{W})$ belong to the normal cone $N_{C_1}(w_N, w^w)$, i.e.

$$\left\{\begin{matrix} p_N \\ \mathcal{W} \end{matrix}\right\} = \lambda \left\{\begin{matrix} 1 \\ 1 \end{matrix}\right\}, \quad \lambda \ge 0, \quad w_N - g - w^w \le 0, \quad \lambda(w_N - g - w^w) = 0.$$

It is seen that the wear driving force $\mathcal{W}$ always will be equal to the normal contact pressure p_N. The entropy defined by (12) becomes

$$S = \frac{\mathcal{C}}{\mathcal{T}_o}(\mathcal{T} - \mathcal{T}_o).$$

Next, a particular choice of pseudo-potential $\Phi(\mathbf{p}_T, \mathcal{W}, \theta^1, \theta^2; \mathcal{P})$, identified as a coupling between Coulomb friction and Archard's law of wear, is proposed. We take

$$\Phi(\mathbf{p}_T, \mathcal{W}, \theta^1, \theta^2; \mathcal{P}) = I_{C_2(p_N, w^w, \mathcal{T})}(\mathbf{p}_T, \mathcal{W}) + \frac{1}{2}\sum_{r=1}^{2} \frac{\vartheta^r}{\mathcal{T}^r}(\theta^r)^2,$$

$$C_2(p_N, w^w, \mathcal{T}) = \{(\mathbf{p}_T, \mathcal{W}) \mid |\mathbf{p}_T| + \frac{k\mathcal{W}^2}{6p} \le \mu p_N\},$$

where $|\cdot|$ denotes the euclidic norm, $k = k(w^w, \mathcal{T})$ is a wear constant, p is the penetration hardness of the softer material, $\mu = \mu(w^w, \mathcal{T})$ is a friction coefficient and $\vartheta^r = \vartheta^r(p_N, w^w, \mathcal{T})$ are thermal contact conductances. Since the subdifferential $\partial I_{C_2(p_N, w^w, \mathcal{T})}(\mathbf{p}_T, \mathcal{W})$ is equal to the normal cone $N_{C_2(p_N, w^w, \mathcal{T})}(\mathbf{p}_T, \mathcal{W})$, (17) can be written as

$$\left\{\begin{matrix} \dot{\mathbf{w}}_T \\ \dot{w}^w \end{matrix}\right\} = |\dot{\mathbf{w}}_T| \left\{\begin{matrix} \frac{\mathbf{p}_T}{|\mathbf{p}_T|} \\ \frac{k\mathcal{W}}{3p} \end{matrix}\right\}, \quad |\mathbf{p}_T| + \frac{k\mathcal{W}^2}{6p} \le \mu p_N, \quad |\dot{\mathbf{w}}_T|(|\mathbf{p}_T| + \frac{k\mathcal{W}^2}{6p} - \mu p_N) = 0$$

and

$$\mathbf{q}^r \cdot \mathbf{n}_c^r = \vartheta^r \theta^r \quad (r = 1, 2),$$

which are interfacial forms of Fourier's heat diffusion law. This together with (7), (8), (12) and (18) give the following expression for the time rate of the internal energy and the evolution of the temperature

$$\dot{E} = \frac{C}{\mathcal{T}_o} \mathcal{T} \dot{\mathcal{T}} = p_N |\dot{\mathbf{w}}_T| \frac{k\mathcal{W}}{3p} + \mathbf{p}_T \cdot \dot{\mathbf{w}}_T + \sum_{r=1}^{2} \vartheta^r \theta^r.$$

Finally, the wear gap $\delta(w_i^w)$ makes it possible to express the total worn away volume V at a time τ as

$$V = \int_0^\tau \int_{\Gamma_c} \delta(\dot{w}_k^w) dA dt = \int_0^\tau \int_{\Gamma_c} |\dot{\mathbf{w}}_T| \frac{k p_N}{3p} dA dt.$$

This is actually a global form of Archard's wear law, which is more familiar in the form

$$V = \frac{kNx}{3p},$$

where k is a constant, which Archard (1953) interpreted as the probability that a fragment will be formed at an adhesive joint, N is the normal force between the bodies, x is the sliding distance and p is the penetration hardness of the softer material.

ACKNOWLEDGEMENT

This work was supported by the Swedish Research Council for Engineering Sciences (TFR).

REFERENCES

Archard, J.F., 1953, Contact and rubbing of flat surfaces, *J. Appl. Phys.* 24:981.

Curnier, A., 1984, A theory of friction, *Int. J. Solids Structures* 20:637.

Fremond, M., 1988, Contact with adhesion, *in:* "Topics in Nonsmooth Mechanics," J.J. Moreau , P.D. Panagiotopoulos and G. Strang, ed., Pirkhäuser Verlag, Bassel.

Germain, P., 1973, The method of virtual power in continuum mechanics. part 2: microstructure, *SIAM J. Appl. Math.* 25:556.

Johansson, L., and Klarbring, A., 1993, Thermoelastic frictional contact problems: modelling, finite element approximation and numerical realization, *Comp. Meth. Appl. Mech. Eng.* 105:181.

Klarbring, A., 1990, Derivation and analysis of rate boundary-value problems of frictional contact, *Eur. J. Mech., A/Solids* 9:53.

Mróz, Z., and Stupkiewicz, S., 1994, An anisotropic friction and wear model, *Int. J. Solids Structures* 31:1113.

Murdoch, A.I., 1976, A thermodynamic theory of elastic material interfaces, *Quart. J. Mech. Appl. Math.* 29:245.

Suh, N.P., 1973, The delamination theory of wear, *Wear* 25:111.

Zmitrowicz, A., 1987, Thermodynamical model of contact, friction and wear: I Governing equations, II Constitutive equations for materials and linearized theories, III Constitutive equations for friction, wear and frictional heat, *Wear* 114:135.

THE JKR-DMT TRANSITION IN THE PRESENCE OF A LIQUID MENISCUS AND THE EXTENSION OF THE JKR THEORY TO LARGE CONTACT RADII

Daniel Maugis

Laboratoire des Matériaux et des Structures du Génie Civil
Unité Mixte CNRS-LCPC, UMR 113
2, allée Kepler, 77420 Champs (France)

INTRODUCTION

In the Hertz theory of frictionless contacting spheres, only compressive stresses in the contact area are assumed. In the JKR theory (Johnson et al., 1971) tensile stresses in the contact area are allowed, whereas in the DMT theory (Derjaguin et al., 1975) adhesion forces act around an unilateral contact. It was shown (Maugis and Barquins, 1978) that the problems of mechanics of contact with adhesion are fracture mechanics problems and that the Sneddon's equations (Sneddon 1965) for axisymmetric frictionless punches are particularly suitable to derive the stress intensity factors (Maugis and Barquins, 1981; Barquins and Maugis, 1982; Maugis and Barquins, 1983). Using a Dugdale model for adhesion force (constant constraining stresses acting around the contact area in an annulus whose size is such that the total stress intensity factor cancels), Maugis (1992) has obtain a general theory with the the JKR and DMT theories as limiting cases. As a liquid meniscus at the edge of a contact is a perfect example of a Dugdale zone, this theory was used to study the change in profile of two crossed cylinders in contact when the size of a surrounding meniscus increases. On the other hand, recent experiments on adhesion of small particles on soft elastic substrates (Rimai et al.,1994) have revealed that the contact radius under zero load can be very large and does not vary as the particle radius to the 2/3 power as required by the JKR theory (which is restricted to small contact radii). These two last points are the subject of this presentation. The corresponding full papers are published elsewhere (Maugis and Gauthier-Manuel, 1994; Maugis, 1994).

Contact Mechanics, Edited by M. Raous *et al.*
Plenum Press, New York, 1995

The Sneddon's Equations

For a frictionless punch whose profile is given by $f(\rho)$ (a is the radius of contact, $\rho = r/a$ and $f(0) = 0$), Sneddon (1965) has shown that the depth of penetration δ of the tip in the elastic half-space (Young's modulus E, Poisson's ratio ν) and the load P that must be applied to achieve this penetration are given by

$$\delta = \int_0^1 \frac{f'(x)dx}{\sqrt{1-x^2}} + \frac{\pi}{2}\chi(1) \tag{1}$$

$$P = \frac{\pi a E}{1-\nu^2}\int_0^1 \chi(t)dt \tag{2}$$

where

$$\chi(t) = \frac{2}{\pi}\left[\delta - t\int_0^t \frac{f'(x)dx}{\sqrt{t^2-x^2}}\right] \tag{3}$$

Furthermore, the distribution of pressure $\sigma(\rho,0)$ under the punch and the displacement $u_y(\rho,0)$ of the surface of the indented plane are given by

$$\sigma_y(\rho,0) = -\frac{E}{2(1-\nu^2)a}\left[\frac{\chi(1)}{\sqrt{1-\rho^2}} - \int_{r/a}^1 \frac{\chi'(t)\,dt}{\sqrt{t^2-\rho^2}}\right], \quad \rho < 1 \tag{4}$$

$$\begin{aligned} u_y(r,0) &= \int_0^1 \frac{\chi(t)dt}{\sqrt{\rho^2-t^2}}, \qquad \rho > 1 \\ &= \chi(1)\sin^{-1}\frac{1}{\rho} - \int_0^1 \chi'(t)\sin^{-1}\frac{t}{\rho}dt \end{aligned} \tag{5}$$

For punches with continuous profile, Sneddon let the arbitrary rigid body displacement $\chi(1)$ equal to zero to have finite stress at the edge of the contact and used this criterion to determine δ. As discussed by Maugis and Barquins (1981) and Barquins and Maugis (1982), if $\chi(1) \neq 0$ these stresses and displacements are those of fracture mechanics with a stress intensity factor

$$K_I = -\frac{E}{2(1-\nu^2)}\chi(1)\sqrt{\frac{\pi}{a}} \tag{6}$$

and an energy release rate

$$G = \frac{1}{2}\frac{1-\nu^2}{E}K_I^2 = \frac{\pi E}{8a(1-\nu^2)}\chi^2(1) \tag{7}$$

Compared with classical fracture mechanics, the factor 1/2 is introduced in G because the punch is undeformable, so that energy is released by the half-space only; we are faced with half a Griffith crack). It is important to note that Eq.(7) is an expression for G which is only valid in the Linear Elastic Fracture Mechanics approximation, i.e. when there is no interaction betwen the crack lips or when the length of the interaction zone is small compared to the crack length or to a characteristic length of the system (here the radius of contact).

The JKR solution

Inserting for the profile of the spherical punch (radius R) the parabolic approximation

$$f(\rho)=\frac{a^2}{2R}\rho^2 \tag{8}$$

one has

$$\chi(t)=\frac{2}{\pi}\left(\delta-\frac{a^2}{R}t^2\right)$$

and Eqs (2), (4)-(7) become

$$\delta=\frac{a^2}{3R}+\frac{2P}{3aK} \tag{9}$$

$$K_I=\frac{\frac{a^3K}{R}-P}{2a\sqrt{\pi a}} \tag{10}$$

$$\sigma_y(r,0)=\frac{K_I}{\sqrt{\pi a}}\frac{1}{\sqrt{1-\rho^2}}-\frac{3aK}{2\pi R}\sqrt{1-\rho^2} \tag{11}$$

$$\left[u_y\right]=\frac{2(1-\nu^2)}{\pi E}K_I\sqrt{\pi a}\cos^{-1}\frac{1}{\rho}+\frac{a^2}{\pi R}\left[\sqrt{\rho^2-1}+(\rho^2-2)\cos^{-1}\frac{1}{\rho}\right] \tag{12}$$

$$G=\frac{3K}{8\pi a}\left(\delta-\frac{a^2}{R}\right)^2=\frac{\left(\frac{a^3K}{R}-P\right)^2}{6\pi a^3K} \tag{13}$$

with

$$\frac{1}{K}=\frac{3}{4}\left(\frac{1-\nu^2}{E}+\frac{1-\nu'^2}{E'}\right)$$

(The primes are for a punch wich is not rigid, and $\left[u_y\right]$ is the discontinuity of displacement).

Writting the equilibrium as $G=w$, where $w=\gamma_1+\gamma_2-\gamma_{12}$ is the *Dupré energy of adhesion* (γ_1 and γ_2 are surface energies, γ_{12} is the interfacial energy), the JKR results (Johnson et al., 1971) are immediately recovered. In particular the radius of contact under zero load, a_0, is given by

$$a_0^3=\frac{6\pi wR^2}{K} \tag{14}$$

and varies as $R^{2/3}$.

The External Crack with Internal Loading

Let us assume that the stresses acting between the lips, around the contact, have a constant value σ_0 up to a radius c and let $d=c-a$. (This stress can be the Laplace pressure in a liquid meniscus). This internal loading leads to a stress intensity factor[1]:

---

[1] As explained in Maugis (1992), the theory of Lowengrub and Sneddon (1965) has been used, giving stresses and displacements for an axisymmetric pressurized external crack. The main difficulty was that the stresses in the neck did not equilibrate the pressure in the crack, due to their implicit assumption of no elastic displacement at infinity. (The stresses vanished at infinity but their integral was finite).

$$K_m = -\frac{\sigma_0 a}{\sqrt{\pi a}}\left[\sqrt{m^2-1} + m^2 \tan^{-1}\sqrt{m^2-1}\right] \tag{15}$$

with $m = c/a > 1$. The stresses distibution in the contact area and the elastic displacement of the plane, due to this loading, are:

$$\sigma_y(r,0) = \frac{K_m}{\sqrt{\pi a}}\frac{1}{\sqrt{1-\rho^2}} + \frac{2\sigma_0}{\pi}\tan^{-1}\sqrt{\frac{m^2-1}{1-\rho^2}}, \qquad \rho < 1 \tag{16}$$

$$u_y = \frac{2(1-v^2)}{\pi E}K_m\sqrt{\pi a}\cos^{-1}\frac{1}{\rho} - \frac{4(1-v^2)}{\pi E}\sigma_0 a$$
$$\times\left[\sqrt{m^2-1}\left(\sqrt{\rho^2-1} - \cos^{-1}\frac{1}{\rho}\right) - m^2\int_1^{\min(\rho,m)}\frac{\sqrt{\rho^2-t^2}}{t^2\sqrt{m^2-t^2}}dt\right] \tag{17}$$

The Dugdale Model

Adding the stresses due to external and internal loadings, eqs. (4) and (16), we see that the non-physical stress singularity disappears if

$$K_I + K_m = 0 \tag{18}$$

This relation gives the ratio c/a as a function of σ_0, P and a and ensures the continuity of stresses at the edge of the contact.

The *crack opening displacement (COD)* δ_t is the air gap at $r = c$, (i.e. $\rho = m$). Adding eqs (12) and (17) and using the relation (18), the COD is

$$\delta_t = \frac{a^2}{\pi R}\left[\sqrt{m^2-1} + (m^2-2)\tan^{-1}\sqrt{m^2-1}\right]$$
$$+\frac{4(1-v^2)}{\pi E}\sigma_0 a\left[\sqrt{m^2-1}\ \tan^{-1}\sqrt{m^2-1} - m + 1\right] \tag{19}$$

According to the Rice J-integral, the energy release rate in a Dugdale model is simply $G = \sigma_0\delta_t$ and the equilibrium is given by the Griffith relation, $G = w$. The problem is thus solved. When $m \to 1$ one recovers the JKR results, whereas when $m \to \infty$, the DMT results are recovered, with $G = \left(\frac{a^3K}{R} - P\right)/2\pi R$.

Introducing the dimensionless parameters:

$$A = \frac{a}{(\pi w R^2/K)^{1/3}}$$

$$\overline{P} = \frac{P}{3\pi w R}$$

$$\Delta = \frac{\delta}{(\pi^2 w^2 R/K^2)^{1/3}}$$

$$\lambda = \frac{2\sigma_0}{\left(\pi w K^2 / R\right)^{1/3}}$$

the Griffith relation gives:

$$\tfrac{1}{2}\lambda A^2\left[\sqrt{m^2-1}+\left(m^2-2\right)\tan^{-1}\sqrt{m^2-1}\right]+\tfrac{4}{3}\lambda^2 A\left[\sqrt{m^2-1}\ \tan^{-1}\sqrt{m^2-1}-m+1\right]=1 \qquad (20)$$

For a given material parameter λ, and a given radius of contact a, one can extract m, and obtain the load and the penetration at equilibrium:

$$\overline{P} = A^3 - \lambda A^2\left[\sqrt{m^2-1}+m^2\tan^{-1}\sqrt{m^2-1}\right] \qquad (21)$$

$$\Delta = A^2 - \tfrac{4}{3}\lambda A\sqrt{m^2-1} \qquad (22)$$

The air gap can be written with the elliptic integrals of the first and second kind $E(\varphi,k)$ and $F(\varphi,k)$. Using the same normalisation as for δ, we have for $r < c$:

$$\overline{[u_y]} = \frac{A^2}{\pi}\left[\sqrt{\rho^2-1}+\left(\rho^2-2\right)\cos^{-1}\frac{1}{\rho}\right]+\frac{8\lambda A}{3\pi}\sqrt{m^2-1}\cos^{-1}\frac{1}{\rho}$$
$$+\frac{8\lambda A}{3\pi}\left[\frac{\sqrt{\rho^2-1}}{\sqrt{m^2-1}}-mE(\varphi,k)\right] \qquad (23)$$

with

$$\varphi = \sin^{-1}\left(\frac{m}{\sqrt{m^2-1}}\frac{\sqrt{\rho^2-1}}{\rho}\right)$$

$$k = \frac{\rho}{m} = \frac{r}{c}$$

and for $r > c$:

$$\overline{[u_y]} = \frac{A^2}{\pi}\left[\sqrt{\rho^2-1}+\left(\rho^2-2\right)\cos^{-1}\frac{1}{\rho}\right]+\frac{8\lambda A}{3\pi}\sqrt{m^2-1}\cos^{-1}\frac{1}{\rho}$$
$$+\frac{8\lambda A}{3\pi}\left[\frac{\sqrt{m^2-1}}{\sqrt{\rho^2-1}}-\rho E(\varphi,k)+\left(1-\frac{m^2}{\rho^2}\right)\rho F(\varphi,k)\right] \qquad (24)$$

with

$$\varphi = \sin^{-1}\left(\frac{\sqrt{m^2-1}}{m}\frac{\rho}{\sqrt{\rho^2-1}}\right)$$

$$k = \frac{m}{\rho} = \frac{c}{r}$$

When $\lambda \to \infty \quad (c/a \to 1)$, one has the JKR air gap, and when $\lambda \to 0 \quad (c/a \to \infty)$, one has the DMT (Hertzian) air gap.

Experiments

Experiments with a Surface Force apparatus (crossed cylinders of mica, with a radius of about 2 cm) have been performed by Maugis and Gauthier-Manuel (1994) with a precision of few nanometers on the profile near the contact area. The contacts are made under zero load. In dry air a JKR profile is observed with its sharp discontinuity of the slope (Fig. 1). With a water meniscus of radius $c \cong 60$ μm, the profile change (Fig.2), in agreement with the theory, and for a large meniscus the DMT profile (Hertzian) is recovered (Fig.3).

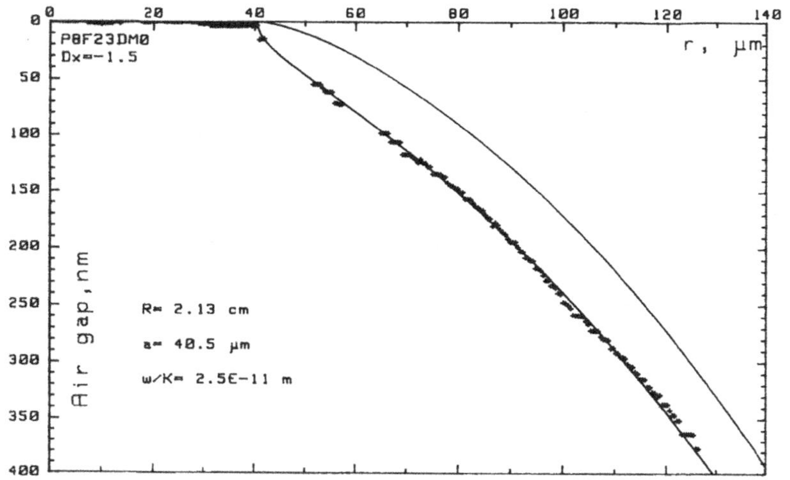

Figure 1. Mica-mica contact in dry air: a JKR profile is observed, in agreement with the theory fitted with w/K=2.5 10^{-11} m.

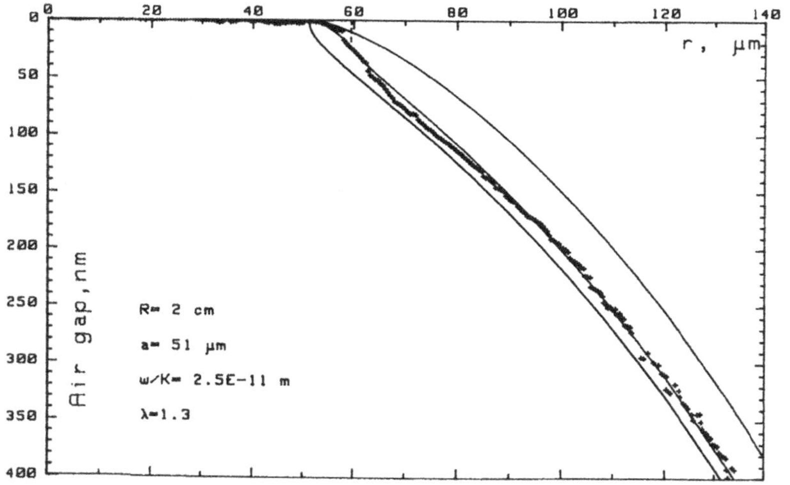

Figure 2. Mica-mica contact in humid air. The Hertzian profile and the JKR profile (with w/K=2.5 10^{-11} m) corresponding to this radius of contact (51 μm) are drawn for comparison.

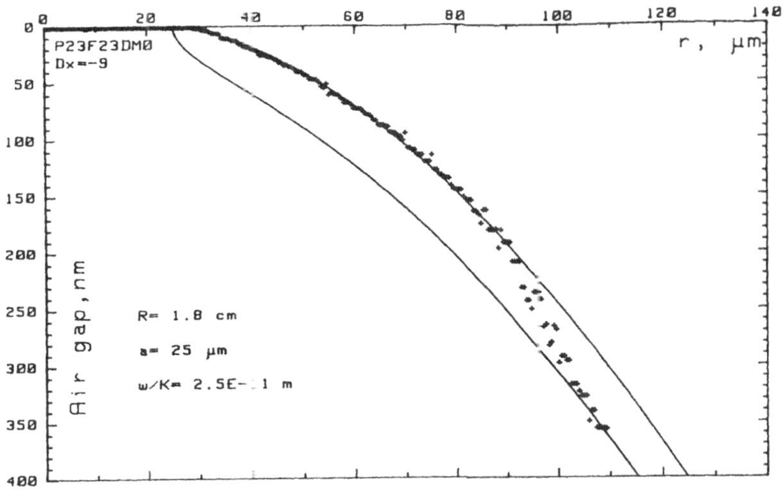

Figure 3. Mica-mica contact with a large water meniscus: a DMT (Hertzian) profile is observed.

EXTENSION OF THE JKR THEORY TO LARGE CONTACT RADII

If instead of the parabolic approximation, the exact profile of the sphere is used:

$$f(\rho) = R - \sqrt{R^2 - a^2\rho^2} \tag{25}$$

one has

$$\chi(t) = \frac{2}{\pi}\left(\delta - \frac{at}{2}\ln\frac{R+at}{R-at}\right)$$

and Eqs (2) and (7) become

$$P = \frac{3aK}{2}\left(\delta - \frac{R}{2} + \frac{R^2 - a^2}{4a}\ln\frac{R+a}{R-a}\right) \tag{26}$$

$$G = \frac{3K}{8\pi a}\left(\delta - \frac{a}{2}\ln\frac{R+a}{R-a}\right)^2 \tag{27}$$

The equilibrium relations $\delta(a)$ and $P(a)$ are obtained by making[2] $G = w$:

$$\delta = \frac{a}{2}\ln\frac{R-a}{R-a} - \sqrt{\frac{8\pi aw}{3K}} \tag{28}$$

$$P = \frac{3aK}{2}\left(\frac{R^2 + a^2}{4a}\ln\frac{R+a}{R-a} - \frac{R}{2} - \sqrt{\frac{8\pi aw}{3K}}\right) \tag{29}$$

and the equilibrium radius of contact under zero load is given by

$$\left(1 + \frac{a_0^2}{R^2}\right)\ln\frac{R+a_0}{R-a_0} - 2\frac{a_0}{R} - \frac{4a_0}{R^2}\sqrt{\frac{8\pi a_0 w}{3K}} = 0 \tag{30}$$

Rimai et al. (1994) have observed in a scanning electron microscope the contact under zero load of small spherical glass particles (radii 0.5 to 100 µm) on soft polyurethane substrate and shown that the contact radius did not vary as the particle radius to the 2/3

[2] If w=0 one obtains the extension of the Hertz theory for large contacts.

power law as required by the JKR theory, Eq.(14), but rather to the 3/4 power . Their data points are displayed in the figure 4. They clearly show a continuous bending from the slope 2/3 to the slope 1 as R decreases, in agreement with the above theory.

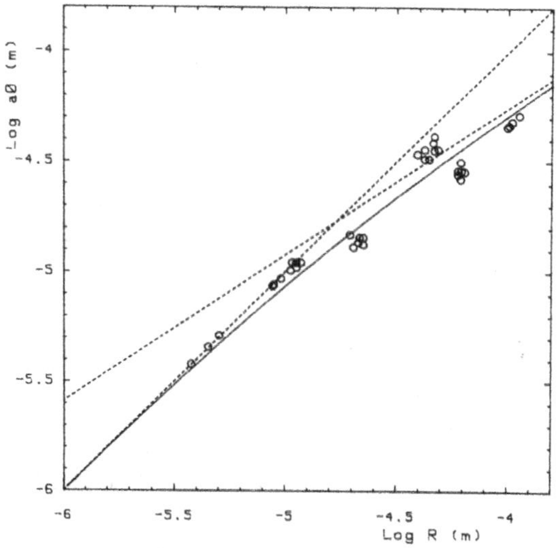

Figure 4. Radius of contact under zero load. Comparison of experimental results of Rimai et al. (1994) for glass beads on soft polyurethane substrates with the theory (w/K=9 10^{-7} m). The dashed curves are for JKR and for $a=R$.

REFERENCES

Barquins, M. and Maugis, D., 1982, Adhesive contact of axisymmetric punches on an elastic half-space: the modified Hertz-Huber stress tensor for contacting spheres, *J. Méc. Théor. Appl.* 1:331.

Derjaguin, B.V., Muller, V.M., Toporov, Yu.P., 1975, Effect of contact deformation on the adhesion of particles, *J. Colloid Interface Sci.* 53:314.

Johnson, K.L., Kendall, K. and Roberts, A.D., 1971, Surface energy and the contact of elastic solids, *Proc. Roy. Soc. A* 324:301.

Lowengrub, M. and Sneddon, I.N., 1965, The distribution of stress in the vicinity of an external crack in an infinite elastic solid, Int. J. Eng. Sci., 3:451.

Maugis, D.,1992, Adhesion of spheres: the JKR-DMT transition using a Dugdale model, *J. Colloid Interface Sci.* 150:243.

Maugis, D.,1994, Extension of the JKR theory of the elastic contact of spheres to large contact radii, *Langmuir* (submitted).

Maugis, D. and Barquins, M., 1978, Fracture mechanics and the adherence of viscoelastic bodies, *J. Phys. D: Appl. Phys.* 11:1989.

Maugis, D. and Barquins, M., 1981, Adhesive contact of a conical punch on an elastic half-space, *J. Phys. Lettres* 42:L95.

Maugis, D. and Barquins, M., 1983, Adhesive contact of sectionally smooth-ended punches on elastic half-space: theory and experiment, *J. Phys. D: Appl. Phys.* 16:1843.

Maugis, D. and Gauthier-Manuel, B.,1994, JKR-DMT transition in the presence of a liquid meniscus, *J. Adhesion Sci. Technol.* (in press).

Rimai, D.S, DeMejo, L.P.,Vreeland, W.B., and Bowen, R.C, 1994, Adhesion induced deformation of a highly compliant elastomeric substrate in contact with rigid particles, *Langmuir* (in press)

Sneddon, I.N., 1965, The relation between load and penetration in the axisymmetric Boussinesq problem for a punch of arbitrary profile, *Int. J. Eng. Sci.* 3:47.

MODELS OF ANISOTROPIC FRICTION DEPENDING ON A SLIDING PATH CURVATURE

Alfred Zmitrowicz

Institute of Fluid-Flow Machinery
Polish Academy of Sciences
ul. J. Fiszera 14, 80-952 Gdask, Poland

Physical properties of solids and surfaces are very often non-homogeneous, and they can form different field singularities in a contact area between two bodies (concentric circles, ellipses, spirals etc.). The phenomenon refers to complex physical properties of materials (e.g. wood, crystals) or to specific techniques of manufacture and finishing. Non-homogeneous physical properties of solids and surfaces can generate a non-homogeneous anisotropic friction in the contact. This is the case where friction depends on the position of a contact point with respect to a singular field center. Then, the frictional anisotropy follows on the one hand from the physical properties of the surface, on the other hand it can additionally depend on a sliding path in this surface. This fact should be included in the mathematical formulation of the friction law. In the contact surface with complex properties besides rectilinear particular friction directions also curved particular directions can exist.

The objective of this study is to present our first research results towards the contribution of the sliding path curvature effects in mathematical models of anisotropic friction and to give a suitable framework allowing a rational and unified formulation of non-homogeneous anisotropic friction.

A trajectory of the material point in the surface R^2 in the range of time $I = (t_o, t_e)$ is defined by an image set of the set I and the following mapping

$$I \ni t \longrightarrow \mathbf{x}(t) \in R^2, \tag{1}$$

where, $\mathbf{x}$ is a radius vector, t is time. With the aid of representation (1) we can introduce an arc length parameter s and a one-dimensional parameterization of the plane curve (motion trajectory). There exists the following relation between the length parameter s and the time t:

$$s(t) = \int_{t_o}^{t} V(\tau)d\tau \equiv \int_{t_o}^{t} \mid \mathbf{V}(\tau) \mid d\tau, \tag{2}$$

$$I \ni t \longrightarrow s(t) \in R^1, \tag{3}$$

Contact Mechanics, Edited by M. Raous *et al.*
Plenum Press, New York, 1995

where, $\mathbf{V}$ is a velocity vector. From the definition of the velocity vector it follows that this vector is tangent to a curve representing the motion path

$$\mathbf{V} = \frac{d\mathbf{x}}{dt} = \frac{d\mathbf{x}}{ds}\frac{ds}{dt} = \mathbf{v}V, \tag{4}$$

where, $\mathbf{V}$ is a velocity value and $\mathbf{v}$ is a unit vector tangent to the trajectory.

According to the Frenet-Serret first formula, we have

$$\frac{d\mathbf{v}}{ds} = \frac{\mathbf{n}}{\rho}, \tag{5}$$

where, $\mathbf{n}$ is a unit vector normal to the sliding path ($\mathbf{n} \cdot \mathbf{v} = 0, \mid \mathbf{n} \mid = 1$), ρ is a sliding path curvature radius. The parameterization s determines uniquely the pair of vectors $\mathbf{v}$ and $\mathbf{n}$, see (4) and (5). Both definitions are always valid, i.e. for any instant of time and any place of the motion trajectory.

Usually it has been assumed that the friction force vector $\mathbf{t}$ at the point $\mathbf{x}$ of the contact has the same value for all sliding trajectories which have the same sliding velocity unit vector $\mathbf{v}$ tangent to the trajectories. In other words, the friction force vector depends on the velocity direction only through the sliding velocity unit vector $\mathbf{v}$ at $\mathbf{x}$, it does not depend on other sliding trajectory parameters. We postulate that the friction force vector at the given point depends on the sliding velocity direction $\mathbf{v}$ and on the sliding path curvature. Therefore, we extend a set of independent variables of a friction force equation taking into account the derivative of the sliding velocity unit vector i.e. $d\mathbf{v}/ds$.

Let us replace the friction force formulation (Zmitrowicz, 1992 and 1992; He and Curnier, 1993)

$$\mathbf{t} = \mathbf{t}(\mathbf{v}, N), \tag{6}$$

by the following formulation

$$\mathbf{t} = \mathbf{t}\left(\mathbf{v}, \frac{d\mathbf{v}}{ds}, N\right), \tag{7}$$

where, N is a normal pressure. The sum of two terms (monomials) can generate a simple realization of the function (7): one term depends on the tangent unit vector $\mathbf{v}$, the other depends on the normal unit vector $\mathbf{n}$

$$\mathbf{t} = \mathbf{t}_{\mathbf{v}} + \mathbf{t}_{\mathbf{n}} = -N\left(\mathbf{C}\mathbf{v} - \mathbf{E}\frac{d\mathbf{v}}{ds}\right). \tag{8}$$

The second order tensor $\mathbf{C}$ defines homogeneous anisotropic friction properties of the contact surface, and these properties are independent on the sliding path. Tensor $\mathbf{E}$ includes the non-homogeneous friction properties of the surface.

The vector $d\mathbf{v}/ds$ defines the effects caused by the sliding path curvature. If the motion trajectory is a curve at a contact point then the contribution of $d\mathbf{v}/ds$ can be taken into account. If the curvature vanishes ($\rho = \infty$), then the dependence between the friction force and $d\mathbf{v}/ds$ vanishes too (the sliding path is a straight line). If the motion trajectory reduces to a point ($\rho = 0$), then there is no motion and no friction (i.e. dynamic friction). In this case the friction force is not defined, in the light of the formulation (8). Therefore, the component $\mathbf{t}_{\mathbf{n}}$ differentiates rectilinear and curved trajectories

$$\mathbf{t}_{\mathbf{n}} \begin{cases} = 0 & for \quad \rho = \infty \\ \neq 0 & for \quad 0 < \rho < \infty. \\ = \infty & for \quad \rho = 0 \end{cases} \tag{9}$$

Friction which depends on the direction of sliding is called the anisotropic friction. A deviation in the friction force from the direction of sliding and a dependence of the friction magnitude on the sliding direction are typical features of contacts with frictional anisotropy. The anisotropic friction coefficient μ_α and the angle β of friction force inclination for any sliding direction can be obtained from the following relations

$$\mu_\alpha = N^{-1} \mid \mathbf{t} \mid, \tag{10}$$

$$sin\beta = \frac{\mathbf{t} \cdot \mathbf{n}}{\mid \mathbf{t} \mid}. \tag{11}$$

The friction equation in the manner described has the following properties.

The friction equation (7) and its variables satisfy the axiom of material objectivity. Adopting the rule proposed by W. Noll, the material objectivity condition for friction force (7) has the following form

$$\mathbf{Rt}\left(\mathbf{v}, \frac{d\mathbf{v}}{ds}, N\right) = \mathbf{t}\left(\mathbf{Rv}, \mathbf{R}\frac{d\mathbf{v}}{ds}, N\right), \quad \forall \mathbf{R} \in O, \tag{12}$$

where, $\mathbf{R}$ is the orthogonal tensor i.e. $\mathbf{RR}^T = \mathbf{R}^T\mathbf{R} = \mathbf{1}$, $det\mathbf{R} = \pm 1$, O is the full orthogonal group.

From the Second Law of Thermodynamics it follows that the power of the friction force in every case of the frictional contact is non-positive

$$\mathbf{t} \cdot \mathbf{V} \leq 0, \quad \forall \mathbf{V}. \tag{13}$$

Substituting the friction force components (8) into (13), assuming that $\mathbf{t_v}$ and $\mathbf{t_n}$ friction components are independent and taking into account that $N, V, \rho \geq 0$, we can replace the inequality (13) by the following two restrictions for the tensors $\mathbf{C}$ and $\mathbf{E}$

$$\mathbf{V}^T\mathbf{CV} \geq 0, \quad \mathbf{n}^T\mathbf{Ev} \leq 0, \quad \forall \mathbf{V}. \tag{14}$$

In experimental works it is observed that: (a) friction of wood is dependent on whether sliding took place parallel to or perpendicular to the wood fibers; (b) when two surfaces of PTFE (Teflon) are oriented with their molecular chains, in parallel, the friction is approximately 30% higher when sliding occurs across the chains than when sliding occurs along them; (c) friction of composites depends on fiber orientation with respect to the sliding direction, it has been shown that the lowest coefficient of friction and rate of wear are obtained when fibers are oriented parallel to the direction of sliding; (d) for surfaces with definite marks (surface roughness) friction is smaller when the machining lines lie in the direction of sliding and it reaches the greatest values when they are perpendicular to the sliding direction. Generally, it is observed that a motion "along the marks" (i.e. wood fibers, molecular chains of PTFE, composite fibers, machining marks) occurs with the lowest resistance to motion, and it has the greatest resistance in the direction "perpendicular to the marks".

Let us consider non-homogeneous friction properties which form concentric circles. In this case, there are two types of privileged sliding directions: the first is radial (i.e. along radial from the center of concentric circles), the second is concentric circular (i.e. along concentric circles). At any point of contact and in any sliding direction, the greatest resistance to motion is in the radial directions, and the lowest resistance is in the concentric circular directions. As a measure of the lowest and greatest resistance to motion we take a value of energy dissipated in friction, i.e. the power of the friction force referred to the unit velocity ($V = 1$).

For sliding along any radius and for any distance from the center of concentric circles, it can be observed that the friction coefficient has a constant value ($\mu_\alpha = const$) and the friction force component normal to the sliding direction does not exist, i.e. the inclination angle of the friction force is equal to zero $\beta = 0$. The energy dissipated in friction for sliding in the radial direction achieves the maximum value

$$D(R, \alpha_v) = \quad -\mathbf{t} \cdot \mathbf{v} = \quad \max_{\substack{R \in <0,\infty) \\ \alpha_v \in <0,2\pi>}} D(R, \alpha_v), \tag{15}$$

for all contact points i.e. for every circle radius $\rho = R \in< 0, \infty)$ and for all sliding directions $\alpha_v \in< 0, 2\pi >$, where α_v is a measure of an oriented angle between a reference direction and the sliding direction $\mathbf{v}$.

The sliding along the concentric circles has the lowest resistance to motion, it does not depend on a position in the circle. The friction force component $\mathbf{t_n}$ is active in the motion along the concentric circles. A value of the component $\mathbf{t_n}$ decreases with the distance from the center of the concentric circles, so the component $\mathbf{t_n}$ vanishes for $R = \infty$. The inclination angle $\beta \neq 0$ depends on the circle radius R, and it is constant for all points of the circle, and for $R \longrightarrow \infty, \beta \longrightarrow 0$. For sliding along the concentric circles the energy dissipated in friction achieves the minimum value

$$D(R, \alpha_v) = \quad -\mathbf{t} \cdot \mathbf{v} = \quad \min_{\substack{R \in <0,\infty) \\ \alpha_v \in <0,2\pi>}} D(R, \alpha_v) \,. \tag{16}$$

The dissipated energy has the minimum value for all concentric circles, if the friction coefficient $\mu_\alpha^{||}$ of the friction force component collinear with the sliding direction is constant

$$\mu_\alpha^{||} = -N^{-1}\mathbf{t} \cdot \mathbf{v} = const, \quad \forall R \in< 0, \infty). \tag{17}$$

In the case of concentric circles, the friction force changes from point to point, and some properties of friction vanish in an infinity ($R \longrightarrow \infty$). However, particular properties of friction (radial and concentric circular directions of extreme friction values, symmetry properties etc.) are the same for all contact points, they are global (homogeneous).

Most of the activity in anisotropic friction investigations is motivated by the interest in fibrous and laminated composites and crystals applied in modern machine parts operating in contact conditions.

REFERENCES

Zmitrowicz, A., 1992, A constitutive modelling of centrosymmetric and non-centrosymmetric anisotropic friction, *Int. J. Solids Structures.* 29:3025.

Zmitrowicz, A., 1992, Illustrative examples of centrosymmetric and non-centrosymmetric anisotropic friction, *Int. J. Solids Structures.* 29:3045.

He, Q.-C., and Curnier, A., 1993, Anisotropic dry friction between two orthotropic surfaces undergoing large displacements, *Eur.J.Mech., A/Solids.* 12:631.

NUMERICAL EXPERIMENTS IN GRANULAR DYNAMICS : VIBRATION-INDUCED SIZE SEGREGATION

J. J. Moreau

Laboratoire de Mécanique et Génie Civil
URA CNRS 1214, Université Montpellier II
34095 Montpellier Cedex 5, France

INTRODUCTION

Numerical techniques have been developed for a few years in our laboratory aimed at computing the motion of collections of rigid or deformable bodies, taking into account the unilateral constraints of *non-interpenetrability* and, at possible contacts, *friction*. These techniques primarily apply to dynamical situations, so inertia terms play a central role in calculations. In the event of collision, *velocity jumps* should be expected. The latter accidents, as well as the nonsmooth character of the law of dry friction and the geometrical roughness of the set of configurations permitted by the non-interpenetrability conditions (say millions of inequalities in usual applications), make the problems in view belong to the field of *Nonsmooth Mechanics*. Our choice is to face nonsmoothness without resorting to regularizing approximation tricks such as artificial elasticity or artificial viscosity. Evolution is naturally treated through time-discretization: a dominant feature of our numerical integration schemes is that they are *implicit*, at least with regard to velocities.

We commonly refer to this approach as the *Contact Dynamics method*, abbreviatively: *CD method*. The above features make it very different from the techniques currently used to the same end -with success- by numerous authors, more or less in the line of the pioneering work of Cundall (1971).

The lecture delivered at the Symposium mainly consisted in displaying computer-generated animations. A videotape has also been played, in which some experiments made in another laboratory, concerning the *heaping* phenomenon in vertically shaken 2-D packs of beads (Clément et al., 1992) were (convincingly...) compared with computed motions.

For the description of the numerical techniques used and the discussion of various topics, the reader is invited to refer to Moreau (1994). However, since *collisions* have been a subject of recurrent interest during the Symposium, we are to give here some precisions about their handling in CD method.

Contact Mechanics, Edited by M. Raous *et al.*
Plenum Press, New York, 1995

After that, some drawings produced by the same numerical experiments as the played animations will be inserted and commented. They solely concern *size-segregation in shaken granular materials*, an essentially dynamical phenomenon.

The simulation of shear bands and the pattern of load chains in slowly deformed assemblies of grains, also presented at the Symposium, are not illustrated in what follows. The latter phenomena are usually viewed as *quasi-static*, i.e. inertia effects are alleged negligible. This does not prevent CD method from being quite effective in the treatment of such situations. Inertia terms have in these cases a mollifying virtue which, like in most other numerical techniques, favor computation speed. In our opinion, the sake of computation speed should not induce to enter artificially large inertia terms. When watching physical experiments on slowly deformed granular assemblies, one is used to hear a noise. In fact, two-dimensional experimental set-ups (Schneebeli materials) allow one to observe that the evolution, however globally slow, takes place as the result of many local rearrangements which are brutal processes (Meftah et al., 1993). These rearrangements are also seen on simulations, but plausibility requires that masses are entered with their true physical values.

COLLISIONS

In the applications presented here, bodies will be treated as perfectly rigid. The bottleneck of such a several centuries old practice may be viewed in the handling of collisions. In the majority of the papers devoted to this topic, the intense effects which take place during a collision are assumed localized in the vicinity of the locus of impact. A multiple scaling analysis of what happens in an 'infinitely small' domain, during the 'infinitely short' shock episode, then allows one to take into account in more or less detail the material behaviour of the colliding bodies.

Physical situations to which such a treatment is relevant certainly exist, but in general the effect of a collision cannot be local. For instance, material dissipation in the vicinity of the impact is not the only cause of the energy loss observable at the macroscopic level of observation. Even if the concerned bodies are assumed perfectly elastic, energy conservation cannot be expected. Disturbances are to propagate from the collision locus to the whole system and also, if the latter is linked with some external support, to the outside world. After contact recedes, vibrations are likely to persist somewhere. At the macroscopic observation level, this does not contradict the rigidity assertion, but the energy involved in microscopic agitation may not be negligible. A precise double-scaling set-up, conciliating the analysis of global deformation with the macroscopic description of bodies as rigid may be found in (Martins and Trabucho, 1989).

Also as a consequence of global deformation, a collision may, at the microscopic time-scale, split into several separate contact episodes: an example of such a double bounce is calculated in closed form in (Timoshenko, 1948, Chap. 12). Finite element computation of the collision of two elastic bodies performed in our laboratory has shown the same. So the conception of a collision as consisting of a compression phase followed by a so-called restitution phase cannot be considered as general.

All this makes the outcome of a collision depend on many factors, in particular on the shape of the concerned bodies.

Still more severe difficulties arise from that collisions are frequently *multiple*, i.e. several contact loci are involved at the same time. This is unavoidable if one of the impacting bodies is part of a cluster of objects already in contact. The propagation of disturbances into such a cluster is a problem similar to that of sound in granular media.

One thus has to accept that any given model of collision can only have a limited scope. Every occasion of comparing its features with calibrated experiments should be seized, in order to estimate this scope as precisely as possible.

The simulations of granular materials presented here, like many others performed in recent decades through various numerical techniques (a review of them may be found in Barker, 1994), are restricted to spherical grains. Fortunately, for bodies having such a simple shape, collisions happen to prove more tractable. After a detailed analysis of the frictional collision process of deformable spherical bodies has been carried out by Maw et al.(1981), several authors (Walton, 1993; Lun and Bent, 1993) have chosen to base their numerical or theoretical developments on a simplified *collision law* which, for a pair of unconstrained balls viewed as rigid and colliding under any angle with any respective velocities and spins, allows one to compute velocities and spins after the collision. This law involves three phenomenological parameters which may be called the *friction coefficient*, the *normal restitution coefficient* and the *tangential restitution coefficient*. Experimental investigation by Foerster et al. (1994) has shown this law to be quite acceptable in a fairly large domain of values of the parameters.

Now it turns out that the CD approach that we are to explain in Sec. 3 below, though based on a completely different rationale, yields for spherical bodies exactly the same three-coefficient collision law. This comforting convergence inspires confidence in the simulations of strongly shaken granular material presented in Secs. 4 and 5, since binary collisions are dominant in such highly dilated assemblies.

The experiments of Clément et al. (1992) or Duran et al. (1993) on the shaking of compact assemblies of beads provide, on the contrary, an example of collisions with very large degree of multiplicity. The success of CD method in simulating these experiments calls for further investigation but seems to stem essentially from that the period of shaking is long with regard to the time that sound waves in the bead pack take to travel the experiment cell.

CONTACT LAW

Let $[t_I, t_F]$ be an interval of the time-discretization (I as 'initial', F as 'final'). In CD computation method, all the contact forces acting on the members of the investigated system during this time-interval are treated in terms of their respective *impulsions*, whether these forces arise from permanent contact of from collisions. For every contact (here assumed to be punctual) detected as effective in the concerned computation step, the a priori unknown contact impulsion R has to be connected with some other elements of the motion, through the phenomenological relations one has accepted as summarizing the contact physics

First of all, a phenomenological relation has to be adopted to describe *friction*. When simulating granular materials one uses to be content with the friction law of Coulomb (with friction coefficient possibly depending on the actual or past cicumstances). This consists in a relation between R and the *sliding velocity* U of the contacting bodies. The latter is of course an unknown of the problem. In an 'explicit' computation scheme, one would introduce into Coulomb's law the value U_I, namely the sliding velocity found for time t_I as a result of the preceding time-step computation. More sophisticated schemes would mix in some way this known value with the value U_F to be computed.

For smooth motions, the vector $U_F—U_I$ is the same order of magnitude as $t_F—t_I$. Such is no more the case if the system dynamics implies a velocity jump in the time-interval, possibly the result of a collision between some of its members (nonnecessarily those involved in the considered contact). Then U_I and U_F are viewed as representatives of the respective velocities before and after the collision. For brevity, we are tackling here directly the algorithmic aspects. If some attention was paid to the mathematical formulation of the problem, prior to numerical techniques, one would assume velocities to be functions of time with *locally bounded variation* (see e.g. Moreau, 1988a, b). This assumption secures at every instant the existence of the left-side limit U^- and of the right-side limit U^+. The difference U_F—U_I actually embodies the sum of all the jumps U^+-U^- occuring in the time interval, added to the continuous variation of U.

The contact impulsion R equals the time-integral of the contact reaction, a function of time expected to assume 'very large' values on the 'very short' duration of a collision. Even if one assumes the friction law of Coulomb valid at every instant, the nonlinear character of such a relation, which connects the instant reaction force with some sliding velocity (actually difficult to identify : it is liable to vary from one point to another of the very small contact zone), in general doesn't allow one to integrate the relation into a law which would connect R with some alleged average velocity.

Since the policy of treating bodies as perfectly rigid involves that collisions are only assessed as generating a new set of values for the velocity parameters, to be computed from the values they had before the shock, we pragmatically decide to connect the unknown R through a Coulomb-like law with some *formal velocity* . The latter is defined by an *averaging procedure* into which velocities before and after shock are entered. Here is the simplest version of such a procedure.

For every contact, the sliding velocity U is decomposed into its normal component $U^N \in \mathbf{R}$ and its tangential component U^T, a vector of the tangent plane $\mathbf{T}$. As before, the subscripts I and F refer to values associated, in the course of computation, with the beginning and the end of the time-step; the subscript 'a', as in 'average', refers to the formal velocity one is to define. Let

$$U^N_a = \frac{\rho}{1+\rho} U^N_I + \frac{1}{1+\rho} U^N_F \in \mathbf{R} \tag{1}$$

$$U^T_a = \frac{\tau}{1+\tau} U^T_I + \frac{1}{1+\tau} U^T_F \in \mathbf{T} \tag{2}$$

where ρ and τ are chosen real numbers, possibly depending on the contact under investigation.

The effectiveness of this trick lies in that it allows one to formulate at the same time the bounce conditions.

To this end the following concepts have to be introduced. With every contact referenced as possible in the considered time-step, some criterium f, a real function of the system configuration variable $q \in \mathbf{R}^{dof}$, is associated in such a way that $f(q) \le 0$ corresponds to the configurations compatible with the constraint of non-interpenetrability, while $f(q)=0$ characterizes effective contact occurring at some point M(q). In computation, a certain amount of violation is also tolerated, so that the definition of the normal unit vector $\mathbf{n}(q)$ to the contacting bodies at point M(q) has to be extended, in a smooth arbitrary way, to configurations with nonzero f . Let us put

$$K(q)=\{U \in \mathbf{R}^3 : \mathbf{n}(q).U \ge 0\} \quad \text{if} \quad f(q) \ge 0 \quad \text{and} \quad K(q)=\mathbf{R}^3 \quad \text{otherwise.}$$

A *contact law*, i.e. a relation between some elements R and U of $\mathbf{R}^3$, is said *complete* if it involves $U \in K$ in all circumstances and $R=0$ if $U \in \text{int}K$.

The interest of this concept lies in that, if complete contact laws are valid for all possible contacts at every instant and if the initial position of the system agrees with the non-interpenetrability constraints, the latter are automatically fullfilled in the sequel (Moreau, 1988b).

Treating the non-interpenetrability constraints as relations between contact forces and velocities (in which the configuration q *usually also appears) is the dominant feature of the CD method.*

The definition of a complete contact law clearly makes that R can be nonzero only if **n**(q).U=0. If the law is applied with U=U_a, the definition of U_a^N then yields

$$U_F^N = -\rho\, U_I^N \tag{3}$$

In other words, if the concerned contact has nonzero reaction in the investigated impulsional process, the parameter ρ coincides with *Newton's restitution coefficient.* and is called for this reason the *normal restitution coefficient* of the contact.

Symmetrically, it may happen that U_a^T=0 (this is not uncommon if a law of dry friction is applied, but can be checked only after computation is completed). In such an event

$$U_F^T = -\tau\, U_I^T$$

so that τ is called the *tangential restitution coefficient.*

The reader may find in (Moreau, 1994) two examples of synthetic formulations of contact laws which automatically turn out to be complete: they respectively concern *frictionless unilateral contact* and *unilateral contact with Coulomb friction*, the latter stated in terms of De Saxcé's *bipotentials* (see his contribution in the present volume).

Though equ. (3) allows one to identify ρ with Newton's coefficient in some cases, the present approach proves richer than the traditional restitution assumption. This is illustrated by the application of what precedes to the problem, familiar in earthquake engineering literature, of the *rocking* of a slender rectangular block supported by a fixed horizontal plane. For simplicity, assume the lower edge slightly concave, so that contact can only occur through corners. Let the left corner remain in contact during an episode where the block rotates to the right, until the right corner collides. If at this time Newton's assumption was applied to both contact points, this would yield zero normal velocity for the left corner, so no rocking could be found. On the contrary, the assumption of a complete contact law leaves the possibility for U_F^N to be nonzero while R vanishes. The result of the analysis depends on the aspect ratio of the block.

A capital issue, when collision laws are proposed, is the *energy balance*: in scleronomic systems (this excludes the case of external boundaries with prescribed motion), collisions should not generate energy and many authors have observed that, for collisions between bodies deprived from the customary symmetry properties, an analysis based on Newton's restitution may break this thermodynamic requirement. When the trick of using an average sliding velocity was first proposed (Moreau, 1988b), this was with ρ and τ assuming for all contacts a single value in the interval [0,1]. Then the dissipativity of collisions was found secured by that of the relation assumed to hold between U_a and R, namely a law of friction.

Contact dynamics calculations, in particular the drawing of energy balance, lead to associate with every possible point of contact a 3×3 linear operator, self-adjoint with regard to the Euclidean metric and positive, say H (see e.g. Moreau, 1994, Sec 7): this operator pictures the system inertia from the viewpoint of the considered contact. For the special case of contact between two homogeneous spherical balls, H is axissymmetric about the common normal and this is found to imply the compatibility of the averaging procedure (1) (2) with the dissipativity requirement. In general, this requirement would induce to relate the choice of a definition of U_a with the operator H. This agrees with the assertion we made in Sec. 2, that collisions cannot be treated as purely local processes. But we don't claim that the above pragmatic approach would constitute a general theory of collisions.

Example

On figure 1, the successive bounces of two balls on a fixed horizontal ground are compared. The two balls differ only in the tangential restitution coefficient which vanishes for the ball on the right, while it equals the normal restitution coefficient, say 0.8, for the ball on the left. The successive ballistic flights of the right ball show regular decrease, while those of the left ball exhibit some alternance, totalizing into stronger damping. The key difference between the two modes of bouncing naturally lies in their effect upon the ball spin.

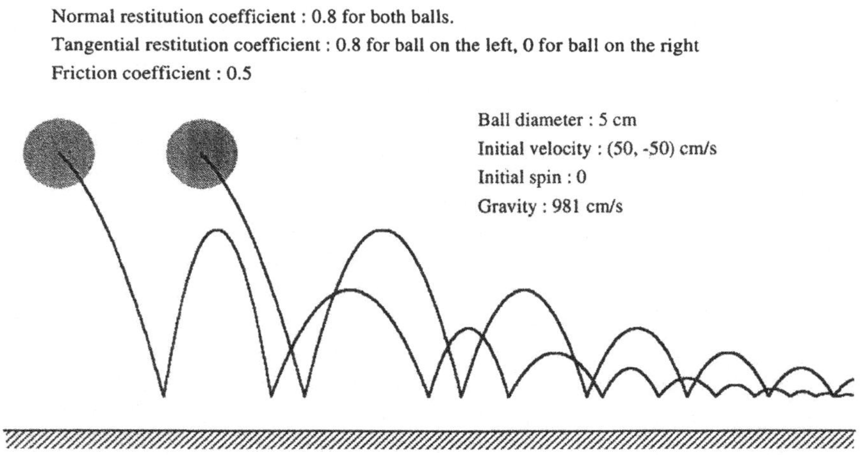

Fig. 1.– Effect of tangential restitution

In this numerical experiment, balls were thrown with zero initial spin. Otherwise, it is well known that sufficient backspin may induce a ball to bounce backward. High friction and high tangential restitution favour this effect which may become spectacular with balls made of special materials. All this can readily be reproduced by CD simulations.

Measurements published by Foerster et al. (1994) for beads of glass or of cellulose acetate show the tangential restitution have an order of magnitude of about half the normal restitution.

SIZE-SEGREGATION AT BOUNDARIES

When a vessel containing grains of different sizes is vertically shaken, one commonly observes a progressive accumulation of larger grains at the top of the pack, even if they are denser than the rest. Similar segregation may also occur in other situations of granular dynamics: flow on a slope, rotating drum (deceptively intended to act as a mixer), etc. and is a nuisance in industrial processes such as the preparation of pharmaceuticals. This is commonly called the 'Brazil nut effect'(Rosato et al., 1987).

The case of vertical shake, to which we restrict ourselves here, has been the subject of a fairly large number of studies tending to explain the segregative effect on the basis of statistical mechanics (Mehta and Barker, 1991) or proposing explicit mechanisms at the scale of individual particles (Jullien et al., 1992)(Duran et al., 1993). The same authors have also made use of computer simulations, but generally not of the properly dynamic sort. These simulations, based on geometry and random drawings (in so-called 'Monte Carlo methods'), are rather meant to test the consequences of tentative conceptions; they may be viewed as an aid to reasoning.

In real situations, several different and possibly antagonistic effects are liable to intervene, so that phenomena may depend qualitatively on the experimental parameters. Computer simulations performed by the CD method (Moreau, 1993) strongly suggested that some of the physical experiments made in the past with a view to investigate size-segregation by shaking, actually produced no proof of any tendency of the larger objects to migrate upwards relatively to the rest of the material. In fact, the created animations of two- or three-dimensional granular motions showed that, under the conditions of the said experiments, the possible tendency was masked by an extremely apparent *convection* effect. If friction is nonzero between the grains and the vertical boundaries of the shaken vessel, strong downward currents appear along these boundaries implying, for material conservation, some upward convection in the central part. This convection brings everythings up, the large grains at the same speed as the smaller ones. Only when large grains reach the top, the boundary layers appear too thin to recycle them down (except for limited incursions produced by small upper-corner vortices). Segregation is thus the result of *filtration* through boundary layers.

Independently of the above simulations, experiments performed by Knight et al. (1993), under somewhat different shaking conditions, have led to the same observation.

There remains to explain the generation of downward boundary currents. We reproduce here, with more detail resulting from new computation (figures 2 and 3), the explanation presented in Moreau (1994).

A cylindrical vessel, the same size as that in the experiments of Knight et al., is submitted to vertical sinusoidal shake with frequency 25 hertz, peak-to-peak amplitude 0.2cm, a motion with maximal acceleration equal to 2.51 g. Due to such large acceleration, the pack of 4000 beads placed in the vessel loses contact with the vessel bottom for a part of each period. When the lowest beads take contact again, the propagation of collisions induce strong agitation in the whole pack. Analogously to a gas, one may say that the *pressure* in the pack is then high. The average pressure experienced by an individual bead during a time-interval (say 1/50 period) can be defined as the total of the normal components of the collisional impulses it receives, divided by the duration and by the bead surface.

For clarity, only the beads with centers in a thin meridian slice are shown on figures 2 and 3. On figure 3, pressures are visualized through levels of gray. The upper image of figure 3 corresponds to the pack having broken contact from the bottom, while its internal agitation is still high. At this time the average pressure exerted by peripheral beads upon the cylindrical boundary is rather large, while their sliding velocity is upward (the average vertical component of the sliding velocity is plotted at the bottom of figure 2; the possible spin of beads is taken into account for its computation, but proves unimportant). Consequently, the *friction force* exerted by beads on the cylindrical boundary is relatively large and directed upward. For five horizontal sections of this boundary, the normal and vertical components of bead-exerted forces are represented by line segments. By equality of action and reaction, peripheral beads are thus found to experience relatively large *downward* forces.

The bottom image of figure 3 corresponds to the pack falling back, at a time its internal agitation has much been damped by repeated slightly inelastic collisions. The average force experienced by peripheral beads is now upward, but definitely smaller than before. The plot in the middle of figure 2 shows that the time average of the vertical force experienced by peripheral beads is downward, hence the circulatory motion shown in the top drawing of the same figure.

Vertically shaken cylindrical vessel

Vessel diameter : 3.5 cm
Shake : frequency 25 hertz, peak-to-peak amplitude 0.2 cm (maximal acceleration 2.51 g)
Beads : 3999 with diameter 0.2 cm, one with diameter 0.5 cm
At all contacts : friction 0.8, normal restitution 0.95, tangential restitution 0.4

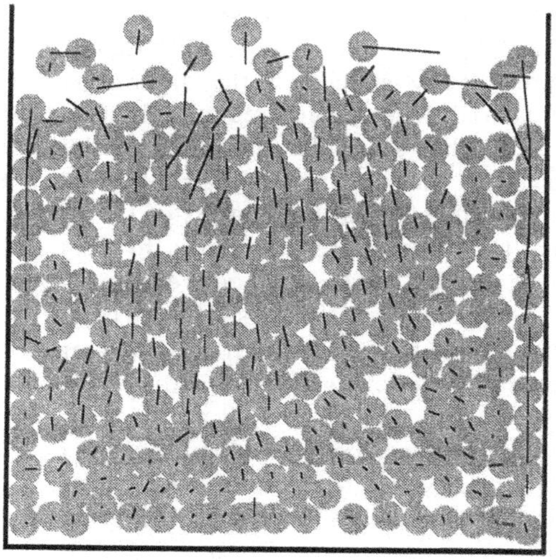

Left : vessel in its lowest position
Beads with centers in a 0.2 cm thick meridian slice are shown.
The line segments figure their respective displacements after 9 shakes.

Below : Time interval of 2 periods, beginning with vessel in lowest position.

Upper curves : Normal and vertical component of the total reaction experienced by beads from the cylindrical boundary. The peak of normal component corresponds to the strong agitation produced in the descending pack of beads when it collides the ascending bottom ; the vertical component has negative time-average.

Lower curve : Vertical component of the sliding velocity averaged over all the beads contacting the cylindrical boundary.

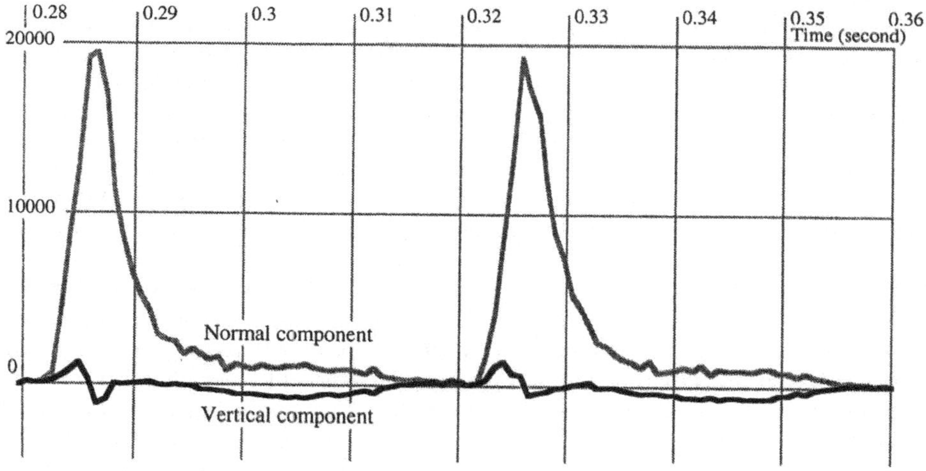

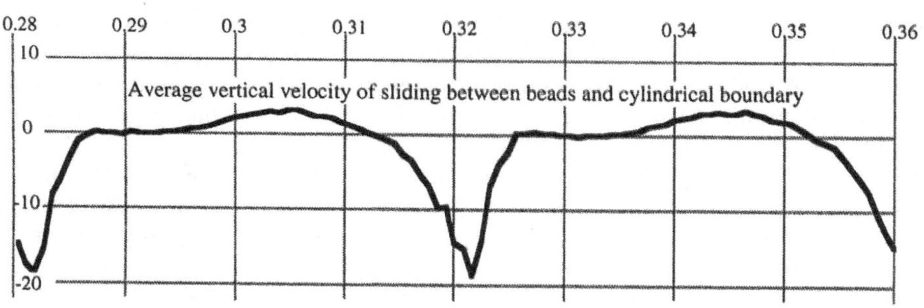

Figure 2

Same computation as on Figure 2

Beads with centers in a 0.2 cm thick meridian slice and their velocities *relative to vessel.*

Levels of gray represent the average pressure experienced by bead during the preceding 1/50 period.

On right side of vessel : for 5 sections of the cylindrical boundary, line segments represent the normal and vertical component of the average force exerted by beads on the considered section during the same 1/50 period.

Time : 0.56 period from lowest position of vessel

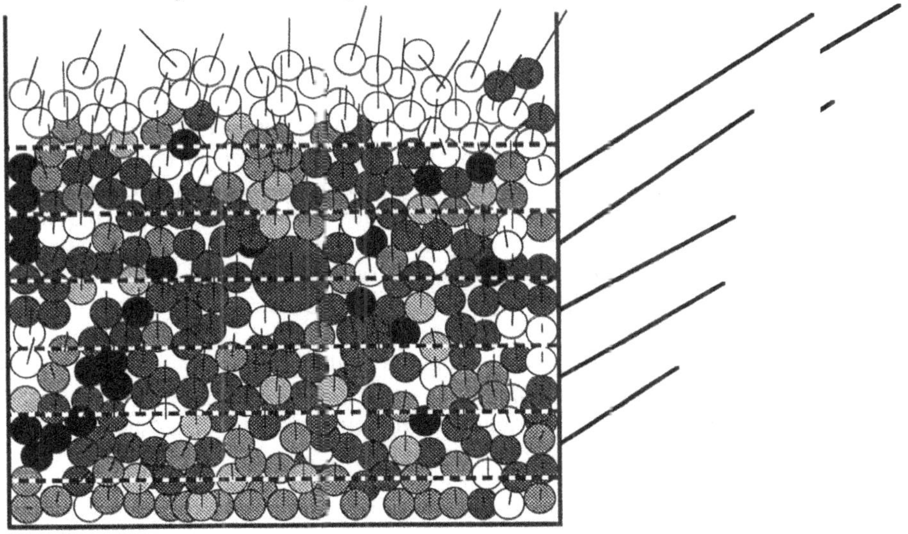

Time : 0.96 period from lowest position of vessel

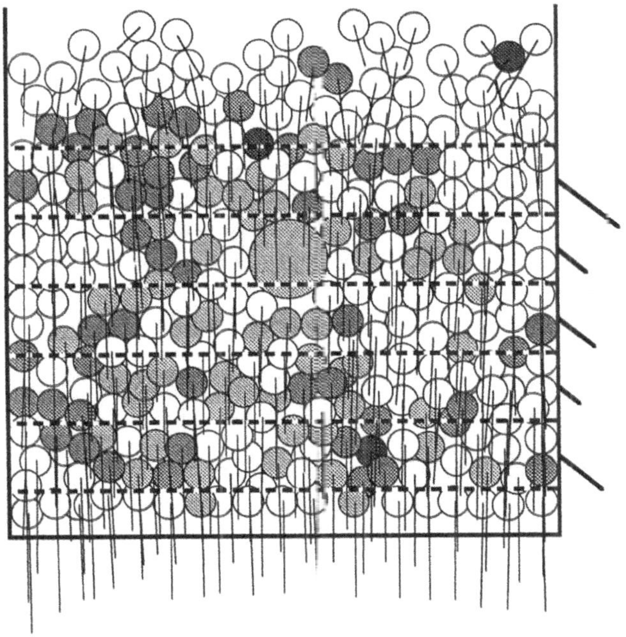

Figure 3

The rapidity of agitation damping, even with normal restitution coefficient as large as 0.95 has to be reminded. Incidentally, this has nothing to do with the *relaxation* of a nearrest pack of grains under gravity, a process which on the contrary is known to be relatively slow.

SIZE-SEGREGATION IN THE BULK

What precedes does not imply that *bulk segregation*, i.e. a tendency of large objects to migrate upward with respect to the surrounding smaller ones, cannot manifest itself in some other range of values of the parameters. Anyway, it must be observed that the relative importance of boundary effects necessarily becomes smaller when larger vessels are considered.

Bulk segregation is apparent in this yet unpublished simulation (figure 4).

Cylindrical vessel 1.5 cm in diameter, with *zero friction* assumed between its boundary and the enclosed beads; normal restitution 0.9 (tangential restitution is immaterial there, since friction vanishes). Vertical sinusoidal vibration with frequency 20 hertz, peak-to-peak amplitude 0.25 cm (making the maximal acceleration equal 2.012 g).

This vessel contains 2200 spherical beads of the same density with sizes as follows :

• a dominant population of 2000 "larger" beads with diameters uniformly distributed between 0.2 and 0.1 cm,

• a minority population of 200 "smaller" ones with diameter 0.02 cm, say one fifth of the smallest diameter in the dominant population.

Between any couple of beads, the friction coefficient equals 0.5, the normal restitution coefficient 0.9, the tangential restitution coefficient 0.4.

Here is the produced evolution:

Smaller beads, initially placed on top, begin performing large ballistic flights as they surf the denser layer constituted by the dominant population. They are progressively captured in this layer, like a gas dissolving into a liquid, and migrate toward the bottom. After 100 shakes or so, a regime is attained where the smaller beads seem *trapped* at bottom.

The cylindrical boundary does not play any part; in fact no boundary currents are expected since friction is zero (Moreau, 1993).

The above is a well known scenario, commonly presented as a gravity-driven trickle of the smaller objects through the voids existing between the larger ones. There probably exist some 'calm' situations to which such a view apply but this is certainly not the case in the present motion. Gravity should be acknowledged as essential in directing the stratification. But agitation is so strong that if at an instant where the 200 smaller beads are concentrated near the bottom, one withdraws the 2000 larger ones out of the computation, the remaining cloud of beads rapidly expand like a gas, a motion which does not resemble a gravity driven spill .

For comparison, another computation have been performed with the same data as above, except that the population of 200 smaller beads has now the diameter 0.05, say half the smallest diameter in the dominant population.

The phenomenon is essentially the same, except that features are not as neat : the capture of smaller beads by the dominant population becomes slower and when a regime seems attained, the small beads make a thicker layer, diffusing higher in the dominant population.

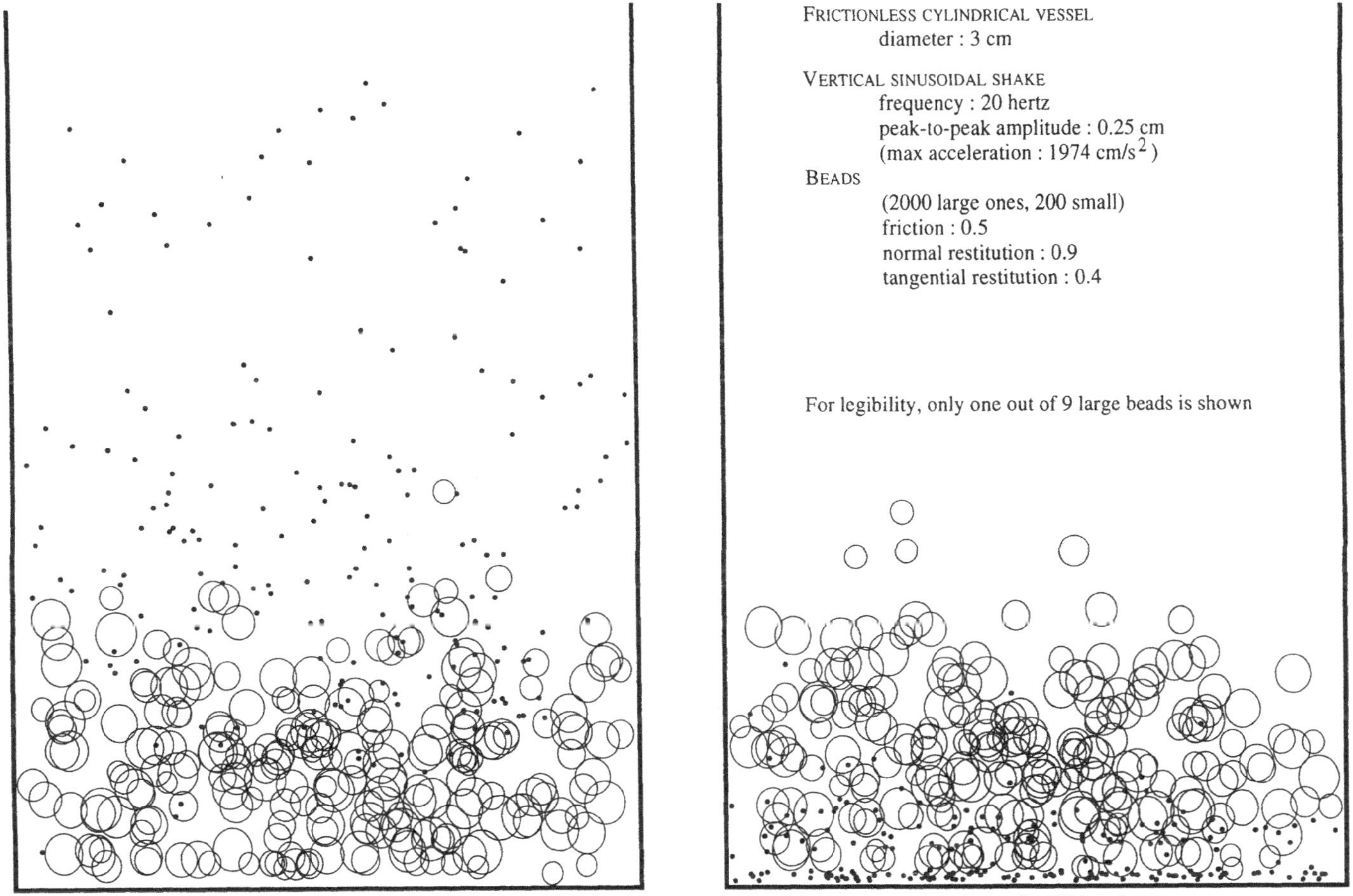

Initial state

State after 104 shakes

Figure 4

REFERENCES

Barker, J. C., 1994, Computer simulations of granular materials, *in* : "Granular Matter", A. Mehta, ed., Springer-Verlag, New York : 35-83.

Bridgwater, J., 1994, Mixing and segregation mechanisms in particle flow, *in* : "Granular Matter", A. Mehta, ed., Springer-Verlag, New York : 161-193.

Clément, E., Duran, J. and Rajchenbach, J., 1992, Experimental study of heaping in a two-dimensional 'sandpile' , *Phys. Rev. Lett.*. 69:1189-1192.

Cundall, P. A., 1971, A computer model for simulating progressive large scale movements of blocky rock systems, *in* : "Proceedings of the Symposium of the International Society of Rock Mechanics", Nancy, France, Vol.1 : 132-150.

Duran, J., Rajchenbach,J. and Clément, E., 1993, Arching effect model for particle size segregation, *Phys. Rev. Lett*, 70 : 2431-2434.

Foerster, S., Louge, M., Chang, H. and Allia, K., 1994, Measurements of the collision properties of small spheres, *Phys. Fluids*, 6 : 1108-1115.

Jean, M., 1994, Frictional contact in collections of rigid or deformable bodies : numerical simulation of geomaterial motions, *in* : "Mechanics of Geomaterial Interfaces", A. P. S. Salvadurai, ed., Elsevier Science Publisher.

Jean, M. and Moreau, J. J., 1992, Unilaterality and dry friction in the dynamics of rigid body collections, *in* : "Contact Mechanics International Symposium", A. Curnier, ed., Presses Polytechniques et Universitaires Romandes, Lausanne: 31-48.

Jullien, R., Meakin, P. and Pavlovitch, A., 1992, Three-dimensional model for particle-size segregation by shaking, *Phys. Rev. Lett.*, 69 : 640-643.

Knight, J. B., Jaeger, H. M. and Nagel, S. R., 1993, Vibration-induced size separation in granular media : the convection connection, *Phys. Rev. Lett.*, 70:3728-3731.

Lun, C. K. K. and Bent, A. A., 1993, Computer simulation of simple shear flow of inelastic frictional spheres, *in*: "Powders and Grains 93", Thornton, C., ed., Balkema, Rotterdam, 301-306.

Martins, J. A. C. and Trabucho, L., 1989, A mathematical formulation for a generalized Hertz impact problem, *in*: "Mathematical Models for Phase Change Problems", J. P. Rodrigues, ed., Birkäuser Verlag, Basel : 337-356.

Meftah, W., Evesque, P., Biarez, J., Sornette, D. and Abriak, N. E., 1993, Evidence of local 'seisms' of microscopic and macroscopic stress fluctuations during the deformation of packings of grains, *in* : "Powders and grains 93", Thornton, C., ed., Balkema, Rotterdam, 173-178..

Mehta, A. and Barker, G. C., 1991, Vibrated powders : a microscopic approach, *Phys. Rev. Lett.*, 67:394-397.

Moreau, J. J., 1988a, Bounded variation in time, *in* : "Topics in Nonsmooth Mechanics", Moreau, J. J., Panagiotopoulos, P. D. and Strang, G., eds., Birkhäuser, Basel, Boston, Berlin : 1-74.

Moreau, J. J., 1988b, Unilateral contact and dry driction in finite freedom dynamics, *in* : "Nonsmooth Mechanics and Applications", Moreau, J. J. and Panagiotopoulos, P. D, eds., CISM Courses and Lectures, n° 302, Springer-Verlag, Wien, New York : 1-82.

Moreau, J. J., 1993, New computation methods in granular dynamics, *in*: "Powders and Grains 93", C. Thornton, ed., Balkema, Rotterdam : 227-232.

Moreau, J. J., 1994, Some numerical methods in multibody dynamics : application to granular materials, *Eur. J. Mech., A/Solids*, 13, n°4 - suppl.: 93-114.

Rosato, A., Strandburg, K. J., Prinz, F. and Swendsen, R. H., 1987, Why the Brazil nuts are on top : size segregation of particulate matter by shaking, *Phys. Rev. Lett.*, 58 : 1038-1040.

Timoshenko, S., 1948, "Théorie de l'Elasticité" , Béranger, Paris, Liège.

Walton, O. R., 1993, Numerical simulation of inelastic, frictional particle-particle interactions, *in*: "Particulate Two-phase Flow", Rocco, M. C., ed., Butterworth-Heinemann, Boston : 884-910.

LOCAL FRICTION EFFECT ON THE GLOBAL BEHAVIOUR OF GRANULAR MEDIA

Abriak N.E.

GRECO GEOMATERIAUX
Ecole des Mines de Douai,
941, Rue Charles Bourseul
Douai - France

INTRODUCTION

If loads oct upon a granular mediam, strains appear at a microstructural level. Many authors (Schlosser, Luong, etc...) have shown that granular mediam deformations are caused by three mechanisms :

- deformations of particles at contact points
- relative sliding and rotations of particles
- bursting of roughness and particles.

As a metter of fact, during loading, relative displacements of particles lead to friction at contact point (local friction).

In order to study the influence of friction between particles (local friction) on the global friction (or angle of internal friction), it hase been made use, in a first time, of the modified CASAGRANDE'S box (ABRIAK, 1991) obtained from the original one by making an aparture on the box and the wagon, in order to can film the tests and so, to observe with more precision (and a posteriori), the behaviour of rolls during rupture.

In a second time, to make this study more complete and have a better understanding of real effects of local frction on global one, triaxial tests have been effected by using a classical triaxial apparatus (ABRIAK, 1993).

EXPERIMENTAL RESULTS

The following analogical materials were used :

- rolls in duralumin (Ecole Centrale de Paris)
- rolls in P.V.C. (Mines de Douai)
- rolls in Copper
- rolls in P.V.C. rolled in Hostun sand.

For each of these materials, many tests at the shearing box have been realized. The curves "tangential stress-deformation drown begin regulary in a restricted domain bounded

Contact Mechanics, Edited by M. Raous *et al.*
Plenum Press, New York, 1995

by null deformation straight line and that relative to a deformation of order 0.8 per cent (Figure 1). From a deformation of 0.8 per cent numerous fluctuations appar and the interpretation of curves becomes difficult. In order to solve this difficulty, the test has, in first time, automatized completely and then a program has been made for the calculus of mean curves (Figure 1). From these last curves, the internal friction angle is deducted (global friction angle)

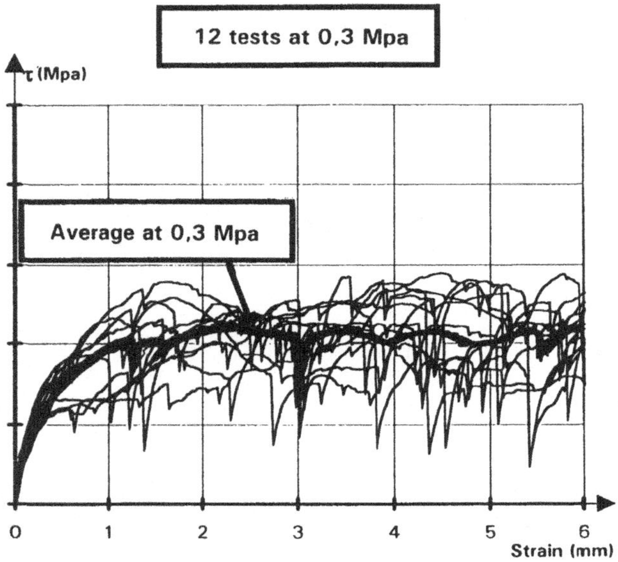

Figure 1. Tangantial stress-deformation
- Casagrande box, rolls in P.V.C.

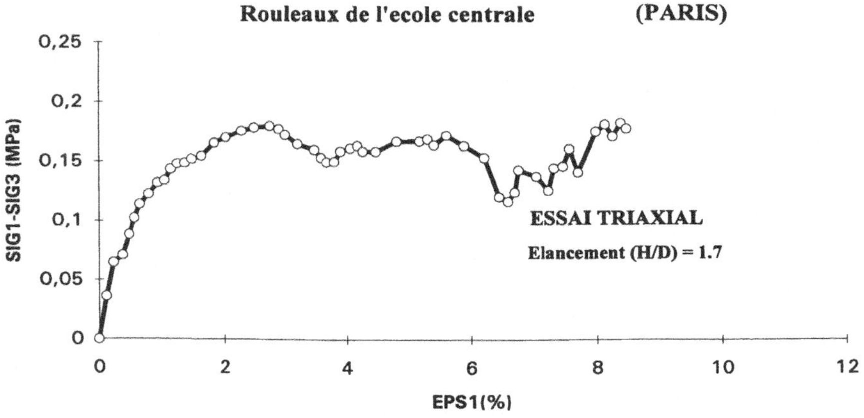

Figure 2. Deviatoric stress-deformation ($\sigma 2 = \sigma 3 = 0,3$ MPa)
- Triaxial apparatus, rolls in P.V.C.

The curves obtained with the triaxial apparatus permit to determine the internal friction angle at the peak (Φ pic) and at the constant level (Φ pp) figure 2, 3, 4.

Table 1. Summarizes the obtained results

		Triaxial	Triaxial	Casagrande Box
Material	Ψ°	Φ PP (mean)	Φ PIC (mean)	Φ PP (mean)
Duralumin	11	20	23	18.9
P.V.C.	18	19.7	25	18
Copper	23	22.5	25.5	20.5
P.V.C. + Sand	30	25.6	27	21

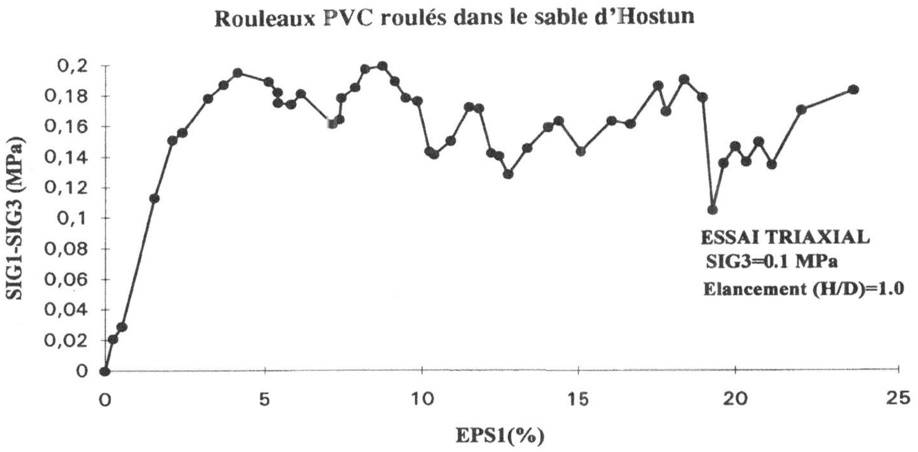

Figure 3. Deviatoric stress-deformation
- Triaxial apparatus, rolls in P.V.C. rolled in Hostun sand (σ2 = σ3 = 0,3 Mpa).

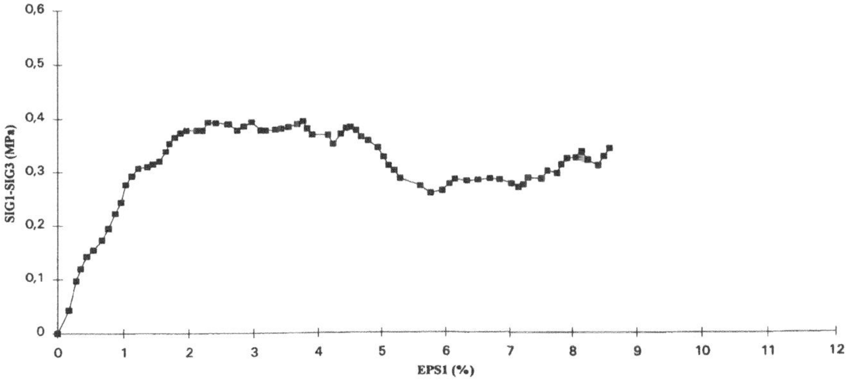

Figure 4. Deviatoric stress-deformation
- Triaxial apparatus, rolls in Copper (σ2 = σ3 = 0,1 MPa).

Results obtained for the internal angle at the level in the CASAGRANDES'S box are slightly different from those measured with the triaxial apparatus. In the case of filmed tests, it has been observed that any important relative motion of roll relative to some other induced a desegregation of the set of rolls at the plane of shearing. It is possible to think that the evolution of perturbations at the shearing plane is influenced, on one part, by the surface state of rolls and otherwise by the orientation of a field of intergranular forces during the shear. In fact, when the inperparticle friction in two materials is different, motions of their particles must be also different.

So, when the relatvie sliding is preponderant during the test, the material exhibits a weak local friction and when the inerparticle friction increases a great part of the particles must be rolling more than sliding each others.

The results of table 1 show that the perfect plasticity internal friction angle and taht at the peak are practically independant from the local friction (Ψ). Moreover, for a material with small angle of internal friction (less than 23°) the global friction angle is greater than the local on and vice-versa.

In the case where the local friction angle is great, this apparently paradoxical can be explained by the existance of important rotations inside the medium.

In fact, if oll intergranular contact could be onaintained in spite of the forces which develop themselves inside the medium, the deformation should be that of the grains. But, at the contrary, it has been observed, on the filmed tests, that a great number of contacts are instable and provoke a reattangement of the rolls in the shear plane, the number, the localization and the nature of these modification leing apparently aleatory.

CONCLUSION

In this study it has been put in evidence firstly, that the global friction is rather independant of local friction and when the local friction of a material is small lower (than 23°) its global friction is greater than the local one and vice-versa.

Secondly, that the behaviour of a granular medium with a great local friction is explained by the rotation mechanism of particles, the behaviour of a medium with small local friction being rather explained by a sliding mechanism. So, in order th shear a given medium, the shear force will depend not only onthe number of contact points between particles but also on their ability to rotate or slide.

REFERENCES

Abriak, N.E., 1991, Flow of granular medium through an orifice : wall effect, Doctorat Thesis. University of sc. and Techno. of Lille, France.

Abriak, N.E., 1993, Etude cinématique et statique des matériaux granulaires, Habilitation à diriger des recherches, Université des Sciences et Technologie de Lille.

Abriak, N.E., and Parsy, F., 1993, Microstructural analysis of granular media, Second International Seminar on Soil Mechanics and Foundations Engineering of Iran.

Abriak, N.E., and Mahboubi, A., 1992, Influence du frottement local sur le frottement global, Rapport scientifique, GRECO-Geomatériaux, Aussois, France.

Schlosser, F., 1974, Influence des déformations des grains dans les milieux granulaires, Bulletin de liaisons des Ponts et Chaussées, N° 69.

MICROMECHANICS MODELLING OF BALLASTED TRACKS

Iyadh Laalaï, Karam Sab and Nathalie Guérin

CERAM-ENPC
1, Avenue Montaigne, 93167 Noisy-le Grand Cedex France

INTRODUCTION

The purpose of this paper is to study the stability of the ballast submitted to a cyclic loading. Naturally, the behaviour of the ballast is linked with the granular nature of the ballast. We propose here a mechanical model at the grain scale.

The "Laboratoire de Mécanique et de Génie Civil" (LMGC) of the University of Montpellier II has developed a software that solves two-dimensional problems involving rigid circular bodies. See for exemple: "Unilateral and dry friction in the dynamics of rigid body collections", Preprint n° 92-2, LMGC, Université de Montpellier II, Montpellier, 1992. A version of this software has been put on the disposal of the "Centre d'Enseignement et d'Analyse des matériaux" (CERAM-ENPC) according to a cooperation agreement between the two laboratories. For the needs of the present study, we have enriched and improved this software by introducing the possibilty of solving two-dimensional problems involving non-circular cylinders.

The study of the stability of the ballast will take into account two parameters: the grains'shape and Coulomb coefficient.

NON CIRCULAR GRAINS

One of the advantages of the LMGC algorithm is its ability to handle granular particles of wide variety of shapes. In the actual configuration, the LMGC software only treats contact between particles of circular shape. The main cause is the difficulty to define an afficient representation of non-circular particles and a straightforward algorithm to detect the contatct points (and the normal vector at the contact points). We propose here a description of the grains by the angle φ between the tangential vector at a current point M of the boundary and the absissa axis. This description is such that $M(\varphi)$ can be simply determined.

1- Desciption of non circular grains

The grain shape is represented by a convex polygon described by a series of vertices $I_1, I_2, ..., I_n$ joined by straight line segments. In what follows, we will conventionnaly note:

Contact Mechanics, Edited by M. Raous *et al.*
Plenum Press, New York, 1995

$I_0 = I_n$ and $I_1 = I_{n+1}$. Let P_i denote the middle point of segment $\underline{I_{i-1}I_i}$. The grain shape is generated by joining point P_i to point P_{i+1} with a convex curve. Let φ denote the angle between the tangential vector $\underline{t}$ at the current point M (see figure 1) of this curve and the abscissa axis. The chosen curve is such that M can be analytically computed as a function of φ. This curve is given by:

$$\underline{OM(\varphi)} = \underline{OI_i} + \left[1 - \cos^{\varepsilon_i}(\theta(\varphi))\right]\underline{I_i P_{i+1}} + \left[1 - \sin^{\varepsilon_i}(\theta(\varphi))\right]\underline{I_i P_i}$$

where ε_i (i=1,n) is a parameter describing the form of the curve connecting P_i and P_{i+1} ($0 \le \varepsilon_i \le 1$). If $\varepsilon_i = 1$, then this curve describes a portion of an elliptical shape. Otherwise, if $\varepsilon_i = 0$, then this curve describes a portion of a rectangular shape.
$\theta(\varphi)$ is determined such that:

$$\frac{d\underline{OM(\varphi)}}{d\varphi} = \varepsilon_i \sin(\theta(\varphi))\cos(\theta(\varphi))\left[\cos^{\varepsilon_i - 2}(\theta(\varphi))\underline{I_i P_{i+1}} + \sin^{\varepsilon_i - 2}(\theta(\varphi))\underline{I_i P_i}\right]$$

is parallal to the tangential vector $\underline{t}$ of the current point M. Therefore:

$$tg(\theta(\varphi)) = \left[\frac{\left|\underline{I_i P_{i+1}}\right|}{\left|\underline{I_i P_{i+1}}\right|} \cdot \frac{\sin(\varphi_{i+1} - \varphi)}{\sin(\varphi - \varphi_i)}\right]^{\frac{1}{\varepsilon_i - 2}}$$

2- Contact points

In the LMGC algorithm, the study of the contatct between two grains A and A' of any shape consists in determining the contact points C and C' of the two bodies and the tangential vectors $\underline{t}$ and $\underline{t}'$ at these points such that:

$$\underline{CC'}.\underline{t} = 0 \quad and \quad \underline{t} = -\underline{t}'$$

As the two bodies are described with the angle φ of any current point M, the previous equation can be written:

$$\underline{C(\varphi)C'(\varphi')}.\underline{t(\varphi)} = 0 \quad and \quad \varphi' = \varphi + \pi$$

or:

$$\underline{C(\varphi)C'(\varphi + \pi)}.\underline{t(\varphi)} = 0$$

The problem is then to determine the angle φ such that the previous equation is satisfied. We propose an algorithm which converges to the solution of the problem noted φ^c:

$$\varphi^{n+1} = \varphi^n + \alpha^n.\underline{C(\varphi^n)C'(\varphi^n + \pi)}.\underline{t(\varphi^n)}$$

with fixed φ^0 and $\alpha^n \succ 0$.

Remark:

This algorithm admits two stationary points: φ^c and $\varphi^c + \pi$. Nevertheless, φ^c is the single stable point:

$$\varphi^n \prec \varphi^c \Rightarrow \underline{C(\varphi^n)C^{'}(\varphi^n+\pi)}.\underline{t(\varphi^n)} \succ 0 \Rightarrow \varphi^n \prec \varphi^{n+1}$$

$$\varphi^n \succ \varphi^c \Rightarrow \underline{C(\varphi^n)C^{'}(\varphi^n+\pi)}.\underline{t(\varphi^n)} \prec 0 \Rightarrow \varphi^n \succ \varphi^{n+1}$$

Indeed:

$$\varphi^n \prec \varphi^c+\pi \Rightarrow \underline{C(\varphi^n+\pi)C^{'}(\varphi^n)}.\underline{t(\varphi^n+\pi)} \prec 0 \Rightarrow \varphi^n \succ \varphi^{n+1}$$

$$\varphi^n \succ \varphi^c+\pi \Rightarrow \underline{C(\varphi^n+\pi)C^{'}(\varphi^n)}.\underline{t(\varphi^n+\pi)} \succ 0 \Rightarrow \varphi^n \succ \varphi^{n+1}$$

So, this algorithm converges to the solution φ^c of the problem.
The description of non-circular bodies and the contact points detection algorithm have been implemented in LMGC software.

SIMULATIONS AND RESULTS

Our purpose is to study the influence of grains' shape and Coulomb coefficient on the macroscopic behaviour of the ballast.

We will present numerical simulations with circular cylinders and elliptical cylinders. It will be established that, even with a very large value of Coulomb coefficient, circular cylinders are unstable under cyclic loads. Whereas, elliptical cylinders are stable if they are well oriented and if the Coulomb coefficient is large enough.

1- Circular cylinders

Perfect lattice

The first simulation concerns a collection of identical circular cylinders occupying a perfect lattice as shown in figure 2 and submitted to cyclic loading. The contact between the cylinders and the rigid right wall is frictionless.The mutual contact between grains obeys Coulomb law.

Simulations show that this lattice is unstable (figure 2). We shall see that this instability is not specific to this lattice. It is due to the circular shape of grains.

Random sample

The second simulation concerns a random sample generated as follows: the cylinders' diameters are randomly picked according to uniform probability distribution (figure 3). At the first step of the simulation the cylinders undergo gravity. This produces a first settlement of the sample. At the second step of the simulation, a load is exerted on the sleeper (Coulomb coefficient is one). It produces a second settlement after some twenty cycles. At the last step of the simulation, we proceed with the same load. We see that the sample is stable for a while, before a new settlement occurs. And so on. Actually, the sample is unstable under cyclic loads (figure 3). It falls down slowly.

We have performed the same simulation with very small Coulomb coefficient (frictionless contact between cylinders). The collapse occurs very quickly after one or two cycles.

Figure 1. Description of non cicular grains.

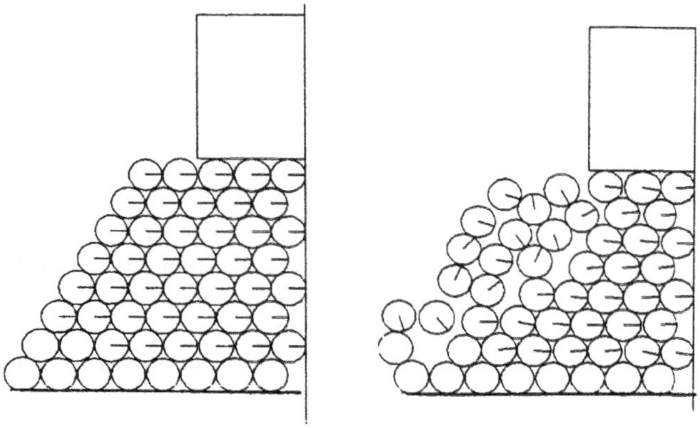

Figure 2. Circular grains - Perfect lattice.

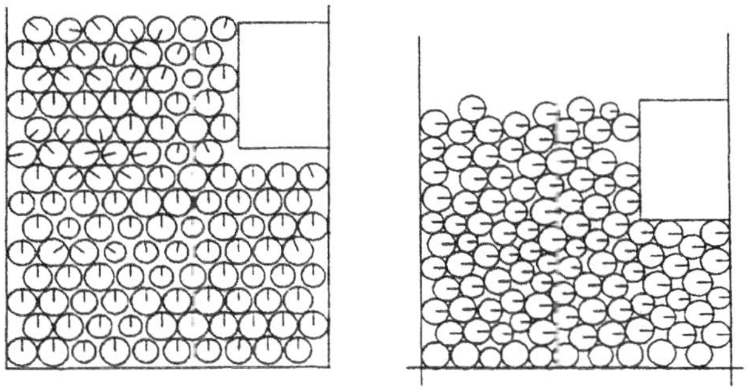

Figure 3. Circular grains - Random sample.

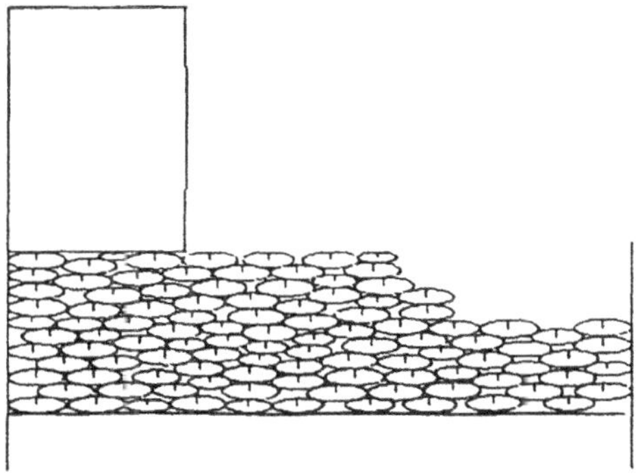

Figure 4. Elliptical grains - Random sample.

In conclusion, we can say that grains with circular shape are unstable under cyclic loading.

2- Elliptical cylinders

Perfect lattice

The first simulation concerns a collection of identical elliptical cylinders obtained by a uniform contraction in the vertical direction of the circular sample of figure 2 (perfect circular lattice). The simulation is performed with the same load and under the same assumptions as for the circular sample. Coulomb coefficient is very small (frictionless contact between cylinders). Simulations show that no collapse occurs. Then we introduce small defects in the lattice as for the circular lattice. We see that, unlike the circular sample, no collapse occurs. Thus, elliptical cylinders are stable.

Random sample

The second simulation concerns a random sample obtained by a uniform contraction in the vertical direction of the sample of figure 3 (see figure 4). With a Coulomb coefficient equal to one, we find that the deflexion curve becomes periodic after two cycles. Another simulation is performed with Coulomb coefficient equal to 0.1. We find that the sample collapse quickly.

CONCLUSION

The main conclusions of this study are:

- The stability of the ballast depends on the grains'shape. If the grains are circular, then the ballast is unconditionally unstable. For non-circular grains, the ballast is unstable if Coulomb coefficient is less than some critical value, and it is stable if Coulomb coefficient is greater than this critical value.

DETECTION OF COLLISIONS OF POLYGONS BY USING A TRIANGULATION

D. Müller and Th. M. Liebling

Département de Mathématiques
École Polytechnique Fédérale de Lausanne
CH-1015 Lausanne (Switzerland)

INTRODUCTION

This paper presents a sketch of a method useful in numerical simulations to efficiently manage motions and collisions of polygonal grains in the plane. This method uses a triangulation of the space between the polygons and it efficiently maintains its coherence in the course of time. This triangulation allows the calculation of exact times of collision and defines a neighborhood that improves simulation time. We use it to study flows of polygonal grains in a hopper and the phenomenon of segregation in granular media.

TRIANGULATION CONSTRUCTION

The triangulation is built in two stages: we first make a *left to right triangulation* that we then modify to obtain a *constrained triangulation.* For a while just keep vertices of polygons and forget polygons themselves. A simple way to triangulate this cluster of points is the following:

1. Sort points lexicographically and number them from 1 to n.
2. **For** i=2 **to** n **do**
   connect the point i with all points j<i if the segment ij does not cross any already constructed segment.

In this way we obtain a left to right triangulation. We want now to build a constrained triangulation where we impose that edges of polygons (which will from now on be called *mandatory edges*) be edges of the triangulation.

Starting from a left to right triangulation, we repeat the following operations for all mandatory edges a_k which are not in the triangulation:

a. Eliminate the edges crossed by a_k
b. Put a_k in the triangulation
c. Triangulate the inner of both induced polygons which adjoin a_k.

Contact Mechanics, Edited by M. Raous *et al.*
Plenum Press, New York, 1995

The following is a simple method for c:

1. Examine the vertices of the polygon counterclockwise and put them into a circular list L
2. Let: a:=beginning of the list
   b:=following(a)
   c:=following(b)
3. **If** the triangle abc is positively oriented **and if** ac is entirely in the polygon **then**:
   - add segment ac
   - remove point b from list L
   - b:=c; c:=following(b)

   else
   - a:=b; b:=c; c:=following(b)
4. **While** following(c)≠a **go to** 3.

After removing edges inside grains, we obtain a triangulation of the space between the polygonal grains (see fig. 1).

Figure 1. A triangulation of the space between the polygons

TRIANGULATION MAINTENANCE IN THE COURSE OF TIME

To illustrate this paragraph in concrete terms we will imagine that polygons are set on a vertical plane and subject to gravitational attraction. Because of the motion of polygons, the triangulation degenerates quickly if it is not maintained by topological operations. We show how to calculate when a triangle becomes flat (hence the time when the triangulation will degenerate) and how to maintain this triangulation coherent by local operations.

In principle, the motion without friction of an object has two aspects: the centre of gravity follows the law of a material point and thus describes a parable, while the objet itself rotates at a constant angular velocity about its centre of gravity. Formulas (1) and (2) are the equations of motion of the polygon centre of gravity.

$$x(t) = x(0) + \dot{x}(0)\cdot t + \frac{1}{2}\ddot{x}\cdot t^2 \tag{1 a}$$

$$y(t) = y(0) + \dot{y}(0)\cdot t + \frac{1}{2}\ddot{y}\cdot t^2 \tag{1 b}$$

$$\dot{x}(t) = \dot{x}(0) + \ddot{x}\cdot t \tag{2 a}$$

$$\dot{y}(t) = \dot{y}(0) + \ddot{y} \cdot t \qquad (2\ b)$$

Herein (x(t),y(t)) are the coordinates at time t, $(\dot{x}(t), \dot{y}(t))$ is the velocity and $(\ddot{x}, \ddot{y})$ is the gravitational attraction. The formulas describing the position of a vertex v are the following:

$$x^v(t) = x(t) + r \cdot \cos(\omega t + \alpha) = x(t) + r_x \cdot \cos(\omega t) - r_y \cdot \sin(\omega t) \qquad (3\ a)$$

$$y^v(t) = y(t) + r \cdot \sin(\omega t + \alpha) = y(t) + r_y \cdot \cos(\omega t) + r_x \cdot \sin(\omega t)\ , \qquad (3\ b)$$

where ω is the rotational velocity, α the angle at time 0, r the distance between the centre of gravity and the vertex v, $r_x = r \cdot \cos(\alpha) = x - x^v$ and $r_y = r \cdot \sin(\alpha) = y - y^v$ (see fig. 2).

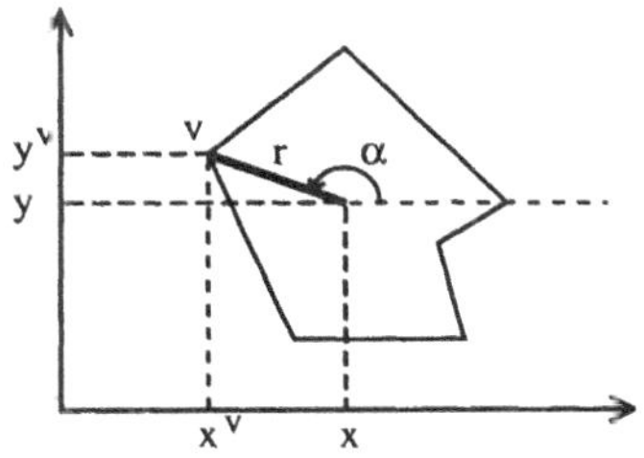

Figure 2. Nomenclature used for a polygon

Trigonometric functions have two drawbacks: computation is slow and relatively inaccurate; moreover finding the roots of a function including trigonometrical functions can be difficult. Instead we develop $\cos(\omega t+\alpha)$ and $\sin(\omega t+\alpha)$ into series of a few power terms by using Mac-Laurin formulas. The number of terms used (two terms for sinus and three for cosinus) allows a good approximation (the maximum error is 10^{-7}) for angles ranging from -0.1 to 0.1 radians. With the help of the Mac-Laurin formulas, we can rewrite equations (3 a) and (3 b) as follows:

$$x^v(t) = x(0) + r_x + (\dot{x}(0) - r_y \cdot \omega) \cdot t + \frac{1}{2}(\ddot{x} + r_x\, \omega^2) \cdot t^2 + \frac{1}{6} r_y\, \omega^3\, t^3 + \frac{1}{24} r_x\, \omega^4\, t^4 \qquad (4\ a)$$

$$y^v(t) = y(0) + r_y + (\dot{y}(0) - r_x \cdot \omega) \cdot t + \frac{1}{2}(\ddot{y} - r_y\, \omega^2) \cdot t^2 - \frac{1}{6} r_x\, \omega^3\, t^3 + \frac{1}{24} r_y\, \omega^4\, t^4 \qquad (4\ b)$$

The aspect of triangles evolves with time, since their vertices move according to (4). Now and then some triangles become flat (area = 0). Two cases are possible:

1. The longest edge of this triangle is not a mandatory edge. In this case there is no collision. We just need to flip the edge in order to still have a coherent triangulation (see fig. 3). This operation consists in exchanging both diagonals of a convex quadrilateral.
2. The longest edge of this triangle is a mandatory edge. There is a collision. We use formulas given by Wang & Mason (1992) to determine the new velocities of the grains.

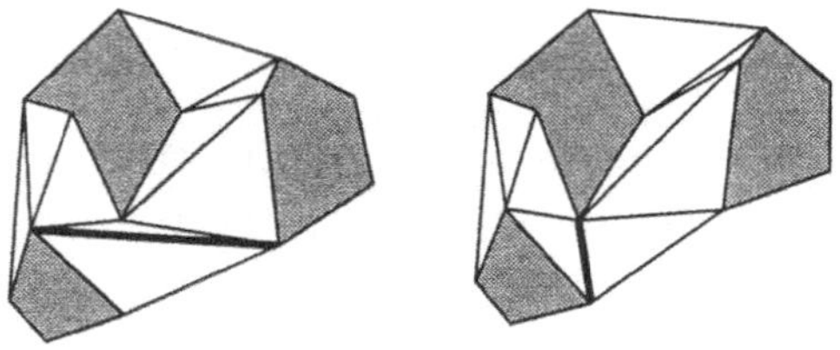

Figure 3. Triangulation maintenance when a polygon goes through a not mandatory edge

We can compute the time where a triangle with vertices **u**(t), **v**(t), **w**(t) will become flat by calculating the 2 by 2 determinant below (**u**(t), **v**(t), **w**(t) are bidimensional vectors):

$$\det(t) = | \mathbf{u}(t) - \mathbf{v}(t) \quad \mathbf{u}(t) - \mathbf{w}(t) | \tag{4}$$

and by solving the equation det(t)=0.

This determinant is a 8th degree polynomial whose roots we must extract. We have to cut time up into small *periods* of one thousandth of second in order to attain sufficient precision because of Mac-Laurin approximations. Actually only the smallest root in the interval $[t_1,t_2]$ interests us, where t_1 is the current time and t_2 the time of the period end. We find this root quickly by using *Sturm sequences* described by Ralston (1978) and a mixture between the Newton method (for rapidity) and the classical bisection method.

EXAMPLE OF APPLICATION

As an example of application, fig. 4 presents a flow of 500 polygonal grains in a hopper. We can study the influence of the shape and the matter of grains (characterized by coefficient of friction and restitution) on the flow. We can modify the slope and the neck of the hopper. The computation time to empty this hopper is very depending on the parameters: it is contained between one and five hours (between one and six seconds in real life). We use a SiliconGraphics workstation with a 100 MHz processor.

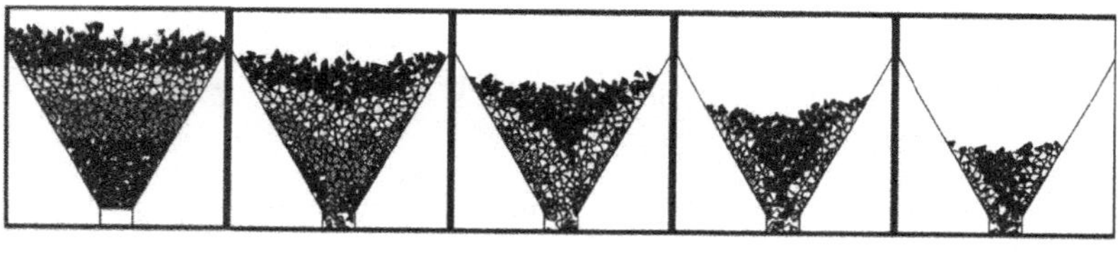

Figure 4. Flow in a hopper

REFERENCES

Ralston A., 1978, "A first course in numerical analysis", McGraw-Hill, New York

Wang Y.,Mason M. T., 1992, Two-dimensional rigid-body collisions with friction, *Journal of Applied Mechanics* 59:635.

DISCRETE MODELS FOR CONTACT PROBLEMS

Bernard Cambou, Marie Chaze, Philippe Dubujet,
Abdelwahab Ghaouti, Yves-Michel Lamidon and François Sidoroff

Laboratoire de Tribologie et Dynamique des Systèmes
Ecole Centrale de Lyon
36, avenue Guy de Collongue
BP 163
69130 Ecully Cedex, France

INTRODUCTION

Contact problems are essentially characterized by the interaction of volume and surface effects. Surface mechanics however is much less clear than volume mechanics and the underlying microscopic physics clearly implies molecular aspects. Adhesion and surface energy in particular are macroscopic manifestation of long range molecular interaction (Maugis, 1992). The increasing interest in molecular dynamics for tribology and contact is therefore not surprising. For instance it has been successfuly used to simulate stick-slip at the molecular level (Thomson and Robbins, 1990) or plastic-like wear in contact (Landmann et al., 1993).

Although very simple in their principles, these simulations are computationally very heavy. One reason for this is the very small time step ($\approx 10^{-13}$s) which is required for taking into account thermal fluctuations which for many problems are not essential, especially in connection with solids. There is a need for simpler simulation tools. The present paper is devoted to the presentation of such simplified approaches and to a brief survey of some results illustrating their application and the physical insight they may provide on various contact problems. Attention will be focused on two dimensional problems.

MOLECULAR DYNAMICS AND GRANULAR MECHANICS

Molecular dynamics is based on the integration of the motion equation

$$m_i \frac{d^2\mathbf{x}_i}{dt^2} = \mathbf{F}_i = \sum_{j \neq i} \mathbf{F}_{ij} + \mathbf{F}'_i \qquad (1)$$

for each particle. The force $\mathbf{F}_i$ acting on particle i is the sum of the forces exerted

Contact Mechanics, Edited by M. Raous *et al.*
Plenum Press, New York, 1995

on i from all other particles j plus some additional external forces F'_i if necessary.

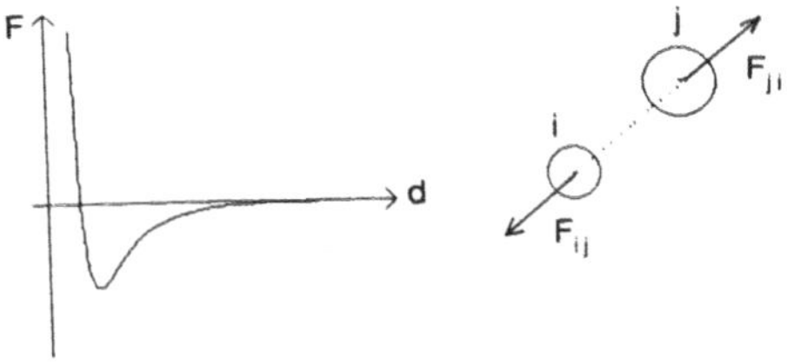

Figure 1. Interaction force.

The interaction force F_{ij} is obtained from an interaction potential w

$$F_{ij} = -\frac{dw}{dr_{ij}} \frac{\mathbf{x}_i - \mathbf{x}_j}{r_{ij}} \qquad r_{ij} = |\mathbf{x}_i - \mathbf{x}_j| \tag{2}$$

The Lennard Jones potential is standard with $n_1 = 12,\ n_2 = 6$

$$w(d) = \frac{w_o}{n_1 - n_2}\left[n_2\left(\frac{d_o}{d}\right)^{n_1} - n_1\left(\frac{d_o}{d}\right)^{n_2}\right] \tag{3}$$

Numerical integration of (1) can then be performed, usually under isothermal condition which means a fixed total kinetic energy. This can be for instance performed through a frequent rescaling of the particle velocities.

A similar situation is encountered in the direct simulation for granular materials (Cundall and Strack, 1979) which has to deal with different interaction laws essentially based on contact and which includes intergranular friction as well as normal stiffness. Often also a viscous dissipative term is added

$$m_i\frac{d^2\mathbf{x}_i}{dt^2} + \mu_i\frac{d\mathbf{x}_i}{dt} = \mathbf{F}_i = \sum_c \mathbf{F}_i^{\,c} \tag{4}$$

where the contact forces $F_i^{\,c}$ depend on the relative positions and velocities of the concerned particles. Different kinds of integration schemes and models have been developed (Cundall and Strack, 1979 ; Moreau, 1994).

Though similar to (1) this is dynamically entirely different. While dynamics is essential in the conservative system (1), both in the underlying physics and in the expected results, it is not in the dissipative system (2) where we are essentially interested in the local motions and global resulting forces and where thermal fluctuations are irrelevant. No matter whether the dynamical term is numerically essential as in (Moreau, 1994) or a mere convenience as in Cundall and Strack (1979), the basic physics remains in the quasi-static case

$$\mathbf{F}_i = 0 \tag{5}$$

Our basic objective is to use such a quasi-static approach together with the interaction law (2)(3) to model molecular process in solids. Bubble raft models also provide a useful experimental background (Mazuyer, Georges and Cambou, 1989).

NUMERICAL ASPECTS

This problem has been approached from two points of view.

First a specific software MPART has been developed using a general interaction law (2) and taking into account long range interaction.

Specific interaction laws with the walls have been incorporated including smooth and particulate walls.

The motion of these walls is imposed inducing displacement, deformation and rearrangement of the particle aggregate. Force control of the wall motion is also implemented. At present time the quasi-static equation (5) is solved through a pseudo-viscosity model :

$$\mu \frac{dx_i}{dt} = F_i = \sum_{j \neq i} F_{ij} + F'_i \tag{6}$$

and a simple Euler algorithm.

The second approach consists in an adaptation of the TRUBAL software as developed for granular materials by Cundall on the basis of the distinct element method (Cundall and Strack, 1979). This method uses a central finite difference scheme for the numerical integration of the motion law (4)

$$\dot{x}_i^{n+1/2} = \left[\dot{x}_i^{n-1/2} \left(1 - \frac{\mu}{m} \frac{\Delta t}{2} \right) + F_i^n \frac{\Delta t}{m} \right] \frac{1}{1 + \frac{\mu}{m} \frac{\Delta t}{2}} \tag{7}$$

$$x_i^{n+1} = x_i^n + \dot{x}_i^{n+1/2} \Delta t$$

The adaptation to our problem specifically includes

- a purely normal contact forces without friction (zero tangential stiffness)
- the introduction of an attractive force

$$F^a_{ij} = - k \left(\frac{r_i + r_j}{r_{ij}} \right)^n n_{ij} \qquad n_{ij} = \frac{x_i - x_j}{r_{ij}} \qquad r_i \text{ radius of particle } i \tag{8}$$

in addition to the usual repulsive contact force described by the normal stiffness k_n

$$Fc_i^n = Fc_i^{n-1} + k_n (\dot{x}_i^{n-1/2} - \dot{x}_j^{n-1/2}) \Delta t \, n_{ij} \otimes n_{ij} \tag{9}$$

- the possibility to impose the motion of some particles for non homogeneous tests.

EXPLOITATION

Two different kinds of particle arrangement can be considered : a crystalline aggregate is obtained with one single kind of particle while an amorphous one results from different particle diameters (Mazuyer et al., 1991) ; the corresponding results will be very different and the amorphous case is probably more relevant.

Generating the initial configuration is quite easy in the crystallographic case but much less obvious in the amorphous case : we usually start from some loose randomly generated arrangement which is then prepared through compaction, relaxation and forming before being loaded by the prescribed wall motion.

The final result of the computation is the evolution of the position of each particle with time from which the interaction forces as well as the external forces to be exerted on the walls can be obtained providing the global structural response of the system together with a "movie" of the deformation.

More quantitative insight into the process can be obtained through the graphical representation at each time of

– The contact forces between neighbouring particles showing the "force lines" along which most of the forces are transmitted

– The velocity of each particle which shows the localization of the deformation process

– A micro stress tensor may be defined on each particle by the relation

$$\boldsymbol{\sigma}_i = \frac{1}{D} \sum_{j \neq i} (\mathbf{r}_i - \mathbf{r}_j) \otimes \mathbf{F}_{ij} \tag{10}$$

Indeed when averaged on all particles in an homogeneous system, this quantity can be shown to be the macroscopic stress tensor (Weber's relation, Cambou and Sidoroff,1985). In case of slow motion it also coincides with the virial expression (Landmann et al., 1993, eq (2)).

Some of these representations will be given in the following examples.

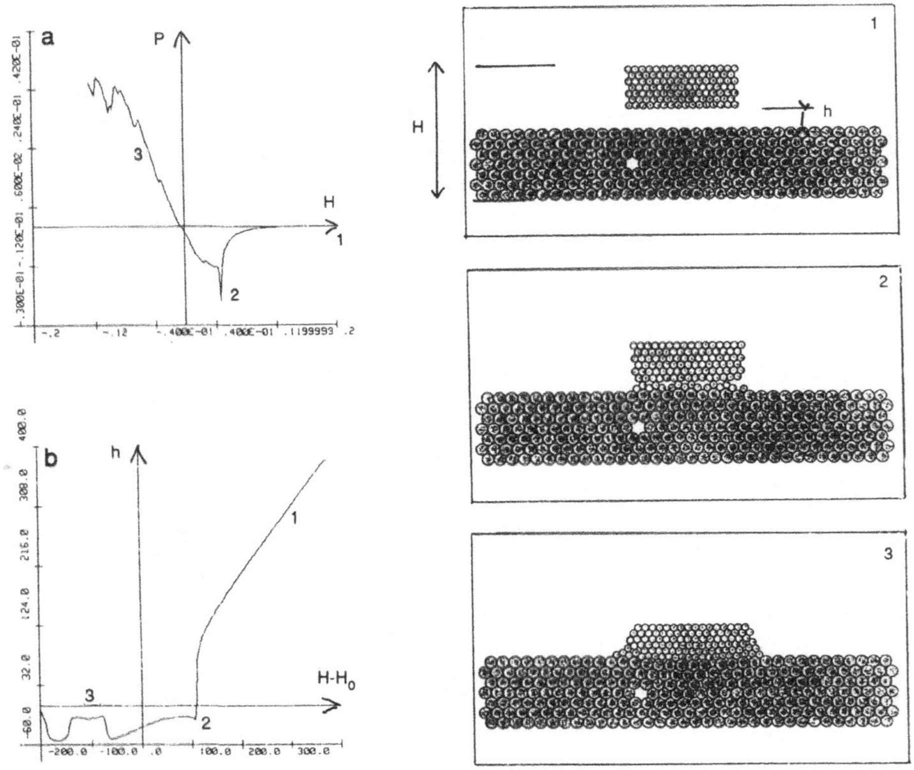

Figure 2. Adhesion.

ADHESION - RUPTURE

Adhesion and rupture phenomena may be simulated by the slow approach of two blocks followed by their separation.

A typical situation is shown in Figure 2 for the adhesion of two crystalline rectangular aggregates, with representation of the normal force P in Figure 2a and of the separating distance h in Figure 2b as a function of the global imposed displacement H (or $H-H_0$). Configuration 2 in particular corresponds to the so-called "adhesive avalanche" which is the instability phenomenon through which the two bodies come into contact. A more precise discussion of this phenomenon using a one dimensional model can be found in Lamidon, Chaze and Sidoroff (1993).

Another instability occurs when the two bodies are separated, leading to rupture. These global instabilities are well known and can be observed in soap bubble experiment (Mazuyer et al., 1991) as well as in complete molecular dynamics simulation.

INDENTATION

Indentation of an half space by a crystalline triangular punch can be analyzed in a similar way. Figures 3, 4 and 5 give two examples of indentation of an amorphous half space. In Figure 3 the load-displacement curve is completed by two representations of a given stage of the indentation process : contact forces in Figure 3b and instantaneous velocities in Figure 3c.

In Figures 4 and 5 on the other hand, attention is focussed on the evolution of the velocity modulus (Figure 4, the gray level is proportional to the instantaneous velocity

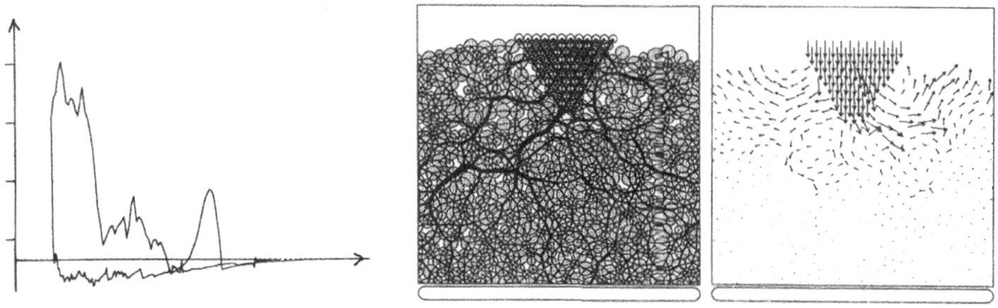

Figure 3. Indentation.

modulus) and of the vertical normal stress as defined by (10) (Figure 5, traction is represented in black, compression in white) for each particle during the identation process.

In both cases a succession of typical configurations are shown starting from equilibrium with separated indentor and substrate. As the indentor is progressively advanced a small adhesive avalanche is obtained for the first contact (second view in Figure 4 and 5) followed by the actual indentation process which proceeds by a

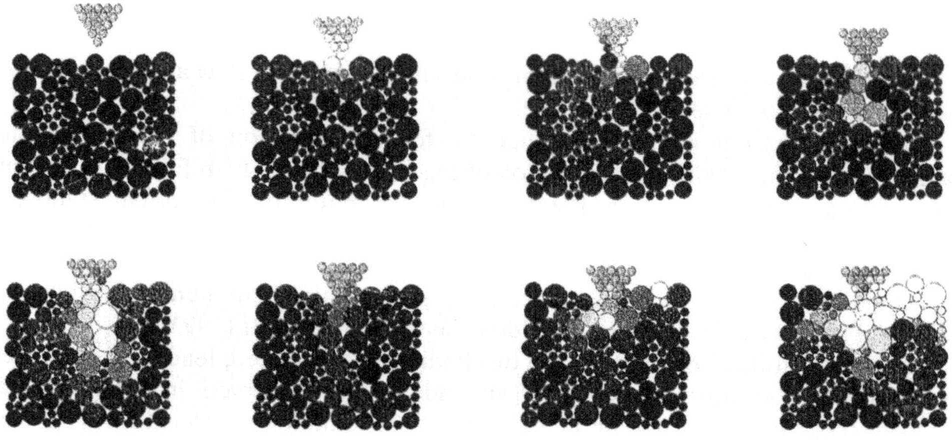

Figure 4. Force amplitude evolution.

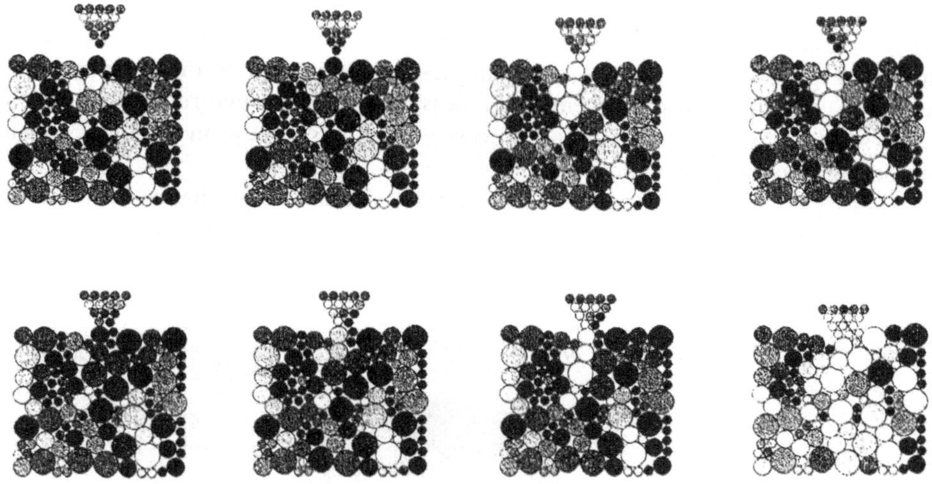

Figure 5. Vertical normal stress evolution.

succession of local instabilities and a global compression superposed on the initial internal microstress distribution resulting from the geometrical arrangement of the particles.

The same velocity representation is given in Figure 6 for a crystalline substrate. In each case the deformation process is characterized by a succession of local instabilities. This is in agreement to what is observed in soap bubbles experiment (Mazuyer et al., 1991), however, these local catastrophes are found sharply isolated both in time and space in the crystallographic case which they are more smoothly distributed in the amorphous case.

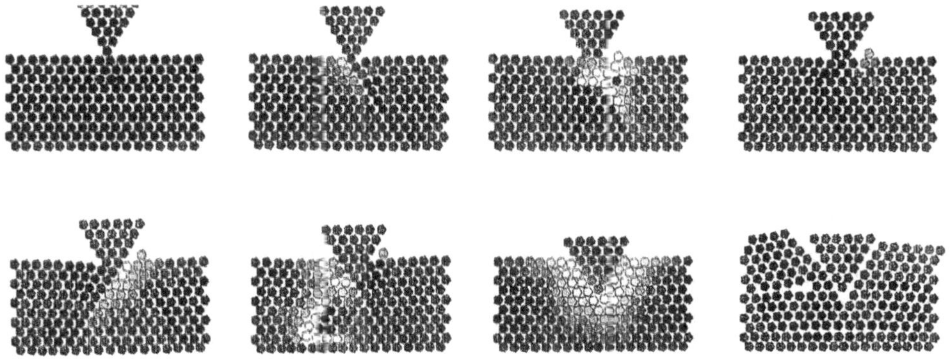

Figure 6. Indentation of a crystalline half space.

SIMPLE SHEAR

Simple shear of a film is represented in Figure 6 : imposed horizontal relative displacement of the upper and lower layers with zero vertical displacement (periodicity conditions on the lateral sides). Evolution of the macroscopic global stress

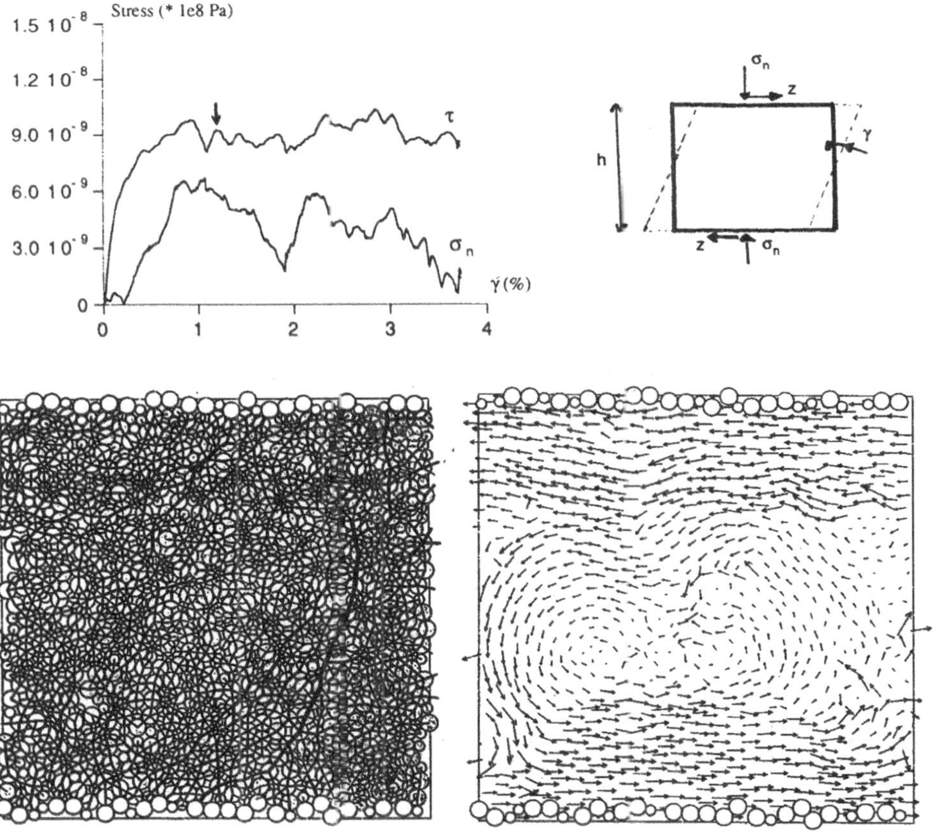

Figure 7. Simple shear.

τ and normal stress σ_n are given together with an example of the instantaneous velocity field and contact force distribution.

WEAR AND FRETTING

Wear can be simulated through the sliding of two crystalline walls separated by an amorphous layer. As the sliding distance increases the thickness of the amorphous layer is observed to increase due to the absorption of particles from the crystalline walls. This corresponds to the wear of these walls, although they do not participate in the shear motion which is limited to the amorphous film.

CONCLUSION

These results shows the interest of this kind of method for the understanding and qualitative analysis of the microscopic physical phenomena involved in contact mechanics.

REFERENCES

Cambou, B. and Sidoroff, F., 1985, Description de l'état d'un matériau granulaire par variables internes statiques à partir d'une approche discrète, *J. Méca. Th. et Appl.*, 4:223-242.

Cundall, P.A. and Strack, O.D.L., 1979, A discrete numerical model for granular assemblies, *Géotechnique*, 29:47-65.

Lamidon Y.M., Chaze, M. and Sidoroff, F., 1993, Mécanique particulaire monodimensionnelle : application à l'avalanche adhésive, 11ème Congrès Francais de Mécanique, Lille.

Landmann, U. et al., 1993, Nanotribology and the stability of nanostructures, *Jpn. J. Appl. Phys.*, 32:1444-1462.

Maugis, D., 1992, Adhesion of spheres : the JKR-DMT transition using a Dudgak model, *Journal of Colloid and Interface Science*, 150:243-269.

Mazuyer, D. et al., 1991, De l'utilisation des bulles de savon, a film, Produced by Lipsis Production, CNRS, MESR.

Mazuyer, D., Georges, J.M. and Cambou, B., 1989, Shear behavior of an amorphous film with bubble soap raft model, *J. Phys.*, France 49:1057-1067.

Moreau, J.J., Some numerical methods in multibody dynamics : application to granular materials, *Eur. J. Mech. /A*, Special issue (2nd ESMC Genoa) 13:93-114.

Thomson, P.A. and Robbins, M.O., 1990, Shear flow near solids : Epitaxial order and fluid boundary conditions, *Phys. Rev. A*, 41(12):6830-6836.

ENERGY CONSIDERATIONS DURING OBLIQUE PARTICLE INTERACTIONS

Colin Thornton[1] and Guoping Lian[2]

[1]Department of Civil Engineering
Aston University
Birmingham, B4 7ET, UK
[2]Unilever Research Colworth Laboratory
Colworth House, Sharnbrook
Bedford MK44 1LQ, UK

INTRODUCTION

The paper considers the oblique interactions of elastic spheres. Mindlin (1949) showed that the application of a tangential force causes relative slip over an annulus of the contact area which spreads radially inwards until sliding occurs. Any reversals in the direction of the relative tangential displacement initiates counterslip at the perimeter of the contact area and, consequently, leads to hysteretic force-displacement curves. For the special case of the tangential force oscillating between $\pm T^*$, with the normal force constant, Mindlin et al (1951) derived an expression for the energy dissipated per cycle.

In this paper, we examine the incremental energy dissipated during the evolution of the microslip annulus directly by considering the slip component of displacement and the distribution of the shear traction over the annulus for the loading, unloading and reloading cases. By subtraction from the total incremental work done the increment of elastic energy is also obtained. We then examine the elastic tangential stiffness and show that the elastic behaviour is hysteretic.

Due to space limitations we restrict our attention to the case when the normal force at the contact remains constant.

Contact Mechanics, Edited by M. Raous *et al.*
Plenum Press, New York, 1995

ENERGY DISSIPATION DUE TO AN OSCILLATING TANGENTIAL FORCE

For two elastic spheres in contact under a constant normal force P, which are subsequently subjected to an oscillating tangential force $\pm T^*$ ($T^* < \mu P$), the tangential force-displacement curve follows a closed symmetrical hysteresis loop after the inital tangential loading stage, as shown in Figure 1. Mindlin et al (1951) calculated the area enclosed by the loop to provide the following equation for the frictional energy dissipated per cycle.

$$\Delta W_{slip} = \frac{9\mu^2 P^2}{10G^* a}\left\{1-\left(1-\frac{T^*}{\mu P}\right)^{5/3} - \frac{5T^*}{6\mu P}\left[1+\left(1-\frac{T^*}{\mu P}\right)^{2/3}\right]\right\} \tag{1}$$

where μ is the coefficient of friction and a is the radius of the contact area, which according to Hertzian theory, is given as

$$a^3 = \frac{3PR^*}{4E^*} \tag{2}$$

and for two spheres of radii R_i and elastic properties E_i, G_i and ν_i ($i = 1,2$)

$$\frac{1}{R^*} = \frac{1}{R_1} + \frac{1}{R_2} \tag{3}$$

$$\frac{1}{E^*} = \frac{1-\nu_1^2}{E_1} + \frac{1-\nu_2^2}{E_2} \tag{4}$$

$$\frac{1}{G^*} = \frac{2-\nu_1}{G_1} + \frac{2-\nu_2}{G_2} \tag{5}$$

Energy Dissipation During Loading

If the tangential force is monotonically increased from zero to T, slip is initiated at the perimeter of the contact area and progresses inward to a radius b. The distribution of the tangential traction is given as $q = q^{(1)} + q^{(2)}$ where

$$q^{(1)} = \frac{3\mu P}{2\pi a^3}\left(a^2 - r^2\right)^{1/2} \qquad r \le a \tag{6}$$

$$q^{(2)} = -\frac{3\mu P}{2\pi a^3}\left(b^2 - r^2\right)^{1/2} \qquad r \le b \tag{7}$$

The tangential force and relative tangential displacement between the two spheres are

$$T = \mu P\left(1 - \frac{b^3}{a^3}\right) \tag{8}$$

$$\delta = \frac{3\mu P}{16G^* a}\left(1 - \frac{b^2}{a^2}\right) \tag{9}$$

The elastic displacement over the slip annulus is

$$u_i = \frac{3\mu P}{16G_i a}(2 - \nu_i)\left\{1 - \frac{b^2}{a^2} - \frac{1}{\pi a^2}\left[g_1(b,r) - \frac{\nu_i}{4 - 2\nu_i} g_2(b,r)\cos 2\psi\right]\right\} \tag{10}$$

and the relative slip is then given as

$$s = \delta - (u_1 + u_2) = \frac{3\mu P}{16\pi G^* a^3}\left[g_1(b,r) - \frac{G^*}{2}\left(\frac{\nu_1}{G_1} + \frac{\nu_2}{G_2}\right) g_2(b,r)\cos 2\psi\right] \tag{11}$$

where

$$g_1(b,r) = \left(r^2 - 2b^2\right)\arccos\frac{b}{r} + b\sqrt{r^2 - b^2} \tag{12}$$

$$g_2(b,r) = r^2 \arccos\frac{b}{r} - \left(\frac{2b^2}{r^2} - 1\right) b\sqrt{r^2 - b^2} \tag{13}$$

If the tangential force increases by an amount ΔT then the relative tangential displacement will increase by an amount $\Delta\delta$ and the radius of the inner stick region will decrease by an amount Δb. Hence, from (8), (9) and (11)

$$\Delta T = -3\mu P \frac{b^2}{a^3}\Delta b \tag{14}$$

$$\Delta\delta = -\frac{3\mu P b}{8G^* a^3}\Delta b \tag{15}$$

$$\Delta s = -\frac{3\mu P}{16\pi G^* a^3}\left[4b\arccos\frac{b}{r} + G^*\left(\frac{\nu_1}{G_1} + \frac{\nu_2}{G_2}\right)\frac{4b^4 - 4b^2 r^2}{r^2\sqrt{r^2 - b^2}}\cos 2\psi\right]\Delta b \tag{16}$$

Combining (15) and (16) we obtain

$$\Delta s = -\frac{1}{2\pi b}\left[4b\arccos\frac{b}{r} + G^*\left(\frac{\nu_1}{G_1} + \frac{\nu_2}{G_2}\right)\frac{4b^4 - 4b^2 r^2}{r^2\sqrt{r^2 - b^2}}\cos 2\psi\right]\Delta\delta \tag{17}$$

We may now define the energy dissipated by friction as

$$\Delta W_{slip} = \int_b^a \int_0^{2\pi} q\Delta s r dr d\psi \tag{18}$$

which, it can be shown, leads to

$$\Delta W_{slip} = \frac{\mu P}{2}\left(2 - 3\theta_L + \theta_L^3\right)\Delta\delta \tag{19}$$

$$\theta_L = \frac{b}{a} = \left(1 - \frac{T}{\mu P}\right)^{1/3} \tag{20}$$

Energy dissipation during unloading and reloading

Following similar procedures it can be shown that, during unloading,

$$\Delta W_{slip} = -\frac{\mu P}{2}\left(2 - 3\theta_U + \theta_U^3\right)\Delta\delta \tag{21}$$

$$\theta_U = \left(1 - \frac{T^* - T}{2\mu P}\right)^{1/3} \tag{22}$$

and during reloading

$$\Delta W_{slip} = \frac{\mu P}{2}\left(2 - 3\theta_R + \theta_R^3\right)\Delta\delta \tag{23}$$

$$\theta_R = \left(1 - \frac{T - T^{**}}{2\mu P}\right)^{1/3} \tag{24}$$

When the tangential force is oscillated between $\pm T^*$, $T^{**} = -T^*$, and unloading and reloading are identical events except for the sign reversal. Therefore, the frictional energy dissipation per cycle may be defined as

$$W_{slip} = -\mu P \int_{\delta*}^{-\delta*} \left(2 - 3\theta + \theta^3\right) d\delta \tag{25}$$

$$d\delta = \frac{3\mu P}{4G^* a}\theta d\theta \tag{26}$$

Hence

$$W_{slip} = \frac{3\mu^2 P^2}{4G^* a}\int_{\theta^*}^{1}\left(2 - 3\theta + \theta^3\right)\theta d\theta \tag{27}$$

$$\theta^* = \left(1 - \frac{T^*}{\mu P}\right)^{1/3} \tag{28}$$

Integrating (27) and rearranging gives

$$W_{slip} = \frac{9\mu^2 P^2}{10G^* a}\left[1 - \theta^{*5} - \frac{5}{6}\left(1 - \theta^{*3}\right)\left(1 + \theta^{*2}\right)\right] \tag{29}$$

Substitution of (28) into (29) leads to (1) confirming that the above procedures are correct.

INCREMENTAL ELASTIC BEHAVIOUR

During loading to T^*, the incremental frictional energy dissipation is given by

$$\Delta W_{slip} = \frac{\mu P}{2}\left(2 - 3\theta_L + \theta_L^3\right)\Delta\delta \tag{30}$$

The incremental elastic energy is obtained by subtracting (30) from the total incremental work done by the tangential force. Thus

$$\Delta W_{elastic} = \left\{T - \frac{\mu P}{2}\left(2 - 3\theta_L + \theta_L^3\right)\right\}\Delta\delta = \frac{3\mu P}{2}\left(\theta_L - \theta_L^3\right)\Delta\delta \tag{31}$$

and the incremental elastic displacement is given by

$$\Delta\delta_{elastic} = \frac{\Delta W_{elastic}}{T} = \frac{3\theta_L\left(1 - \theta_L^2\right)}{2\left(1 - \theta_L^3\right)}\Delta\delta \tag{32}$$

The corresponding tangential force increment is

$$\Delta T = 8G^* a\theta_L \Delta\delta \tag{33}$$

therefore the elastic compliance, using (8) and (9), is found to be

$$\frac{\Delta\delta_{elastic}}{\Delta T} = \frac{3\left(1 - \theta_L^2\right)}{16G^* a\left(1 - \theta_L^3\right)} = \frac{\delta}{T} \tag{34}$$

which shows that,during loading, the elastic stiffness is equal to the current secant stiffness given by the total force-displacement curve.

During unloading from T^* to $-T^*$,the incremental elastic energy is obtained in a similar way using (21) and (22):

$$\Delta W_{elastic} = -\mu P\left[\frac{3}{2}\theta_U\left(1-\theta_U^2\right)+\left(1-\theta_U^3\right)\right]\Delta\delta \tag{35}$$

The incremental elastic displacement is

$$\Delta\delta_{elastic} = \frac{\Delta W_{elastic}}{T-T^*} = \left[\frac{1}{2}+\frac{3\theta_U\left(1+\theta_U\right)}{4\left(1+\theta_U+\theta_U^2\right)}\right]\Delta\delta \tag{36}$$

and the elastic compliance is given by

$$\frac{\Delta\delta_{elastic}}{\Delta T} = \frac{\Delta\delta}{2\Delta T}+\frac{3\left(1-\theta_U^2\right)}{32G^*a\left(1-\theta_U^3\right)} = \frac{1}{2}\left[\frac{\Delta\delta}{\Delta T}+\frac{\delta-\delta^*}{T-T^*}\right] \tag{37}$$

Hence, the elastic compliance during unloading is equal to the average of the tangential and secant compliances given by the total force-displacement curve. This can be shown to be true also during reloading.

From the above, it is possible to construct the elastic force-displacement curves for loading, unloading and reloading. Figure 2 shows that the elastic behaviour exhibits a hysteresis loop. However, it can be shown that, by integrating the elastic work increment over a clomplete unload-reload cycle, the total elastic work done per cycle is zero. Calculating the area enclosed by the loop we find that it is equal to $2T^*(\delta^* - \delta^*_{elastic})$.

CONCLUSIONS

It has been found that, when the contact between two elastic spheres is compressed by a constant normal force and then subjected to the application of a tangential force, the elastic behaviour is non-linear and hysteretic. This means that the elastic behaviour is path dependent due to the way in which the evolution of the inner stick region of the contact area occurs.

It has been shown that, during tangential loading, the elastic stiffness is equal to the secant stiffness of the total force-displacement curve. During unloading and reloading, the elastic compliance is equal to the average of the tangential and secant compliances of the total force-displacement curve. These relationships suggest a possible simple way to incorporate non-linear elasticity into elasto-plastic continuum models for soils.

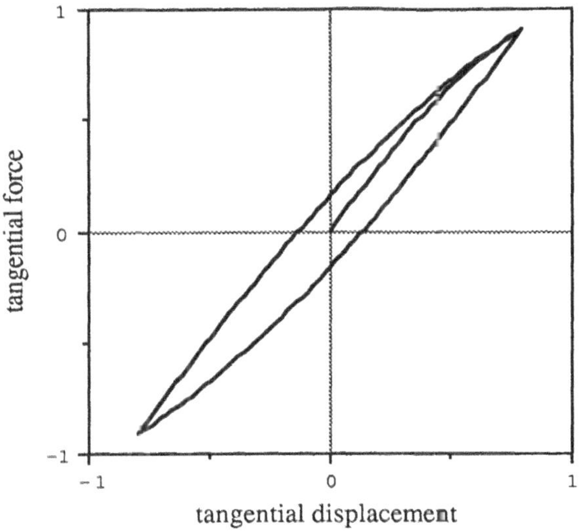

Figure 1. Force displacement curves for an oscillating tangential force. Normalised force ($T/\mu P$) plotted against normalised displacement ($16G^*a\delta/3\mu P$)

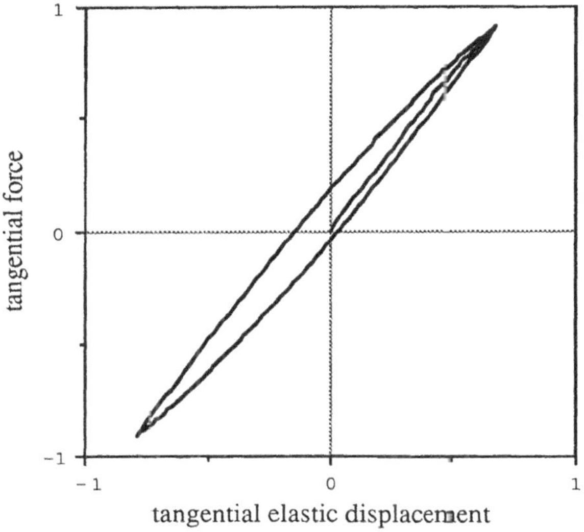

Figure 2. Elastic force displacement curves for an oscillating tangential force. Normalised force ($T/\mu P$) plotted against normalised displacement ($16G^*a\delta/3\mu P$)

There are no absolute checks on contemporary computer simulations of large non-linear, dissipative particle systems. It is, therefore, important to monitor how energy is distributed and dissipated in such systems. Motivated by such a need, expressions have been obtained for both the slip and elastic work increments during oblique interactions between elastic spheres subjected to a constant normal force. Further work, taking into account a varying normal force, will be presented in a future paper.

REFERENCES

Mindlin, R.D., 1949, Compliance of elastic bodies in contact,J. Appl. Mech.,Trans. ASME 71:259.

Mindlin, R.D., Mason, W.P., Osmer, T.F. and Deresiewicz, H., 1951, Effects of an oscillating tangential force on the contact surfaces of elastic spheres, in: Proc. 1st. US Nat. Cong. Appl. Mech., 203.

AN EXPERIMENTAL STUDY OF RIGID BODY IMPACTS

Martin Fandrich and Caroline Hogue

Department of Engineering
University of Cambridge
Cambridge, England

INTRODUCTION

The modelling of rigid body impacts in two and three dimensions is a research area of growing interest. Several models have recently been proposed, some at a microscopic level, where each impact is modelled in detail, and others at a macroscopic level, where individual impacts are modelled approximatively but the overall behaviour of an assembly of rigid bodies is of primary concern. One of the main problems arising in these investigations is the difficulty of validating the models against experimental data. This is primarily due to the complexity of obtaining accurate data for all but the simplest cases of contact experiments.

This paper considers the problem of validating two-dimensional macroscopic models of rigid body collisions by presenting the results of a basic set of experimental impact studies performed on an air table. The cases considered range from a single body hitting a fixed wall with varying frictional characteristics to multiple bodies in simultaneous impact. The data obtained from these experiments include pre and post-impact velocities and the coefficient of restitution as well as material characteristics such as the coefficient of friction.

Since the primary application of this data is to validate impact models, a rigid body model treating collisions as a sequence of instantaneous binary impacts [Hogue 94] is tested as an example.

SCOPE OF EXPERIMENT

The term 'rigid body impact' refers to any collision in which deformation is small and limited to the immediate area of contact without affecting any other part of the body. A rigid body impact can be subclassed as either a direct or oblique impact and as either a central or eccentric impact. A direct impact is one in which the bodies' relative velocity is purely normal to the plane of impact, or local tangent at the point of impact, with no tangential component. In order to consider the general two dimensional case of most use for validating impacts, the experiments performed were oblique

Contact Mechanics, Edited by M. Raous *et al.*
Plenum Press, New York, 1995

impacts and thus had velocity components both parallel and tangential to the plane of impact. Central impacts occur when the centers of mass of the bodies both lie on a line perpendicular to the plane of impact and passing through the point of impact; eccentric impacts occur otherwise. The easiest way to experimentally produce central impacts is to use two circular bodies or a circular body and a flat wall. All such impacts are central, yielding data useful for testing a theoretical model with a simplified impact. Almost all impacts between non-circular objects are oblique, eccentric impacts and the different geometries of impacting pucks used in this research allow both central and eccentric impacts and impacts between corners, curves and planes.

Frictional effects are specifically considered by creating central impacts between circular objects and flat, fixed surfaces and repeating the impact with different surface preparations and different incident angles. This simple contact geometry has the desirable effect of isolating the effects of friction from other effects created by more complicated impacts.

The simultaneous impact of multiple rigid bodies is studied by impacting two touching bodies, initially at rest, with a third body.

EXPERIMENTAL APPARATUS

In order to approximate two dimensional impacts, an air table was used, providing a frictionless base on which pucks, manufactured from a Perspex sheet of constant thickness, may slide [Fandrich 94]. The air table operates by ejecting air through regularly spaced holes in its surface so that the pucks do not touch the surface of the table but are supported at a uniform height by a cushion of air.

The motion of the pucks on the air table was recorded using a strobe light and camera. By setting the camera on extended exposure, images of the pucks were captured at the regular intervals of the strobe flash. Due to the near instantaneous nature of an impact, the actual instant of impact could not be filmed but the motion preceeding and following the impact was captured on a single photograph. A large print of this photograph was developed and computer scanned at high resolution to determine the coordinates of the marked centers of mass and rotation lines of the pucks. Since the time interval between the images was given by the strobe frequency, the linear and angular velocities of the pucks both before and after the impact could be found.

Experimental constants are also required in order to provide all the information necessary to validate a theoretical model; important constants are listed in Table 1. The coefficient of friction μ between two impacting pucks was found by placing the pucks on an inclined plane with their impacting edges in contact and finding the angle of the plane at which initiating sliding between the pucks yielded a constant relative speed; the tangent of the inclination angle is the coefficient of friction [Fandrich 94]. This experiment was repeated over several trials to account for varying conditions and uncertain aspects of friction.

The mass moments of inertia were calculated from the measured dimensions.

RESULTS

The first experimental geometry is that of a circular body hitting a flat wall as shown in Figure 1. This was repeated, as shown in Table 2, at different incident angles and with frictional coefficients μ differing from those given in Table 1. While

Table 1. Impact Puck Data

Puck	Geometry	Major Axis [mm]	Minor Axis [mm]	Area [mm^2]	Mass [g]	Mass Moment of Inertia [$g \cdot mm^2$]	μ
circ	circular	102	102	8 170	94	122 200	0.25
obl	oblong	178	63	10 360	125	327 000	0.25
rect	rectangular	151	75	11 330	136	322 200	0.25

Table 2. Velocity Data for Frictional Study Impacts

#	Pre-Impact Velocities			Post-Impact Velocities			e	μ
	V_{norm} [$\frac{mm}{s}$]	V_{tan} [$\frac{mm}{s}$]	ω [$\frac{rad}{s}$]	V_{norm} [$\frac{mm}{s}$]	V_{tan} [$\frac{mm}{s}$]	ω [$\frac{rad}{s}$]		
1	-1283±12	-187±12	0.07±0.03	551±2	-5±5	8.46±0.03	0.43	0.15±0.01
2	-1169±5	-521±3	0.31±0.03	581±2	-380±5	8.01±0.05	0.50	0.15±0.01
3	-831±20	-769±25	1.50±0.12	356±5	-656±7	7.33±0.09	0.43	0.15±0.01
4	-1154±12	-129±8	-0.79±0.03	608±3	-23±5	4.69±0.05	0.53	0.25±0.04
5	-1085±6	-498±5	-0.23±0.03	541±7	-293±6	8.08±0.05	0.50	0.25±0.04
6	-873±2	-870±30	-1.22±0.10	467±7	-741±22	6.63±0.14	0.53	0.25±0.04
7	-1100±6	-492±3	-1.12±0.03	337±2	-388±3	5.34±0.05	0.31	0.39±0.09
8	-842±8	-843±13	0.07±0.05	280±5	-598±7	11.01±0.07	0.33	0.39±0.09

the standard coefficient of friction was achieved with a milled surface finish, the low coefficient corresponded to a sandpapered finish and the high coefficient was gained by covering the puck edge with masking tape.

The coefficient of restitution e was found from the ratio of the relative normal velocity of the bodies at the point of contact after and before impact, as defined by Newton.

Table 3 gives some geometrical information, at the instant of impact, for the two and three body impacts and must be used in conjunction with Tables 4 and 5 which give the velocities. Most impact models require velocities and displacements in components normal and tangential to the plane of impact, as given in Tables 2, 3 and 4 for one and two body collisions, but for three body impacts the velocities are presented in the fixed XY coordinate system, based on the orientation of the photograph when it was computer scanned, in which the velocities were originally calculated. The angles of the planes of impact, clockwise from the positive X axis, have been given to facilitate the transition between coordinate systems in the manner individually suited to a particular model. A sketch of the geometry of the impact, given the information in Table 3, will make the transition straightforward and will distinguish between the case where the moving body hits both of the other bodies directly and the case where it only hits one of the bodies directly, moving the second body through the first hit.

For the bodies which were initially stationary, the center of mass was read directly from the photograph, but for the pucks which were moving before impact, a different technique was required. The coordinates of the pucks were used to derive the equations of the straight lines traced by the puck before and after impact. Geometrical rules were then used to determine the intersection of these two lines which corresponds with the coordinates of the puck's center of mass at the instant of impact.

For impacts between circular pucks, the point of impact is halfway between the locations of the centers of mass at impact. Impacts involving other bodies were con-

Table 3. Two and Three Body Impact Data for the Instant of Impact

#	Type of Impact	Body	Puck	Center of Mass [mm, mm]	Puck Rotation [deg]	Point of Impact [mm, mm]	Impact Plane [deg]
9	round on round	1	circ	(51±1,0±0)		(0,0)	0.0
		2	circ	(-51±1,0±0)			
10	round on round	1	circ	(51±1,0±0)		(0,0)	0.0
		2	circ	(-51±1,0±0)			
11	round on round	1	circ	(51±1,0±0)		(0,0)	0.0
		2	circ	(-51±1,0±0)			
12	corner to plane	1	rect	(-72±5,33±5)	138.2	(0,0)	0.0
		2	rect	(39±5,-20±5)	0.0		
13	corner to plane	1	rect	(85±3,-6±2)	121.0	(0.0)	0.0
		2	rect	(-38±3,20±2)	0.0		
14	round on round	1	circ	(473±2,514±2)		(423±2,520±2)	83.3
		2	circ	(374±1,526±1)			
	round on round	3	circ	(274±1,560±1)		(324±1,543±1)	71.1
15	round on round	1	circ	(498±1,644±1)		(460±1,611±1)	-49.6
		2	circ	(421±1,578±1)			
	round on round	3	circ	(318±1,598±1)		(370±1,588±1)	78.8
16	round on round	1	circ	(411±1,585±1)			
	on round	2	circ	(314±1,567±1)		(363±1,576±1)	-79.3
		3	circ	(386±1,499±1)		(398±1,542±1)	-16.1
17	round on round	1	circ	(412±1,593±1)			
	on round	2	circ	(317±1,565±1)		(364±1,579±1)	-74.0
		3	circ	(394±1,494±1)		(403±1,543±1)	-10.1
18	round on plane	1	circ	(276±2,663±2)	0.0	(257±2,620±2)	-23.2
		2	rect	(231±1,550±1)	246.8	(171±1,494±1)	
	plane on plane	3	rect	(226±1,431±1)	156.5	(237±1,467±1)	-23.5

sidered individually. The distance from the center of mass to the impacting point of the puck can be found on the last image before impact and added to the location of the center of mass at impact. Rotation of the puck between the time of the last pre-impact image and the instant of impact was also taken into account. The location of the center of mass at impact was used to determine the fraction of a strobe period between the last pre-impact image and the instant of impact.

Table 4. Velocity Data for Two Body Impacts

#	Body	Pre-Impact Velocities			Post-Impact Velocities			e
		V_{norm} $[\frac{mm}{s}]$	V_{tan} $[\frac{mm}{s}]$	ω $[\frac{rad}{s}]$	V_{norm} $[\frac{mm}{s}]$	V_{tan} $[\frac{mm}{s}]$	ω $[\frac{rad}{s}]$	
9	1	-891±2	-884±2	-2.43±0.02	-15±3	-809±3	1.69±0.03	0.96
	2	0	0	0	-874±1	-90±1	4.59±0.05	
10	1	-972±3	-631±1	-2.88±0.03	-12±3	-496±1	2.60±0.03	0.96
	2	0	0	0	-949±4	-148±2	5.81±0.10	
11	1	-972±16	431±4	1.97±0.09	403±18	540±1	6.67±0.12	0.94
	2	451±20	1083±5	1.06±0.10	-930±12	947±1	5.41±0.17	
12	1	1047±8	135±1	1.08±0.05	262±5	196±1	9.91±0.05	0.91
	2	0	0	0	766±8	-58±2	4.52±0.09	
13	1	-888±42	612±10	0.17±0.07	173±20	645±1	5.08±0.14	0.87
	2	367±33	953±5	0.26±0.07	-689±26	902±2	9.44±0.09	

Table 5. Velocity Data for Three Body Impacts

#	Body	Pre-Impact Velocities V_x $[\frac{mm}{s}]$	V_y $[\frac{mm}{s}]$	ω $[\frac{rad}{s}]$	Post-Impact Velocities V_x $[\frac{mm}{s}]$	V_y $[\frac{mm}{s}]$	ω $[\frac{rad}{s}]$
14	1	-741±10	-1204±7	2.20±0.05	-78±4	-1167±7	-1.31±0.05
	2	0	0	0	-188±5	-166±1	-4.47±0.05
	3	0	0	0	-491±2	151±2	0.0±0.3
15	1	-824±3	-1107±13	0.65±0.09	120±5	-186±2	-1.97±0.05
	2	0	0	0	-199±8	-928±6	-7.80±0.14
	3	0	0	0	-763±4	29±5	-4.56±0.10
16	1	-946±10	-1084±5	-0.37±0.03	-67±1	221±3	5.24±0.02
	2	0	0	0	-349±1	-118±1	0.0±0.28
	3	0	0	0	-525±5	-1174±3	4.29±0.12
17	1	-926±20	-970±10	-0.35±0.03	224±6	-79±3	-4.97±0.05
	2	0	0	0	-1066±14	-494±3	-4.92±0.07
	3	0	0	0	-114±4	-400±1	0.68±0.02
18	1	-894±5	-949±3	-0.70±0.03	-234±2	375±3	3.74±0.02
	2	0	0	0	-325±2	-335±2	0.65±0.03
	3	0	0	0	-137±13	-576±2	9.04±0.17

The angle of the plane of impact was similarly found. For impacts with flat sided pucks, the angle was found by measuring the angle of the last image before impact and adding the rotation until impact. In the case of impacts between circular pucks, the plane of impact is defined as being perpendicular to the line joining the centers of mass at impact.

Since impact # 18 contains a planar impact as shown in Figure 2, Table 3 lists two points of impact for the second plane of impact. These correspond to the corners of the contacting plane and may be dealt with in different ways by different models.

No coefficient of restitution is given in Table 5 because there is no accepted method of defining the coefficient for simultaneous multiple body impacts. It is assumed that enough information has been given to determine this coefficient according to the specific definition used by a particular theory.

UNCERTAINTY ANALYSIS

In order to check the accuracy of the method of data analysis, the procedure of scanning, reading and analyzing a print was performed on a print for which the velocities were pre-determined. This print consisted of the graphical output of a computer impact simulation [Hogue 94], printed in a form similar to the experimental photographs. Although the velocities obtained through this analysis were very close to the known velocities, differing by less than 1% in most cases, there was a large variability in the readings taken to compose the average. Since the variability is considered in the uncertainty parameter of the velocity, the velocity data from the experimental prints can be taken to be accurate within this uncertainty, which is determined from both the variability of the velocities determined at different time steps, taken to compose the average velocity given, and the uncertainty of the individual measurements.

Another method used to check the accuracy of the data was to test the data against the laws of conservation of linear and angular momentum, relations independent of any impact theory. The data were found to correspond within 2%.

VALIDATION OF A THEORETICAL MODEL

Several of the impact cases were used to test a computer-based impact simulation [Hogue 94]. The data was useful in debugging the computer code as well as in checking the validity of the theory. Due to the uncertainties of the data, it is expected that during the verification of any theoretical work the analysis will be repeated, varying the coefficient of friction and other values within their uncertainties. In the validation of the computer model the theoretical output was found to closely match the experimental data for very low values of the coefficient of friction, up to half the value determined experimentally and well out of the range of certainty listed. This could be attributed to a non-Coulomb friction behavior occurring during impact or to a dynamic coefficient of friction, determined as described, which is not representative of friction during impact.

DISCUSSION

The impact situation of impacts # 16 and # 17 was repeated 47 times with roughly the same initial conditions, serving as a sensitivity analysis. The resulting impacts were remarkably similar; without detailed analysis of each photo, the post-impact trajectories of each case were qualitatively the same and followed the pattern shown in Figures 3 and 4. Some impacts were expected to show the moving body reversing direction after impact, indicating that the other two bodies were hit simultaneously. Other trials were expected to show the puck with a post-impact trajectory angled greatly to one side or the other, indicating that one puck was hit slightly before the other. Experimentally, all the results indicated the sequential collisions with no simultaneous collisions shown. The collisions were evenly distributed with regard to the side of rebound indicating that the initial trajectories were distributed about a mean that should have created a simultaneous impact.

CONCLUSIONS

Although the data obtained from the experiments has been found to be accurate, each further calculation increases the uncertainty of the results. Therefore, to accurately predict the outcome of an impact, a model should be as computationaly simple as possible.

The most common friction law used in the modelling of impacts is that developed by Coulomb. This approach has been found to be valid during impact [Routh 77] but the usefulness of using a coefficient experimentally determined under relatively light normal loads, unlike the actual impact situation, should be investigated further.

The production of impacts in which three independently moving objects meet at exactly the same moment in time takes great precision and care and is beyond the scope of this experiment. Because this type of impact didn't occur despite a number of simple attempts to cause it, it is possible that such a simultaneous multiple body impact occurs so rarely that it is only worth considering in extreme cases.

The experimental data has been tested at its primary purpose of validating theoretical models and found to be useful. It had been presented in a form intended to be useful regardless of the model being validated.

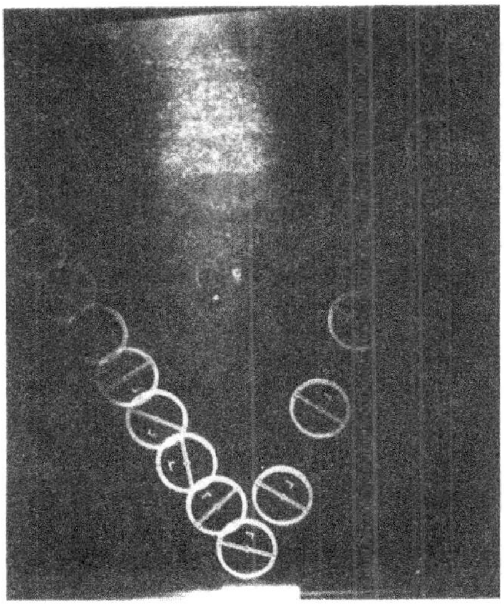

Figure 1. The photograph of impact # 2 - A typical circular body on wall impact. The circular puck starts at the top right and rebounds from the wall at the bottom of the picture.

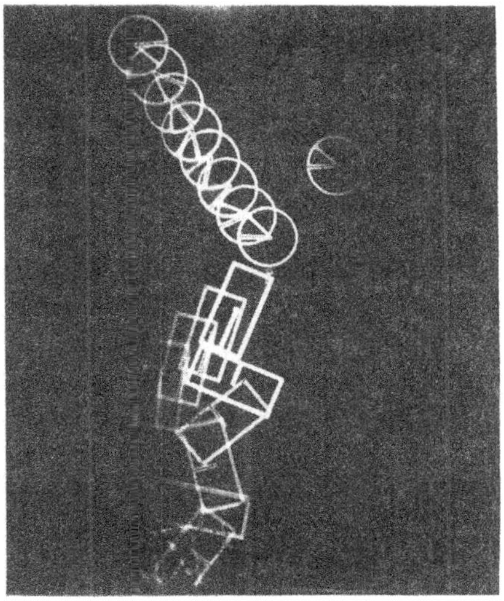

Figure 2. The photograph of impact # 18 - A circular body on rectangular body on rectangular body simultaneous impact. The circular puck starts at the top right and the initial positions of the rectangular pucks can be seen from their thicker outlines.

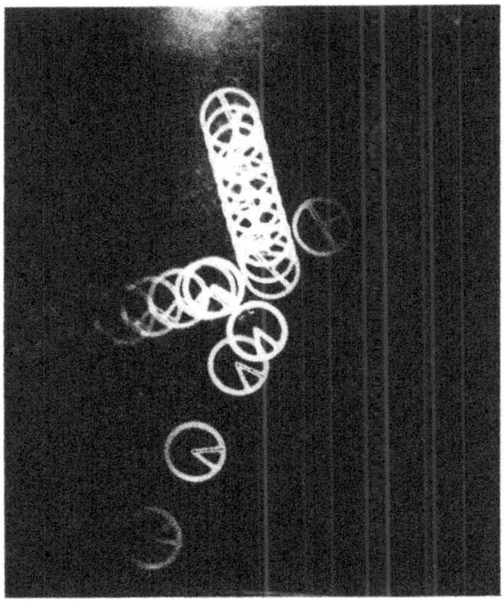

Figure 3. The photograph of impact # 16 - A rough attempt to impact three bodies simultaneously

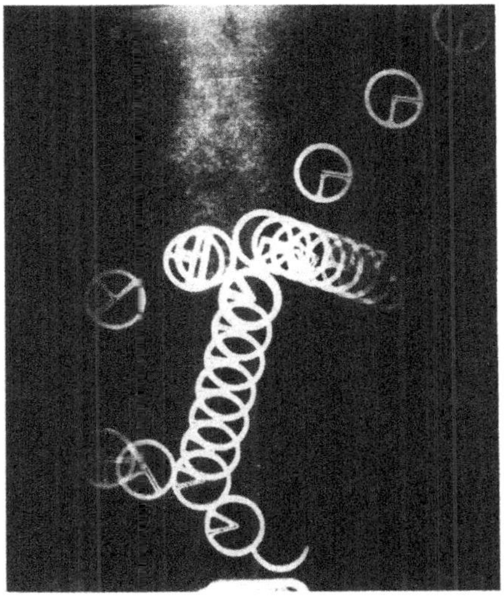

Figure 4. The photograph of impact # 17 - A rough attempt to impact three bodies simultaneously

REFERENCES

[Fandrich 94] M. E. Fandrich. Impact dynamics of rigid bodies. Technical report, University of Cambridge, 1994.

[Hogue 94] C. Hogue and D. Newland. Efficient computer-simulation of moving granular particles. *Powder Technology*, 78(1):51–66, 1994.

[Routh 77] E. J. Routh. *An Elementary Treatise on the Dynamics of a System of Rigid Bodies.* MacMillan and Co., London, 3rd edition, 1877.

COLLISIONS OF RIGID BODIES

Michel Frémond

Laboratoire des Matériaux et des Structures du Génie Civil
Unité Mixte CNRS-LCPC, UMR 113
2, allée Kepler, 77420 Champs (France)

ABSTRACT

Consider a point moving with respect to a rigid body. The system made up of these two elements is deformable since the distance of the point and the body changes. Because the system is deformable, we define strain rates and interior forces. The latter are percussions and forces defined by their virtual work. The equations of motion are derived from the principle of virtual work. The constitutive laws for the interior percussions and forces are the other equations which describe the evolution of the system in particular the collisions which occur between the point and the body. Examples illustrate the ability of the theory to analyse practical situations.

THE SYSTEM. ITS VELOCITIES AND STRAIN RATES

Let there be a system composed of a point and a rigid body. This system is deformable because the distance d_1 of the point $\mathbf{x}_1$ to the body Ω may change:

$$d_1(t) = (\mathbf{x}_1(t) - \text{proj}\mathbf{x}_1(t)).\mathbf{N}(\text{proj}\mathbf{x}_1(t)),$$

where $\mathbf{N}$ is the outward normal vector to Ω at the projection $\text{proj}\mathbf{x}_1$ of $\mathbf{x}_1$ on the boundary $\partial\Omega$ of Ω (figure 1). Let us note that the point does not interpenatrate the solid if and only if

$$\forall t,\ d_1(t) \geq 0. \tag{1}$$

Contact Mechanics, Edited by M. Raous *et al.*
Plenum Press, New York, 1995

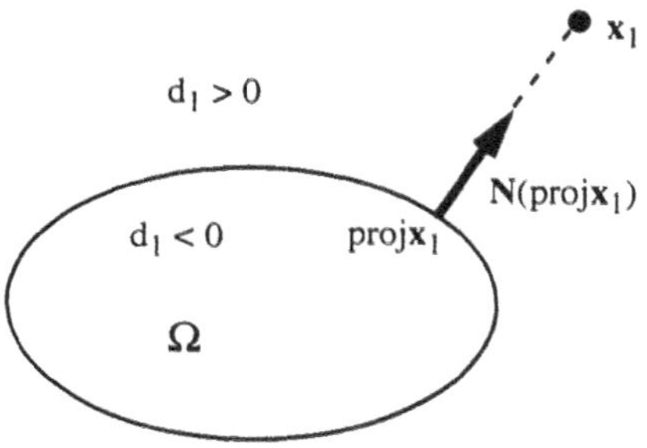

Figure 1. The point $\mathbf{x}_l(t)$ and the solid $\Omega(t)$ move and collide. The outwards normal to $\Omega(t)$ is $\mathbf{N}$, proj$\mathbf{x}_l$ is the projection of $\mathbf{x}_l$ on the boundary $\partial\Omega$ of Ω.

Let $\mathbb{V}_1$ and $\mathbb{V}_2$ be the linear spaces of the virtual velocities of the point (vectors which depend on t) and the rigid body (vectors which depend on t and $\mathbf{x}\in\Omega(t)$) ; let $\mathbb{R}$ be the linear subset of $\mathbb{V} = \mathbb{V}_1 \mathrm{x}\ \mathbb{V}_2$ of the rigid system velocities ($\mathbf{V}_l(t) = \mathbf{W}(\mathbf{x}_l(t),t)$, $\mathbf{V}_2(\mathbf{x},t) = \mathbf{W}(\mathbf{x},t)$; $\mathbf{W}(\mathbf{x},t) = \mathbf{V}_0(t)+\omega(t)\wedge(\mathbf{x}-\mathbf{x}_0)$), defined by the same twist, $\{\mathbf{V}_0(t),\omega(t)\}$, because in such a movement the distances of the points of the system remain constant). Let $\mathbb{D} = \mathbb{V}/\mathbb{R}$, be the linear space of the velocities defined up to a rigid system velocity: it is also the space of the velocities of the point with respect to the rigid body. The strain rate of the deformable system is defined by

$$\mathbf{V} = (\mathbf{V}_l,\mathbf{V}_2)\in\mathbb{V} \rightarrow \mathbf{D}(\mathbf{V})\in\mathbb{D} \text{ with } \mathbf{D}(\mathbf{V})(t) = \mathbf{V}_l(t)-\mathbf{V}_2(\mathbf{x}_l(t),t).$$

Due to collisions, the velocities can be discontinuous. The indices – and + refer to states before and after collisions.

THE INTERIOR FORCES. THE EQUATIONS OF MOTION

Interior forces are defined by their work. They are forces and percussions. The latter are forces concentrated in time. The virtual work of the acceleration forces is

$$\mathcal{T}_a(\mathbf{V},t_1,t_2) = \int_{t_1}^{t_2} \Big\{ m\frac{d\mathbf{U}_l}{dt}(t).\mathbf{V}_l(t) + \int_{\Omega(t)} \rho_2 \frac{d\mathbf{U}_2}{dt}(\mathbf{x},t).\mathbf{V}_2(\mathbf{x},t)d\Omega \Big\} dt$$

$$+ \sum_{t\in Z(\mathbf{V},\mathbf{U},t_1,t_2)} \Big\{ m[\mathbf{U}_l](t).\frac{\mathbf{V}_l^+(t)+\mathbf{V}_l^-(t)}{2} + \int_{\Omega(t)} \rho_2[\mathbf{U}_2](\mathbf{x},t).\frac{\mathbf{V}_2^+(t)+\mathbf{V}_2^-(t)}{2} d\Omega) \Big\}$$

where $\mathbf{U} = (\mathbf{U}_l,\mathbf{U}_2)$ are the actual velocities, $[\mathbf{U}_l](t) = \mathbf{U}_l^+(t)-\mathbf{U}_l^-(t)$ and $Z(\mathbf{V},\mathbf{U},t_1,t_2)$ is the set of the instants of the time interval $[t_1,t_2]$ where the actual or virtual velocities are discontinuous; m is the mass of the point and ρ the density of the rigid body. This expression is meaningful if the velocities are bounded variation functions (Moreau, 1988). The virtual work of the acceleration forces is such that the actual work is the variation of the kinetic energy between the times t_1 and t_2.

The virtual work of the interior forces $\mathcal{T}_{int}$ is a linear function on $\mathbb{D}$ which is zero for any rigid system velocity . The expression we choose is

$$\mathcal{T}_{int}(\mathbf{V},t_1,t_2) = -\int_{t_1}^{t_2} \mathbf{R}^{int}(t).\mathbf{D}(\mathbf{V})(t)dt - \sum_{t\in Z(\mathbf{V},\mathbf{U},t_1,t_2)} \left\{\mathbf{P}^{int}(t).\frac{\mathbf{D}(\mathbf{V}^+)(t)+\mathbf{D}(\mathbf{V}^-)(t)}{2}\right\},$$

where $\mathbf{R}^{int}(t)$ is a force and $\mathbf{P}^{int}(t)$ a percussion. It is obvious that $\mathcal{T}_{int}(\mathbf{V},t_1,t_2) = 0$ for $\mathbf{V}\in\mathbb{R}$.

The virtual work of the exterior forces $\mathcal{T}_{ext}$ is a linear function on $\mathbb{V}$. We choose,

$$\mathcal{T}_{ext}(\mathbf{V},t_1,t_2) = \int_{t_1}^{t_2}\left\{\mathbf{f}_1(t).\mathbf{V}_1(t) + \int_{\partial\Omega(t)}\mathbf{T}_2(x,t).\mathbf{V}_2(\mathbf{x},t)d\Gamma\right\}dt$$

$$+ \sum_{t\in Z(\mathbf{V},\mathbf{U},t_1,t_2)}\left\{\mathbf{P}_1^{ext}(t).\frac{\mathbf{V}_1^+(t)+\mathbf{V}_1^-(t)}{2} + \int_{\partial\Omega(t)}\mathbf{P}_2^{ext}(\mathbf{x},t).\frac{\mathbf{V}_2^+(t)+\mathbf{V}_2^-(t)}{2}d\Omega\right\},$$

where $\mathbf{P}_1^{ext}(t)$ is the exterior percussion applied to the point and $\mathbf{P}_2^{ext}(\mathbf{x},t)$ the surfacic density of exterior percussion applied to the solid, $\mathbf{f}_1(t)$ is the exterior force applied to the point and $\mathbf{T}_2(\mathbf{x},t)$ the exterior surfacic force applied to the solid.

The equations of motion result from the principle of virtual work. The virtual work of the acceleration forces between times t_1 and t_2 is equal to the sum of the virtual works of the interior and exterior forces between the same instants:

$$\forall t_1,\ \forall t_2,\ \forall \mathbf{V}\in\mathbb{V},\quad \mathcal{T}_a(\mathbf{V},t_1,t_2) = \mathcal{T}_{ext}(\mathbf{V},t_1,t_2) + \mathcal{T}_{int}(\mathbf{V},t_1,t_2).$$

The equations of motion of the point are: at any time t, in particular when there is a collision,

$$m[\mathbf{U}_1](t) = -\mathbf{P}^{int}(t) + \mathbf{P}_1^{ext}(t); \tag{2}$$

and almost everywhere,

$$m\frac{d\mathbf{U}_1}{dt} = -\mathbf{R}^{int} + \mathbf{f}_1. \tag{3}$$

The equations of motion of the rigid body are equations satisfied at any time t involving $\mathbf{P}^{int}$ and equations satisfied almost everywhere involving $\mathbf{R}^{int}$.

A MATHEMATICAL RESULT

Let the movements of a point and a solid such that they do not interpenetrate. Let Δt be a positive time increment. Let I be the indicator function of $\mathbf{R}^+$ ($I(x) = 0$, if $x \geq 0$; $I(x) = +\infty$, if $x < 0$). We have

$$I(d_1(t-\Delta t)) \geq I(d_1(t)) + (d_1(t-\Delta t) - d_1(t))A(t), \text{ for any } A(t) \in \partial I(d_1(t)),$$
$$I(d_1(t-\Delta t)) \geq I(d_1(t+\Delta t)) + (d_1(t-\Delta t) - d_1(t+\Delta t))B(t+\Delta t), \text{ for any } B(t+\Delta t) \in \partial I(d_1(t+\Delta t)),$$

with $\partial I(x) = \{0\}$ if $x > 0$, $\partial I(0) = \mathbf{R}^-$, $\partial I(x) = \varnothing$ if $x < 0$. By dividing the relations by $\Delta t > 0$ and letting Δt tend to 0, we get

$$A(t)\frac{d^-d_1}{dt} \geq 0, \text{ for any } A(t) \in \partial I(d_1(t)), \tag{4}$$
$$C(t)\left\{\frac{d^-d_1}{dt} + \frac{d^+d_1}{dt}\right\} \geq 0, \text{ for any } C(t) \in \partial I_+(d,t), \tag{5}$$

where $\frac{d^+d_1}{dt} = \lim_{\Delta t \to 0,\ \Delta t > 0} \frac{d_1(t+\Delta t) - d_1(t)}{\Delta t}$ is the right derivative,

$\frac{d^-d_1}{dt} = \lim_{\Delta t \to 0,\ \Delta t > 0} \frac{d_1(t) - d_1(t-\Delta t)}{\Delta t}$ is the left derivative;

d is the function : $t \to d_1(t)$ and $\partial I_+(d,t) = \lim_{\Delta t \to 0,\ \Delta t > 0} \{\partial I(d_1(t+\Delta t))\}$ is the right limit of $\partial I(d_1(t))$. This set is $\mathbf{R}^-$ if $d_1(\tau)$ remains equal to 0 in some time interval after the time t and $\{0\}$ if $d_1(\tau)$ becomes strictly positive after the time t. The function C(t) is negative if contact is maintained after the time t and zero if contact is not maintained afterwards. It is posible to prove that

$$\frac{dd_1}{dt} = \mathbf{D}(\mathbf{U}).\mathbf{N}(\text{proj}\mathbf{x}_1).$$

It results from (4) and (5) that: at any time t, in particular when there is a collision,

$$C(t)\{\mathbf{D}(\mathbf{U}^+) + \mathbf{D}(\mathbf{U}^-)\}.\mathbf{N}(\text{proj}\mathbf{x}_1(t)) \geq 0, \text{ for any } C(t) \in \partial I_+(d,t), \tag{6}$$

and almost everywhere, $A(t)\mathbf{D}(\mathbf{U}).\mathbf{N}(\text{proj}\mathbf{x}_1(t)) \geq 0$, for any $A(t) \in \partial I(d_1(t))$. (7)

THE CONSTITUTIVE LAWS

For the sake of simplicity, we assume the temperature to be constant. Thus the second principle of thermodynamics imposes the interior forces $\mathbf{P}^{int}$ and $\mathbf{R}^{int}$ to satisfy:

at any time t, in particular when there is a collision,

$$-[\Psi(t)] + \mathbf{P}^{int}(t).\frac{\mathbf{D}(\mathbf{U}^+)(t) + \mathbf{D}(\mathbf{U}^-)(t)}{2} \geq 0, \tag{8}$$

and almost everywhere, $-\frac{d\Psi}{dt} + \mathbf{R}^{int}.\mathbf{D}(\mathbf{U}) \geq 0,$ (9)

for any actual evolution, where Ψ is the free energy of the system assumed to depend on the distance d_1. The interior forces $\mathbf{R}^{int}$ and $\mathbf{P}^{int}$ are defined in the following way: we let

$$\mathbf{R}^{int}(t) = \{R^r(t)+R^{nd}(d_1(t))\}\mathbf{N}(\text{proj}\mathbf{x}_1(t)+\mathbf{R}^d(d_1(t),\mathbf{D}(\mathbf{U}(t)), \quad (10)$$

$$\mathbf{P}^{int}(t) = P^r(d,t)\mathbf{N}(\text{proj}\mathbf{x}_1(t))+\mathbf{P}^d(d_1(t),\mathbf{D}(\mathbf{U}^+(t),\mathbf{D}(\mathbf{U}^-(t)). \quad (11)$$

The interior force is the sum of the non-interpenetration reaction force $R^r\mathbf{N}$, of the non dissipative force $R^{nd}\mathbf{N}$ and of the dissipative force $\mathbf{R}^d$. The interior percussion is the sum of the non-interpenetration reaction percussion $P^r\mathbf{N}$ and of the dissipative percussion $\mathbf{P}^d$.

The Non-Interpenetration Reactions

They are defined by

$$R^r(t) \in \partial I(d_1(t)),\ P^r(d,t) \in \partial I_+(d,t). \quad (12)$$

These relations have two meanings: on the one hand that the internal constraint (1) is satisfied or that there is no interpenetration because the subdifferentials ∂I are not empty; on the other hand they give the values of the reactions R^r and P^r. For instance, in a collision the non-interpenetration percussion P^r is negative if contact is maintained in the following instants and zero if it is not.

The Dissipative Force and Impulse

They are functions which are assumed to satisfy

$$\forall d_1, \forall \mathbf{D} \in \mathbb{D},\ \mathbf{P}^d(d_1,\mathbf{D}^+,\mathbf{D}^-).\frac{\mathbf{D}^+ + \mathbf{D}^-}{2} \geq 0,\ \ \mathbf{R}^d(d_1,\mathbf{D}).\mathbf{D} \geq 0. \quad (13)$$

The dissipative impulse $\mathbf{P}^d$ and force $\mathbf{R}^d$ take into account the dissipations which result from the surface and volume characteristics of the body and the point. It is easy to prove that the properties (6), (7), the definitions (10), (11), (12) and the assumption (13) insure that the Clausius-Duhem inequalities (8) and (9) are satisfied (Frémond, 1990).

From now on we focus on collisions and no longer deal with smooth evolutions of the system. Let us say briefly that the dissipative force $\mathbf{R}^d$ describes friction when the point slides on the solid and that the non-dissipative force $R^{nd}(d_1(t)) = \frac{\partial \Psi}{\partial d_1}(d_1(t))$ describes actions at a distance between the point and the solid. As an example, one can think of an elastic string linking the point and the solid. We assume that there are no actions at a distance in collisions, i.e. $\mathbf{P}^d = 0$ for $d_1 > 0$.

EXAMPLES OF DISSIPATIVE PERCUSSIONS

First Example

Let us assume that the exterior actions applied to the rigid body are such that in a collision its velocity is continuous. We denote $\mathbf{W} = \mathbf{D}(\mathbf{U}) = \mathbf{U}_1 - \mathbf{U}_2(\mathbf{x}_1)$ the relative velocity of the point with respect to the solid and $W_N = \mathbf{W}.\mathbf{N}$. We choose

$$\mathbf{P}^d(\mathbf{D}(\mathbf{U}^+),\mathbf{D}(\mathbf{U}^-)) = k(W_N^+ + W_N^-)\mathbf{N}, \text{ with } k > 0. \quad (14)$$

The inequality (13) is satisfied. Let us compute the velocity after a collision without exterior percussion ($\mathbf{P}^{ext} = 0$):

if the point is not heavy ($m \le k$), equations (2) and (14) give,

$$W_N^+ = \frac{m-k}{m+k} W_N^-, \text{ and } \mathbf{W}_T^+ = \mathbf{W}_T^-,$$

(the subscript $_T$ denotes the tangential component).

The point bounces with a velocity slightly lower than the incident velocity (we retrieve a classical relation (Jean et al., 1991)). The non-interpenetration percussion of reaction is zero; if the point is heavy ($m \ge k$), the equations give $W_N^+ = 0$: the contact persists after the collision. The non-interpenetration percussion of reaction is not zero: $P^r = (m-k)W_N^- < 0$. One can separate the point and the solid abruptly by applying a normal percussion P_N^{ext} large enough to have

$$W_N^+ = \frac{m-k}{m+k} W_N^- + \frac{P_N^{ext}}{m+k} > 0.$$

Second Example

Let us choose the following function which satisfies the inequality (13),

$$\mathbf{P}^d(\mathbf{D}(\mathbf{U}^+),\mathbf{D}(\mathbf{U}^-)) = -pn\{c+k(W_N^+ + W_N^-)\}\{|c+k(W_N^+ + W_N^-)|\}^{p-2}\mathbf{N},$$

with $pn\{x\} = \sup\{0,-x\}$ and $p > 1$, $c \ge 0$, $k > 0$.

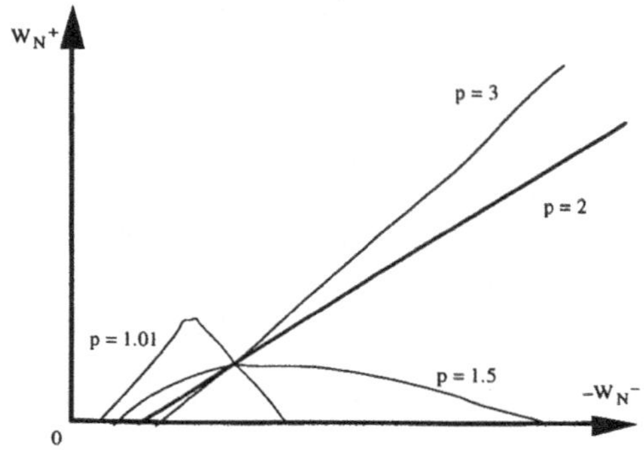

Figure 2. The normal velocity W_N^+ after the collision versus the normal velocity before $-W_N^-$ for different values of p (Cholet, 1994).

Figure 2 shows W_N^+ versus $-W_N^-$ for different values of p and $m \leq k$. One can note that for some values of p, the point bounces only if the incident velocity is not too large or too low. This phenomenon occurs in some practical situations. The figure 2 curve can be obtained from experiments. Thus the normal dissipative reaction can be obtained from experiments. Many other functions $\mathbf{P}^d$ can be used to fit with the experimental curves.

COLLISION WITH AN EXTERIOR PERCUSSION

One can investigate a collision of a light point ($m \leq k$) concomitant with an exterior positive normal percussion P_N^{ext}. The results of figure 3 describe what occurs in a pin-ball machine where an electrical percussion is applied whenever a ball hits an obstacle (Moreau, 1988). If the ball assumed to be a point arrives slowly the future movement depends only on the electrical action and if it arrives quickly it depends also on the mechanical properties.

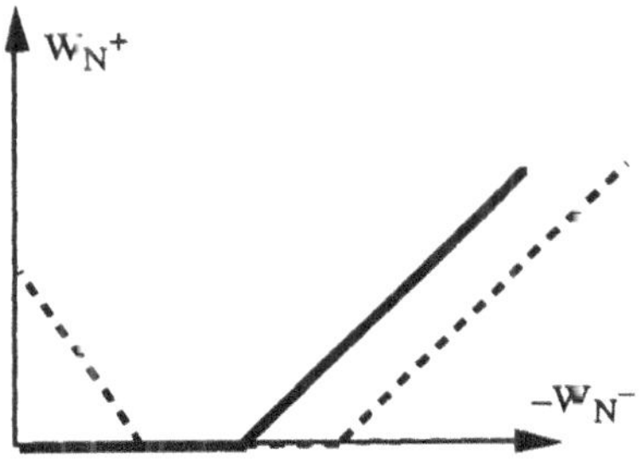

Figure 3. The velocity W_N^+ after the collision versus the velocity before $-W_N^-$ with (dotted line) and without (bold line) an exterior positive normal percussion (p = 2).

EVOLUTION OF TWO RIGID BODIES. MULTIPLE COLLISIONS

Let us consider the movement of an irregular wheel evolving over a plane (figure 4). The points A and B can collide the plane at the same time : it is a multiple collision.

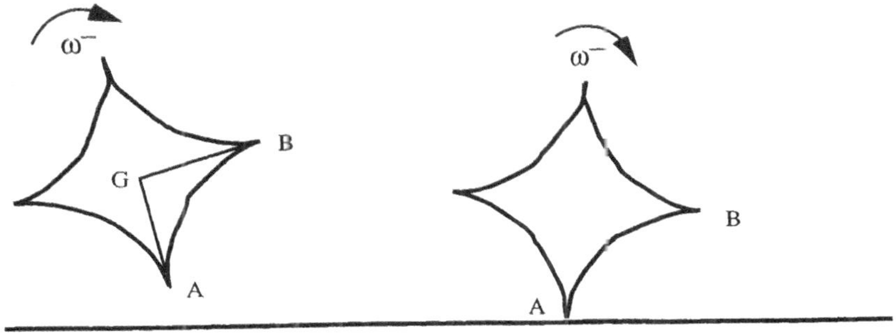

Figure 4. The irregular wheel moves above the plane. We investigate the movement of the wheel rotating around the point A which is in contact with the plane.

Let us investigate the movement of the wheel rotating with velocity ω^- around point A in contact with the plane. We assume the constitutive laws of the collisions occuring at the points A and B to be those of the first example. We have

$$m[\mathbf{U}_G] = -\mathbf{P}(\mathbf{U}_A{}^+ + \mathbf{U}_A{}^-) - \mathbf{P}(\mathbf{U}_B{}^+ + \mathbf{U}_B{}^-), \tag{15}$$

$$I[\omega] = -\mathbf{GA}\Lambda\mathbf{P}(\mathbf{U}_A{}^+ + \mathbf{U}_A{}^-) - \mathbf{GB}\Lambda\mathbf{P}(\mathbf{U}_B{}^+ + \mathbf{U}_B{}^-), \tag{16}$$

with $\mathbf{U}_A = \mathbf{U}_G + \omega\Lambda\mathbf{GA}$, $\mathbf{U}_B = \mathbf{U}_G + \omega\Lambda\mathbf{GB}$, where $\mathbf{U}_G$ is the velocity of the centre of mass G , ω is the rotation vector, m is the mass and I the mass moment of inertia at point G. Equations (15) and (16), the constitutive law (14) give a unique $(\mathbf{U}_G{}^+,\omega^+)$ depending on $(\mathbf{U}_G{}^-,\omega^-)$ solution of

$$\mathsf{A}(\mathbf{U}_G{}^+,\omega^+) = \begin{vmatrix} m\mathbf{U}_G{}^- \\ I\omega^- \end{vmatrix},$$ where the operator A is defined by

$$\begin{vmatrix} \mathbf{V}_G \\ \theta \end{vmatrix} \rightarrow \begin{vmatrix} m\mathbf{V}_G + \mathbf{P}(\mathbf{V}_G + \theta\Lambda\mathbf{GA} + \mathbf{U}_A{}^-) + \mathbf{P}(\mathbf{V}_G + \theta\Lambda\mathbf{GB} + \mathbf{U}_B{}^-) \\ I\theta + \mathbf{GA}\Lambda\mathbf{P}(\mathbf{V}_G + \theta\Lambda\mathbf{GA} + \mathbf{U}_A{}^-) + \mathbf{GB}\Lambda\mathbf{P}(\mathbf{V}_G + \theta\Lambda\mathbf{GB} + \mathbf{U}_B{}^-) \end{vmatrix} = \mathsf{A}(\mathbf{V}_G,\theta).$$

The solution of equations (15) and (16) is unique because the operator A is strictly monotone. This monotonicity property is general in the setting we have chosen . It ensures coherence in terms of mechanics and mathematics. The different possible evolutions after the multiple collision are shown in figure 5.

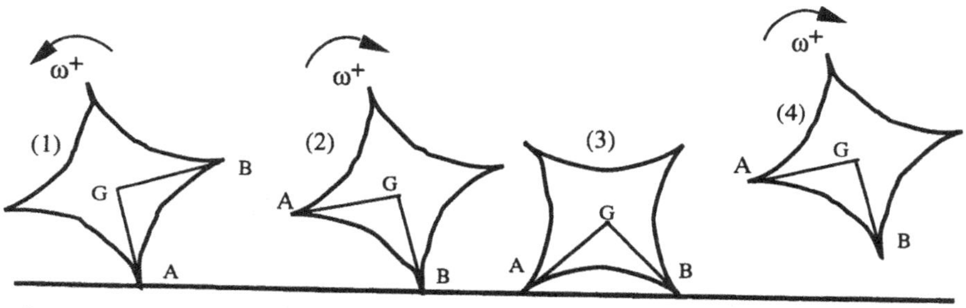

Figure 5. The possible evolutions (depending on the values of k, m and I) of the irregular wheel rotating around point A before the collision of point B with the plane: (1) point A remains fixed, point B bounces; (2) point A moves, point B stops; (3) points A and B stop; (4) points A and B move.

REFERENCES

Cholet, C., Chocs de solides. Mémoire de DEA., Université P. et M. Curie, 1994.

Frémond, M., Sur l'inégalité de Clausius-Duhem, C.R.Acad.Sci.Paris, 311, Série II, 1990, p.757-762.

Moreau, J.J., Bounded variation in time in Topics in non smooth mechanics, J.J.Moreau, P.D.Panagiotopoulos, G.Strang éditeurs, Birkhauser, Basel, 1988, Chapitre I, p.1-71.

Moreau, J.J., Unilateral contact and dry friction in finite freedom dynamics, in Non smooth mechanics and applications, J.J.Moreau, P.D.Panagiotopoulos éditeurs, Springer,Wien, 1988, p.1-81.

Jean, M., J.J.Moreau, J.J., Dynamics of elastic or rigid bodies with frictional contact : numerical methods. Mécanique, modélisation et dynamique des matériaux, Publication du Laboratoire de Mécanique et d'Acoustique, 124, R.H. Blanc, M. Raous, P. Suquet, Editeurs, Marseille 1991, p.9-29.

SEISMIC RESPONSE OF STRUCTURES WITH UNILATERAL CONTACT CONDITIONS AT THE FOUNDATION

E.N. Mitsopoulou, I.N. Doudoumis and P.A. Zervas

Department of Civil Engineering,
Aristotle University of Thessaloniki, Greece

INTRODUCTION

In the paper the dynamic response of framed structures with surface foundation supported on soil which is capable of supporting only compressive stresses (unilateral frictionless contact conditions) is studied. The structures are supposed to be elastic with infinitesimal displacements and strains.

A method is presented for the numerical solution of this problem. Two or three dimensional structures, discrete or discretized by the finite element method are solved using also a temporal discretization. For each time step an elimination of the internal degrees of freedom gives rise to a Linear Complementarity Problem (L.C.P.) on the boundary for only a small number of unknowns, which are the unknown unilateral displacements of the time step under consideration.

A parametric numerical analysis of 2D and 3D structures under seismic excitations is made with unilateral and bilateral contact conditions. Some characteristic results of this analysis are presented in the paper.

FORMULATION OF THE PROBLEM

We consider first the system of the bodies Ω_o for which only the kinematical constraints on Γ_u hold. The dynamic equations of equilibrium of this system at any time are written as:

$$\boldsymbol{M}\ddot{\boldsymbol{u}} + \boldsymbol{C}\dot{\boldsymbol{u}} + \boldsymbol{K}\boldsymbol{u} = \boldsymbol{p}(t) \qquad (1)$$

In (1) $\boldsymbol{K}$ is the $6n\times 6n$ stiffness matrix, $\boldsymbol{M}$ is the $6n\times 6n$ positive definite mass matrix, $\boldsymbol{C}$ is the $6n\times 6n$ damping matrix, $\boldsymbol{p}(t)$ is the $6n$ vector of the nodal forces, and $\boldsymbol{u}$, $\dot{\boldsymbol{u}}$, $\ddot{\boldsymbol{u}}$ the $6n$ vectors of the nodal displacements, velocities and accelerations

As long as contact-impact does not take place at any boundary point, the motion is fully described by the linear differential equations (1) and by the "initial"

Contact Mechanics, Edited by M. Raous *et al.*
Plenum Press, New York, 1995

conditions at time $t=t_A$: $u=u_A$ and $\dot{u}=\dot{u}_A$. The "initial" conditions are introduced at time $t=0$ and at any time step t_A for which discontinuity of the velocities occurs due to contact-impact.

For the solution of the problem also an appropriate time discretization algorithm is used. Here the simple one-step algorithm of implicit type proposed by Zienkiewicz, Wood and Taylor (1980) is used. On the basis of this algorithm the differential equations of motion (1) of the system Ω_0 are converted to a set of algebraic equations which for the discrete time interval $(t-\Delta t)\div t$ takes the form:

$$\overline{\boldsymbol{K}} \cdot \boldsymbol{u}_t = \overline{\boldsymbol{P}}_t \tag{2}$$

After the above discretization, the unilateral contact conditions at the boundaries Γ_S are introduced. At the m pairs of the discrete contact points, fictitious semi-rigid unilateral bonds with infinitesimal size are introduced, which can carry only compressive stresses. Each bond i has a direction normal to the contact surface and connects the adjacent nodes k and l. Denoting by:

s_i the stress of the bond i (normal contact reaction at the pair i),

e_i the imposed strain of the bond i corresponding to s_i (relative displacements of the node pair k and l), and

h_i the initial gap between the node pair i,

the following relations hold:

$$-s_i \geq 0, \qquad \varepsilon_i = e_i + h_i = \boldsymbol{G}_i^T\, \boldsymbol{u} + h_i \geq 0, \qquad -s_i \cdot \varepsilon_i = 0. \tag{3}$$

where

$\boldsymbol{G}_i^T = [\boldsymbol{o} \ldots \boldsymbol{g}_{ik} \ldots \boldsymbol{g}_{il} \ldots \boldsymbol{o}]$

is the $1\times n$ strain-displacement matrix of the bond i connecting nodes k and l,

$-\boldsymbol{g}_{ik} = \boldsymbol{g}_{il} = \boldsymbol{g}_i = [n_{ix}, n_{iy}, n_{iz}, 0, 0, 0]$

and $n_i=[n_{ix}, n_{iy}, n_{iz}]$ is the unit vector of the direction cosines which is normal to the bodies A and B at the location of the bond i. For the total number of the m unilateral bonds, at the time step t, the following matrix relations can be written:

$$\boldsymbol{\varepsilon} = \boldsymbol{G}^T\boldsymbol{u} + \boldsymbol{h} \geq \boldsymbol{0}, \qquad -\boldsymbol{s} \geq \boldsymbol{0}, \qquad \boldsymbol{s}^T \cdot \boldsymbol{\varepsilon} = 0 \tag{4}$$

using the vectors: $\mathbf{s}_t = [\, s_{1t} \,..\, s_{mt} \,]$, $e_t = [\, e_{1t} \,..\, e_{mt} \,]$, $h_t = [\, h_{1t} \,..\, h_{mt} \,]$, and $\boldsymbol{G}^T = [\, \boldsymbol{G}_1^T \;\; \boldsymbol{G}_2^T \; . \; . \; . \; \boldsymbol{G}_m^T \,]$.

The unilateral contact kinematic conditions $\boldsymbol{\varepsilon}_t=\boldsymbol{G}^T\boldsymbol{u}_t+\boldsymbol{h}$ will be taken into account together with the equations of dynamic equilibrium (2), throught the technique of Langrange multipliers. Accordingly to this technique if (2) are the dynamic equilibrium equations without any constrains, $\boldsymbol{G}^T\boldsymbol{u}_t-\boldsymbol{\varepsilon}_t+\boldsymbol{h}=\boldsymbol{0}$ are the kinematic constraints, and $\boldsymbol{s}_t$ are the reactions corresponding to the kinematic constraints, then (see e.g. Washizu 1975):

$$\overline{\boldsymbol{K}}\boldsymbol{u}_t + \boldsymbol{G}\boldsymbol{s}_t = \overline{\boldsymbol{p}}_t, \qquad \boldsymbol{G}^T\boldsymbol{u}_t = \boldsymbol{\varepsilon}_t - \boldsymbol{h} \tag{5}$$

Since the matrix $\overline{\boldsymbol{K}}$ is always non-singular, $\boldsymbol{u}$ can be eliminated from equations (5), and since also relations (4) hold, the following relations are obtained:

$$\boldsymbol{\varepsilon}_t = -(\boldsymbol{G}^T\overline{\boldsymbol{K}}^{-1}\boldsymbol{G})\boldsymbol{s}_t + (\boldsymbol{G}^T\overline{\boldsymbol{K}}^{-1}\overline{\boldsymbol{p}}_t + \boldsymbol{h}) = \boldsymbol{F}\boldsymbol{s}_t + \boldsymbol{\varepsilon}_{0t},$$

$$\boldsymbol{\varepsilon}_t \geq \boldsymbol{0}, \qquad -\boldsymbol{s}_t \geq \boldsymbol{0}, \qquad \boldsymbol{s}_t^T\boldsymbol{\varepsilon}_t = \boldsymbol{0} \tag{6}$$

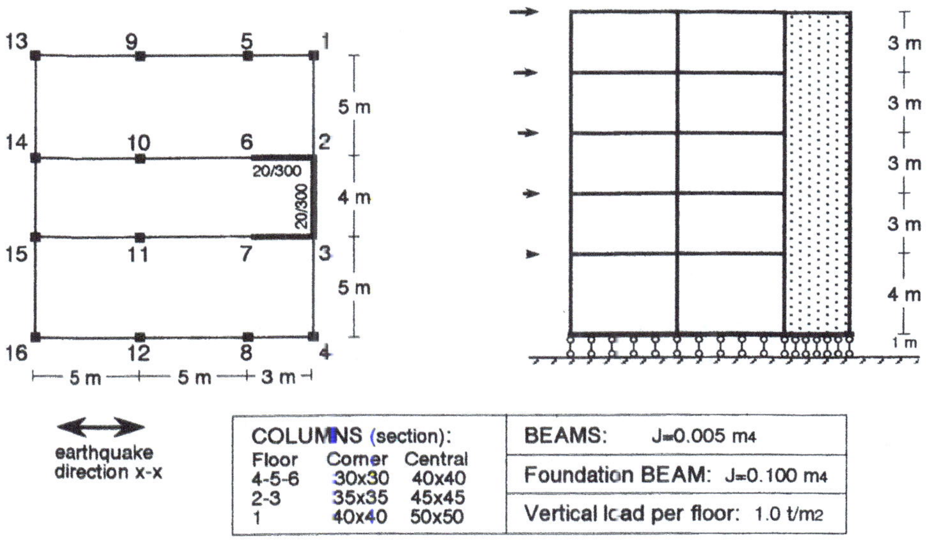

Figure 1. Plan view and vertical section of a five-storey building

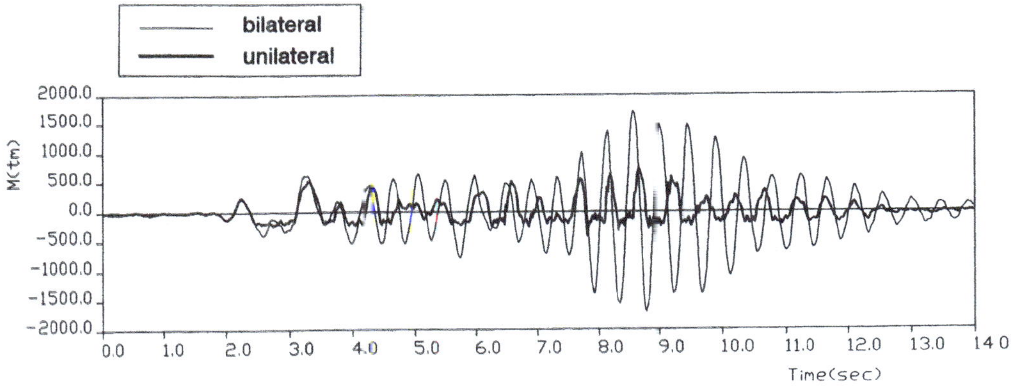

Figure 2. Moments at the base of the core, stiff soil.

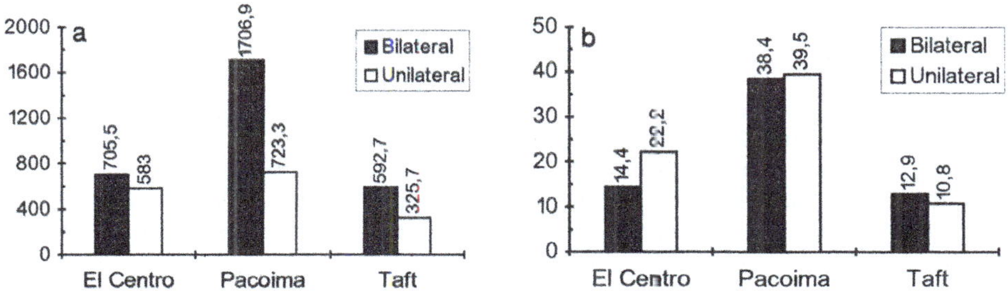

Figure 3. a) Maximum moments (tm) at the base of the core
b) Maximum moments (tm) at the base of the columns 14 and 15

where $\boldsymbol{F}$ is the $m \times m$ influence matrix of the reactions of the bonds to their corresponding fictitious strains, and $\boldsymbol{\varepsilon}_{ot}$ is the m-vector of the fictitious strains of the bonds due to the external loading. The matrix $\boldsymbol{F}$ is non-singular, thus $\boldsymbol{D} = \boldsymbol{F}^{-1}$ exists, and relations (6) can be written as:

$$\boldsymbol{s}_t = \boldsymbol{D}\boldsymbol{\varepsilon}_t + \boldsymbol{D}\boldsymbol{\varepsilon}_{ot} = \boldsymbol{D}\boldsymbol{\varepsilon}_t + \boldsymbol{s}_{ot}, \quad \boldsymbol{\varepsilon}_t \geq \boldsymbol{0}, \quad -\boldsymbol{s}_t \geq \boldsymbol{0}, \quad \boldsymbol{s}_t^T \boldsymbol{\varepsilon}_t = 0 \tag{7}$$

where $\boldsymbol{D}$ is the $m \times m$ influence matrix of fictitious strains of the bonds to their corresponding reactions, and $\boldsymbol{s}_{ot}$ is the m-vector of the reactions of the bonds due to the external loading.

The relations (6) (resp. the relations 7), which constitute a Linear Complementarity Problem (L.C.P.) give the solution of the problem.

Here for the numerical solution the relations (7) are used. The matrix $\boldsymbol{D}$ and the vector $\boldsymbol{s}_{ot}$ are not calculated through matrix invertion, but by using their physical meaning (Doudoumis and Mitsopoulou 1988) and the L.C.P. is solved by the Lemke's algorithm.

NUMERICAL EXAMPLES

As an example of unilateral behaviour of a structure under seismic loading, the five-storey building of figure 1 was studied. The strucrure is founded on stiff soil through continoous foundation beams. The structure is subjected to static vertical loads in conjuction with dynamic seismic excitation for which the accelerograms of Pacoima, Taft and El Centro earthquakes were used.

In figure 2 the bending moments at the base of the core during the Pacoima earthquake are shown for unilateral and bilateral structural behaviour. In figure 3 the maximum values of the bending moments at the base of the core and the columns 14 and 15 are shown, for the above mentioned earthquakes. The presented here sample of results shows that the structures with unilateral contact support conditions (if partial uplift occurs during an earthquake) take special dynamic characteristics and the response values could drastically change from the response values of the corresponding bilateral structures.

A systematic parametric analysis has been carried out (Zervas 1993), by which it has become clear that generally there is a decrease of the response values at the unilateral structure. But it should be noted that for some special cases a significant increase of the response values of the unilateral structure may also occur.

REFERENCES

Zienkiewicz O., Wood W.L., and Taylor R.L., 1980, An alternative single-step algorithm for dynamic problems, *Earthq. Eng. Struct. Dyn., Vol. 8*, 31-40.

Washizu K., 1975, Variational Methods in Elasticity and Plasticity, Pergamon Press, Oxford, 2nd ed.

Doudoumis I.N. and Mitsopoulou E.N., 1988, On the solution of the unilateral contact frictional problem for general static loading conditions, *Computers and Structures, Vol 30, No 5*, 1111-1126.

Zervas P.A., 1993, Seismic behaviour of frames with unilateral contact support conditions, Ph.D. Thesis, Aristotle University of Thessaloniki.

THE IMPACT OF RIGID BODIES WITH A FINITE LINE OF CONTACT. AN EVOLUTIVE METHOD FOR FRICTION PERFORMANCE

A. Sinopoli

Dipartimento di Scienza e Tecnica del Restauro
Venice Architecture University
30135 Venice - Italy

INTRODUCTION

The impact problem, though a classical subject of mechanics, seems to be till now an open problem, as tested by the numerous works produced in the last few years.

In this paper, a brief recall of the rigid body theory for the single point impact is given, bringing out some aspects of the question which may induce controversial results. These contradictory aspects regard the role of the restitution's coefficient; they are put in evidence in the case of a block hitting a rigid ground.

A preliminary elastic impact model is investigated to identify the main characteristics of the contact. Then, the inelastic impact is considered and the formulation for the planar impact of a rigid block, with a finite line of contact, is briefly illustrated; no reference is made to any restitution coefficient and the dry friction is taken into account by means of a generalized Routh method. Different behaviours concerning the post-impact motion are found as a function of the geometric sizes and the friction coefficient.

RIGID BODY THEORY OF IMPACT

It can be stated that an impact occurs between two bodies, which enter suddenly into contact, only when the relative velocities of the contact points violate the impenetrability law of the solids mechanics; that is, when the relative velocities normal to the tangent plane at the contact points are negative. However, in the scientific literature also the case named tangential impact (Wang and Mason, 1992) has been included in the category of impacts; such a case occurs when a body is sliding along the contact surface in presence of a dry friction large enough, but with zero relative velocity in the normal direction.

Here, this situation is not included among the impact phenomena. In fact, in such a case the impulsive dynamics are due only to the Coulomb friction performance; the normal velocity remains always zero and therefore consistent with the normal contact law.

An impact is a very complex problem; in fact during its very brief duration different phenomena can occur: elastic vibrations, plastic deformations, activation of both frictional and plastic mechanisms of energy dissipation. Nevertheless, a common and practical way to investigate the problem of the impact has generally been to use the dynamics equations of rigid bodies. Such an approach, in the case of a single-point impact, is based on the assumptions that the impact is instantaneous, the positions of the two bodies during the shock remain unvaried and the motions of the two bodies are always described (before, during and after the shock) by the rigid-body dynamics equations. According to the assumptions above, the effects of the impact consist only on instantaneous discontinuities of the velocities, due to the action of the contact impulse $\boldsymbol{I}_j$ on each *j-th* body *(j=1,2)* ; further, the effect of any finite force during the impact is neglected.

Assuming for each body the coordinates of the centre of mass and the angular displacements as lagrangian coordinates, the classical and well-known rigid-body dynamics equation for the impulsive motion is:

$$\boldsymbol{K}_j - \boldsymbol{M}_j \, \Delta\dot{\boldsymbol{q}}_j = 0 \qquad j = 1,2 \tag{1}$$

where $\boldsymbol{M}_j$ and $\dot{\boldsymbol{q}}_j$ are respectively the mass matrix and the generalized velocity of the *j-th* body, while $\boldsymbol{K}_j$ is the covariant component of the impulse $\boldsymbol{I}_j$.

According to (1), the impact is formulated as an algebraic problem; but, eq.(1) is generally not sufficient to identify the unique solution of the impulsive motion. In fact, the number of the unknown quantities (namely, the components of the impulse $\boldsymbol{K}_j$ and the generalized velocities components after the shock) is larger than the number of equations. For example, for a single contact point and a planar impact, two further relationships are necessary; they can be obtained by taking into account the friction performance and any other source of energy dissipation. These aspects of the problem have been extensively investigated and discussed. The several papers produced in the last few years on rigid bodies impact basically refer to the classical models defined:

- by Routh, for Coulomb friction. By means of this simple and efficient graphical method different kinds of impulsive contacts can be analyzed, depending on the existence, persistence and stop of either sticking or sliding motions, including reversal slips.
- by Newton or Poisson, for the other sources of energy dissipation. Both made use of the so-called restitution coefficient e^N; but, e^N is defined differently in the two models.

The Newton restitution coefficient e^N is defined as the ratio between the relative velocities $\dot{r}_i^N$ normal to the contact surface at the contact point P_i , respectively after and before the impact:

$$e^N = \frac{\dot{r}_i^{N+}}{\dot{r}_i^{N-}} \tag{2}$$

On the contrary, in the Poisson's hypothesis, the impact is divided in two phases, respectively named *compression* and *restitution* phase. Two impulses $I_{c,i}^N$ and $I_{r,i}^N$, both positive, are then assumed; their ratio defines the restitution's coefficient:

$$e^N = \frac{I_{r,i}^N}{I_{c,i}^N} \tag{3}$$

The aspect of the impact problem concerning the amount of dissipated energy is extremely delicate. Stronge (1990) demonstrated that the Newton's impact law (2) can imply, in non-collinear impacts with friction, a value for the post-impact kinetic energy increased with respect to the corresponding one of the pre-impact motion; further, in the case of reversal slips, the Poisson's hypothesis also can fail, if the motion is not evolutively determined, as by means of the Routh-Poisson method proposed for a single point impact by Wang and Mason (1992). Anyway, the dissipated energy depends on both the restitution coefficient e^N and the dry friction coefficient μ ; they are independent coefficients connected to two different mechanisms of dissipation: the viscous, plastic or thermoelastic internal dissipation and the frictional one.

But, though the coefficient e^N had been initially defined only for direct and central impacts, since then it has been used in the formulation of any kind of impacts, with the implicite meaning that it depends generically on the nature of the material. In the same sense, its use has been sometimes extended (Brach, 1989) also to take into account the variations of the tangential relative velocities; the tangential restitution coefficient e^T is thus defined as the ratio between the post-impact and pre-impact tangential velocity.

Therefore, the rigid body formulation attributes to the restitution coefficients the role to describe all the effects of the contact; they depend on many aspects, as: the shapes of the bodies, the surface area of the contact, the velocities field before the shock, the constitutive law of the material, the history of the loading process, the local frictional dissipation and other aspects, as the thermoelastic ones. Consequently, it is not surprising that few values are available in the technical-scientific handbooks for e^N *and* e^T, except the extreme cases of $e^N = 0$ and $e^N = 1$, respectively named *inelastic* and *perfectly elastic* impact.

Furthermore, although some experimental papers have been produced in the last few years (e.g., Sondergaard et al., 1990), the lack of an experimental tradition on this subject does not permit the validation of the rigid body model for the impact, neither in the extreme case of $e^N = 0$ nor for $e^N = 1$. Analytical comparisons have been made between the Poisson's assumption versus the Newton's law (Wang and Mason, 1992), stating that, from both a philosophical and practical point of view, Poisson's hypothesis is preferable to Newton's law of restitution; but, it seems that nothing can be said about the validity of the rigid body model to describe actual impacts.

In the author's opinion, the restitution coefficients e^N and e^T can represent misleading parameters of the problem. As an example, consider the case of a perfectly elastic impact ($e^N=1$); in accordance to both the Newton's law and Poisson's hypothesis, the normal post-impact velocity of the contact point is greater than zero, still better, it is always positive during the restitution phase. Then, how is it possible to take into account the frictional effects, if the contact is lost ? Another relevant example is discussed in the next section.

THE CASE OF THE BLOCK

Consider a rigid block of base length b and height h (Fig. 1a) which, rotating clockwise around its corner edge A, impacts the ground with all the base. According to the Newton law, for $e^N = 0$, to which point of the base must the coefficient of restitution be applied? If the answer is to all the points of the base, then the block cannot rotate after the impact, in complete disagreement with experimental observations. The same indetermination is obtained by using the Poisson hypothesis.

Then, in order to simplify the question, we can assume that the base of the block is shaped in such a way (Fig.1b) that only point B is involved in the impact; further, assume that the friction coefficient is large enough in order to prevent any sliding.

If $\dot{x}_G$, $\dot{y}_G$ and $\dot{\theta}$ are respectively the horizontal and vertical velocity of the mass's centre and the angular velocity of the block, while the apex $^-$ and $^+$ refer to their values respectively before and the after the impact, according to eqs. (1) and (2), the post-impact motion as a function of the restitution's coefficient e^N is:

$$\dot{x}_G{}^{+} = \frac{2 - b^2/h^2\,(1 + 3\,e^{N})}{2\,(1 + b^2/h^2)}\,\dot{x}_G{}^{-}$$

$$\dot{y}_G{}^{+} = -\frac{2 - b^2/h^2 + e^{N}\,(4 + b^2/h^2)}{2\,(1 + b^2/h^2)}\,\dot{y}_G{}^{-} \qquad (4)$$

$$\dot{\theta}^{+} = \frac{2 - b^2/h^2\,(1 + 3\,e^{N})}{2\,(1 + b^2/h^2)}\,\dot{\theta}^{-}$$

If the bloch is slender (for example, $b/h = 0.2$) and $e^{N} = 0$, after the shock the block rotates clockwise and with a reduced angular velocity around the corner B, as first stated by Housner (1963); but, for $e^{N} = 1$, the post-impact vertical velocity of the centre of mass is opposite in sign and about three times larger with respect to the corresponding pre-impact velocity. A more surprising result is obtained if the block is stocky, as for $b/h = 2$. In this case, if $e^{N} = 0$, all the post-impact velocity components are opposite in sign with respect to the corresponding pre-impact ones, with the evident violation of the impenetrability law for all the points of the base; for $e^{N} = 1$, the violation reduces to the base points belonging to a region close to the corner A .

Now, the main question is: why violations are obtained? Shenton et al., (1991) proposed to solve the question by assuming that, only in case of violation, an additional impulse must be considered to act at point A ; but this is a particular trick, not a general answer to the question.

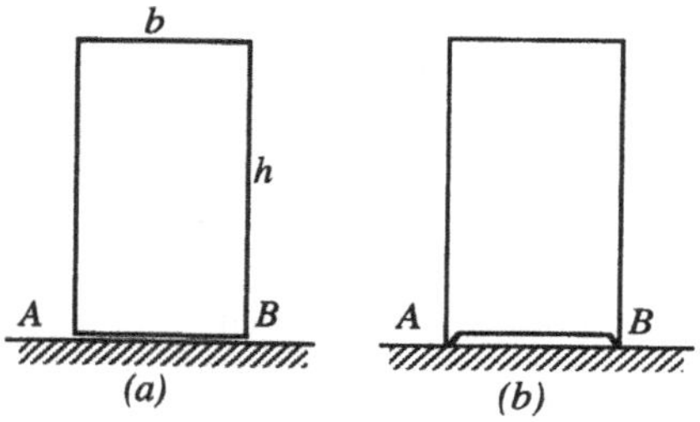

Figure 1.

In the author's opinion, there are two sources of the contradictory results above. The first concerns the necessity of a general formalism for the impulsive rigid body dynamics. The second follows from the fact that, depending on the features of the body and pre-impact velocities, either the whole or a part of the initial rigid body kinetic energy is transformed in elastic energy during the contact. Then, the elastic energy is either completely or partially dissipated so that, when the contact breaks off, the residual elastic energy is partly restituted in the form of rigid body kinetic energy and partly maintained in the form of free vibrations.

PRELIMINARY ELASTIC IMPACT MODEL

In order to understand the elastic features of the contact and to identify the elastic modes started by the impact, consider the pre-impact velocities distribution of the block. It can be decomposed according to Fig.2; the distribution *(a)* starts longitudinal vibrations, while the distributions *(b)* and *(c)* the flexural ones. Under the assumption that, during the motion, the cross sections remain planar and neglecting the effects of shear and rotatory inertia, the equations governing the coupled longitudinal-flexural vibrations are:

$$\frac{\partial N}{\partial y} = \mu \frac{\partial^2 v}{\partial t^2}$$
$$\frac{\partial^2}{\partial y^2} [E\,J(y,t) \frac{\partial^2 u}{\partial y^2}] + N \frac{\partial^2 u}{\partial y^2} + \frac{\partial N}{\partial y}\frac{\partial u}{\partial y} = -\mu \frac{\partial^2 u}{\partial t^2} \tag{5}$$

where $u(y,t)$ and $v(y,t)$ are the displacements in the x and y direction and μ is the mass per unit length; further, $N = E\,A(y,t)\,\partial v/\partial y$, with E Young s modulus, $A(y,t)$ cross section area and $J(y,t)$ inertia moment of $A(y,t)$. Eqs. (5) are subject to the unilateral boundary conditions at the base. Consequently, A is a function of y and t; in fact, during the motion, it is possible that the area of the base section reduces and the solid deforms, varying all the cross sections areas.

Anyway, let us assume that the vibrations are unccupled and evaluate when this assumption is acceptable. The first uncoupled flexural frequency ω_f is:

$$\omega_f = \frac{(1.875)^2}{h^2} \sqrt{\frac{EJ}{\rho A}} \tag{6}$$

where ρ is the mass density. The first longitudinal frequency ω_l, on the contrary, is:

$$\omega_l = \frac{\pi}{2h} \sqrt{\frac{E}{\rho}} \tag{7}$$

By comparing eq.(6) and (7) it results that, for $\omega_f << \omega_l$, and then, for $b/h << 1.55$, the vibrations are uncoupled. Therefore, during the impact of a thin block, the longitudinal vibrations are the most important. After the shock, the distribution *(a)* is opposite in sign and reduced, in dependence of the internal longitudinal damping. On the contrary, the distributions *(b)* and *(c)* maintain the sign and are only slightly reduced; this reduction corresponds to the transformation of the proper rigid body kinetic energy in the potential flexural one, during a time interval equal to the half-period of the longitudinal vibrations.

Following such an approach and neglecting the potential flexural energy after the impact, it is found that, for $b/h = 0.2$, the block rotates around B with a reduction of the angular velocity of about *8 %* ; the reduction obtained by the rigid body inelastic model was *6 %* .

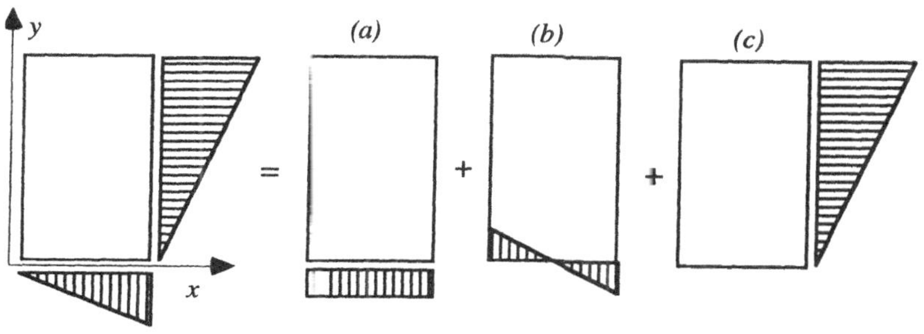

Figure 2. Decomposition of the pre-impact velocity distributions of the block.

On the contrary, the flexural vibrations become more and more important at the increase of the ratio *b/h* ; therefore, it is expected that, for very stocky blocks, only flexural elastic behaviour determines the motion, in the sense that the block continues to oscillate elastically with the bearing section alternately lying around corner *A* and *B* .

From the oservations above, it seems that the restitution's coefficient e^N, for $0 < e^N \leq 1$, does not represent a reliable parameter of the impact, above all in the case of $e^N = 1$. I mean that the rigid body theory of perfectly elastic impact, given the pre-impact velocities, assumes that the whole pre-impact kinetic energy is transformed in elastic one and then restituted as rigid body kinetic energy; furthermore, that occurs by assigning the same importance to different vibrations, characterized by different propagation's velocities and wave-lengths.

As a consequence, it seems that the case when reliability would be given to the rigid body model is only for either a material point or inelastic impact ($e^N = 0$) ; in this last case, only rigid body kinetic energy is taken into account both before and after the impact. That does not mean necessarily that the whole elastic energy is dissipated during the shock; it can be partly preserved and transformed in rigid body energy and partly dissipated in the post-impact motion. Then, the inelastic impact is the one which ends when the normal velocity of any point still in contact is zero; the identification of the points maintaining the contact in the post-impact motion follows from the solution of the impact problem. From an energetic point of view, the inelastic impact neglects any elastic energy, which can be taken into account by means of experimental, numerical or analytical investigations.

INELASTIC IMPACT WITH A FINITE LINE OF CONTACT

In this section a synthetic review of the analysis, recently proposed (Sinopoli, 1994b) to investigate the planar impact of rigid bodies, is given. During the impact, no reference has been made to any restitution coefficient; the performance of dry friction has been taken into account by means of a generalized Routh method.

Consider again the rigid body (Fig.1a) impacting the ground. Let $\dot{\boldsymbol{r}}_P^N$ indicate the velocity vector of any point *P* belonging to the base *AB*, in the direction of the outward normal **N** to the contact surface. The impenetrability condition imposes that the admissible velocity $\dot{\boldsymbol{r}}_P^{N+}$ of any point *P* , in the post-impact motion, is:

$$\dot{\boldsymbol{r}}_P^{N+} \geq 0 \qquad\qquad \forall\ P \in AB \tag{8}$$

Due to the velocities distribution in the pre-impact motion, if the solution exists, it is expected that a resultant positive impulse $\boldsymbol{I}^N$ acts on the body, in order that rel.(8) is satisfied. The point *Q* where the resultant impulse $\boldsymbol{I}^N$ is applied is unknown; anyway it is internal to *AB* , so that, if * means transposition, the contact law for admissible motions is:

$$\boldsymbol{I}^{N*}\, \dot{\boldsymbol{r}}_Q^{N+} \geq 0 \tag{9}$$

But, the problem is equivalent to the one of two impulses, positive or zero, applied, respectively, $\boldsymbol{I}_A^N$ on *A* and $\boldsymbol{I}_B^N$ on *B* ; further, it is possible to show that (Sinopoli, 1994) the contact law for actual post-impact motion is such that:

$$\boldsymbol{I}_A^{N*}\, \dot{\boldsymbol{r}}_A^{N+} = \boldsymbol{I}_B^{N*}\, \dot{\boldsymbol{r}}_B^{N+} = 0 \tag{10}$$

Then, by assuming the metric of the kinetic energy, the frictionless inelastic impact can be formulated as: among all the kinematically admissible velocities such that:

$$\dot{\underline{t}}_A^{N+} \geq 0 \qquad \text{and} \qquad \dot{\underline{t}}_B^{N+} \geq 0 \tag{11}$$

and, among all the dynamically possible ones such that:

$$\Delta\dot{u}^{\,*}\,\dot{\underline{u}}^{\,+} \geq 0 \tag{12}$$

the actual post-impact motion corresponds to:

$$\Delta\dot{u}^{\,*}\,\dot{u}^{\,+} = 0 \tag{13}$$

where * means transposition and $\dot{u} \equiv (\dot{u}_x, \dot{u}_y, \dot{u}_\theta)$ is the generalized rigid body velocity in the linear mapping which transforms the configuration space in itself, with the new metric. Rels.(11) define in the plane $(\dot{u}_y, \dot{u}_\theta)$ the tangent (t_B^N, t_A^N) and normal (n_A^N, n_B^N) cones. The kinematical admissible velocities belong to the cone (t_B^N, t_A^N) ; the dynamical possible velocity variations $\Delta\dot{u}^N$ in the plane (u_y, u_θ) belong to the cone (n_A^N, n_B^N) .

The actual motion is such that $\Delta\dot{u}^N$ is either tangent or internal to the cone $(n_A^N\, n_B^N)$, depending on $\dot{u}^{N+}$ is either different from or equal to zero. In the first case, the contact is maintained on one point, either A or B; in the second, in both the points.

Eq.(13) coincides with the formulation given by Moreau (1983) for frictionless inelastic impact; but, it coincides also with the results obtained by Almansi (1916), who investigated the frictionless impact as a dynamics problem with unilateral constraints.

In presence of Coulomb dry friction, the dynamics formulation corresponds to the normal contact law (10) and rels.(11), together to:

$$[\Xi_A^T + \Xi_B^T - \Delta\dot{u}]^{*}\,\dot{\underline{u}}^{\,+} \geq 0 \tag{14}$$

$$[\Xi_A^T + \Xi_B^T - \Delta\dot{u}]^{*}\,\dot{u}^{\,+} = 0 \tag{15}$$

where Ξ_A^T and Ξ_B^T are the covariant components of the friction performance; they are governed by the Coulomb law of minimal dissipated power:

$$[\Xi_A^T + \Xi_B^T]^{*}\,\dot{\underline{u}}^{\,+} \geq [\Xi_A^T + \Xi_B^T]^{*}\,\dot{u}^{\,+} \tag{16}$$

Thus, an evolutive method is necessary, to respect the behaviour of the friction, in accordance with (16), and to avoid incorrect energetic balance. The method is a generalization of the one suggested first by Routh: two Coulomb cones and two sticking lines, corresponding to the maintenance of the contact on either point A or B , can be defined. Different solutions are then obtained, depending on the initial conditions and on the geometrical and mechanical features of the block.

According to the formulation above, a parametric and analytical investigation has been performed on the impact of a block, rotating around the base corner edge A ; symmetrical results are obtained in the case of rocking around point B. Three situations, corresponding to zero, negative and positive tangential pre-impact velocity of point A, have been analyzed.

Here only the results related to the identification of the mechanisms, as a function of the ratio b/h and the kinetic friction coefficient μ_k, in the case of zero tangential pre-impact velocity are presented (Fig.3). Each mechanism has been labeled by means of the initial letters of the particular motion; then, R means rocking, S sliding, Re block at rest and S-R sliding-rocking. If the attention is restricted to values of μ_k larger enough and b/h lower than $\sqrt{2}$, a wide region of rocking can be identified. This region defines the boundaries of the model proposed by Housner (1963) to investigate the behaviour of inverted pendulum-like structures. The upper limit of this region coincides with the result of a preliminary paper (Sinopoli, 1987), where the dry friction had been simulated roughly; so that, for: $b/h < \sqrt{2}$, the motion was a rocking, while for: $b/h \geq \sqrt{2}$, only sliding might have been expected.

A further confirmation of the generality of the formulation proposed concerns the results for $\mu_k = 0$; they coincide with the results of the standard inelastic impact by Moreau (1983) and a previous paper by the author (Sinopoli, 1989), where a first attempt of a contact theory had been tackled, by neglecting friction; there, $b/h = \sqrt{2}/2$, was the separation value between sliding-rocking and sliding motions.

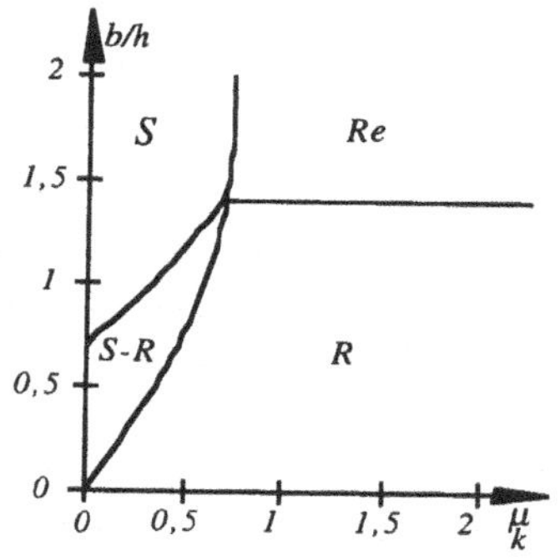

Figure 3. Mechanisms for zero tangential velocity.

REFERENCES

Almansi, E., 1916, "Sulla Teoria degli Impulsi", *Rendiconti Accademia Nazionale dei Lincei*, **25**, 410-416.

Brach, R. M., 1989, "Rigid Body Collisions," ASME *Journal of Applied Mechanics*, **56**, 133-138.

Housner, W.G., 1963, "The Behavior of Inverted Pendulum Structures during Earthquakes," *Bull. of Seism. Society of America*, **53**, 403-417.

Moreau, J.J., 1983, "Liaisons Unilaterales sans Frottement et Chocs Inelastiques," *C. R. Acad. Sc. Paris*, 296, Serie II, 1473-1476.

Shenton III, H.W., Jones, N.P., 1991, "Base Excitation of Rigid Bodies. I. Formulation', *Journal of Engineering Mechanics, ASCE*, **117** (10), 2286-2306.

Sinopoli, A., 1987, "Dynamics and Impact in a System with Unilateral Constraints. The Relevance of Dry Friction," *Meccanica*, **22**, 210-215.

Sinopoli, A., 1989, "Kinematic Approach in the Impact Problem of Rigid Bodies," *Applied Mechanics Reviews*, **42**(11), 233-244.

Sinopoli, A., 1994, "Unilaterality and Dry Friction: A Geometric Formulation for Smooth Rigid Body Dynamics," Submitted to ASME *Journal of Applied Mechanics*.

Sinopoli, A., 1994, "Impact Between Rigid Bodies. The Case with a Finite Line of Contact," Submitted to ASME *Journal of Applied Mechanics*.

Sinopoli, A., 1995, "A Geometric Formulation for Rigid-Body Dynamics: The Locking of the Rod," *Proceedings of IV Pan American Congress of Applied Mechanics, Buenos Aires*.

Sondergaard, R., Chaney, K. and Brennen, C. E., 1990, "Measurements of Solid Spheres Bouncing Off Flat Plates," *Transactions of the ASME*, **57**, 694-699.

Stronge, W. J., 1990, "Rigid Body Collisions with Friction," *Proceedings of Royal Society of London*, **A431**, 169-181.

Wang, Y., and Mason, M., 1992, "Two-Dimensional Rigid-Body Collisions with Friction," ASME *Journal of Applied Mechanics*, **59**, 635-642.

COUPLING OF FRICTION AND INTERNAL DISSIPATION IN PLANAR COLLISION OF COMPLIANT BODIES

W.J. Stronge

Department of Engineering
University of Cambridge
Cambridge CB2 1PZ, U.K.

INTRODUCTION

For collisions between hard bodies, the energetic *coefficient of restitution* has been defined by Stronge (1990) as a direct measure of energy dissipated in internal inelastic deformations. This energy can be calculated from work done on the bodies during the collision by the normal component of contact force if the contact region has negligible tangential compliance. This calculation is based on separating the contact period into an initial period of compression and a subsequent period of restitution. During compression the normal component of relative velocity at the contact point is negative whereas during restitution the bodies are moving apart so that this component of relative velocity is positive. The normal component of relative velocity vanishes at the transition from compression to restitution. The contact force **F** has a normal component F_2 that does work $-W_c$ on the colliding bodies during compression. This work transforms some kinetic energy of relative motion to internal energy of deformation; a part of this internal energy (the elastic strain energy) is available to drive the bodies apart during restitution.

For collisions between hard bodies with non-conforming surfaces, the contact area remains small and deformations are concentrated around the point of initial contact C. In this case a theory of deformation can be employed to relate normal contact force to deformation and subsequently to obtain the maximum contact force (or maximum indentation) corresponding to work W_c done on the contact region by the normal force during compression — this work transforms part of the *kinetic energy of normal relative motion* into internal energy. For elastic bodies with spherical contact surfaces, Hertz (1882) developed such a quasi-static theory of 'rigid body' impact that has proved to be extremely useful. This theory of contact between non-conforming surfaces has been extended into the range of small elastic-plastic deformations by Johnson (1985). While elastic collisions finally experience no change in kinetic energy of relative motion due to work of normal force, an elastoplastic collision suffers an energy loss that depends on maximum depth of indentation δ_c (i.e. indentation at termination of compression). This energy loss due to internal plastic deformation depends primarily on the kinetic energy of

Contact Mechanics, Edited by M. Raous *et al.*
Plenum Press, New York, 1995

normal relative motion, the curvature of surfaces in the contact region and material properties (elastic modulus and flow stress). For non-collinear impact configurations where the centre of mass of at least one of the colliding bodies is not on the line parallel to the common normal which passes through the contact point, we will find that the energy dissipated by internal plastic deformation also depends on friction and sliding in the contact region. The effect of friction on the coefficient of restitution is a secondary effect caused by coupling between normal and tangential forces in collisions between rough bodies in non-collinear configurations.

COMPLIANCE AND WORK TO DEFORM CONTACT REGION

Elastic Indentation

Suppose colliding bodies B and B′ initially come into contact at point C where the surfaces of the bodies are non-conforming. In a neighbourhood of C the two bodies have surfaces with curvatures R_{B}^{-1} and $R_{\mathrm{B'}}^{-1}$ respectively, as illustrated in Fig. 1. If these bodies

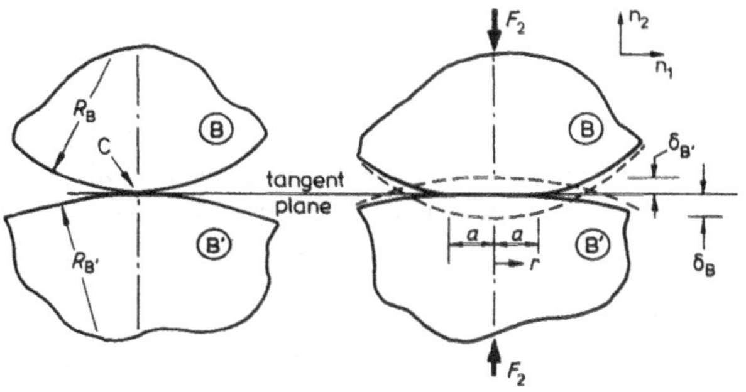

Figure 1. Compression and indentation of spherical contact surfaces.

are elastic and they are compressed in the normal direction by a force F_2, Hertz showed that the contact area spreads to radius a. Within the contact area there is an elliptical distribution of contact pressure. At the centre of the contact patch this pressure compresses each body by $\delta_i (i = \mathrm{B}, \mathrm{B'})$. If an effective surface curvature R^{-1} is defined as

$$R^{-1} \equiv R_{\mathrm{B}}^{-1} + R_{\mathrm{B'}}^{-1}$$

then if $R >> a$ the indentation or normal displacement of one body relative to the other $\delta \equiv \delta_{\mathrm{B}} + \delta_{\mathrm{B'}}$ is related to the contact radius by

$$\frac{\delta}{R} = \frac{a^2}{R^2} \tag{1}$$

For elastic bodies with Young's moduli E_i and Poisson's ratios ν_i an effective elastic modulus E has been defined as

$$E \equiv [(1-\nu_B^2)E_B^{-1} + (1-\nu_{B'}^2)E_{B'}^{-1}]^{-1}$$

With these expressions we can relate the mean pressure $\bar{p}$ in the contact area and the normal component of contact force F_2 to normal indentation δ,

$$\frac{\bar{p}}{E} = \frac{4}{3\pi}\sqrt{\delta/R}, \qquad \frac{F_2}{ER^2} = \frac{4}{3}\left(\delta/R\right)^{3/2} \tag{2}$$

This force can be integrated to obtain the work W done by the normal contact force in compressing the small region of significant deformation around the small contact patch.

$$\frac{W}{ER^3} = \int_0^{\delta/R} \frac{F_2(\delta'/R)}{ER^2} d(\delta'/R) = \frac{8}{15}\left(\delta/R\right)^{5/2} \tag{3}$$

Indentation and Work at Yield in Elastic-Plastic Bodies

Elastic indentation continues until a body begins to plastically deform. If plasticity initiates at a uniaxial yield stress Y, the elliptical (Hertz) contact pressure distribution gives solely elastic deformation if the mean pressure $\bar{p} < 1.1Y$. Further indentation requires an increase in contact pressure and involves elastic-plastic deformations. With increasing indentation beyond the yield value δ_Y, contact pressure rapidly approaches an upper limit $\bar{p} \approx 3Y$ where the pressure is uniformly distributed and plastic flow is fully developed. (The factor 3 relates indentation pressure to flow stress in a Brinell hardness test.) Here for simplicity this transition region of elastic-plastic deformation for $1.1 < \bar{p}/Y < 3$ is neglected and we assume that fully plastic indentation occurs at pressure $\bar{p} = \kappa Y$ where $1.1 < \kappa < 3$. With this model the transition pressure $\bar{p}_Y$ is given by

$$\frac{\bar{p}_Y}{Y} \equiv \kappa = \frac{4}{3\pi}\frac{E}{Y}\sqrt{\delta_Y/R} \tag{4}$$

Thus *indentation and work required to initiate yield are material properties*,

$$\frac{\delta_Y}{R} = \left(\frac{3\pi}{4}\right)^2\left(\frac{\kappa Y}{E}\right)^2, \qquad \frac{W_Y}{ER^3} = \frac{8}{15}\left(\frac{\delta_Y}{R}\right)^{5/2}, \qquad \frac{W_Y}{\kappa YR^3} = \frac{2\pi}{5}\left(\frac{3\pi}{4}\right)^4\left(\frac{\kappa Y}{E}\right)^4 \tag{5}$$

Uncontained Plastic Indentation

Perfectly plastic indentation takes place at pressure $\bar{p} = \kappa Y$ which is uniformly distributed across the contact area. Assuming that there is negligible elastic deformation in material surrounding the contact patch, the edge of the contact area neither sinks in nor piles up and consequently indentation into a plastically deforming core is given by

$$\frac{\delta}{R} = \frac{\delta_Y}{2R} + \frac{a^2}{2R^2}, \qquad \delta > \delta_Y \tag{6}$$

This relation gives continuity of both indentation and contact area at the transition from elastic to fully plastic behaviour. Then for perfectly plastic materials with negligible strain hardening, the contact force F_2 can be expressed as

$$\frac{F_2}{\kappa Y R^2} = \pi\left(\frac{2\delta}{R} - \frac{\delta_Y}{R}\right) \tag{7}$$

while work required to indent the bodies is

$$\frac{W}{\kappa Y R^3} = \frac{W_Y}{\kappa Y R^3} + \pi\left(\frac{\delta^2}{R^2} - \frac{\delta\delta_Y}{R^2}\right), \qquad \delta > \delta_Y \tag{8}$$

The maximum indentation δ_c and maximum normal force $F_2(\delta_c)$ occur simultaneously at the end of compression. At the same instant, the work W_c done by the normal force in compressing the deforming region is a maximum,

$$\frac{W_c}{W_Y} = 1 + \frac{5}{2}\left(\frac{\delta_c^2}{\delta_Y^2} - \frac{\delta_c}{\delta_Y}\right), \qquad \delta_c > \delta_Y \tag{9}$$

Elastic Unloading

As a consequence of plastic deformation during loading, curvature R^{-1} of contacting surfaces is changed to $\overline{R}^{-1}$. Chang and Ling (1992) assumed that this transition occurs at maximum indentation. They use continuity of contact area and indentation depth at maximum indentation δ_c to obtain the change in indentation δ_r that takes place during elastic unloading.[1]

$$\frac{\delta_r}{\delta_c} = \frac{\delta_Y}{\delta_c}\left(2\frac{\delta_c}{\delta_Y} - 1\right)^{1/2} \tag{10}$$

After complete unloading the final indentation δ_f is given by $\delta_f = \delta_c - \delta_r$.

With an elastic compliance relation, we obtain the work W_r removed from the deforming region during unloading — this work is equivalent to the elastic strain energy removed from the deforming region as the normal contact force is reduced during restitution or unloading.

$$\frac{W_r}{W_Y} = -\left(2\frac{\delta_c}{\delta_Y} - 1\right)^{3/2} \tag{11}$$

---

[1] At the transition from loading to unloading expression (10) does not provide continuity of force for elastic-plastic deformations. It is inevitable that any simple model will result in some imbalance at this transition since in practice the contact pressure distribution does not change instantaneously from uniform during fully plastic loading to elliptic during elastic unloading (Johnson 1968). An alternative expression that at transition satisfies continuity of force (but not contact area) is

$$\frac{\delta_r}{\delta_c} = \left(\frac{3}{2}\right)^{1/2}\frac{\delta_Y}{\delta_c}\left(2\frac{\delta_c}{\delta_Y} - 1\right)^{1/2}, \qquad \text{giving} \qquad \frac{W_r}{W_Y} = -\left(\frac{3}{2}\right)^{3/2}\left(2\frac{\delta_c}{\delta_Y} - 1\right)^{3/2}$$

The outcome of the difference from (10) is only a small decrease in the normal impact speed to initiate yield v_Y so the former relation is retained in order to provide continuity at the elastic limit.

RESOLVED DYNAMICS OF PLANAR RIGID BODY IMPACT

The rigid body theory of impact assumes that the collision period is sufficiently brief that during this period there are changes in velocity but no movement. This asymptotic limit of a very brief time dependent process is appropriate for collisions between hard bodies where the contact area remains small in comparison with both length and cross-sectional dimensions of the body; this condition is also required by the contact deformation theory in the first section. When this is satisfied, during the brief contact period continuous changes in relative velocity across the small deforming region that surrounds the contact area can be obtained by supposing that it is represented by an *infinitesimal deformable particle* located at the contact point C between colliding rigid bodies; the particle is assumed to have negligible tangential compliance. Consequently if $\mathbf{V}_C$ and $\mathbf{V}_{C'}$ are velocities of the two bodies at C, the relative velocity $\mathbf{v} \equiv (v_1, v_2)$ across the deformable particle is defined as $\mathbf{v} \equiv \mathbf{V}_C - \mathbf{V}_{C'}$. This relative velocity is resolved into a component v_1 in the common tangent plane and a component v_2 normal to this plane; the relative velocity v_1 is termed slip. The coordinate system is oriented such that at incidence $v_1(0) \geq 0$ and $v_2(0) < 0$. Equations of motion for this system can be expressed in terms of components of impulse $\mathbf{p} \equiv (p_1, p_2)$ of the reaction force $\mathbf{F} \equiv (F_1, F_2)$ at the contact point C.

$$dv_1 = B_1\, dp_1 - B_3\, dp_2 \tag{12a}$$

$$dv_2 = -B_3\, dp_1 + B_2\, dp_2 \tag{12b}$$

where inertia parameters B_1, B_2, B_3 were defined by Wang and Mason (1993). These parameters depend on the mass of each body M, M', the radius of gyration of each about its centre of mass k, k' and the locations of the centres of mass $\mathbf{r}, \mathbf{r}'$ relative to the contact point C, $\mathbf{r} \equiv (x_1, x_2)$.

$$B_1 = M^{-1}(1 + x_2^2 / k^2) + M'^{-1}(1 + x_2'^2 / k'^2) \tag{13a}$$

$$B_2 = M^{-1}(1 + x_1^2 / k^2) + M'^{-1}(1 + x_1'^2 / k'^2) \tag{13b}$$

$$B_3 = M^{-1} x_1 x_2 / k^2 + M'^{-1} x_1' x_2' / k'^2 \tag{13c}$$

The coordinate system for describing the configuration is illustrated in Fig. 2.

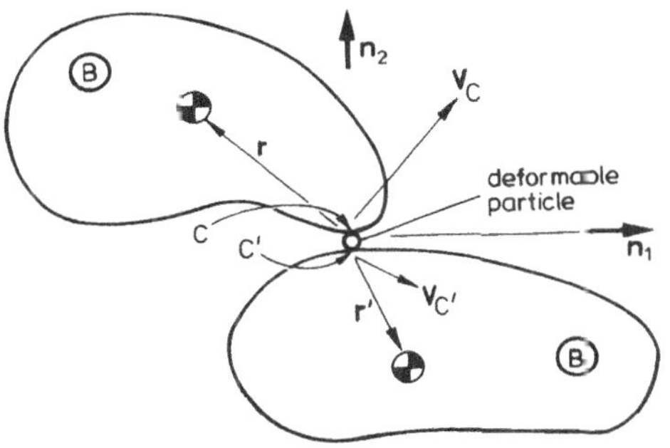

Figure 2. Colliding rigid bodies with contact points C and C' separated by deformable particle.

Coulomb's Law of Friction

If the contact is sliding ($v_1 \neq 0$) the tangential force F_1 has magnitude proportional to that of the normal force F_2 and a direction opposed to sliding. Alternatively, if the contact sticks ($v_1 = 0$) the tangential force is just sufficient to maintain sticking. Since $d\mathbf{p} = \mathbf{F}dt$, Coulomb's law can be expressed as a relation between components of impulse,

$$|dp_1| < \mu dp_2 \qquad \text{if } v_1 = 0$$

$$dp_1 = -s\mu dp_2 \qquad \text{if } v_1 \neq 0$$

where $s = \mathrm{sgn}(v_1)$ is the direction of sliding and μ is the coefficient of friction. (For simplicity the static and kinetic coefficients are assumed to be equal.)

Equations of Relative Motion

With Coulomb's law of friction the equations of motion (12) can be expressed in terms of a monotonously increasing independent variable $p \equiv p_2$,

$$dv_1 / dp = -s\mu B_1 - B_3 \tag{14a}$$

$$dv_2 / dp = B_2 + s\mu B_3 \tag{14b}$$

These equations can be integrated to give the relative velocity at any impulse during an initial period of unidirectional slip.

$$v_1(p) = v_1(0) - (s\mu B_1 + B_3)p \tag{15a}$$

$$v_2(p) = v_2(0) + (B_2 + s\mu B_3)p \tag{15b}$$

Furthermore the equations of motion can be examined to identify the range of inertia and friction parameters for different contact processes.

Contact Processes during Collision

For all impact configurations the definitions (13a–c) give $B_1 > 0,\ B_2 > 0$ and $B_1 B_2 > B_3^2$. Moreover, in order for normal force to oppose indentation, Eq. (14b) requires[2]

$$-B_2 B_3^{-1} < \mu < B_2 B_3^{-1} \qquad \text{to give} \quad dv_2 / dp > 0 \tag{16}$$

The process of slip is described by (14a). For $v_1(p) > 0$ slip is retarded if $dv_1 / dp < 0$ whereas for $v_1(p) < 0$ retardation requires $dv_1 / dp > 0$. If initial retardation causes slip to vanish at impulse p_s before separation, reversal in direction of slip requires $dv_1 / dp > 0$ during $p > p_s$. Conditions of this type give the range of parameters for any particular slip process; e.g. the tangential direction of acceleration is constant so if sliding vanishes before separation, subsequently the contact either sticks or slip reverses:

---

[2] For eccentric impact configurations ($B_3 \neq 0$) inequality (16) provides an upper bound on coefficient of friction in order to avoid jam or equivalently Painleve's paradox (Stronge 1990, Lötstedt 1981).

sticks if $\quad -\mu < B_3 B_1^{-1} < \mu \quad$ (17)

slip reverses if $\quad \mu < B_3 B_1^{-1} \quad$ (18)

Reversal can occur only if $B_3 > 0$. Integration of (14) indicates that sliding halts or reverses during compression if

$$-\frac{v_1(0)}{v_2(0)} < \frac{B_3 + \mu B_1}{B_2 + \mu B_3} \tag{19a}$$

while sliding halts or reverses during restitution if

$$\frac{B_3 + \mu B_1}{B_2 + \mu B_3} < -\frac{v_1(0)}{v_2(0)} < \left(\frac{B_3 + \mu B_1}{B_2 + \mu B_3}\right)\frac{p_f}{p_c} \tag{19b}$$

where p_c is the normal impulse at transition from compression to restitution and p_f is the terminal normal impulse at separation. When sliding vanishes, $v_1(p_s) = 0$, the normal impulse p_s can be obtained from (15b),

$$p_s = (B_3 + \mu B_1)^{-1} v_1(0) \tag{20}$$

and the normal component of relative velocity is $v_2(p_s) = v_2(0) + (B_2 + \mu B_3)p_s$. If sliding vanishes during compression an additional impulse $p_c - p_s$ occurs before compression terminates. For parameters that give slip reversal during compression, the impulse at compression-restitution transition is

$$p_c = -\frac{v_2(0)}{B_2 - \mu B_3}\left\{1 + \left(\frac{2\mu B_3}{B_3 + \mu B_1}\right)\frac{v_1(0)}{v_2(0)}\right\}, \qquad 0 < p_s < p_c \tag{21a}$$

whereas for unidirectional slip

$$p_c = -(B_2 + \mu B_3)^{-1} v_2(0), \qquad p_s > p_c \tag{21b}$$

Work and Indentation when Compression Terminates

The contact force does work on the 'rigid' bodies that is the negative of the work W_c done on the deforming region during the same period. For each separate period of unidirectional sliding this work can be calculated by a theorem (Stronge 1992) that goes back to Kelvin. Here for example we consider initial sliding that reverses direction during compression.[3] Before sliding reverses the normal force does work $-W_s$ on the bodies and thereby reduces their kinetic energy of normal relative motion. After reversal and before the end of compression the normal force imparts additional normal impulse $p_c - p_s$ that does further work $-W_c + W_s$ on the colliding rigid bodies.

---

[3] Expressions for work during other possible slip processes were obtained by Hogue and Newland (1994).

$$2W_s = (B_2 + \mu B_3)p_s^2 + 2(B_2 - \mu B_3)(p_c - p_s)p_s$$

$$2W_c = (B_2 - \mu B_3)p_c^2 + 2\mu B_3 p_s^2$$

Thus with (5) we obtain for compression

$$\frac{W_c}{W_Y} = \frac{(B_2 - \mu B_3)p_c^2 + 2\mu B_3 p_s^2}{(4\pi/5)(3\pi/4)^4(\kappa Y/E)^4 \kappa Y R^3} \tag{22}$$

For continuous slip during compression, (21b) and (22) give

$$\frac{W_c}{W_Y} = \frac{1}{m(B_2 + s\mu B_3)} \frac{v_2^2(0)}{v_Y^2} \tag{23}$$

where

$$v_Y^2 \equiv \frac{4\pi}{5}\left(\frac{3\pi}{4}\right)^4 \left(\frac{\kappa Y}{E}\right)^4 \kappa Y R^3 m^{-1} \qquad \text{and} \qquad m^{-1} \equiv M^{-1} + M'^{-1} \tag{24}$$

For collinear collisions $B_2 = m^{-1}$ and $B_3 = 0$ so that effects of normal and tangential force are decoupled. Thus irrespective of the initial slip velocity, at the end of compression $p_c = -mv_2(0)$ and

$$\frac{W_c}{W_Y} = \frac{5}{4\pi} \frac{mv_2^2(0)}{(3\pi/4)^4(\kappa Y/E)^4 \kappa Y R^3} = \frac{v_2^2(0)}{v_Y^2} \tag{25}$$

Expression (22), (23) or (25) can be equated with (9) to obtain maximum indentation δ_c / δ_Y for any particular impact condition.

COEFFICIENT OF RESTITUTION

The energetic coefficient of restitution is a ratio of work done on the small deforming region during compression to the work done by this region on the adjacent 'rigid' bodies during restitution; i.e. it is a measure of the part of the kinetic energy of normal relative motion transformed to energy of deformation during compression which is recoverable during restitution. Stronge (1991) gave the following definition:

The square of coefficient of restitution e_*^2 is the ratio of elastic strain energy released at the contact point during restitution to the energy absorbed by internal deformation during compression.

This energy ratio can be calculated from the work done by the normal component of contact force if tangential compliance is negligible,

$$e_*^2 = -\frac{W_r}{W_c} = \frac{W_Y}{W_c}\left(\frac{8}{5}\frac{W_c}{W_Y} - \frac{3}{5}\right)^{3/4} \tag{26}$$

This final expression combines (9) and (11) with the energy transformed in compression W_c / W_Y obtained from Eq. (22) or (23). Together with (24a) it relates e_* to the 'damage number' defined by Johnson (1972). The result is similar to that obtained by Adams and Tran (1993) but here my expression explicitly incorporates the effect of friction in various possible slip processes. In the limit of $W_c / W_Y >> 1$ equation (26) indicates that $e_* \approx (v_2(0)/v_Y)^{-1/4}$ where v_Y depends on effective mass contact curvature and material properties but is *independent of impact configuration*. This functional relation between coefficient of restitution and normal impact speed agrees with measurements on a wide range of metals reported by Goldsmith (1960).

ECCENTRIC IMPACT OF RIGID ROD ON ELASTIC-PLASTIC HALF-SPACE

To illustrate the effect of friction on the coefficient of restitution, consider a slender rod inclined at angle θ when it strikes a massive elastic-plastic half-space. At incidence the impact end of the rod C has tangential and normal components of relative velocity, $v_1(0)$ and $v_2(0)$. The process of slip that evolves during contact depends on initial sliding speed as well as the impact configuration.

Figure 3 shows the calculated coefficient of restitution e_* as a function of normal relative speed at incidence for direct ($\theta = \pi/2$) as well as two eccentric impact configurations ($\theta = \pi/4$ and $\theta = 3\pi/4$). If the contact surfaces are smooth, there is only a normal component of contact force. As a function of angle of inclination θ, the inertia opposing normal acceleration at C is a maximum for direct or collinear collision; hence, for any particular mass and normal impact speed, maximum indentation and a minimum coefficient of restitution occur for collinear impact, $\theta = \pi/2$.

In eccentric impact configurations with $x_1 > 0$, friction supplements the normal force in accelerating C in the normal direction if $v_1 > 0$, and it diminishes the effect of normal force if $v_1 < 0$. Thus for $\theta = \pi/4$, friction from initial sliding that persists throughout the contact period slightly increases the coefficient of restitution whereas if sliding reverses at incidence, $v_1(0) = 0$, friction decreases e_*. If direction of sliding reverses, the value of e_* for eccentric impact of rough bodies is in the range between bounds obtained from unidirectional sliding. For inclination $\theta = 3\pi/4$ the correspondence is reversed between direction of sliding and changes in the coefficient of restitution (with the exception that in this configuration if slip vanishes the contact subsequently sticks rather than slip reversing).

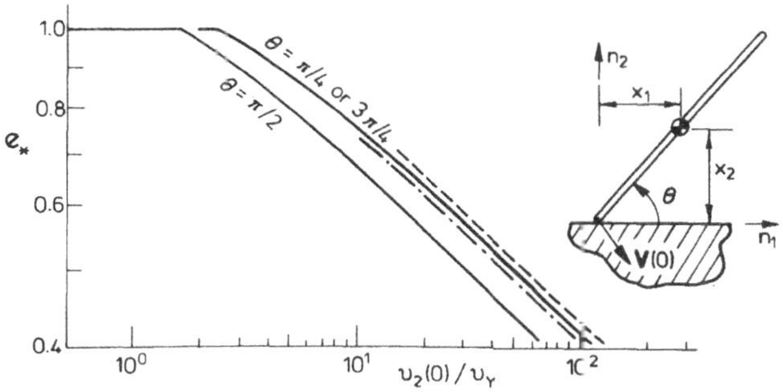

Figure 3. Coefficient of restitution for impact between end of rod inclined at $\theta = \pi/4, \pi/2$ or $3\pi/4$ and half space. Continuous curves are for smooth surfaces, $\mu = 0$, while dashed curves give the range of values resulting from all initial speeds of slip on a rough surface, $\mu = 0.5$. Friction has no effect on e_* for direct or collinear impact, $\theta = \pi/2$.

In all cases, if coefficient of friction is not too large, $\mu \leq 1$, the effect of friction on the coefficient of restitution is insignificant. There is only a second order effect that is entirely due to coupling between normal and friction forces in eccentric collisions of rough bodies. This second order effect is caused by friction slightly affecting maximum indentation in eccentric collisions. Friction also dissipates energy due to sliding between rough surfaces; this is in addition to and entirely separate from the energy loss represented by the energetic coefficient of restitution.

REFERENCES

Adams, G.G. and Tran, D.N., 1993, The coefficient of restitution for a planar two-body eccentric impact, *ASME J. Appl. Mech.* 60:1058.

Chang, W.R. and Ling, F.F., 1992, Normal impact model of rough surfaces. *ASME J. Appl. Mech.* 114:439.

Goldsmith, W., 1960, "Impact, the Theory and Physical Behaviour of Colliding Solids," Edward Arnold Pub., London.

Hertz, H., 1882, Uber die beruhrung fester elastischer korper (On the contact of elastic solids), *J. reine und angewundte Mathematik* 92:156.

Hogue, C. and Newland, D., 1994, Efficient computer simulation of moving granular particles, *Powder Tech.* 78:51.

Johnson, K.L., 1968, Experimental determination of contact stresses between plastically deformed cylinders and spheres, *in*: "Engineering Plasticity," J. Heyman and F. Leckie eds., Cambridge Univ. Press, UK.

Johnson, K.L., 1985, "Contact Mechanics," Cambridge University Press, UK.

Johnson, W., 1972, "Impact Strength of Materials," Edward Arnold Pub., London.

Lötstedt, P., 1981, *Z. angew. Math. Mech.* 61:605.

Stronge, W.J., 1990, Rigid body collisions with friction, *Proc. R. Soc. Lond.* A431:169.

Stronge, W.J., 1991, Unraveling paradoxical theories for rigid body collisions, *ASME J. Appl. Mech.* 58:1049.

Stronge, W.J., 1992, Energy dissipated in planar collision, *ASME J. Appl. Mech.* 59:681.

Wang, Y. and Mason, M.T., 1992, Two-dimensional rigid-body collisions with friction, *ASME J. Appl. Mech.* 59:635.

NUMERICAL SENSITIVITY OF A DYNAMICAL SYSTEM WITH DRY FRICTION AND UNILATERAL AND INTERMITTENT CONSTRAINTS

Matthias Storz and Peter Vielsack

Universität Karlsruhe
Institut für Mechanik

MECHANICAL SYSTEM

Vibratory driving of piles into granular soil is a common procedure in engineering practice. Rotating unbalances produce a harmonic driving force $F = F_0\ sin\Omega t$. The amplitude F_0 contains the square of the exciting frequency Ω. As an adequate model the total process can be described by a simple 1 DOF system with the coordinate x (see fig. 1) and the weight G (Verspohl, 1990). In the following only such types of motions are of interest when the pile tip can hit a rigid obstacle in a certain depth of the soil. As shown in Storz (1992) the wall resistance against longitudial motion is the Coulomb friction force with a constant value R_0 and the impact is assumed to be ideally plastic.

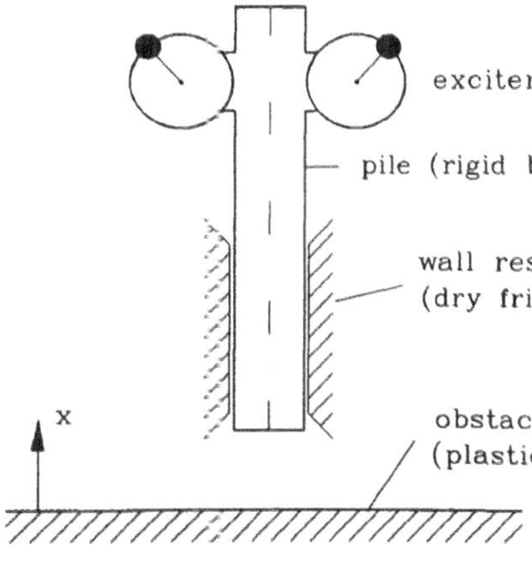

Figure 1. Mechanical system

Contact Mechanics, Edited by M. Raous *et al.*
Plenum Press, New York, 1995

EQUATION OF MOTION

The problem is governed by two specific control-parameters

$$\begin{aligned} A &= G/R_0 , \\ B &= F_0/R_0 , \end{aligned} \tag{1}$$

a dimensionless coordinate

$$\xi = \frac{G\Omega^2}{gR_0} x \tag{2}$$

and a normalized time scale

$$\tau = \Omega t . \tag{3}$$

The period of excitation is 2π. The equation of motion

$$\xi'' + sgn\xi' = B\ sin\tau - A\ ; \quad \xi > 0 \tag{4}$$

is only defined for positive values of ξ. Two analytical solutions for $\xi' > 0$ and $\xi' < 0$ can easily be given, but switching conditions are needed to combine both solutions.

If the pile moves downwards and strikes the obstacle, the conditions $\xi' = 0$ and $\xi = 0$ hold. Two cases are possible: Either there exists a sudden change to an upward motion or the system stops for a certain time interval as long as the inequality

$$B\ sin\tau - A < 1 \tag{5}$$

is valid. Then the motion equation degenerates to the description of a state with static equilibrium. Moreover a third case is possible: A reversal of motion can also take place without impact under the conditions $\xi > 0$ and $\xi' = 0$.

Solutions of the problem can be derived by piecewise adding the two analytical solutions of equation (4) and considering the switching conditions between three possible states of motion, namely $\xi' > 0$, $\xi' < 0$ and $\xi' \equiv 0$. The values of ξ and ξ' of a foregoing state at a certain switching time give the initial conditions for the following state. Then the whole problem is reduced to successive numerical calculations of zeros of transcendental equations to get the sequence of switching times during motion.

MODES OF MOTION

It is straight forward to imagine, that the nonlinear equation (4) in combination with unilateral and intermittent constraints and additionally the possibility of static equilibrium has an infinite number of solutions. They only depend on the two control-parameters A and B (Storz, 1992). As an example in fig. 2 a bifurcation diagram with the amplitude ξ plotted versus the exciting force B is shown, keeping the weight $A = 0.1$ fixed.

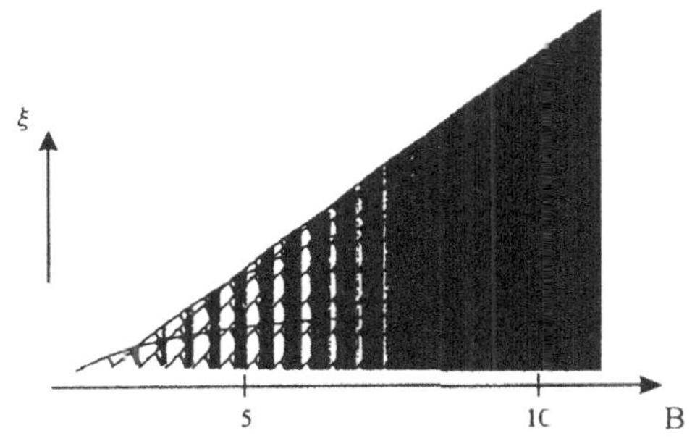

Figure 2. Bifurcation diagram

The response of the system has the same frequency as the excitation for small values of the exciting force B. Increasing the excitation leads to period doublings and other types of bifurcation. Windows with simple modes of motion (for which the system period is a low multiple of the exciting period) and complicated modes of motion (system period is much larger than the exciting period) follow one another. For large values of B the windows for simple modes become smaller and finally vanish.

NUMERICAL SENSITIVITY

It is still an open question how to achieve an accurate result for complicated modes of motion. The numerical calculation is rather unsatisfactory because of limited accuracy of computer algorithmic (hard- and software) non-optimal algorithms and non-exact determination of the sequence of switching times. In many cases it cannot be proved whether a numerically calculated chaotical behaviour really exists or successive transient motions induced by numerical errors are artifically created.

An example for a miscalculation is given in fig. 3. Here 200 000 Poincaré-sections with distance 2π (exciting period) are plotted for arbitrary chosen values $A = 0.5$ and $B = 16.3$ (Storz, 1994).

The exact result for the same values of A and B can be seen in fig. 4.

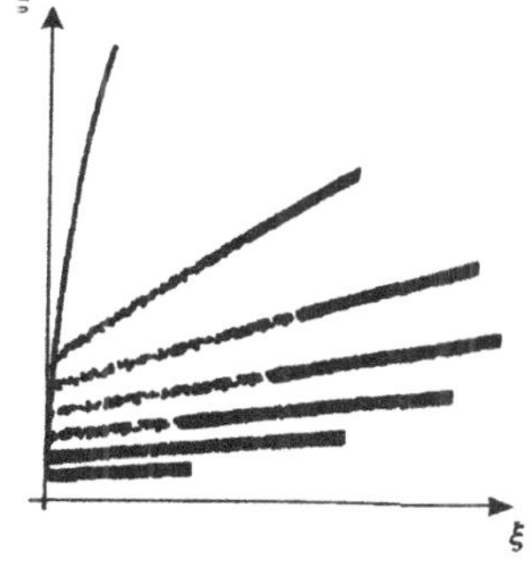

Figure 3. Poincaré-map: miscalculation

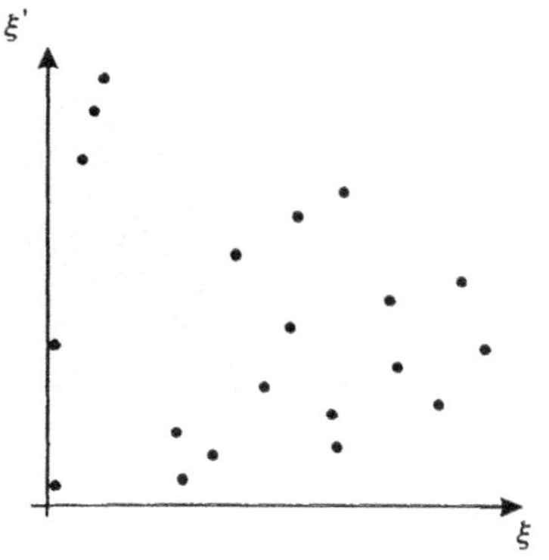

Figure 4. Poincaré-map: correct result

The difference between the diagrams is created by using different accuracy to calculate the sequence of switching times. In fig. 3 a time step $\Delta\tau = 4 * 10^{-4}$ is used which seems very small compared with the overall integration time of $200\,000 * 2\pi$. Speculations are possible in the interpretation of this result. Either the map shows a motion with extremely long period or even a strange attractor. Raising the accuracy to smaller values $\Delta\tau \leq 1 * 10^{-4}$ leads to the correct map in fig. 4. The motion has a period twenty times larger than the exciting period, which is indicated by 20 dots.

In fact, a long-time calculation over many thousands of exciting periods can lead to unavoidable small errors which nevertheless have great influence on the result. There is no clear distinction between numerically produced and real chaos. In addition numerical problems can arise even for simple modes of motion. If both the coordinate ξ and the velocity ξ' become zero within the same time step $\Delta\tau$ no correct decision about the mode of motion can be achieved. In the case if $\Delta\tau$ reaches the order of accuracy of numbers in the computer, then no reasonable calculation is possible.

REFERENCES

Verspohl, J., 1990, "Ungefesselte hysteretische Systeme unter besonderer Berücksichtigung des Vibrationsrammens", Thesis, Universität Karlsruhe.

Storz, M., 1992, Chaotic motions in pile-driving, in: "Soil Dynamics and Earthquake Engineering", Computational Mechanics Publications, Southampton, Boston, 503-512.

Storz, M., 1994, "Stabilität eines Reibschwingers mit Stoß am Beispiel des Vibrationsrammens", Thesis, Universität Karlsruhe.

VARIATIONAL PRINCIPLES FOR CONTACT PROBLEMS INCLUDING IMPACT PHENOMENA

P.D. Panagiotopoulos

Dept. of Civil Eng., Aristotle University
54006 Thessaloniki, Greece
Faculty of Mathematics and Physics, RWTH
52062 Aechen, F.R.G.

INTRODUCTION

Dynamic problems for deformable bodies include usually impact phenomena which play an important role in the whole evolution of the body behavior, within the time interval under consideration. This is obvious due to the abrupt change of the velocity because of the impact. The impact phenomenon was studied first by the ancient Greek engineer and mathematician Heron and by his students in the Alexandria School, following previous ideas by Aristotle and his students. In modern time Wallis, Wren and Huygens study the impact phenomenon in the framework of a competition of the Royal Society of London in 1668. Newton makes reference to the experimental results by Wren later in 1687. The results of Huygens concerning the conservation of momentum are based only on the Galilean principle of relativity and no use is made in these results of the principles of dynamics. At this point we should mention that the impact phenomenon is described by means of Newton's conjecture, which for two bodies subjected to onedimensional motion with velocities v_1 and v_2 before and v_1' and v_2' after impact reads

$$e(v_1 - v_2) = -(v_1' - v_2') \tag{1}$$

Here e is called the recovery coefficient, it is a dimensionless parameter and takes values in the interval $[0, 1]$. For $e = 0$ (resp. $e = 1$) we have an absolutely inelastic impact (resp. elastic impact). Note that in the reality e depends slightly on the velocities of the colliding bodies as it is experimentally verified. In the present paper we assume that Newton's conjecture holds for the impact of a mass point with a rigid support and ignoring (for the moment) the frictional effects in the tangential direction, which may cause discontinuity of the tangential velocities (cf. in this respect Painlévé paradoxon (1895) and the corresponding investigations of Klein, Prandtl et al. in the Zeitschrift

Contact Mechanics, Edited by M. Raous *et al.*
Plenum Press, New York, 1995

für mathematische Physik Vol. 58 (1909)) we may write the following relations

$$v_{T_+} = v_{T_-} \text{ and } v_{N_+} = -ev_{N_-} \tag{2}$$

Here v_T (resp. v_N) denotes the tangential (resp. the normal) velocity with respect to the support and the indices (+) and (−) denote that the velocity is considered just after and just before the impact moment. Concerning the mathematical treatment of the impact problem we refer to Monteiro Marques (1993), Paoli and Schatzman (1993), Panagiotopoulos and Liolios (1988) and Moreau (1988).

In the solution of a problem related to the motion of a deformable body subjected possibly to impact(s) with a support the exact determination of the moment(s) of impact(s) and of the point(s) of impact is of great importance. Especially in the numerical solution of dynamic problems of large structures the inaccurate determination of the time of impact may lead to erroneous final results. A well known "recipe" in Mechanics is that in order to determine a quantity one should formulate a variational expression containing the variation of this quantity. The aim of the present paper is the derivation of variational "principles" holding for deformable bodies subjected to impact effects. Thus we are obliged to consider variation of the time of impact, and of the place of impact analogously to Panagiotopoulos and Liolios (1988), Kozlov and Treshchëv (1991). The variational expressions here are derived by means of simple tools of the calculus of variations. We refer in this respect also to the doctoral thesis of Caratheodory (1904), who dealt with externals having discontinuous tangents (cf. the case of the motion of a material point subjected to impact).

THE IMPACT PROBLEM FOR A DISCRETE SYSTEM

We consider a discrete system of material particles having r degrees of freedom. The system is moving in the admissible domain

$$C = \{u \in \mathbb{R}^r,\ f(u) \geq 0\} \tag{3}$$

and may be subjected to an impact with the surface

$$\Sigma = \{u \in \mathbb{R}^r,\ f(u) = 0\}\ . \tag{4}$$

The functional f is assumed to be regular in the sense that grad$f \neq 0$ for $f = 0$. The kinetic energy of the system is given by the bilinear form

$$T = \frac{1}{2}\sum_{i=1}^{r} m_i \dot{u}_i^2\ , \tag{5}$$

where m_i denotes the mass corresponding to the i-th degree of freedom and $\dot{u}_i$ is the velocity $\frac{du_i}{dt}$. The geometric parameters of the system u_i are refereed to a fixed Cartesian orthogonal coordinate system, which is assumed as have the metric induced by the bilinear form (5). Possible impacts with the surface Σ are assumed to be absolutely elastic. We denote further by $v = \{\dot{u}_1, ..., \dot{u}_r\}$ the velocity of the system, by $v_- = \lim_{t\to o_-} \frac{du}{dt}$ the velocity before the impact and we denote by $(.,.)$ the scalar product for the metric (5), i.e.

$$(a, a^*) = \frac{1}{2}\sum_{i=1}^{r} m_i a_i a_i^* \tag{6}$$

We denote by n the outward unit normal vector to Σ at the point of the impact and let u_0 be the configuration of the system and t_0 the time of the impact.

We introduce further the normal and the tangential velocities of the system before impact v_{N_-} and v_{T_-} with respect to the surface Σ and the metric (6), i.e.

$$v = v_{T_-} + v_{N_-} \ , \quad v_{N_-} = (v_-, n)n \ , \quad v_{T_-} = v_- - v_{N_-} \tag{7}$$

and let v_{N_+} and v_{T_+} be the corresponding velocities after impact. One can write again the generalized impact law in the form (2), where now $e = 1$. We consider now the motion of the system in the admissible domain (non necessarily convex) C within a time interval $[t_1, t_2]$ which includes the time of the first impact of the system with Σ. We assume first that the forces acting on the system have a potential V, i.e. $\frac{\partial V}{\partial u_i} = -f_i$, $i = 1, ..., r$. From classical analytical Mechanics it is well known that the motion of the present system in the interior of C satisfies Hamilton's principle: the action integral $\int_{t_1}^{t_2} L(u, \dot{u})dt$ becomes stationary, i.e.

$$\delta \int_{t_1}^{t_2} L(u, \dot{u})dt = 0 \quad \delta u(t_1) = \delta u(t_2) = 0 \ , \tag{8}$$

where $L = T - V$. Here δ denotes the symbol of variations in the sense of the classical calculus of variations. Roughly speaking the variations ∂u are appropriately small deviations of the real trajectory $t \to u(t)$. Classical Hamilton' s principle holds as a stationarity principle. Additional assumptions about convexity (e.g. through the consideration in V of forces derived by a convex potential - or even superpotential) permit that one can derive Hamilton's principle as a minimum principle (Panagiotopoulos 1982). Note also at this point that in the int$C$ the energy $T + V$ is conserved. Now we assume that an impact occurs with $e = 1$. In this case energy is also conserved during the impact due to the relation (2) for $e = 1$. We recall here that one can prove the reflection law of the light by means of Fermat Principle of Optics which is analogous to Maupertuis Principle of Mechanics. Therefore it seems to be quite obvious that one should try to incorporate the case of impact into the Maupertuis "principle", or furthermore into Hamilton's "principle". Recall here the relation between the above two "principles" for classical trajectories, i.e. without impact phenomena. We prove now the following result.

Proposition 1: Let us consider a time interval (t_1, t_2) including one impact on Σ of the system considered. A trajectory $t \to \tilde{u}(t) \in \mathbb{R}^r$ is a motion of the system, if and only if it is a solution of the stationarity problem (8) of the action functional.

Proof: Let $\tilde{t}$ be the time of impact, $\tilde{t} \in (t_1, t_2)$. The trajectory of the system $t \to \tilde{u}(t)$ satisfies the relations $f(\tilde{u}(t)) > 0$ for all $t \in (t_1, \tilde{t})$ and $t \in (\tilde{t}, t_2)$, $t \neq \tilde{t}$ and $f(\tilde{u}(\tilde{t})) = 0$. We consider now a set of trajectories depending on a parameter ε, $\varepsilon \in (-\alpha, +\alpha)$ $\alpha > 0$, i.e., $t \to u_\varepsilon(t) \in \mathbb{R}^r$, such that

i) they are well-defined for all $\varepsilon \in (-\alpha, +\alpha)$,

ii) they have the property that $u_0(t) = \tilde{u}(t)$ for all $t \in (t_1, t_2)$,

iii) $u_\varepsilon(t_1) = \tilde{u}(t_1), u_\varepsilon(t_2) = \tilde{u}(t_2)$ for all $\varepsilon \in (-\alpha, +\alpha)$.

Moreover we consider that the time of impact $\tilde{t}$ is subjected to a variation depending on ε and let t_ε be an appropriately smooth function of ε such that $t_0 = \tilde{t}$.

Accordingly we will have that $f(u_\varepsilon(t_\varepsilon)) = 0$. Finally we assume that the function $(\varepsilon, t) \to u_\varepsilon(t) : (-\alpha, \alpha) \times [t_1, t_2] \to \mathbb{R}^r$ is appropriately smooth in the domains $(-\alpha, \alpha) \times [t_1, t_\varepsilon]$ and $(-\alpha, \alpha) \times [t_\varepsilon, t_2]$. Further let us define the perturbed action integral

$$I(\varepsilon) = \int_{t_1}^{t_2} L(u_\varepsilon(t), \dot{u}_\varepsilon(t))dt = \int_{t_1}^{t_2} L_\varepsilon dt \tag{9}$$

We will calculate its variation (classical)

$$\delta I = \left.\frac{dI(\varepsilon)}{d\varepsilon}\right|_{\varepsilon=0} . \tag{10}$$

At the time of impact (the perturbed one) t_ε, $t \to u_\varepsilon(t)$ has only left and right time derivatives. Thus we write

$$\left.\frac{dI}{d\varepsilon}\right|_{\varepsilon=0} = \frac{d}{d\varepsilon}\left(\int_{t_1}^{t_\varepsilon} L_\varepsilon dt + \int_{t_\varepsilon}^{t_2} L_\varepsilon dt\right)\Big|_{\varepsilon=0} . \tag{11}$$

We have now that

$$\frac{d}{d\varepsilon}\int_{t_1}^{t_\varepsilon} L(u_\varepsilon(t), \dot{u}_\varepsilon(t))dt = \int_{t_1}^{t_\varepsilon}\left[\frac{\partial L}{\partial \dot{u}_\varepsilon}\frac{\partial \dot{u}_\varepsilon}{\partial \varepsilon} + \frac{\partial L}{\partial u_\varepsilon}\frac{\partial u_\varepsilon}{\partial \varepsilon}\right] dt + L_\varepsilon \frac{dt_\varepsilon}{d\varepsilon} =$$

$$= \frac{\partial L}{\partial \dot{u}_\varepsilon}\frac{\partial u_\varepsilon}{\partial \varepsilon}\Big[_{t_1}^{t_\varepsilon} - \int_{t_1}^{t_\varepsilon}\frac{\partial}{\partial t}\left(\frac{\partial L}{\partial \dot{u}_\varepsilon}\right)\frac{\partial u_\varepsilon}{\partial \varepsilon}dt + L_\varepsilon\frac{dt_\varepsilon}{d\varepsilon} + \int_{t_1}^{t_\varepsilon}\frac{\partial L}{\partial u_\varepsilon}\frac{\partial u_\varepsilon}{\partial \varepsilon}dt . \tag{12}$$

Moreover

$$\frac{\partial u_\varepsilon}{\partial \varepsilon}(t_\varepsilon) = \frac{d}{d\varepsilon}(u_\varepsilon(t_\varepsilon)) - \frac{\partial u_\varepsilon}{\partial t_\varepsilon}\frac{dt_\varepsilon}{d\varepsilon} \tag{13}$$

and

$$\frac{\partial u_\varepsilon}{\partial \varepsilon}(t_1) = 0 . \tag{14}$$

From (13) (14) we obtain that

$$\frac{\partial L}{\partial \dot{u}_\varepsilon}\frac{\partial u_\varepsilon}{\partial \varepsilon}(t_\varepsilon) + L_\varepsilon\frac{dt_\varepsilon}{d\varepsilon} = \frac{\partial L}{\partial \dot{u}_\varepsilon}\frac{d}{d\varepsilon}(u_\varepsilon(t_\varepsilon)) + \left[L(u_\varepsilon, \dot{u}_\varepsilon) - \frac{\partial L}{\partial \dot{u}_\varepsilon}\frac{du_\varepsilon}{dt_\varepsilon}\right]\frac{dt_\varepsilon}{d\varepsilon} \tag{15}$$

From (12) and (15) we obtain that (cf. also eq. (6))

$$\frac{d}{d\varepsilon}\int_{t_1}^{t_\varepsilon} L(u_\varepsilon, \dot{u}_\varepsilon)dt\Big|_{\varepsilon=0} = (\tilde{v}_-, w) - \left[\frac{(\tilde{v}_-, \tilde{v}_-)}{2} + V(\tilde{u}(\tilde{t}))\right]\eta +$$

$$+ \int_{t_1}^{\tilde{t}}\left[\frac{\partial L}{\partial u} - \frac{d}{dt}\left(\frac{\partial L}{\partial \dot{u}}\right)\right]_{\tilde{u}(t)} \delta\tilde{u}(t)dt \tag{16}$$

where $\tilde{v} = \dot{\tilde{u}}$, and

$$\delta\tilde{u}(t) = \frac{\partial u_\varepsilon}{\partial \varepsilon}(t)\Big|_{\varepsilon=0}, \quad \eta = \left.\frac{dt_\varepsilon}{d\varepsilon}\right|_{\varepsilon=0} = \delta\tilde{t} , \quad w = \frac{d}{d\varepsilon}(u_\varepsilon(t_\varepsilon))\Big|_{\varepsilon=0} = \delta\tilde{\tau} . \tag{17}$$

Here $\delta\tilde{\tau}$ denotes the variation (classical) of the tangential vector to Σ at the point of impact, (recall the geometrical meaning of $\frac{du_\varepsilon}{dt_\varepsilon}$ at $\varepsilon = 0$) and $\delta\tilde{t}$ the variation (classical) of the time of impact. Moreover in order to derive (16) we have used the relation

$$L_\varepsilon - \frac{\partial L_\varepsilon}{\partial \dot{u}_{\varepsilon_i}} \dot{u}_{\varepsilon_i} = -(T+V)(u_\varepsilon(t)) \ , \tag{18}$$

i.e. for the perturbed trajectory, and then we take $\varepsilon = 0$. Analogously we find that

$$\frac{d}{d\varepsilon} \int_{t_\varepsilon}^{t_2} L(u_\varepsilon, \dot{u}_\varepsilon) dt \Big|_{\varepsilon=0} = -(\tilde{v}_+, \delta\tilde{\tau}) + \left[\frac{(\tilde{v}_+, \tilde{v}_+)}{2} + V(\tilde{u}(\tilde{t}))\right] \delta\tilde{t} +$$

$$+ \int_{\tilde{t}}^{t_2} \left[\frac{\partial L}{\partial u} - \frac{d}{dt}\left(\frac{\partial L}{\partial \dot{u}}\right)\right]_{\tilde{u}(t)} \delta\tilde{u}(t) dt \ . \tag{19}$$

Accordingly

$$\delta I = (\tilde{v}_- - \tilde{v}_+, \delta\tilde{\tau}) - (T(\tilde{v}_-) - T(\tilde{v}_+))\delta\tilde{t} + \int_{t_1}^{\tilde{t}} \left[\frac{\partial L}{\partial u} - \frac{d}{dt}\left(\frac{\partial L}{\partial \dot{u}}\right)\right]_{\tilde{u}(t)} \delta\tilde{u}(t) dt +$$

$$+ \int_{\tilde{t}}^{t_2} \left[\frac{\partial L}{\partial u} - \frac{d}{dt}\left(\frac{\partial L}{\partial \dot{u}}\right)\right]_{\tilde{u}(t)} \delta\tilde{u}(t) dt \ . \tag{20}$$

From (20) we may prove that the equations of motions and the impact low imply that $\delta I = 0$ and conversely that $\delta I = 0$ implies the equations of motion and the impact law. Indeed from (20) we obtain by assuming that $\delta\tilde{t} = 0$, i.e. $t_\varepsilon = \tilde{t}$, and $\delta\tilde{\tau} = 0$, i.e. $u_\varepsilon(t_\varepsilon) = u_\varepsilon(\tilde{t}) = \tilde{u}(\tilde{t})$, for all ε, that $\delta I = 0$ implies the validity of the equations of motion

$$\frac{\delta L}{\delta u} - \frac{d}{dt}\left(\frac{\partial L}{\partial \dot{u}}\right) = 0 \quad \text{for} \quad t \in [t_1, \tilde{t}) \cup (\tilde{t}, t_2] \ . \tag{21}$$

Moreover we consider variations with $\delta\tilde{t} = 0$. They imply that

$$\delta I = (\tilde{v}_- - \tilde{v}_+, \delta\tilde{\tau}) = 0 \tag{22}$$

for all vectors $\delta\tilde{\tau}$ tangent to Σ at the impact point. Thus (22) implies that

$$\tilde{v}_{T_+} = \tilde{v}_{T_-} \ . \tag{23}$$

From (21)(23) and $\delta I = 0$ we obtain that

$$T(\tilde{v}_-) = T(\tilde{v}_+) \ ; \tag{24}$$

(23),(24) yield that $v_{N_+} = v_{N_-}$, q.e.d.

Analogous proposition can be proved for the case of k (finite) times of impact. We can show easily that (20) may be replaced by

$$\delta I = \sum_{j=1}^{k} \left[(\tilde{v}_{k_-} - \tilde{v}_{k_+}, \delta\tilde{\tau}_k) - (T(\tilde{v}_{k_-}) - (T(\tilde{v}_{k_+}))\delta\tilde{t}_k\right] +$$

$$+ \int_{t_1}^{\tilde{t}_1} \ldots \delta\tilde{u}(t) dt + \int_{\tilde{t}_1}^{\tilde{t}_2} \ldots \delta\tilde{u}(t) dt + \ldots + \int_{\tilde{t}_k}^{t_2} \ldots \delta\tilde{u}(t) dt \tag{25}$$

Note also that the term under the integrals $\int_{t_1}^{\bar{t}} \ldots$ and $\int_{\bar{t}_1}^{t_2} \ldots$ in (20) is actully the classical expression of d'Alembert's principle, i.e.,

$$\int_{t_1}^{\bar{t}} \left[\frac{\partial L}{\partial u} - \frac{d}{dt}\left(\frac{\partial L}{\partial \dot{u}}\right) \right]_{\tilde{u}(t)} \delta\tilde{u}(t)dt = \int_{t_1}^{\bar{t}} (f_i - m_i\ddot{\tilde{u}})\delta\tilde{u}(t)dt \quad \text{etc.} \tag{26}$$

THE IMPACT PROBLEM FOR DEFORMABLE BODIES

Let us consider in the framework of small strain and small displacement theory a linear elastic body occupying an open bounded subset $\Omega \in \mathbb{R}^3$. Let Γ be the boundary of the body, assumed to be appropriately regular. The boundary is divided into three nonoverlapping parts Γ_F, Γ_U and Γ_I. On Γ_F the boundary forces $S_i = \sigma_{ij}n_j = F_i$ are given (here $\sigma = \{\sigma_{ij}\}$ $i,j = 1,2,3$ is the stress tensor and $n = \{n_i\}$ is the outward unit normal vector to Γ), on Γ_U the boundary displacements are prescribed (we take for the sake of simplicity that $u_i = 0$), and on Γ_I absolutely elastic impact may occur with a rigid support. On Γ_I do not act any prescribed forces.

Again we begin by considering the action functional of the body, which now has the form

$$I = \int_{t_1}^{t_2} [\int_\Omega \frac{1}{2}\rho\dot{u}^2 d\Omega - \Pi(u)]dt = \int_{t_1}^{t_2} \mathcal{L}(u,\dot{u})dt\ . \tag{27}$$

Here $\Pi(.)$ is the potential energy of the body

$$\Pi(u) = \frac{1}{2}a(u,u) - \int_\Omega f_i u_i d\Omega - \int_F F_i u_i d\Gamma \tag{28}$$

with $a(.,.)$ the bilinear form of linear elasticity, i.e.

$$a(u,v) = \int_\Omega C_{ijhk}\varepsilon_{ij}(u)\varepsilon_{ij}(v)d\Omega\ , \tag{29}$$

where $C = \{C_{ijhk}\}$ the Hooke's linear elasticity (anisotropic) tensor having the well-know symmetry and ellipticity properties and $\varepsilon = \{\varepsilon_{ij}\}$ is the strain tensor. Moreover $f = \{f_i\}$ are the given body forces, and ρ is the density of the body assumed for simplicity to be constant. The following proposition holds.

Proposition 2: Proposition 1 holds for the defined linear elastic body in the case of absolutely elastic impact with a rigid support.

Proof: We consider appropriate variations of the displacement field $(t,x) \to u(x,t)$ of the body such that all points of Γ_I have perturbed trajectories like the ones in the proof of Prop. 1 and the remaining points of $\bar{\Omega} - \Gamma_I$ have arbitrary perturbed trajectories $u^* - u = \delta u$ compatible with the constraints on Γ_U. Moreover the time of the first impact $\tilde{t}$ is subjected to a variation and let t_ε be its value which is an appropriately smooth function of ε as in the previous section. We obtain as before that

$$\frac{d}{d\varepsilon}\int_{t_1}^{t_\varepsilon} \mathcal{L}(u_\varepsilon, \dot{u}_\varepsilon)dt\bigg|_{\varepsilon=0} = \int_{\Gamma_I} \rho\tilde{v}_{i-}\delta\tilde{\tau}_i d\Gamma - [T(\tilde{v}_-) + \Pi(\tilde{u})]\delta\tilde{t} -$$

$$- \int_{t_1}^{\tilde{t}} \left\langle \left(\frac{d}{dt}\frac{\partial \mathcal{L}}{\partial \dot{u}} - \frac{\partial \mathcal{L}}{\partial u} \right)_{\tilde{u}(t)}, \delta\tilde{u}(t) \right\rangle dt \tag{30}$$

and

$$\frac{d}{d\varepsilon}\int_{t_\varepsilon}^{t_2}\mathcal{L}(u_\varepsilon,\dot{u}_\varepsilon)dt\Big|_{\varepsilon=0} = -\int_{\Gamma_I}\rho\tilde{v}_{i+}\delta\tilde{\tau}_i d\Gamma - [T(\tilde{v}_+)+\Pi(\tilde{u})]\delta\tilde{t}-$$

$$-\int_{\tilde{t}}^{t_2}\left\langle\left(\frac{d}{dt}\frac{\partial\mathcal{L}}{\partial\dot{u}}-\frac{\partial\mathcal{L}}{\partial u}\right)_{\tilde{u}(t)},\delta\tilde{u}(t)\right\rangle dt \qquad (31)$$

where now $\dot{u}=v$, and v_+ and v_- denote the velocity after and before the impact at $t=\tilde{t}$. Accordingly we may write that

$$\delta I = \int_{\Gamma_I}\rho(\tilde{v}_{i-}-\tilde{v}_{i+})\delta\tilde{\tau}_i d\Gamma - [T(\tilde{v}_-)-T(\tilde{v}_+)]\delta\tilde{t}-$$

$$-\int_{t_1}^{\tilde{t}}\left\langle\left(\frac{d}{dt}\frac{\partial\mathcal{L}}{\partial\dot{u}}-\frac{\partial\mathcal{L}}{\partial u}\right)_{\tilde{u}(t)},\delta\tilde{u}(t)\right\rangle dt - \int_{\tilde{t}}^{t_2}\left\langle\left(\frac{d}{dt}\frac{\partial\mathcal{L}}{\partial\dot{u}}-\frac{\partial\mathcal{L}}{\partial u}\right)_{\tilde{u}(t)},\delta\tilde{u}(t)\right\rangle dt . \qquad (32)$$

In the above formulas $\langle,\rangle$ denotes the linear form (or if one introduces function spaces, the duality paining). Thus we have that

$$\left\langle\left(\frac{d}{dt}\frac{\partial\mathcal{L}}{\partial\dot{u}}-\frac{\partial\mathcal{L}}{\partial u}\right)_{\tilde{u}(t)},\delta\tilde{u}(t)\right\rangle = \int_\Omega\rho\ddot{\tilde{u}}_i\delta\tilde{u}_i d\Omega + a(\tilde{u},\delta\tilde{u})-$$

$$-\int_\Omega f_i\delta\tilde{u}_i d\Omega - \int_{\Gamma_F}F_i\delta\tilde{u}_i d\Gamma . \qquad (33)$$

By means of the Green-Gauss theorem we obtain that the right hand side of (33), denoted by A, takes the form

$$A=\int_\Omega(-\sigma_{ij,j}-f_i+\rho\ddot{\tilde{u}}_i)\delta\tilde{u}_i d\Omega + \int_{\Gamma_F}(S_i-F_i)\delta\tilde{u}_i d\Gamma + \int_{\Gamma_I}S_i\delta\tilde{u}_i d\Gamma , \qquad (34)$$

where the well-known Hooke's law and the linear strain-displacement relations hold. Accordingly (32) and (34) imply that

$$\delta I = \int_{\Gamma_I}\rho(\tilde{v}_{i-}-\tilde{v}_{i+})\delta\tilde{\tau}_i d\Gamma - [T(\tilde{v}_-)-T(\tilde{v}_+)]\delta\tilde{t} + \int_{t_1}^{t_2}\int_\Omega(\sigma_{ij,j}+f_i-\rho\ddot{\tilde{u}}_i)\delta\tilde{u}_i d\Omega dt-$$

$$-\int_{t_1}^{t_2}\int_{\Gamma_F}(S_i-F_i)\delta\tilde{u}_i d\Gamma dt - \int_{t_1}^{\tilde{t}}\int_{\Gamma_I}S_i\delta\tilde{u}_i d\Gamma dt - \int_{\tilde{t}}^{t_2}\int_{\Gamma_I}S_i\delta\tilde{u}_i d\Gamma dt , \qquad (35)$$

The last two integrals concerning the part of the boundary Γ_I on which impacts may occur at $t=\tilde{t}$ are equal to zero because on Γ_I do not act any given forces for $t<\tilde{t}$ and $t>\tilde{t}$. Moreover the integrals $\int_{t_1}^{\tilde{t}}$ ... and $\int_{\tilde{t}}^{t_2}$ ... of (33) are written as an one integral over (t_1,t_2) in the remaining expressions of (33) because $\ddot{\tilde{u}}_i$ is well defined in the interior of the body. It is not defined at the points of Γ_I at $t=\tilde{t}$ which are subjected to an impact with the support.

The proof of the proposition is completed as in prop. 1 by considering that $\delta I=0$ and by taking special variations as in prop. 1. q.e.d.

Certain remarks are important .

i) In our proofs we have considered separately the singular part of the problem resulting from the impact phenomenon.

ii) We have assumed that the functions of the problem are appropriately smooth. Since

we consider separately the singularity one could work more rigourously within the Sobolev space framework. These considerations justify the vanishing of the two last terms in (35).
iii) Again $\delta\tilde{\tau}$ denotes the variation of the tangent vector to the support at the point of impact.
iv) The admissible area defined by the support is not necessarily convex. Moreover the support is not necessarily a convex or a concave surface.

EXTENSIONS AND GENERALIZATIONS

Now from the expression of action variation

$$\delta I = \int_{\Gamma_I} \rho(\tilde{v}_{i-} - \tilde{v}_{i+})\delta\tilde{\tau}_i d\Gamma - [T(\tilde{v}_-) - T(\tilde{v}_+)]\delta\tilde{t} + \int_{t_1}^{t_2}\int_{\Omega}(\sigma_{ij,j} + f_i - \rho\ddot{\tilde{u}}_i)\delta\tilde{u}_i d\Omega dt -$$

$$- \int_{t_1}^{t_2}\int_{\Gamma_F}(S_i - F_i)\delta\tilde{u}_i d\Gamma dt \ , \tag{36}$$

where σ_{ij} and ε_{ij} are connected with Hooke's law and ε_{ij} and u_i with the classical strain-displacement relations. We may deduce the following results.
1) Relation (36) implies that during the impact the kinetic energy is conserved. This fact results from $\delta I = 0$ by taking $\delta\tilde{u}_i = 0$, $\delta\tilde{\tau}_i = 0$.
2) One could consider the Newton's conjecture as a subsidiary condition for the problem. Relation (36) implies that d' Alembert's principle, i.e. the two last terms in $\delta I = 0$, is modified by adding via Lagrange multipliers the two subsidiary conditions describing the absolutely elastic impact. Thus one can build up the theory by starting from d' Alembert's principle. A dimensional analysis shows that the Lagrange multipliers introducing the impact relations have the meaning of $\delta\tilde{\tau}$ and $\delta\tilde{t}$ as in (36).
3) One can build up the theory by considering the equations of motion and the impact relations in the time interval $[t_1, t_2]$, i.e.

$$\sigma_{ij,j} + f_i - \rho\ddot{u}_i = 0 \quad \text{in} \quad \Omega \times [t_1, t_2] \tag{37}$$

$$\varepsilon_{ij} = C_{ijhk}\sigma_{ij} \quad \text{in} \quad \Omega \times [t_1, t_2] \tag{38}$$

$$\varepsilon_{ij} = \frac{1}{2}(u_{i,j} + u_{j,i}) \quad \text{in} \quad \Omega \times [t_1, t_2] \tag{39}$$

$$u_i = 0 \quad \text{on} \quad \Gamma_U \times [t_1, t_2] \tag{40}$$

$$S_i = F_i(x, t) \quad \text{on} \quad \Gamma_F \times [t_1, t_2] \tag{41}$$

$$S_i = 0 \quad \text{on} \quad \Gamma_I \times ([t_1, t_2] - \{\tilde{t}\}) \tag{42}$$

$$v_{T+} = v_{T-} \quad \text{on} \quad \Gamma_I \quad \text{at} \quad t = \tilde{t} \tag{43}$$

$$T(v_-) = T(v_+) \quad \text{at} \quad t = \tilde{t} \tag{44}$$

Both the point(s) $\tilde{x}$ of impact and the time $\tilde{t} \in [t_1, t_2]$ of impact are not given. Now we can multiply (37) by a variation $u^* - u$ compatible with the kinematic constraint on Γ_U and by taking into account (38), (39) and the boundary conditions we obtain by means of Green-Gauss theorem the equivalent formulation: Find $u = u(x,t)$ such as to satisfy (40), the variational equality

$$a(u, v-u) = \int_\Omega \tilde{f}_i(v_i - u_i)d\Omega + \int_{\Gamma_F} F_i(v_i - u_i)d\Gamma, \quad \tilde{f}_i = f_i - \rho\ddot{u}_i \tag{45}$$

for every v kinematically admissible and for every $t \in [t_1, t_2]$ and then at $t = \tilde{t}$ the impact relation (43) (44).
Integrating in the time interval the above formulation leads to $\delta I = 0$, where δI is given by (32), with the only difference that instead of $\delta\tilde{u}_i$ we will have $v_i - u_i$.
4) Let us suppose now that on a part Γ_S of the boundary a boundary condition of the type

$$-S \in \partial j(u) \tag{46}$$

holds, where j is a convex superpotential (i.e. a l.s.c., convex, proper functional on $\mathbb{R}^3$) and ∂ denotes the subdifferential. We introduce $J(u) = \{\int_{\Gamma_S} j(u)dx$ if $j(u(.))$ is integrable, ∞ otherwise $\}$ and then we have instead of (44) the variational inequality (cf. P.D.Panagiotopoulos (1985), (1993))

$$a(u, v-u) + J(v) - J(u) \geq \int_\Omega \tilde{f}_i(v_i - u_i)d\Omega + \int F_i(v_i - u_i)d\Gamma, \quad \tilde{f}_i = f_i - \rho\ddot{u}_i \tag{47}$$

It is easy now to obtain that the dynamic impact problem is governed by the variational inequality

$$\delta I = \int_{t_1}^{t_2} [a(\tilde{u}, \delta\tilde{u}) + J(\tilde{u} + \delta\tilde{u}) - J(\tilde{u}) - \int_\Omega f_i \delta\tilde{u}_i d\Omega - \int_{\Gamma_F} F_i \delta\tilde{u}_i d\Gamma] dt +$$

$$+ \int_{t_1}^{t_2} \int_\Omega \rho\ddot{\tilde{u}}_i \delta\tilde{u}_i d\Omega dt + \int_{\Gamma_I} \rho(\tilde{v}_{i_-} - \tilde{v}_{i_+})\delta\tilde{\tau}_i d\Gamma - [T(\tilde{v}_-) - T(\tilde{v}_+)]\delta\tilde{t} \geq 0 \tag{48}$$

which holds for every $\delta\tilde{u}, \delta\tilde{\tau}_i$ compatible with the constraints (kinematic constraints and admissible domain) and $\delta\tilde{t}$. We call (48) the action variational inequality. An analogous one has been derived in Panagiotopoulos (1982) for the case of systems of material points without impact effects.
5) Action inequalities are also obtained in the case of nonconvex superpotentials (Panagiotopoulos 1985, 1993). Suppose that instead of (46) a relation of the type

$$-S \in \bar{\partial} j(u) \quad \text{on } \Gamma_S \tag{49}$$

holds, where now j is a locally Lipschitz functional and $\bar{\partial}$ denotes the generalized gradient of Clarke. Then instead of (48) the impact problem is governed by the hemivariational inequality for the action

$$\delta I = \int_{t_1}^{t_2} [a(\tilde{u}, \delta\tilde{u}) + \int_{\Gamma_S} j^0(\tilde{u}, \delta\tilde{u})d\Gamma - \int_\Omega f_i \delta\tilde{u}_i d\Omega - \int_{\Gamma_F} F_i \delta\tilde{u}_i d\Gamma] dt +$$

$$+ \int_{t_1}^{t_2} \int_\Omega \rho\ddot{\tilde{u}}_i \delta\tilde{u}_i d\Omega dt + \int_{\Gamma_I} \rho(\tilde{v}_{i_-} - \tilde{v}_{i_+})\delta\tilde{\tau}_i d\Gamma - [T(\tilde{v}_-) - T(\tilde{v}_+)]\delta\tilde{t} \geq 0 \tag{50}$$

which holds for every $\delta\tilde{u}_i$ and $\delta\tilde{\tau}_i$ compatible with the constraints and for all $\delta\tilde{t}$.

6) Analogous results we obtain if Hooke's law of elasticity is replaced by a monotone holonomic law derived by a convex superpotential i.e. by

$$\sigma \in \partial w(\varepsilon) \tag{51}$$

where w is a convex, l.s.c, proper functional. Then $a(\tilde{u}, \delta\tilde{u})$ in (48) and (50) is replaced by $W(\varepsilon(\tilde{u} + \delta\tilde{u})) - W(\varepsilon(\tilde{u}))$, where $W(\varepsilon) = \{\int_\Omega w(\varepsilon)d\Omega$ if $w(\varepsilon(.))$ is integrable, ∞ otherwise $\}$.

7) In the case of large displacements and for any type of reference system we place ourselves in the framework of Truesdell and Toupin (1960) p.604 and we write the action variation in the corresponding form. Then we proceed as in the small deformation problem.

8) For a discretized structure, e.g. by the F.E.M., the variational equality (35) when written for the discretized problem constitutes a criterion for the determination of the impact time and place especially if an initial rough estimation is known. Indeed we put it in the form

$$\int_{t_1}^{t_2} (Ku - p - M\ddot{u})^T (u^* - u)dt + (v_{T_-} - v_{T_+})^T diag(m_i)(\tau^* - \tau) -$$

$$-[T(v_-) - T(v_+)](t^* - t) = 0 \tag{52}$$

where K is the stiffness matrix, M is the mass matrix, p is the loading vector, u is the displacement vector, m_i denote the mass related to the degree of freedom i on Γ_I, τ is the corresponding boundary segment length vector and v_T denotes the vector of the tangential velocities. Here u, τ and t correspond to the impact configuration.

REFERENCES

Monteiro Marques.M.D.P, 1993, " Differential Inclusions in Nonsmooth Mechanical Problems. Shocks and Dry Friction", Birkhäuser Verlag, Basel.

Paoli.L, Schatzman.M, 1993, Mouvement a un nombre fini de degrés de liberté avec contraintes unilatérales: cas avec perte d'énergie, *Math. Modelling and Numerical Analysis* 27:673.

Panagiotopoulos.P.D, Liolios.A, 1989, On the dynamics of inelastic shocks, in: "Proc. Greek-German Seminar on Struct. Dyn. and Earthq. Eng. Athens Dec. 16- 19, 1988", A.N.Kounadis and W.B.Krätzig, ed., pp. 12-18, Hellenic Soc. Theor. Appl. Mech., Athens.

Moreau.J.J, 1988, Bounded variation in time, in: "Topics in Nonsmooth Mechanics", J.J.Moreau P.D.Panagiotopoulos, G. Strang, ed., Birkhäuser Verlag, Basel.

Kozlov.V.V, Treshchër.D.V, 1991, Billiards. A genetic introduction to the dynamics of systems with impacts, *Tranl. of Math. Monographs* 89, A.M.S, Providence.

Caratheodory.C, 1904, Ueber die diskontinuirlichen Lösungen in der Variationsrechnung, Göttingen.

Panagiotopoulos.P.D, 1982, Ungleichungsprobleme und Differentialinklusionen in der analytischen, Mechanik, in: "Annual Proc. School of Technology, Aristotle University, Vol. Θ' (1982) 100-140" and Appendix in: "Analytische Mechanik", C.Heinz, RWTH Aachen.

Panagiotopoulos.P.D, 1985, "Inequality Problems in Mechanics and Applications. Convex and Nonconvex Energy Functions", Birkhäuser Verlag, Basel, Boston.

Panagiotopoulos.P.D, 1993, "Hemivariational Inequalities. Applications in Mechanics and Engineering", Springer Verlag, Berlin.

Truesdell.C and Toupin.R.A, 1960, The classical field theories, in: "Handbuch der Physik", S.Flügge, ed., Springer Verlag, Berlin.

A VELOCITY FORMULATION FOR THE CONTACT PROBLEMS

Claude Bohatier

LMGC, URA CNRS 1214
Université Montpellier II, place E. Bataillon
34195 Montpellier Cedex FRANCE

INTRODUCTION

The aim of this paper is to present a way to integrate the boundary contact conditions in a velocity formulation. The control of the penetration is considered during the evolution process within each step of time. When a point come into contact two stage are defined the first is the point is coming into contact, the second is the point can stay or leave the contact.

Several choice of numerical modelling are classically available to take into account the boundary contact conditions without or with friction. The numerical simulations demonstrate that some of them are less or more efficient according to the type of problem. The methods which consist to an implicit projection of any point coming into contact leads to good results by a short time of computation. However, in the cases where the contact conditions evolve, it is often necessary to control the sign of the normal forces which does not involves to an easy implementation and is not compatible with a reliability objective.

This paper propose another way based on a velocity formulation coming from the virtual work principle that control the geometric evolution within any step of time. The considered solid can be rigid or deformable with any kind of behavior. The tests examples are chosen in dynamic solicitation of elastic or viscoplastic solids.

PROBLEM MODELLING

Large deformations or large displacements can be considered. The main variable is the velocity field. The velocity field $\mathbf{v}$ is solution of the equation (1) involving from the virtual work principle (3).

Problem :

To find $\mathbf{v}$ that satisfy the evolution equation (1) on the time interval [t0,t1] :

$$\phi(\mathbf{v}, \bar{\Omega}) = 0 \qquad (1)$$

Contact Mechanics, Edited by M. Raous *et al.*
Plenum Press, New York, 1995

with the initial conditions at any step of time :

$$\mathbf{v}(t_0) = \mathbf{v}_0$$

$$\bar{\Omega}_0 = (\{\mathbf{x}(t_0)\} \; ; \; \mathbf{x}(t_0)= \mathbf{x}_0) \; : \text{solid domain}$$

The domain and its boundary $\bar{\Omega}$ at time t are estimated by

$$\mathbf{x} = \mathbf{x}_0 + \int_{t_0}^{t} \mathbf{v} \; dt \tag{2}$$

$\phi(\mathbf{v}, \bar{\Omega})$ is defined by the integration :

$$\phi(\mathbf{v}, \bar{\Omega}) = \int_{t_0}^{t_1} P^* \; dt \tag{3}$$

where :

$$P = P^*i + P^*e - P^*a \tag{4}$$

$$P^*e = \int_{\Omega} \rho \; \mathbf{f} \, . \, \mathbf{v}^* \; d\Omega \quad + \int_{\partial\Omega} \mathbf{F} . \mathbf{v}^* \; da \tag{5}$$

$$P^*a = - \int_{\Omega} \rho \frac{d\mathbf{v}}{dt} . \mathbf{v}^* \; d\Omega \tag{6}$$

when the solid is deformable :

$$P^*i = - \int_{\Omega} \sigma : \mathbf{D}^* \; d\Omega \tag{7}$$

when the solid is rigid then $P^*i = 0$

σ is the Cauchy stress tensor

$\mathbf{v}^*$ is the virtual velocity field and $\mathbf{D}^*$ is the associated strain rate tensor.

$\mathbf{v}^*$ is chosen as a compatible field with the contact conditions,

Therefore:
$\mathbf{F}_{nij} \cdot \mathbf{v}^*_{nji} = 0$ for any time t within the considered step of time

The normal force of the solid S_i applied onto the solid S_j and the relative velocity of the solid S_j in regard from the solid S_j are complementary vectors:

$$F_{nji} \cdot v_{nij} = 0 \tag{8}$$

The geometric condition which limits the variation domain of the normal velocity during a time interval Dt=h from a gap d :

$$\int_{\Delta t} v_{n_{ij}} \, dt + d \geq 0 \tag{9}$$

When a point comes into contact, then the normal velocity vanishes after a discontinuity at the end of a first part of the time interval. During the second part of the step of time the normal velocity either keep the contact surface or leave it.

NUMERICAL DISCRETIZATION

The discretization is develcped by the space and time finite element method. The interpolation functions are function of the space variable and the time. On the reference element it is assumed that :

$$\mathbf{v}(y,t) = \Sigma\, P_{ij}(y,t)\, \mathbf{V}(y_i,t_j) \tag{10}$$

with the form of the interpolation functions :

$$P_{ij}(y,t) = N_i(y)\, g_j(t) \tag{11}$$

The proposed algorithm is called GSWM (Geometrical Scft Way Method).
The position vector of any point is calculated by the integration of the velocity.

When a point comes into contact, then a first stage on the step of time is carried out by an explicit scheme since the relative velocity at the contact point is zero. The second stage is an implicit scheme comparable to the regular implicit scheme within the point can keep or leave the contact surface. The consequence is a local remeshing on the space-time mesh.

Therefore :
- At each iteration the position vector is control in order to keep the geometric compatibility condition
- When a point comes into contact the local space-time element is decomposed

Notice :
- for the regular point : $\mathbf{x}_1{}^n = \mathbf{x}_0 + (\mathbf{V}_0 + \mathbf{V}_1{}^n)h/2$
- The remeshing can be adapted in regard from the lengh of the impact wave

The evolution algorithm is :

$t_0 = t_{in}$
For all the steps of time from t_{in} to t_{lim}, repeat :

> $\mathbf{V}_0$, $\mathbf{X}_0$: discretized velocity and position vector at time t_0
> $\mathbf{V}_1{}^0 = \mathbf{V}_0$; $\mathbf{X}_1{}^0 = \mathbf{X}_0$; first iteration at time t_1
> $n = 1$
>
> solve iteratively
>
> > $f(\mathbf{V}_1{}^n, \tilde{\Omega}^{n-1}) = 0$
> > $\mathbf{X}_1{}^n = \mathbf{X}_0 + I(\mathbf{V}_0, \mathbf{V}_1{}^n, h)$
>
> until convergence of $\mathbf{V}_1{}^n$ to $\mathbf{V}_1$
>
> t_1 becomes t_0

until $t_1 > t_{lim}$

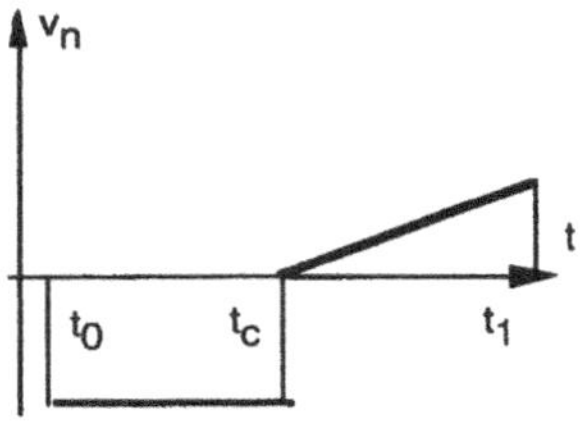

Figure 1. velocity interpolation

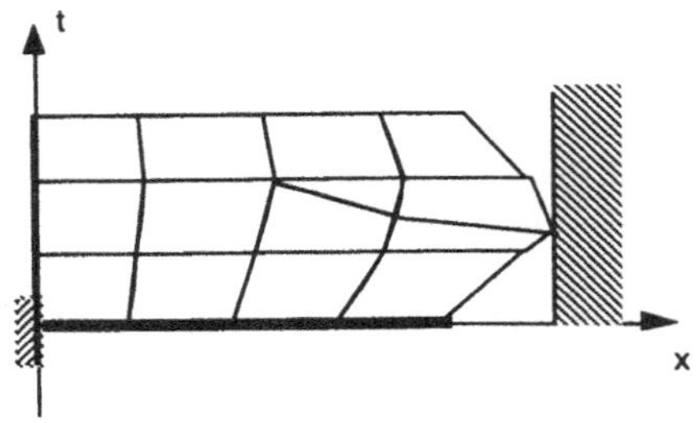

Figure 2. space-time remeshing

APPLICATIONS

The application domain can be in elastic bodies or other type of behavior. See [Bajer & Bohatier 93] where elastic and viscoplastic bodies are considered.

CONCLUSIONS

This velocity formulation is very efficient for the contact problems. It leads to a robust algorithm that permits to choose larger step of time at equal accuracy in regards of classical methods and that saves computer time.

REFERENCES

Alart P., 1991, "A mixed formulation for frictional contact problems prone to Newton like solution methods", Comp.Meth.in Appl. Mech. and Eng., pp. 353-375, vol.92, N°3

Bajer C. et Bohatier, 1993, "The Soft Way Method and the velocity formulation", Computer & Structures, Soumis à publication 3 dec 93

Bajer C. & Bohatier C., 1992, "Solution of thermomechanical problems by the space-time finite elements", International conference NUMIFORM92.

Bohatier C., 1992, "A large deformation formulation from thermomechanical point of view and space-time finite element method", ASME , proceedings Dynamics and Vibration, ESDA 1992, Istambul.

Bohatier C., 1992, "A large deformation formulation and solution with space-time finite elements", Archives of Mechanics., vol. 44, pp. 31-34 , Warszawa

Bohatier C., Chenot J.L., 1989, Finite element formulation for non-steady state large deformations with sliding or evolving contact boundary conditions, Int. J. for Num. Meth. in Eng., vol 28, pp. 753-768.

Jean M., Moreau J.J., 1992, "Unilaterality and dry friction in the dynamics of rigid body collections, Proceedings of contact Mechanics International Symposium, Lausane, Switzerland, Oct. 7-9

Heegaard J.H., Curnier A., 1990, "An augmented lagrangian method for large slip contact problems", pp. 69-72, proceedings Euromech 273 "unilateral contact and dry friction", La Grande Motte

Richelsen A. B., 1990, "Friction modelling in a finite strain viscoplastic analysis of a rolling process", pp 121-124, proceedings Euromech 273 "unilateral contact and dry friction", La Grande Motte, France

DELAMINATION OF LAMINATES UNDER IMPACT LOADING : CONTACT TECHNIQUE AND CONSTITUTIVE RELATION

Francis Collombet, Jérôme Bonini, Valérie Martin, and Jean-Luc Lataillade

LAMEF - Ecole Nationale Supérieure d'Arts et Métiers
Esplanade des Arts et Métiers
33405 Talence Cedex (France)

INTRODUCTION

During transverse low velocity impact of laminates, two kinds of damage - matrix cracking and delamination - are coupled and depend on the evolution of the loading versus time. The optimization of the behaviour of the composite structures in dynamic imposes the development of an hybrid approach synergy between experimental, theoretical and numerical aspects (Collombet & al, 1993). The aim of this paper is to model two kinds of contact, the impact loading on the macroscale (projectile-structure) and the delamination on the mesoscale (between plies of different orientations).

EXPERIMENTAL OBSERVATIONS

The modelling of contact needs experimental data to identify the mechanical behaviour and to give some comparaisons for the numerical development. A drop-weight set-up, using the Hopkinson Pressure Bars principle, allows a precise measurement of the response of the structure at the impacted point versus time (Martin & al, 1994). The projectile is a long rod instrumented with strain gauges (length = 600 mm, diameter Ø = 25 mm). The range of impact energy is from 1 to 50 J (velocity lower than 6 m/s and mass lower than 3 kg). A numerical treatment of the recorded signals gives the normal resultant force F(t) and the normal resultant displacement d(t) of the extremity of the projectile versus time. The plates tested are glass-epoxy ten plies laminates [0n/90m/0n] (diameter Ø = 200 mm and thickness e = 2 mm).

An experimental investigation by means of impact-tests with increasing energy is proposed to observe the appearance of the first damage. Matrix cracking appears before delamination. The cracks first appear in the lower ply (from the impact face). They cross the thickness of the ply and propagate in the direction of the fibres of the ply. The delamination is located in the cracked area at the interface of plies of different orientations. The lower ply (from the impacted face) is saturated with cracks in the delamination area. The delamination propagates in mode II in the direction of the fibers of the lower ply.

The delamination can not be modelled by means of a numerical contact technique without a prior description of the matrix cracking damage within the adjacent plies.

THEORETICAL AND NUMERICAL DEVELOPMENTS

From the experimental results, damage scenarios are developed in the dynamic explicit finite element code PLEXUS (Collombet & al, 1994). This code is based on the resolution of the following fundamental equation :

Contact Mechanics, Edited by M. Raous *et al.*
Plenum Press, New York, 1995

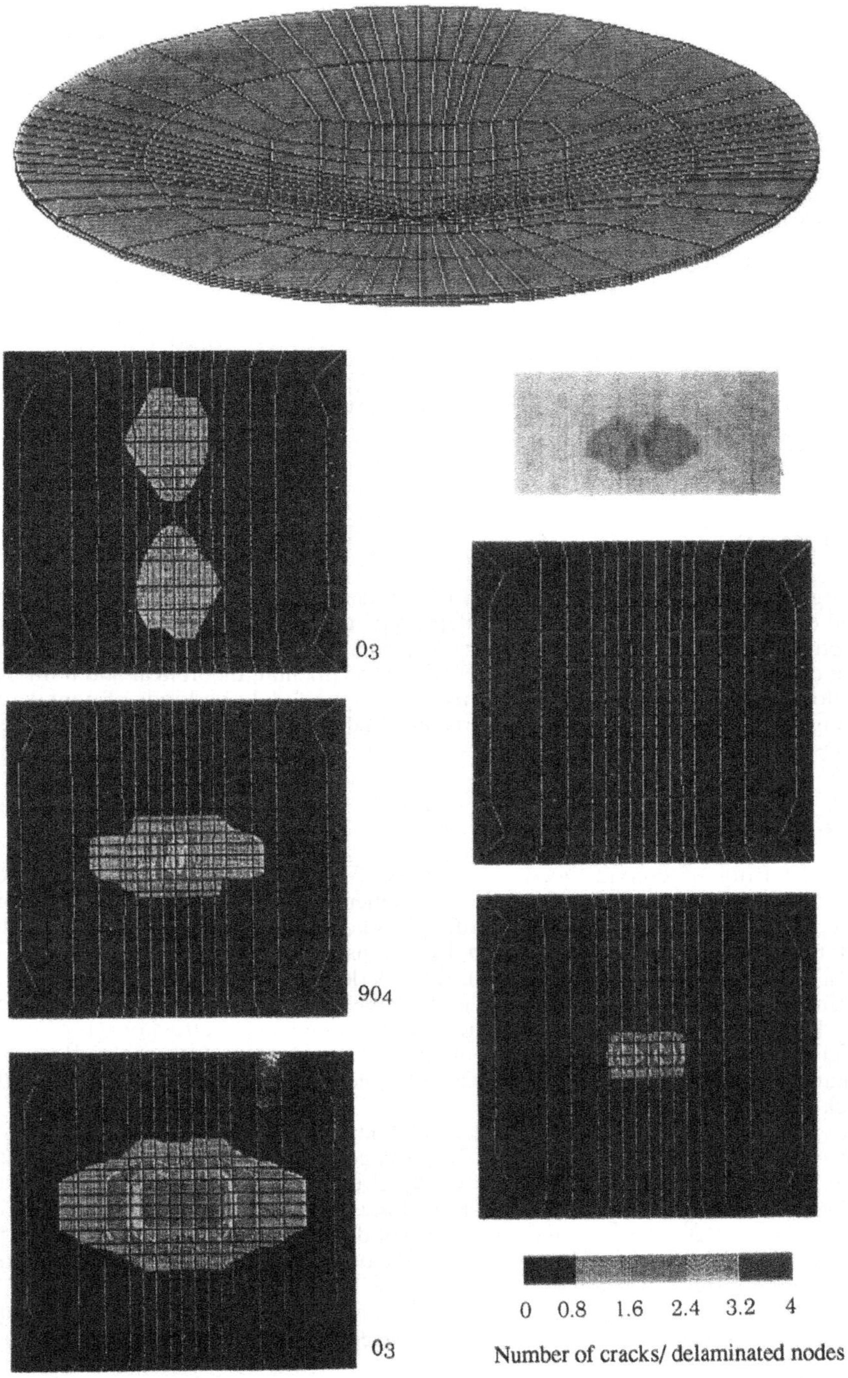

Figure 1. Matrix cracking (left) and delamination area (right) for an impact of 10J on a [03/904/03] glass-epoxy - Comparaison experiments (photography of the final damage state - view from the non-impacted face) vs. calculations. Identical scale of representation of the damage.

$$[M]\{\ddot{u}\} = \{Fext\} - \{Fint\} \quad (1)$$

with [M] the diagonal mass matrix, {Fint} the internal force vector, contribution of each element via the constitutive relation of the material and {Fext} the external force vector applied on the nodes of the mesh. The Newmark integration explicit scheme is used. Coupled with (1), this scheme allows a virtual propagation of the loading, element by element, within the structure. The algorithm carries out a specific calculation of the internal forces, without the assembly of neither a local, nor an overall rigidity matrix [K].

Macroscopical Contact-Impact Modelling

A contact model is first developped to represent on the macroscale, the impact loading versus time (Bonini & al, 1993). If falls into two parts : the contact searching and the calculation of a contact force. The contact searching is based on the well-known master-slave nodes technique. It concerns three kinds of contact : punctual, linear or surface. The calculation of the contact force uses the Lagrange multiplier method in explicit mixed with a defending node technique. Whatever the kind of contact, all the master nodes are concentrated in a fictitious defending node, located on the normal to the master surface going through the slave node. As the algorithm is based on an explicit integration scheme, the condition of stability imposes little time steps and small displacements during a time step. The slave node is close to the master surface. The following contact condition could be written : $u_{defending} = u_{slave}$. Then, the Lagrange Multiplier technique consists in adding to equation (1) a link force {Flink}, external contact force applied on the nodes :

$$[M]\{\ddot{u}\} = \{Fext\} - \{Fint\} + \{Flink\} \quad (2)$$

with {Flink} = ± λ and λ is the Lagrange multiplier

$$\text{and} \quad \lambda = \frac{M_s\left(F^d_{ext} - F^d_{int}\right) - M_d\left(F^s_{ext} - F^s_{int}\right)}{M_s + M_d} \quad (3)$$

The link force is then distributed on the master nodes of the mesh. It is calculated as long as the contact-rupture test is not satisfied. It consists in a displacement condition identified from experimental observations and leads to a cancellation of the link force :

If $u_s > u_{d0}$ then contact rupture : {Flink} = 0
with u_{d0} zero position of the defending node

Mesoscopical Contact : Representation of the laminate

Each ply is considered as an independent orthotropic structure meshed with hexaedric brick elements. A three-dimensionnal node on node stacking of the laminate is proposed from the previous Lagrange multiplier contact technique. The link force, calculated to keep the equality of the displacements of the interfacial double nodes, is given by (2). The continuity of the displacements and a perfect transmission of the loading are ensured in each direction of the space without the classical laminate theory. It allows a physical representation of the delamination by means of an evolution of the contact conditions.

Matrix Cracking Modelling

A model of matrix cracking damage is developed by means of an averaging technique from the results of (Laws & al, 1983; Collombet & al, 1992). The evolution of the matrix cracking is coupled on an average in each elementary volume of the mesh, with the mechanical characteristics of the composite in the perpendicular direction (fibres). The dilution parameter β is the ratio of the thickness 2a of the ply to the average distance between cracks 2x. The new characteristics M tilde of the compliances in each volume could be written as :

$$\tilde{M} = M + \frac{\Pi}{4}\beta\Lambda \qquad \text{with} \qquad \beta = \frac{2a}{2x} \quad (4)$$

with M as the previous compliance, Λ is a tensor defined from a localization tensor Q. The matrix cracking is set off by a maximal mixed strain-stress cracking criterion. A condition on β concerning a saturated cracking state is defined according to the experimental results (depends on the position of the ply in the thickness of the laminate) and the evolution of the size of the elements of the mesh.

Delamination Modelling

A condition of localisation is based on the cracked state in the plies from the experimental observations. The delamination is localised if both the plies at the interface are cracked, and if the lower ply is saturated with cracks. The double node assembly of the laminate leads to suggest an initiation criterion based on the component of the interfacial forces {Flink} in the direction of the fibres of the lower ply.

RESULTS

An original method for the evaluation of the matrix cracking and delamination criteria is developed from the hybrid approach (Collombet & al, 1993). The following values are found, with 2 the perpendicular direction to the fibers :

$$\sigma_{22}^{max} = \frac{E_2}{100} = 51.5 \text{ MPa} \quad \text{and} \quad F_{\alpha}^{crit} = 250 \text{ N}$$

$$\beta^{limit\,/\,0^\circ} = 0.3 \quad \text{and} \quad \beta^{limit\,/\,90^\circ} = 1$$

One application is then presented concerning the impact of a [03/904/03] glass-epoxy laminate. The impact energy is 10 J. The numerical results are in good agreement with experimental observations (Fig 1).

CONCLUSION

The Lagrangian Multiplicator technique is here applied, in a dynamic explicit frame, to represent on the macroscopical scale the interaction between the impactor and the plate, and, on the mesoscale, the coupling between the plies. It presents an original modelling of the delamination from a node-force approach mixed with the craked state of the plies. The contact conditions depends on the evolution of the constitutive relation of the material.

REFERENCES

Bonini J, Collombet F, Lataillade J.L, 1993, "Numerical modelling of contact for low velocity impact damage in composite laminates", Comp. Meth. Contact Mech., pp 453-461, Southampton (UK).

Collombet F, Espinosa C, 1992, "Simulation of through-ply cracking damage history sustained by laminated composite during transverse impact", ISIE, pp 412-418, Sendaï (Japan).

Collombet F, Bonini J, Martin V, Lataillade J.L, 1993, "Hybrid method for the evaluation of composite plates during transverse low velocity impact", Euromech 306, Mech of Contact-Impact, Prague (Czech Republic).

Collombet F, Bonini J, Lataillade J.L, 1994,"A three-dimensional modelling of the low velocity impact damage in composite laminates", Submitted to publication in *Num. Meth. Eng.*

Laws N, Dvorak G.J., Hejazi M, 1983, "Stiffness changes in unidirectional composites caused by crack systems", *Mechanics of Materials*, 2, pp 123-137

Martin V., Collombet F., Moura M., Lataillade J.L., 1994, "A standardized experimental set-up : laminated composite structures under low velocity impact loading", Experimental Mechanics, Lisboa (Portugal)

A DYNAMICAL CONTACT MODEL WITH FRICTION USING COMPLIANCE LAWS OF ROCK MECHANICS. COMPUTATIONAL METHOD AND APPLICATION

A. Lakhal[1], G. Bayada[2], M. Chambat[3] and L. Rochet[1]

[1]C.E.T.E, Labo. des Ponts et Chaussées, 102 av.S.Allende, 69674 Bron
[2]C.N.R.S. URA 740 et 856, I.N.S.A. Math.,Bt. 401, 69621 Villeurbanne
[3]C.N.R.S. URA 740, Univ. Lyon I, L.A.N., Bt 101, 69622 Villeurbanne

INTRODUCTION

A massive where an experimental survey of cracking is possible, is often studied as an assembly of blocks. In the discontinuities several phenomena come into play (friction, dilatance, abrasion ...). To take into account the morphology of contacting surfaces and their behaviour, the compliance laws applied to a fictitious interface seem to be the best adapted; for more explanation see Oden et al (1985) and Rochet (1990). This approach is also used by Klarbring et al (1989). They are explicit relations between stresses and displacements at the contact level, and they can be determined experimentally :
-The one side condition (impenetrability of blocks) can be described by a compliance law between normal stress and normal displacement at the contact level.
-Friction can also be taken into account via a law between tangential stress, normal stress and the tangential displacement. This makes the apparent coefficient of friction (ratio of tangential stress and normal stress) appear as a function of the normal displacement and tangential displacement. The dependence of this coefficient on the tangential displacement is used (Ionescu et al 1994) to explain the "stick-slip" motion phenomen.

The main topic of this work is to present a dynamical frictional contact model using experimental compliance laws of Rock Mechanics. To take into account both the elastic deformations and the motion of a body in contact with a rigid obstacle, the non coercive variational formulation is projected on the rigid body displacement space and on the orthogonal space. This leads to a nonlinear elliptic partial differential equation (governing the elastic behaviour) coupled with a nonlinear ordinary differential equation (governing the motion). After implicit time discretization, a fixed point coupled with preconditioned projected gradient algorithm is used to solve the finite elements problem at each time step. A computational program has been performed from this mathematical algorithm. Applications of the present model can be impact problems and sliding of rock masses. The static case has been presented in a previous paper (Bayada et al 1992).

Contact Mechanics, Edited by M. Raous et al.
Plenum Press, New York, 1995

CLASSICAL AND VARIATIONAL FORMULATIONS

Let us consider a domain Ω in $\mathbb{R}^N$, $N=1,2,3$, with the Lipschitz boundary Γ. Let Γ_c and Γ_F be open and disjoint parts of Γ which do not depend on time $t \in [0,T]$ such

that $\bar{\Gamma} = \bar{\Gamma}_c \cup \bar{\Gamma}_F$, Γ_c being candidate to unilateral contact with a rigid fixed support. At time $t=0$ the body occupies the domain Ω. We shall use the following notations for the normal and tangential components of the displacement and of the stress vector : $u_n = u_i n_i$, $u_{ti} = u_i - u_n n_i$, $\sigma_n = \sigma_{ij} n_i n_j$, $\sigma_{ti} = \sigma_{ij} n_j - \sigma_n n_i$
where $i,j=1,\dots,N$, $n=(n_i)$ is the outward unit normal vector on Γ and the summation convention is used. We assume that the body is subjected to volume forces of density $f=(f_1,\dots,f_N)$ on Ω, to surface traction of density $F=(F_1,\dots,F_N)$ on Γ_F :

$$\sigma_{ij} n_i = F_{i(t)} \quad \text{on } \Gamma_F \tag{1}$$

Following Martins and Pires (1990), we split the displacement u into a purely elastic displacement $u_d \in D$ and a rigid dynamic displacement $u_r \in R$, where :

$$R = \left\{ \varphi \in (H^1(\Omega))^N,\ \varphi(x) = a + b \wedge x \text{ on } \Omega\ ,\ a \text{ and } b \in \mathbb{R}^N \right\} \tag{2}$$

and D is its topological orthogonal subspace in $V=(H^1(\Omega))^N$: $V = R \oplus D$
Concerning u_d, we restrict it to small linear elastic displacements so that :

$$\sigma(u_d) = A.\varepsilon(u_d)\ ;\ \varepsilon(u_d) = ((\nabla(u_d) + \nabla^T(u_d))/2 \tag{3}$$

and making the classical Hertz hypothesis which allows us to neglect $\ddot{u}_d$, the equations of momentum become :

$$\operatorname{div} \sigma(u_d) + f = \rho\, \ddot{u}_r \quad \text{in } \Omega \tag{4}$$

The interface between contacting bodies is an hypothetical medium of vanishing thickness; we wish to characterize the response of such interface to normal deformation in a way that is consistent with experimental results. The initial gap g between Γ_c and the obstacle that may come in contact with Ω is defined as the distance, measured along a line normal to Γ_c, between the highest asperities of the body in the reference configuration. To take into account the normal deformability of the interface asperities, we use the compliance law :

$$\sigma_n = \psi_n(u_n - g) \quad \text{on } \Gamma_c \tag{5}$$

were ψ_n is a continuous function issued from the increasing monotone part of experimental tests (Barton et al), this implies in fact a bound for the normal stress assuming plastic deformations negligable .
The shear tests of Rock Mechanics (Bandis et al) present a nonlinear behaviour characterized by a rapid increase of the shear stresses to an accentuated peak followed by a decrease to a residual value which represents the friction between the surfaces in contact. This behaviour is actually due to the asperities and their irregularities. Indeed, for a given level of normal stress the sharpest asperities are sheared (in this phase the dissipate energy is due to elastic displacements) and the rigid displacement is then imposed by the most resistant asperities (in this phase the dissipate energy is actually due to rigid displacements). To take such a behaviour into account, the following friction law is proposed :

$$-\sigma_t \in \left(\mu\, \psi_n(u_n - g)\, \Psi_t(|u_t|) \operatorname{sgn}(\dot{u}_t) \right) \tag{6}$$

For computational purpose, the multivalued relation sgn(.) is approximated by the function :

$$\psi_{\varepsilon'}(\xi) = \xi/(\varepsilon' + |\xi|)\ ;\ \forall\, \xi \in \mathbb{R}^3 \tag{7}$$

where the regularization parameter ε' is a small positive real number. With this regularization the variational formulation for (1), (3), (4), (5) and (6) is:

$$(Q)\begin{cases}\text{Find } u_r \in C^2([0,T],R) \text{ and } u_d \in C^1([0,T],D) \text{ such that :}\\ a(u_d,v) + (\rho\, \ddot{u}_r,v) + <\Pi_n(u),v>_{\Gamma_c} + <\Pi_t(u,\dot{u}),v>_{\Gamma_c} = L(v)\ ;\ \forall\, v \in V\\ u_r(0) = u_0\ ,\ \dot{u}_r(0) = u_1 \text{ et } u_d(0)=u_{d0}\end{cases}$$

where

$$a(u,v) = \sum_{i,j} \int_\Omega a_{ijkl}\varepsilon_{ij}(u)\varepsilon_{kl}(v)d\Omega \quad ;\quad L(v) = \int_\Omega f.v d\Omega + \int_{\Gamma_F} F.v\, ds \quad ;\quad \forall\, u\ ,\ v \in V$$

$$<\Pi_n(u),v>_{\Gamma_c} = \int_{\Gamma_c} \psi_n(u_n-g)v_n ds;\ <\Pi_t(u,\dot{u}),v>_{\Gamma_c} = \int_{\Gamma_c} \mu\psi_n(u_n-g)\Psi_t(|u_t|)\psi_\varepsilon{}'(\dot{u}_t)v_t ds;\ \forall\, u,v \in V$$

The variational formulation is projected on R and D. this leads to the following equivalent formulation :

$$(D)\begin{cases}\text{Find } Y=(a,b)\ C^2([0,T],(\mathbb{R}^3)^2) \text{ and } u_d \in C^1([0,T],D) \text{ such that :}\\ a(u_d,v_d) + <\Pi_n(u),v_d>_{\Gamma_c} + <\Pi_t(u,\dot{u}),v_d>_{\Gamma_c} = L(v_d)\ ;\ \forall\, v_d \in D \quad (8)\\ \ddot{Y}(t) = \mathbf{F}(Y(t),u_d(t),\dot{Y}(t),\dot{u}_d(t)) = \begin{pmatrix} \frac{1}{mes(\Omega)}\left(\int_\Omega f.d\Omega + \int_{\Gamma_F} F.ds - \int_{\Gamma_c}(\Pi_n(u)+\Pi_t(u,\dot{u})).ds\right) \\ \tilde{J}_G^{-1}\left(\int_\Omega \overrightarrow{GX}\wedge f d\Omega + \int_{\Gamma_F}\overrightarrow{GX}\wedge F ds - \int_{\Gamma_c}\overrightarrow{GX}\wedge(\Pi_n(u)+\Pi_t(u,\dot{u})ds\right)\end{pmatrix} \quad (9)\\ Y(0)=Y_0\ ,\ \dot{Y}(0)=Y_1 \text{ et } u_d(0)=u_{d0}\end{cases}$$

where

$u_r(t,x) = a(t)+b(t)\wedge \overrightarrow{GX}$; $\tilde{J}_G$ being the inertial torser of Ω at the mass center.

For this problem, we have (see Lakhal) an existence and uniqueness theorem for the case without friction. The same result is obtained for the problem with friction and $\Psi_t \equiv 1$ (using a regularized Coulomb law with a small coefficient) after implicit time discretization.

COMPUTATIONAL METHOD

For computational purpose, the interval [0,T] is discretized with a time step Δt. At each time t_i, $\dot{u}(t_i)$ is approximated by $(u(t_i)-u(t_{i-1}))/\Delta t$ in equation (8), and (9) is approximated by an implicit Euler Scheme. We get the problem :

$$\begin{cases} Y(t_{i-2}), Y(t_{i-1}) \text{ et } u_d(t_{i-1}) \text{ known,}\\ \text{find } Y(t_i)=(a(t_i),b(t_i)) \in (\mathbb{R}^3)^2 \text{ and } u_d(t_i) \in D, \text{ such that : } \forall\, v_d \in D\\ a(u_d(t_i),v_d) + <\Pi_n(u(t_i)),v_d>_{\Gamma_c} + <\Pi_t(u(t_i),\frac{u(t_i)-u(t_{i-1})}{\Delta t}),v_d>_{\Gamma_c} = L(v_d)\\ Y(t_i)-2Y(t_{i-1})+Y(t_{i-2}) = (\Delta t)^2\mathbf{F}\left(Y(t_i),u_d(t_i),\frac{Y(t_i)-Y(t_{i-1})}{\Delta t},\frac{u_d(t_i)-u_d(t_{i-1})}{\Delta t}\right)\\ \text{where } u(t_i) = u_r(t_i)+u_d(t_i)\ ;\ u_r(t_i) = a(t_i)+b(t_i)\wedge\overrightarrow{GX}\end{cases}$$

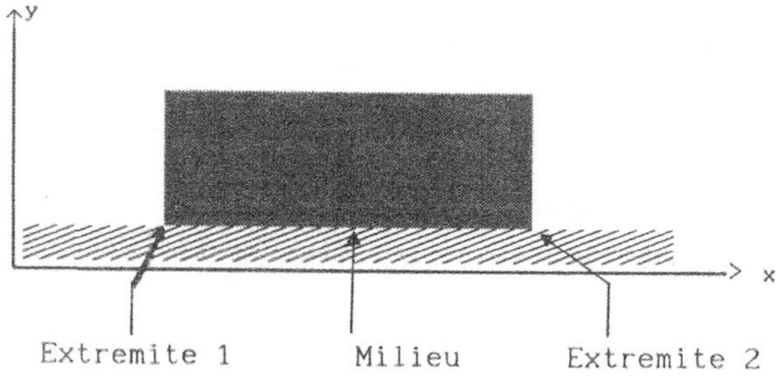

Figure 1. Model problem

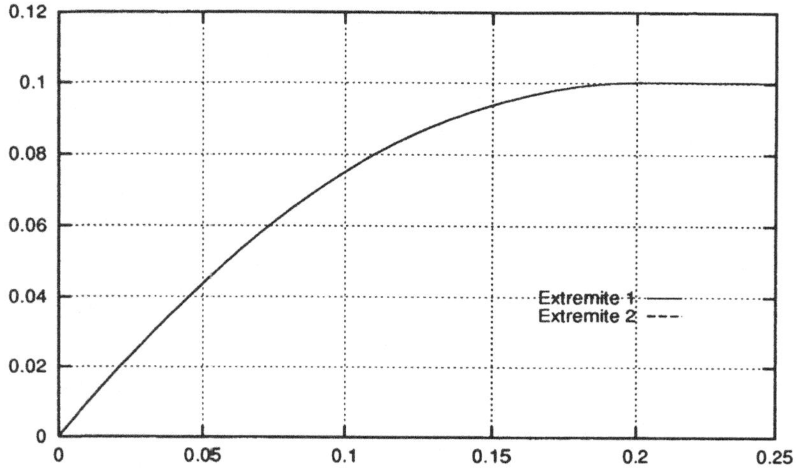

Figure 2. Tangentiel displacement (m) versus times (s)

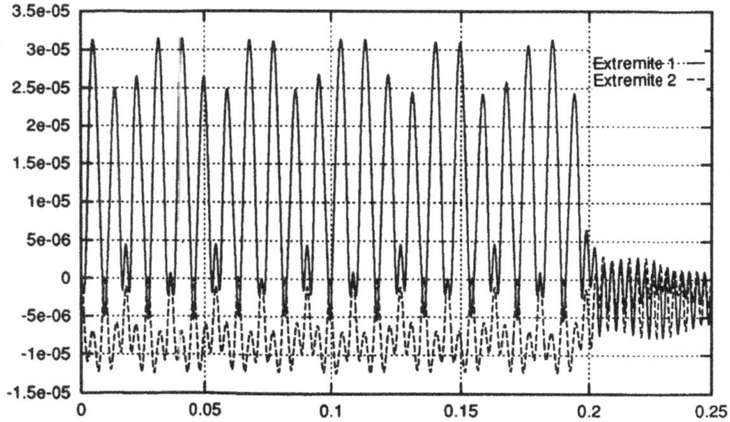

Figure 3. Normal displacement (m) versus times (s)

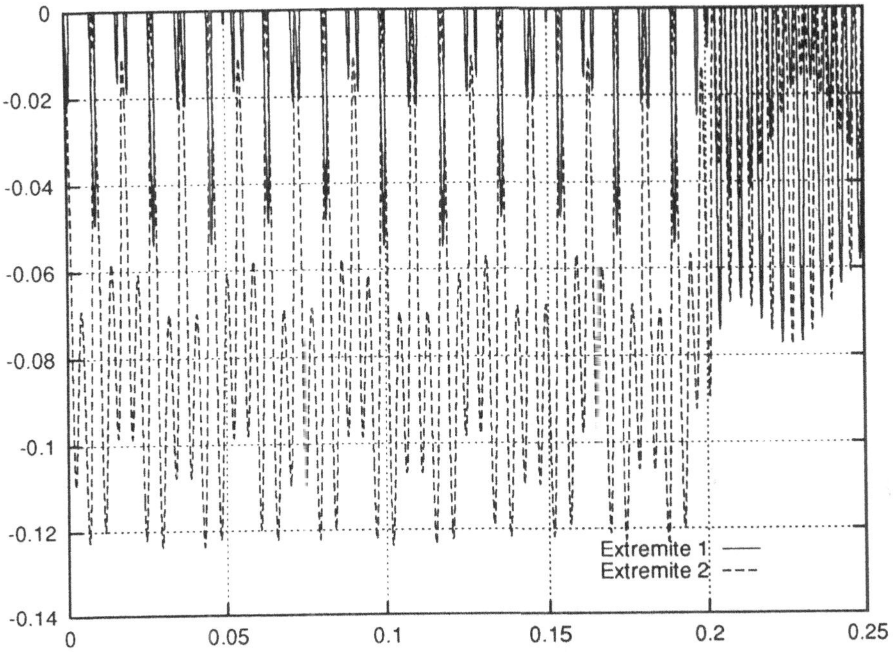

Figure 4. Normal stress (MPa) versus times (s)

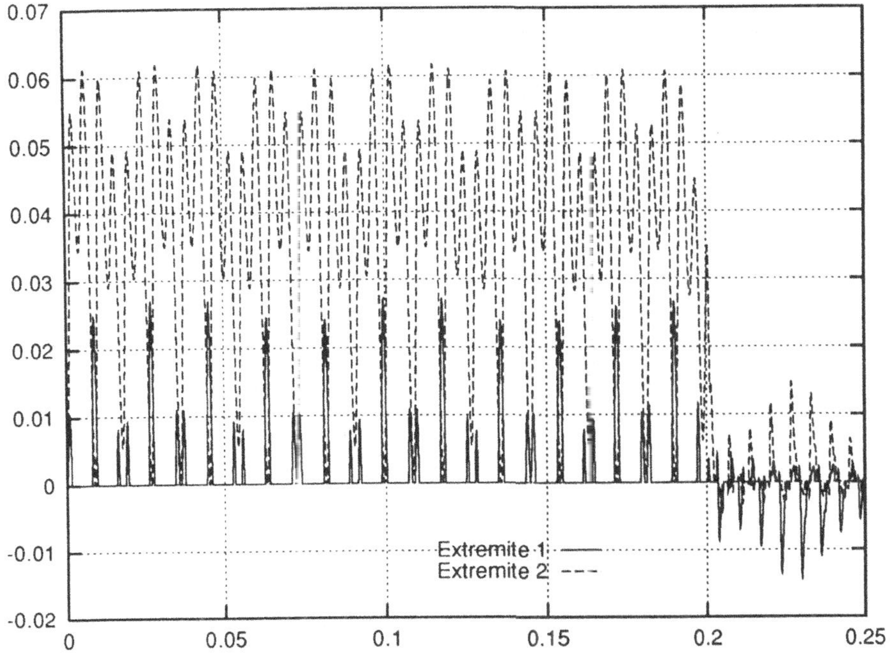

Figure 5. Tangentiel stress (MPa) versus times (s)

At each time step, we have to solve two nonlinear coupled equations :

Loop $(t_{i-2},t_{i-1})\longrightarrow t_i$

$Y(t_{i-2})$, $Y(t_{i-1})$ et $u_d(t_{i-1})$ being known, we compute $(Y(t_i), u_d(t_i))$ by :

Loop $(k-1,k)\longrightarrow k+1$

$(Y^{k-1}(t_i), u_d^{k-1}(t_i))$ et $(Y^k(t_i), u_d^k(t_i))$ known, compute $(Y^{k+1}(t_i), u_d^{k+1}(t_i))$ by :

$$\mathcal{A}(u_d^k(t_i)) = P_D\left[S_m^{-1}(A_{Y^k(t_i)}(u_d^k(t_i)))\right]$$

where P_D is the orthogonal projection onto D.

the operator $A_{Y^k(t_i)} : V \longrightarrow V'$ defined by :

$$<A_{Y^k(t_i)}(u_d^k(t_i)),v> = a(u_d^k(t_i),v) + <\Pi_n(u^k(t_i)),v>_{\Gamma_c} + <\Pi_t(u^{k-1}(t_i), \frac{u^k(t_i)-u(t_{i-1})}{\Delta t}),v>_{\Gamma_c} - L(v)$$

$$u^k(t_i) = u_r^k(t_i)+u_d^k(t_i) \; ; \; u_r^k(t_i) = a^k(t_i)+b^k(t_i)\wedge \overrightarrow{GX} \; ; \; Y^k(t_i)=(a^k(t_i),b^k(t_i))$$

$$<S_m(u),\varphi> = a(u,\varphi) + m\int_\Omega u\,\varphi\, d\Omega; \; \forall\, u \text{ et } \varphi\in V \; ; \; m>0$$

compute : $\rho^k = \dfrac{(\mathcal{A}(u_d^k(t_i))-\mathcal{A}(u_d^{k-1}(t_i)),\,u_d^k(t_i)-u_d^{k-1}(t_i))}{||S_m^{-1}(\mathcal{A}(u_d^k(t_i))-\mathcal{A}(u_d^{k-1}(t_i)))||^2}$

then set : $u_d^{k+1}(t_i) = u_d^k(t_i) - \rho^k\mathcal{A}(u_d^k(t_i))$ and

$$Y^{k+1}(t_i) = 2Y(t_{i-1}) - Y(t_{i-2}) + (\Delta t)^2 \left[F(Y^k(t_i),u_d^k(t_i), \frac{Y^k(t_i)-Y(t_{i-1})}{\Delta t}, \frac{u^k(t_i)-u(t_{i-1})}{\Delta t})\right]$$

NUMERICAL RESULT

Ω is an elastic block (fig.1) discretized by finite elements P_1. External forces are restricted to the gravity. The obstacle is the horizontal rigid plane.

Contact laws are : $\sigma_n = -k'(u_n)^+$; $\sigma_t = -\mu\sigma_n \mathrm{sgn}(\dot{u}_t)$; $k' > 0$ large enough.

In some sence both laws (11) can be considered as regularization of classical Signorini and Coulomb laws. For the sliding test ($\dot{u}_{rt}(0)=0.1$ m/s), the contact law used is (11). We observe the sliding (fig.2) accompanied with micro normal oscillation (fig.3) at the interface. At the equilibrium, the time and the distance obtained are in good accordance with the analytical results and we can also notes (fig.4 and fig.5) that the equilibrium of total stresses is respected. More details and other tests related to impact problems can be found in Lakhal.

REFERENCES

Bandis S.C., Lumsden A.C and Barton N. R., Experimental studies of scale effects on the shear behaviour of rock joints, Int. J. Rock Mech. and Min. Sci. & Geomeck, Abstr., 18, pp. 249-268, (1981).

Barton N.,Bandis S.and Bakhtar K., Strength, Deformation and Conductivity coupling of rock joint, Int.J. Rock Mech. and Min. Sci. & Geomeck, Abstr.Vol. 22, 3, pp 121-140, Pergamon Ltd, London, (1985).

Bayada G., Chambat M. & Lakhal A., Homographic approximation of dynamical unilateral contact with or without friction in Rock Mechanics, Contact Mechanics International Sympos., Alain Curnier, Presses Polytechniques et universitaires Romandes, Lausanne, Suisse, pp. 361-368, (1992).

Ionescu I.R & Paumier J.C, On the contact problem with slip rate dependent friction in elastodynamics. Europ. J. of Mech. Solids, to appear, (1994).

Klarbring A., Mikelic A. & Shillor M, On friction problems with normal compliance, Nonlinear Analysis, Theory, Methods & Application, 13:8, pp. 935-955, (1989).

Lakhal A., Modélisation du contact unilatéral avec frottement en dynamique. Application aux problèmes d'impact et de glissement en Mécanique des Roches, Thèse Maths, Lyon I, N° 218-94, (1994).

Martins J.A.C & Pires E.B., A class of impact problems in linear elasticity, CMEST, Lisboa, (1990).

Oden J.T., and Martins J.A.C., Models and computational methods for dynamic friction phenomena, Computer Methods in Applied Mechanics and Engineering, 52, pp.527-634, (1985).

Rochet L., Mouvements de terrains, Univ. Européenne d'été, Edition Pôle Grenoblois d'Etudes et de Recherches pour la Prevention des Risques Naturels, Grenoble, France, (1990).

THEORETICAL AND NUMERICAL STUDY FOR A MODEL OF VIBRATIONS WITH UNILATERAL CONSTRAINTS

Michel Panet[1], Laetitia Paoli[2] and Michelle Schatzman[3]

[1]Département M.T.C, Centre E.D.F des Renardières,
Route de Sens, 77250 Moret-sur-Loing
[2]Laboratoire d'Analyse Numérique, URA 740 CNRS, Université de St Etienne
[3]Laboratoire d'Analyse Numérique, URA 740 CNRS, Université Lyon 1

INTRODUCTION

We consider a mechanical system with N degrees of freedom. Its motion is described by

$$\ddot{u} = f(t, u, \dot{u}), \tag{1}$$

where $u \in \mathbb{R}^N$ denotes the representative point of the system.

We suppose that the system is subject to convex constraints, i.e:

$$u(t) \in K, \tag{2}$$

where K is a closed convex set of $\mathbb{R}^N$, with a smooth boundary.

We model the loss of energy when constraints are saturated as follows: the tangential component of the velocity is conserved, while the normal component is reversed and multiplied by a restitution coefficient $e \in]0, 1]$, i.e:

$$\dot{u}(t+0) = \dot{u}_T(t-0) - e\dot{u}_N(t-0), \quad \text{if } u(t) \in \partial K. \tag{3}$$

The equations (1)-(2)-(3) can be replaced by the following system:

$$\begin{cases} \ddot{u} + \partial\psi_K(u) \ni f(t, u, \dot{u}), \\ \dot{u}(t+0) = \dot{u}_T(t-0) - e\dot{u}_N(t-0), \quad \text{if } u(t) \in \partial K, \end{cases} \tag{4}$$

where ψ_K is the indicatrix function of the convex of constraints K.

EXISTENCE THEOREM

We have given a notion of solution for problem (4). with initial conditions, under the following assumptions:

(H1) the set K is a closed convex set of $\mathbb{R}^N$, with $\text{Int}(K) \neq \emptyset$, and a C^3 boundary,

Contact Mechanics, Edited by M. Raous *et al.*
Plenum Press, New York, 1995

(H2) the function f is a continuous function from $[0,T] \times \mathbb{R}^N \times \mathbb{R}^N$ to $\mathbb{R}^N$, Lipschitz continuous in its two last arguments uniformly in the first one.

Then, we have the

Definition 1. Let us assume (H1) and (H2). A function u from $[0,T]$ to $\mathbb{R}^N$ is a solution of (4), with initial conditions $u_0 \in K$ and $u_1 \in \mathbb{R}^N$, if and only if, the following assertions are verified:

(i) u is Lipschitz continuous and $\dot{u}$ is a function of bounded variation,

(ii) $u(t) \in K$, for all $t \in [0,T]$,

(iii) for all continuous functions v from $[0,T]$ to K,

$$0 \geq \langle f(t,u,\dot{u}) - \ddot{u}, v - u \rangle,$$

(iv) the initial conditions are satisfied in the following sense

$$u(0) = u_0$$

and

$$\begin{cases} v_0 = u_1 & \text{if } u_0 \in \text{Int}(K) \text{ or if } u_0 \in \partial K \text{ and } \big(u_1, \nu(u_0)\big) \leq 0, \\ v_0 = u_1 - (1+e)\big(u_1, \nu(u_0)\big)\nu(u_0) & \text{if } u_0 \in \partial K \text{ and } \big(u_1, \nu(u_0)\big) > 0. \end{cases}$$

We use a penalty method to obtain the existence result. The idea is to replace the rigid boundary of K by a viscoelastic one (see Paoli and Schatzman, 1993a, 1993c).

We have proved the

Theorem 2. Under hypotheses (H1) and (H2), system (4), with initial conditions $u_0 \in K$ and $u_1 \in \mathbb{R}^N$, has a solution in the sense of definition 1.

Remark: The existence result is still true if we assume only ∂K of class C^2 in (H1).

NUMERICAL METHODS

To solve numerically problem (4), we can use different methods:

- naïve method: to solve the equation of the free flight $\ddot{u} = f(t,u,\dot{u})$, to detect the impact times and to use the impact law (3) to determine new initial conditions;
- penalty method: to solve the penalised equation (5) for λ small enough;
- specific method.

The first method is based on a simple idea, but is not so simple to implement because the precise detection of each time of shock may become very expensive if the impacts are very close.

The second method uses the penalisation used in the proof of existence. Its implementation is also very expensive because we have shown (in Paoli and Schatzman, 1993c) that the penalised solution leaves the convex of constraints during a time interval whose length is an $O(\sqrt{\lambda})$, where λ is the penalisation parameter, intended to converge to zero. So we will have to choose a time step $h \leq O(\sqrt{\lambda})$ to approximate correctly this solution, if we use a fixed time step.

So, it is crucial to propose a numerical method especially suited to this kind of problems, with a good efficiency, which implies to avoid the precise determination of impact times. The details of the analysis can be found in Paoli and Schatzman (1994).

Let F be a function satisfying the following hypothesis:

(H3) F is a continuous function from $[0,T] \times \mathbb{R}^N \times \mathbb{R}^N \times \mathbb{R}^N \times [0,h^*]$, with $h^* > 0$, to $\mathbb{R}^N$, Lipschitz continuous with respect to its second, third and fourth arguments, uniformly in the first one, and such that:

$$F(t,u,u,v,0) = f(t,u,v) \quad \forall (t,u,v) \in [0,T] \times \mathbb{R}^N \times \mathbb{R}^N.$$

To solve numerically problem (4), with initial conditions $u_0 \in K$ and $u_1 \in \mathbb{R}^N$, we propose the following scheme:

$$U^0 = u_0, \quad U^1 = u_0 + hv_0 + \frac{h^2}{2} f(0, u_0, v_0), \tag{5}$$

with

$$\begin{cases} v_0 = u_1 \quad \text{if } u_0 \in \mathrm{Int}(K) \text{ or if } u_0 \in \partial K \text{ and } (u_1, \nu(u_0)) \leq 0, \\ v_0 = u_1 - (1+e)(u_1,\nu(u_0))\nu(u_0) \quad \text{if } u_0 \in \partial K \text{ and } (u_1,\nu(u_0)) > 0, \end{cases} \tag{6}$$

and

$$\frac{U^{n+1} - 2U^n + U^{n-1}}{h^2} + \partial\psi_{(1+e)K}(U^{n+1} + eU^{n-1}) \ni F^n \quad \text{for} \quad n \geq 1, \tag{7}$$

with

$$F^n = F\left(nh, U^n, U^{n-1}, \frac{U^{n+1} - U^{n-1}}{2h}, h\right). \tag{8}$$

This scheme is built by adding, to a centered discretisation of equation $\ddot{u} = f(t,u,\dot{u})$, the term $\partial\psi_{(1+e)K}(U^{n+1} + eU^{n-1})$, which takes into account both the constraint $u(t) \in K$ and the impact law.

Equation (7) is equivalent to:

$$U^{n+1} = -eU^{n-1} + P_{(1+e)K}(2U^n - (1-e)U^{n-1} + h^2 F^n) \quad \text{for} \quad n \geq 1, \tag{9}$$

where $P_{(1+e)K}$ denotes the projection under the convex set $(1+e)K$.

The proof of convergence of this scheme is very technical. The main difficulty comes from curvature effects when U^n is in a neighborhood of the boundary.

We denote $N(h)$ the bigest integer smaller than or equal to T/h, and we define a function u_h by:

$$\begin{cases} u_h(t) = U^n + (t - nh)\dfrac{U^{n+1} - U^n}{h} \quad \text{for } t \in [nh, (n+1)h[,\ 0 \leq n \leq N(h) - 1, \\ u_h(t) = U^{N(h)} \quad \text{for } t \in [N(h)h, T]. \end{cases} \tag{10}$$

We have then the

Theorem 3. Under hypotheses (H1), (H2) and (H3), there exist subsequences of the sequence $(u_h)_{h>0}$ uniformly converging on $[0,T]$, and the limit of any converging subsequence is a solution of problem (4) with initial conditions, in the sense of definition 1.

NUMERICAL RESULTS

We have performed two kinds of numerical experiments, one about a model problem with one degree of freedom, and another one about a more realistic problem, with many degrees of freedom. describing the motion of a guided beam.

The first problem is described by the system:

$$\begin{cases} \ddot{u} + 2\alpha\dot{u} + u + \partial\psi_{[u_{\min},+\infty[}(u) \ni a\cos(\omega t), \\ \dot{u}(t+0) = -e\dot{u}(t-0), \quad \text{if } u(t) = u_{\min}. \end{cases} \tag{11}$$

It is a mechanical system mass-spring-obstacle. We have studied this system only when $u_{\min}$ is a positive number: in this case, it is a modelisation of tight joint.

For this very simple system, it is possible to use the naïve method: we have explicit formulae for the solution of the free flight equation and we detect precisely the impact times. We obtain, by this way, an almost analytical solution of system (11).

We have also used the scheme described in the previous section to solve the system. We have chosen the function F as:

$$F(t, U_1, U_2, V, h) = a\cos(\omega t) - 2\alpha V - U_1.$$

We have observed that, for certain values of the parameters, the almost analytical solution has a periodic permanent regime and, according to the values of these parameters, the periodic motion would have one or many impacts per period (see figures 1 and 2). In this case, the scheme gives a good approximation of the solution, even for not very small time steps (see figure 3).

With other values of the parameters, we have observed, for the almost analytical solution, attractors which look strange (see figure 4). This is linked to sensitivity to initial data. Figure 5 shows the graphs of the almost analytical solutions obtained with the same values of a, α, ω and e, and with the following initial conditions:

$$(A)\begin{cases} u_0 = u_{\min} = 0.54, \\ v_0 = 0.1, \end{cases} \quad \text{and} \quad (B)\begin{cases} u_0 = u_{\min} = 0.54, \\ v_0 = 0.1 + 10^{-8}. \end{cases}$$

The graphs are quite different.

Consequently, it is impossible to obtain, with the numerical scheme, a good accuracy in the approximation of the motion for large times. Nevertheless, the scheme gives a very good approximation of the attractor, and thus, the approximate solution looks the same as the almost analytical one (see figure 6). Lyapunov exponents have been computed by F.Nqi for the attractor of figures 4 and 6, and their numerical values ($\lambda_1 = 3.051$ and $\lambda_2 = -15.083$) justify the chaotic behavior.

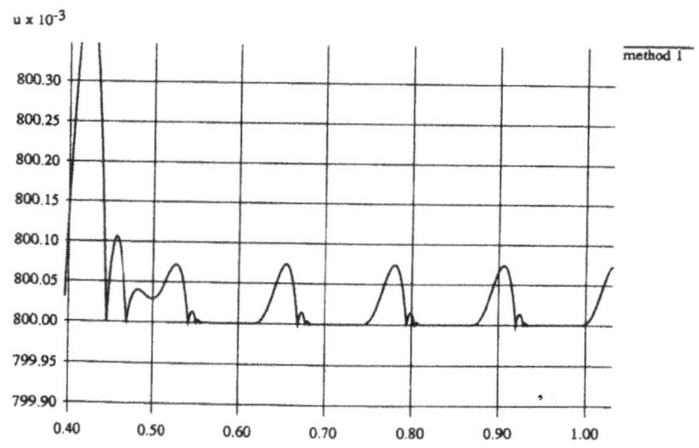

Figure 1. Graph of u with $a = 1$, $\alpha = e = 0.5$, $\omega = 50$, $u_0 = u_{min} = 0.80$, $v_0 = 0.1$.

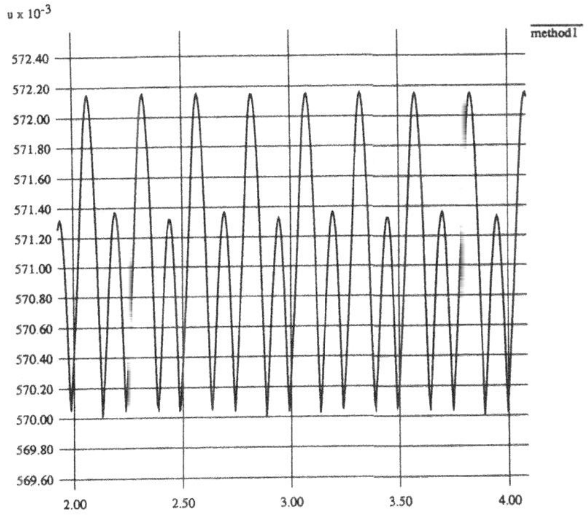

Figure 2. Graph of u with $a = 1$, $\alpha = e = 0.5$, $\omega = 50$, $u_0 = u_{min} = 0.57$, $v_0 = 0.1$.

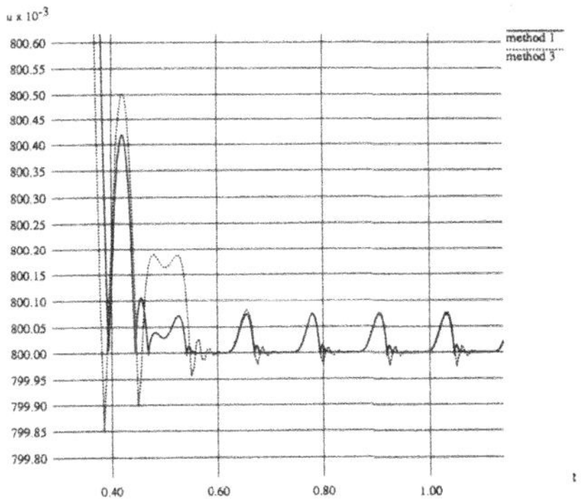

Figure 3. Graphs of the almost analytical solution (full line) and the numerical solution (broken line) with $a = 1$, $\alpha = e = 0.5$, $\omega = 50$, $u_0 = u_{min} = 0.80$, $v_0 = 0.1$ and a time step $h = 0.005$.

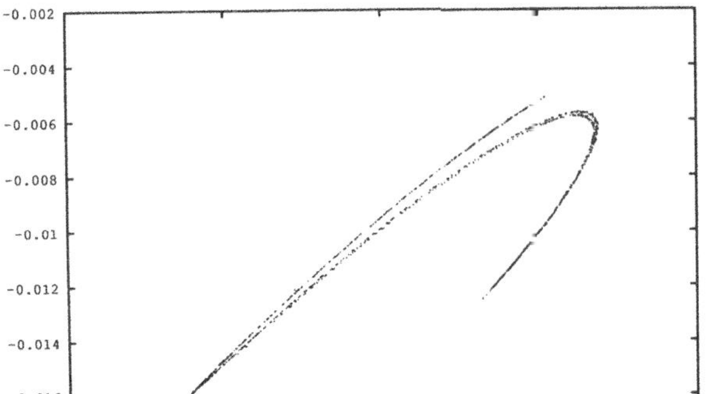

Figure 4. Poincaré's map ($a = 1$, $\alpha = e = 0.5$, $\omega = 50$, $u_0 = u_{min} = 0.54$, $v_0 = 0.1$).

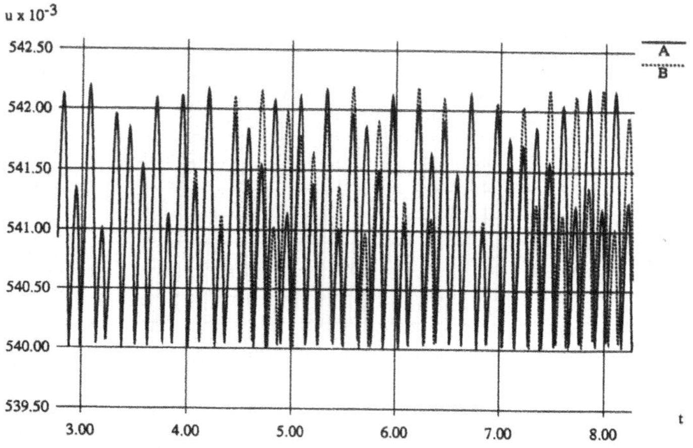

Figure 5. Graphs of the almost analytical solutions with initial data A (full line) and B (broken line) ($a = 1$, $\alpha = e = 0.5$, $\omega = 50$).

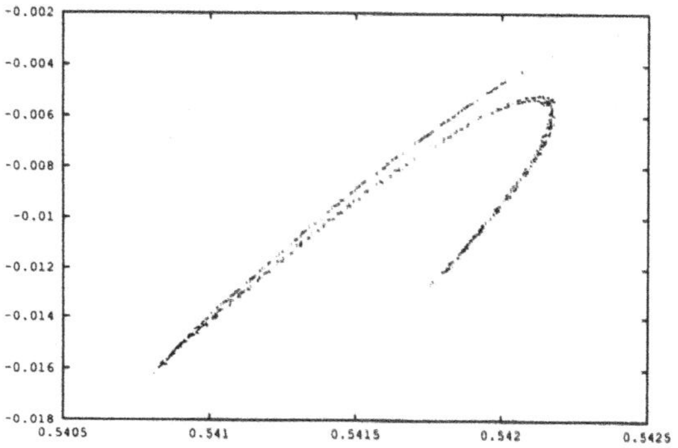

Figure 6. Numerical attractor (initial data A and $a = 1$, $\alpha = e = 0.5$, $\omega = 50$).

Our second set of experiments is about the problem of a beam, clamped at one end, free at the other end and guided. The equations of motion lead to an evolution problem of second order in time, and we have discretised it in space by finite differences. So, the convex set of constraints is

$$K = \left\{ u \in \mathbb{R}^d; u_{i_0} \in [\phi^-, \phi^+] \right\},$$

the matrix A is

$$A = \frac{1}{\Delta x^4} \begin{pmatrix} 6 & -4 & 1 & 0 & \dots & \dots & \dots & \dots & 0 \\ -4 & 6 & -4 & 1 & 0 & \dots & \dots & \dots & 0 \\ 1 & -4 & 6 & -4 & 1 & 0 & \dots & \dots & 0 \\ 0 & 1 & -4 & 6 & -4 & 1 & 0 & \dots & 0 \\ \vdots & & & & & & & & \vdots \\ 0 & \dots & 0 & 1 & -4 & 6 & -4 & 1 & 0 \\ 0 & \dots & \dots & 0 & 1 & -4 & 6 & -4 & 1 \\ 0 & \dots & \dots & \dots & 0 & 1 & -4 & 5 & -2 \\ 0 & \dots & \dots & \dots & \dots & 0 & 1 & -2 & 1 \end{pmatrix},$$

the forcing is given by

$$G_i(t) = \begin{cases} \frac{c}{\Delta x}\cos(\omega t), & \text{if } i = i_1, \\ 0, & \text{if } i \neq i_1, \end{cases}$$

and the problem to solve is

$$\ddot{u} + k^2 Au + \partial\psi_K(u) \ni G(t),$$

with an impact law of type (3).

The scheme is defined by:

$$F(t, U_1, U_2, V, h) = \frac{G(t-h) + 2G(t) + G(t+h)}{4} - \frac{k^2}{4}A(3U_1 + hV + U_2),$$

which corresponds to a Newmark's scheme of second order for the free flight.

We have compared our results with experimental results, obtained by Jacquart et al, EDF, Clamart (see figures 7 and 8).

There is, for this system too, sensitivity to initial data. We have plotted the graph of $t \mapsto |u_{\Delta t_1}(x_{ob}, t) - u_{\Delta t_2}(x_{ob}, t)|$, where Δt_1 and Δt_2 are two different but close values of the time step. We have observed that, at the beginning, the two solutions merge and draw aside exponentially afterwards: this phenomenon is caracteristic of a sensitivity to initial data (see figure 9).

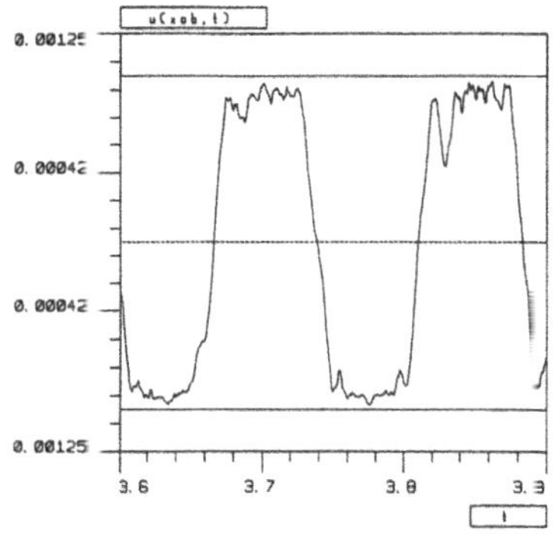

Figure 7. Motion at the abscissa x_{ob}: numerical results with $\Delta t = 5.10^{-5}$s.

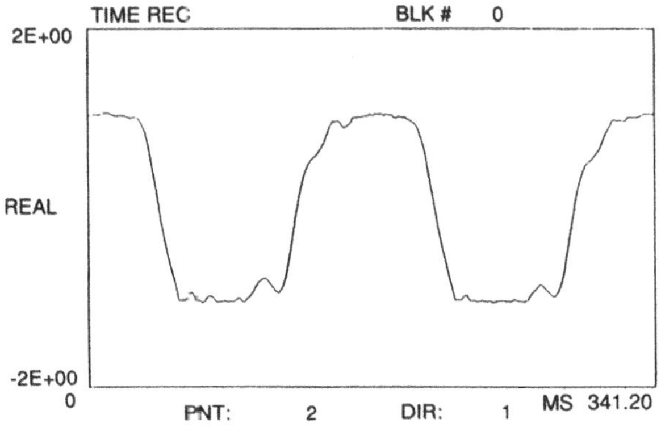

Figure 8. Motion at the abscissa x_{ob}: experimental results.

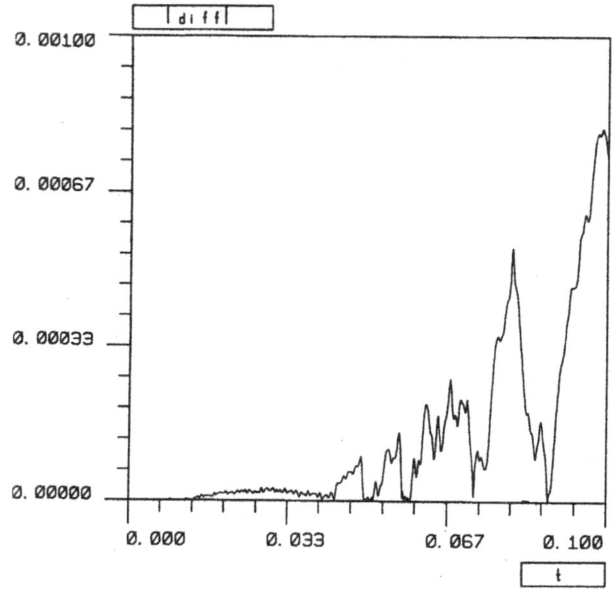

Figure 9. Graph of $t \mapsto |u_{\Delta t_1}(x_{ob}, t) - u_{\Delta t_2}(x_{ob}, t)|$, for $\Delta t_1 = 2.10^{-5}$s and $\Delta t_2 = 5.10^{-5}$s.

Conclusion: We have obtained a set of theoretical and numerical results on finite time intervals. New directions of research include asymptotic behavior for large times and identification of mechanical characteristics of the system from dynamical measurements.

REFERENCES

Berger M. and Gostiaux B., 1987, "Géométrie Différentielle: Variétés, Courbes et Surfaces", PUF, Paris.

Brezis H., 1973, "Opérateurs Maximaux Monotones et Semi-Groupes de Contraction dans les Espaces de Hilbert", North Holland, Amsterdam, Londres, New-York.

Jacquart G. and Gay N., 1992, Computation of impact-friction interaction between a vibrating tube and its loose supports, *in* "Proceedings of ASME 92, Symposium on Flow Induced Vibration and Noise".

Panet M., Paoli L. and Schatzman M., 1992, Vibrations with an obstacle and a finite number of degrees of freedom, *in* "Proceedings of International Symposium on Identification of Nonlinear Mechanical Systems from Dynamics Tests, Euromech 280", L.Jezequel & C.H. Lamarque ed., Balkema, Rotterdam.

Paoli L. and Schatzman M., 1993a, Vibrations avec contraintes unilatérales et perte d'énergie aux impacts, en dimension finie, *C. R. Acad. Sci. Paris*, 317-97:101.

Paoli L. and Schatzman M., 1993b, Schéma numérique pour un modèle de vibrations avec contraintes unilatérales et perte d'énergie aux impacts, en dimension finie, *C. R. Acad. Sci. Paris*, 317-211:215.

Paoli L. and Schatzman M., 1993c, Mouvement à nombre fini de degrés de liberté avec contraintes unilatérales: cas avec perte d'énergie, *Modél. Math. Anal. Numér.*, 27-673:717.

Paoli L. and Schatzman M., 1994, A numerical scheme for a dynamical impact problem with loss of energy, in finite dimension, *submitted to Math. of Comp., Rapport interne de l'équipe d'analyse numérique Lyon-St Etienne.*

Rockafellar R.T., 1970, "Convex Analysis", Princeton University Press, Princeton.

DILATATIONAL EFFECTS IN NONMONOLITHIC STRUCTURES

H. Parland

Department of Civil Engineering, Structural Mechanics
Tampere University of Technology, Finland

The structural analysis of reinforced concrete structures and stone structures has been based on a continuum-mechanical treatment adopted from monolithic structures such as steel structures. The nonlinear discontinuous effects induced by cracks and dry joints have been smeared out and interpreted as softening material nonlinearities. Reinforced concrete and stone structures being nonmonolithic remain in working condition inspite of the gaps and cracks still in the elastic stage where the reversibility of displacements is preserved. In their elastic stage the effect of displacement discontinuities on the internal equilibrium is more pronounced. Because of the impenetrability condition they induce strong dilatational effects which discontinuously change the internal equilibrium.

The solution $\{\sigma,u\}$ of a boundary value problem of an elastic nonmonolithic structure can be decomposed into two components
$\{\sigma,u\}^T = \{\sigma,u\}_e^T + \{\sigma,u\}_h^T$
where $\{\sigma,u\}_e^T$ is the solution of the monolithic structure and $\{\sigma,u\}_h^T$ is the solution induced by the state of selfstress $\sigma_h = \sigma - \sigma_e$ induced by the edge effect of the joints. Any generalized displacement U^i correponding to a unit force $P_i = 1$ which induce $\{\sigma^i,u^i\}_e$ can be determined by work equations and Betti's rule

$$U^i = U_e^i + U_h^i\ ; \tag{1}$$

where U_e^i corresponds to the monolithic structure and

$$U_h^i = \Sigma\int \mathbf{p}_e^i \cdot [\mathbf{u}]d\Gamma \tag{1'}$$

equals the work of the stress σ_e^i in the cracks and dry joints and $[\mathbf{u}]_{\nu\mu} = \gamma_{\nu\mu}$ is the displacement discontinuity in the joint. Applying a unit normal force N = 1 and a unit moment M = 1 to a cracked beam with vertical cracks the mean elongation of the beam
v and the mutual rotation ω of the end faces are, respectively

$$v = \frac{Nl}{EA} + \Sigma\frac{V_h}{A}; \omega = \int\frac{Mdx}{EJ} + \Sigma\frac{y_h V_h}{J} \tag{2}$$

where $V_h = \int_A \gamma_x dA$ is the local crack-volume and y_h is the coordinate of its centroid. Because every volume $V_h > 0$, the crack induces an elongation of the centroid axis of the beam also if the tip of the crack doesn't reach the axis.

Contact Mechanics, Edited by M. Raous *et al.*
Plenum Press, New York, 1995

Let a sequence of identical rectangular panels with thickness t, depth d, length $l = d\lambda$ and $A = td$ be loaded by a compressive force P with constant eccentricity e. The dry joints between the panels remain because of symmetry plane within the contact area.

These contact planes determine the generalized deformations v, ω and the contraction u_p of the panel at the level $y_p = -e$ of the compressive force P. Being functions of P and the moment $M = Pe$, v, ω and u_p can be expressed by deformation parameters $\beta(m,\lambda)$, $\alpha(m,\lambda)$, $\delta(m,\lambda)$

$$v = \frac{Pd}{EA}\beta\,;\ \omega = \frac{Pd}{EAk}\alpha\,;\ u_p = \frac{Pd}{EA}\delta \qquad (3)$$

where k is the kernpoint distance and $m = e/k$. The parameters of the monolithic state β_e, α_e, δ_e and of the edge effect β_h, α_h, δ_h correspond to v_e, ω_e, u_{pe} and v_h, ω_h, u_{ph}, respectively.

Because the strain energy of a panel is $W = P\,u_p/2 = P^2 d\delta/2EA$ there holds

$$\alpha = \frac{1}{2}\frac{\partial\delta}{\partial m}\,;\ \beta = m\alpha - \delta \qquad (4)$$

The parameters α_h, β_h, δ_h are linked to the crack volume V_h, its centroid distance y_h and the edge distance $y_c = d/2$

$$\rho_h = \frac{y_h}{k} = \frac{3\alpha_h}{\beta_h}\,;\ \rho_h + m \cong 1 + \frac{y_c}{k} \cong 4 \qquad (5)$$

This analysis, based on dry joints between the panels, is also applicable to cracks if we assume that due to repeated loading the beam attains the limit state of free contact where no tensile stresses occur in the cracked cross-sections.

There are the following limit cases characterized by the parameters δ, α, β:

I The panels behave monolithically, the end faces remain in complete plane contact.

$\delta_e(m,\lambda) = \lambda(1 + m^2/3)$; $\alpha_e(m,\lambda) = m\lambda/3$; $\beta_e(m,\lambda) = -\lambda$ (6)

II Tensile stresses completely eliminated (corresponding to $\lambda \to 0$), which represents a triangular compression field. The deformation parameters for this case α_o, β_o, γ_o and for the corresponding edge effect $\alpha_{oh} = \alpha_o - \alpha_e$, $\beta_{oh} = \beta_o - \beta_e$, $\delta_{oh} = \delta_o - \delta_e$ are proportional to λ. Thus with m = _ m _

$$\alpha_{ho} = \lambda\frac{(4-m)(m-1)^2}{3(3-m)^2}\,;\ \beta_{ho} = \lambda\frac{(m-1)^2}{(3-m)^2}\,;\ \delta_{ho} = \lambda\dot{\delta}_{ho}(m) = \lambda\frac{(m-1)^3}{3(3-m)} \qquad (7)$$

III If $\lambda \to \infty$ the edge effect of (σ_h) dies away within the length l of the ashlar if $\lambda \geq 1$ and therefore the parameters α_h, β_h, δ_h become independent of λ. Their values are approximately determined by

$$\delta_h(m,\infty) = \sqrt{r}((m-1)(3-m))^r \delta_{ho}(m), \alpha_h(m,\infty) = \frac{(4-m+r(2-m))\delta_h(m,\infty)}{(3-m)(m-1)};$$

$$\beta_h(m,\infty) = \frac{(3+r(2-m)m)}{(3-m)(m-1)}\delta_h(m,\infty)r \cong 0,55 \tag{8}$$

IV If $1/\lambda \to \infty$ and the eccentricity e of P, is very close to the compressed edge d/2 the stresses induced by the compressive force die away at a distance l from the compressed edge as in limit case III. Therefore if $\lambda \leq 1$ the deformation parameter δ is a unique function of $m_a = ((d/2) - e)/l = (3-m)/6\lambda$ only.

Because the cases III and IV are valid for $\lambda = 1$, we obtain for $\lambda \leq 1$ with $m_a(m,\lambda) = m_a(m_1,1)$ where $m_1 = 3 - (3-m)/\lambda$ the approximate values
$\delta_h(m,\lambda) = \delta_e(m_1,1) - \delta_e(m,\lambda) + \delta_h(m_1,\infty)$;
$\alpha_h(m,\lambda) = \alpha_e(m_1,1)/\lambda - \alpha_e(m.\lambda) + \alpha_h(m_1,\infty)/\lambda$; $\beta_h(m,\lambda) = m\alpha_h(m,\lambda) - \delta_h(m,\lambda)$ (9)
If $\lambda > 1$ then applies 8).

Whereas the parameters α_e β_e δ_e determine the deformations of the rectangular ashlar in the monolithic phase the parameters α_h, β_h can be considered as concentrated deformations, dilatations v_h and rotations ω_h at the joints. The present analysis is confined to eccentric compression because the shearforces induce small deformations in joints, compared with the rotations and dilatations caused by moments M and normal forces N.

An interesting problem is the behaviour of clamped beams and plates. In the uncracked state there are no normal forces but cracking induces strong arch- and dome-actions if the supports are fixed.In a clamped beam using as basic system the simply supported beam, the deformation of the beam is determined by eq. 2) and 2'). The first cracks occur at the supports (x = 0 and x = l). The bending moment M depends on the moment of the vertical load M_o and the moment caused by the horizontal thrust. The rigid support provides according to 1) the conditions

$$v = -\frac{Hl}{EA} + \frac{Hd}{2EA}(\beta_h(0) + \beta_h(l)) = 0; \omega = \frac{\int M_o}{EI}dx + \frac{Hd}{2EAk}(\alpha_h(o) + \alpha_F(l)) = 0$$

If the load is symmetric $\beta(0) = \beta(l) = \lambda$; $\alpha(0) = \alpha(l) \leq 0$ then because of 5).

$$H = \frac{3}{2}\int\frac{M_o dx}{ld}$$

If the vertical load q is uniform then $\int_o^l M_o dx = (ql^3)/12$ and

$$H = \frac{M_{omax}}{d} = H_{\min} \tag{10}$$

This H-value is the smallest permissible to which corresponds a linear arch within the beam. Thus the beam behaves as a straight arch. Prof. H. Moseley 1833 proposed a new principle of least pressure. The principle was much disputed but apparently it holds in the case described above. A corresponding analysis for nonstraight arches show that the ratio $H/H_{min} > 1$ and increases with the elevation of the arch.

The procedure used above can be applied also to plates if in the expressions 4) E is replaced by $E/1-v^2$. In a clamped circular disc with diameter 2a and thickness d loaded by a uniform load q the first cracks develop along the support. Using as basic system the simply supported plate and proceeding as above we get with $\beta_h(a) = 2d/a = \lambda$ the unit thrust

$$p_r = \frac{3}{2} \frac{qa\lambda}{\left(|m| + \frac{3\alpha_h}{\lambda} \right)}$$

Taken into consideration $M_{omax} = (3 + v)qa^2/16$ and $m + 3\alpha/\lambda = 4$ we obtain

$$p_r = \frac{M_{o\max}}{(1 + v/3)d} \approx \frac{M_{o\max}}{d} = p_{r\min} \tag{11}$$

This value corresponds closely to the minimum permissible thrust surface. In these cases the thrust is proportional to λ and the load q. The real dependence of the trust on the load is not linear because the number of cracks increases the ratio H/ql. The transition from one H(ql) line to another happens nearly discontinuously and the curve H(ql) is a stepwise rising graph.

Another dilatational effect is the uplift caused by the voids developed at horizontal joints and cracks. If the load is within the cone of the elastic kern E_k the structures remains in the monolithic stage. Loads acting outside this cone induce because of the uplift an increase of the potential energy of gravity. This increase is proportional mainly to the gap volume V_h of the horizontal cracks and joints.

Comparison with tests carried out with beams and frames, confirms qualitatively the theoretical results. On the other hand, there are considerable quantitative discrepancies between theory and test. These are mainly due to the difficulty of a reliable assessment of the effect of the surface roughness of the joint interfaces. Another reason is the difficulty of an accurate measurement of the real deformation at cracks and dry joints. The roughness affects desicively the dilatations and rotations in the joints especially at low stresses. Only in tests with polishes plane interfaces the experimental results conform with the theoretically deduced deformation parameters.

REFERENCES

Moseley H. On a New Principle of Statics. Principle of Least Pressure. Phil.Magazine, Ser. 3., Vol. 3.

Schlaich J. Die Gewölbwirkung in durchlanfenden Stahlbetonplatten. Dissertation. Techn. Hochshule, Stuttgart 1963.

INDEX

The manufacturer's authorised representative in the EU is Springer Nature Customer Service Centre GmbH, Europaplatz 3, 69115 Heidelberg, Germany. If you have any concerns regarding our products, please contact ProductSafety@springernature.com

Printed and bound by CPI Group (UK) Ltd, Croydon, CR0 4YY

23/04/2026

02095593-0013